The Mycota

Edited by
K. Esser

Springer
Berlin
Heidelberg
New York
Barcelona
Hong Kong
London
Milan
Paris
Singapore
Tokyo

The Mycota

The Mycota

A Comprehensive Treatise
on Fungi as Experimental Systems
for Basic and Applied Research

Edited by K. Esser

VIII

*Biology
of the Fungal Cell*

Volume Editors:
R.J. Howard and N.A.R. Gow

With 68 Figures, 2 in Color, and 16 Tables

Springer

Series Editor

Professor Dr. Dr. h.c. mult. KARL ESSER
Allgemeine Botanik
Ruhr-Universität
44780 Bochum, Germany
Tel.: +49(234)32-22211
Fax: +49(234)32-14211
e-mail: Karl.Esser@ruhr-uni-bochum.de

Volume Editors

Professor Dr. Richard J. Howard
DuPont Experimental Station E402/2231
Powder Mill Road
Wilmington, DE 19880-0402, USA
Tel.: +1(302)695-1492
Fax: +1(302)695-4509
e-mail: Richard.J.Howard@usa.dupont.com

Professor Dr. Neil A.R. Gow
Department of Molecular and Cell Biology
Institute of Medical Sciences
University of Aberdeen
Aberdeen AB25 2ZD, Scotland, UK
Tel.: +44(1244)273179
Fax: +44(1224)273144
e-mail: n.gow@abdn.ac.uk

ISBN 3-540-60186-4 Springer-Verlag Berlin Heidelberg New York

Library of Congress Cataloging-in-Publication Data.

The Mycota. Includes bibliographical references and index. Contents: 1. Growth, differentiation, and sexuality/editors, J.G.H. Wessels and F. Meinhardt – 2. Genetics and biotechnology. 1. Mycology. 2. Fungi. 3. Mycology – Research. 4. Research. I. Esser. Karl, 1924– . II. Lemke, Paul A., 1937– . QK603.M87 1994 589.2 ISBN 3-540-57781-5 (v. 1: Berlin: alk. paper) ISBN 0-387-57781-5 (v. 1: New York: alk. paper) ISBN 3-540-58003-4 (v. 2: Berlin) ISBN 0-387-58003-4 (v. 2: New York)

Springer-Verlag Berlin Heidelberg New York
a member of BertelsmannSpringer Science + Business Media GmbH

http://www.springer.de

© Springer-Verlag Berlin Heidelberg 2001
Printed in Germany

Production Editor: PRO EDIT GmbH, Heidelberg, Germany
Cover design: Springer-Verlag, E. Kirchner

Typesetting by Best-set Typesetter Ltd., Hong Kong

Printed on acid-free paper   SPIN 10509616   31/3130/Di   5 4 3 2 1 0

Series Preface

Mycology, the study of fungi, originated as a subdiscipline of botany and was a descriptive discipline, largely neglected as an experimental science until the early years of this century. A seminal paper by Blakeslee in 1904 provided evidence for self-incompatibility, termed "heterothallism", and stimulated interest in studies related to the control of sexual reproduction in fungi by mating-type specificities. Soon to follow was the demonstration that sexually reproducing fungi exhibit Mendelian inheritance and that it was possible to conduct formal genetic analysis with fungi. The names Burgeff, Kniep and Lindegren are all associated with this early period of fungal genetics research.

These studies and the discovery of penicillin by Fleming, who shared a Nobel Prize in 1945, provided further impetus for experimental research with fungi. Thus began a period of interest in mutation induction and analysis of mutants for biochemical traits. Such fundamental research, conducted largely with *Neurospora crassa*, led to the one gene: one enzyme hypothesis and to a second Nobel Prize for fungal research awarded to Beadle and Tatum in 1958. Fundamental research in biochemical genetics was extended to other fungi, especially to *Saccharomyces cerevisiae*, and by the mid-1960s fungal systems were much favored for studies in eukaryotic molecular biology and were soon able to compete with bacterial systems in the molecular arena.

The experimental achievements in research on the genetics and molecular biology of fungi have benefited more generally studies in the related fields of fungal biochemistry, plant pathology, medical mycology, and systematics. Today, there is much interest in the genetic manipulation of fungi for applied research. This current interest in biotechnical genetics has been augmented by the development of DNA-mediated transformation systems in fungi and by an understanding of gene expression and regulation at the molecular level. Applied research initiatives involving fungi extend broadly to areas of interest not only to industry but to agricultural and environmental sciences as well.

It is this burgeoning interest in fungi as experimental systems for applied as well as basic research that has prompted publication of this series of books under the title *The Mycota*. This title knowingly relegates fungi into a separate realm, distinct from that of either plants, animals, or protozoa. For consistency throughout this Series of Volumes the names adopted for major groups of fungi (representative genera in parentheses) are as follows:

Pseudomycota

| Division: | Oomycota (*Achlya, Phytophthora, Pythium*) |
| Division: | Hyphochytriomycota |

Eumycota

Division:	Chytridiomycota (*Allomyces*)
Division:	Zygomycota (*Mucor, Phycomyces, Blakeslea*)
Division:	Dikaryomycota

Subdivision:	Ascomycotina
Class:	Saccharomycetes (*Saccharomyces*, *Schizosaccharomyces*)
Class:	Ascomycetes (*Neurospora*, *Podospora*, *Aspergillus*)
Subdivision:	Basidiomycotina
Class:	Heterobasidiomycetes (*Ustilago*, *Tremella*)
Class:	Homobasidiomycetes (*Schizophyllum*, *Coprinus*)

We have made the decision to exclude from *The Mycota* the slime molds which, although they have traditional and strong ties to mycology, truly represent nonfungal forms insofar as they ingest nutrients by phagocytosis, lack a cell wall during the assimilative phase, and clearly show affinities with certain protozoan taxa.

The Series throughout will address three basic questions: what are the fungi, what do they do, and what is their relevance to human affairs? Such a focused and comprehensive treatment of the fungi is long overdue in the opinion of the editors.

A volume devoted to systematics would ordinarily have been the first to appear in this Series. However, the scope of such a volume, coupled with the need to give serious and sustained consideration to any reclassification of major fungal groups, has delayed early publication. We wish, however, to provide a preamble on the nature of fungi, to acquaint readers who are unfamiliar with fungi with certain characteristics that are representative of these organisms and which make them attractive subjects for experimentation.

The fungi represent a heterogeneous assemblage of eukaryotic microorganisms. Fungal metabolism is characteristically heterotrophic or assimilative for organic carbon and some nonelemental source of nitrogen. Fungal cells characteristically imbibe or absorb, rather than ingest, nutrients and they have rigid cell walls. The vast majority of fungi are haploid organisms reproducing either sexually or asexually through spores. The spore forms and details on their method of production have been used to delineate most fungal taxa. Although there is a multitude of spore forms, fungal spores are basically only of two types: (i) asexual spores are formed following mitosis (mitospores) and culminate vegetative growth, and (ii) sexual spores are formed following meiosis (meiospores) and are borne in or upon specialized generative structures, the latter frequently clustered in a fruit body. The vegetative forms of fungi are either unicellular, yeasts are an example, or hyphal; the latter may be branched to form an extensive mycelium.

Regardless of these details, it is the accessibility of spores, especially the direct recovery of meiospores coupled with extended vegetative haploidy, that have made fungi especially attractive as objects for experimental research.

The ability of fungi, especially the saprobic fungi, to absorb and grow on rather simple and defined substrates and to convert these substances, not only into essential metabolites but into important secondary metabolites, is also noteworthy. The metabolic capacities of fungi have attracted much interest in natural products chemistry and in the production of antibiotics and other bioactive compounds. Fungi, especially yeasts, are important in fermentation processes. Other fungi are important in the production of enzymes, citric acid and other organic compounds as well as in the fermentation of foods.

Fungi have invaded every conceivable ecological niche. Saprobic forms abound, especially in the decay of organic debris. Pathogenic forms exist with both plant and animal hosts. Fungi even grow on other fungi. They are found in aquatic as well as soil environments, and their spores may pollute the air. Some are edible; others are poisonous. Many are variously associated with plants as copartners in the formation of lichens and mycorrhizae, as symbiotic endophytes or as overt pathogens. Association with animal systems varies; examples include the predaceous fungi that trap nematodes, the microfungi that grow in the anaerobic environment of the rumen, the many insectassociated fungi and the medically important pathogens afflicting humans. Yes, fungi are ubiquitous and important.

There are many fungi, conservative estimates are in the order of 100000 species, and there are many ways to study them, from descriptive accounts of organisms found in nature to laboratory experimentation at the cellular and molecular level. All such studies expand our knowledge of fungi and of fungal processes and improve our ability to utilize and to control fungi for the benefit of humankind.

We have invited leading research specialists in the field of mycology to contribute to this Series. We are especially indebted and grateful for the initiative and leadership shown by the Volume Editors in selecting topics and assembling the experts. We have all been a bit ambitious in producing these Volumes on a timely basis and therein lies the possibility of mistakes and oversights in this first edition. We encourage the readership to draw our attention to any error, omission or inconsistency in this Series in order that improvements can be made in any subsequent edition.

Finally, we wish to acknowledge the willingness of Springer-Verlag to host this project, which is envisioned to required more than 5 years of effort and the publication of at least nine Volumes.

Bochum, Germany
Auburn, AL, USA
April 1994

KARL ESSER
PAUL A. LEMKE
Series Editors

Addendum to the Series Preface

In early 1989, encouraged by Dieter Czeschlik, Springer-Verlag, Paul A. Lemke and I began to plan *The Mvcota*. The first volume was released in 1994, other volumes followed in the subsequent years. Also on behalf of Paul, I would like to take this opportunity to thank Dieter Czeschlik, his colleague Andrea Schlitzberger, and Springer-Verlag for their help in realizing the enterprise and for their excellent cooperation for many years.

Unfortunately, after a long and serious illness, Paul A. Lemke died in November 1995. Without his expertise, his talent for organization and his capability to grasp the essentials, we would not have been able to work out a concept for the volumes of the series and to acquire the current team of competent volume editors. He was an outstanding scientist interested in many fields. Together with the volume editors, authors, and Springer-Verlag, I mourn the loss of a very good and reliable friend and colleague.

Since the first Volumes of *The Mycota* were well accepted by the scientific community, the publisher suggested to extend this series, which now will comprise 12 Volumes.

Bochum, Germany   KARL ESSER
April 2001

Volume Preface

Research in cell biology has exploded over the past decade, rendering impossible the task of mortals to stay abreast of progress in the entire discipline. Anyone interested in the biology of the fungal cell has most certainly noticed this trend, even in this fringe field of the larger subject. Indeed, to understand the biology of the fungal cell is to understand its interactions with the environment and with other cells, encompassing a tremendously broad array of subdisciplines. In fact, the Mycota represent one of the last, largely unexplored gold mines of biological diversity. From cellular morphogenesis to colony formation and pathogenesis, this volume provides examples of the breadth and depth of fungal cell biology. Of course, there are many topics that could not be addressed in such limited space, but no matter. Our primary aim has been to provide a selected sampling of contemporary topics at the forefront of fungal cell biology to facilitate the dissemination of information across and between the many enclaves of researchers who study fungal cell biology. These include cell biologists, cytologists, developmental biologists, ecologists, geneticists, medical mycologists, microbiologists, molecular biologists, plant pathologists, and physiologists – many of whom would never consider themselves *mycologists*, and what a pity. We hope that the current volume will, in some ways, serve to bridge the gaps and inequalities that exist between these mycologists and to unite their efforts toward the advancement of our science.

This volume is divided into two parts. The first part considers a sampling of *behavioral* topics – how, or in what manner and to what effect, do cells of fungi behave in various environments; how does environment influence cell biology; how do the cells affect their surroundings, animate and inanimate? Topics include invasive growth, a defining characteristic of the Mycota; controls of cell polarity and shape, and morphological changes that are essential for the virulence of many pathogenic fungi; and a detailed consideration of the ways in which groups of cells of the same species form an individualistic coordinated organism.

The second part of the volume looks at the fungal cell as a structural continuum – from proteins, e.g. hydrophobins, that manage patterns of growth and development in space, to extracellular matrices, molecular connections between extra- and intracellular domains, including the cytoskeleton, to the molecular patterns of genomes that dictate things we do not yet know exist. All of these topics are perfused by recent advances in molecular genetics and are written at a time when fungal genome databases are just becoming established as a tool for the future. We hope that this volume will not only demonstrate that fungal cell biology is useful in representing accessible systems for exploration of biological systems as a whole, but also in illuminating aspects of fungal biology that are unique and fascinating in their own right. We are challenged by an amazing universe of fungal cell biology waiting to be explored.

Wilmington, Delaware, USA   RICHARD J. HOWARD
Aberdeen, Scotland   NEIL A.R. GOW
March 2001   *Volume Editors*

Contents

List of Contributors

JONATHAN ARNOLD, Department of Genetics, University of Georgia, Athens, Georgia 30602, USA

ANNE E. ASHFORD (e-mail: A.Ashford@unsw.edu.au, Tel.: +61-2-93852068, Fax.: +61-2-93851558) School of Biological Science, The University of New South Wales, Sydney 2052, Australia

J.W. BENNETT (e-mail: jbennett@tulane.edu) Department of Cell and Molecular Biology, Tulane University, New Orleans, Louisiana 70118, USA

I. BRENT HEATH (e-mail: brent@yorku.ca, Tel.: +1-416-7365511, Fax.: +1-416-7365698) Biology Department, York University, 4700 Keele Street, Toronto, Ontario M3J JP3, Canada

ALISTAIR J.P. BROWN (e-mail: gen069@abdn.ac.uk, Tel.: +44-1224-273173, Fax.: +44-1224-273144) Department of Molecular and Cell Biology, Institute of Medical Sciences, Foresterhill, University of Aberdeen, Aberdeen AB25 2ZD, UK

LOUISE COLE, School of Biological Science, The University of New South Wales, Sydney 2052, Australia

HANS DE NOBEL, Institute of Molecular Cell Biology, University of Amsterdam, Faculty of Biology, Nieuwe Achtergracht 166, 1018 WV, The Netherlands

NEIL A.R. GOW (e-mail: n.gow@abdn.ac.uk, Tel.: +44-1224-273179, Fax.: +44-1224-273144) Department of Molecular and Cell Biology, Institute of Medical Sciences, University of Aberdeen, Foresterhill, Aberdeen AB25 2ZD, UK

ADRIENNE R. HARDHAM (e-mail: hardham@rsbs-central.anu.edu.au, Tel.: +61-6-2494168, Fax.: +61-6-2494331) The Australian National University, Research School of Biological Sciences, Plant Cell Biology Group, G.P.O. Box 475, Canberra City ACT 2601, Australia

HARVEY C. HOCH, Department of Plant Pathology, Cornell University, New York State Agricultural Experiment Station, Geneva, New York 14456, USA

GEOFFREY J. HYDE, School of Biological Science, The University of New South Wales, Sydney 2052, Australia

FRANS M. KLIS (e-mail: klis@bio.uva.nl, Tel.: +31-20-5257834, Fax.: +31-20-5257934/7662) Institute of Molecular Cell Biology, University of Amsterdam, Faculty of Biology, Nieuwe Achtergracht 166, 1018 WV, The Netherlands

W. LaJean Chaffin (e-mail: micwlc@ttuhsc.edu, Tel.: +1-806-7432545,
Fax.: +1-806-742334) Department of Microbiology and Immunology,
Texas Tech University Health Sciences Center, Lubbock, Texas 79430, USA

In Hyung Lee, School of Biological Sciences, University of Missouri-Kansas City,
5100 Rockhill Road, Kansas City, Missouri 64110-2499, USA

Nicholas P. Money (e-mail: moneynp@muohio.edu, Tel.: +1-513-5292140,
Fax.: +1-513-5294243) Department of Botany, Miami University, Oxford,
Ohio 45056, USA

Stefan Olsson (e-mail: sto@kvl.dk, Tel.: +45-35282646, Fax.: +45-35282606)
Department of Ecology, Royal Veterinary and Agricultural University,
Thorvaldsensvej 40, 1871 Frederiksberg C, Copenhagen, Denmark

Michael Plamann (e-mail: plamannm@umkc.edu, Tel.: +1-816-2352593,
Fax.: +1-816-2351503) School of Biological Sciences, University of Missouri-
Kansas City, 5100 Rockhill Road, Kansas City, Missouri 64110-2499, USA

Brian D. Shaw (e-mail: bshaw@dogwood.botany.uga.edu, Tel.: +1-706-5426026,
Fax.: +1-706-5421805) Department of Botany, 2502 Miller Plant Sciences,
University of Georgia, Athens, Georgia 30602, USA

Yi-Jun Sheu, Department of Molecular, Cellular and Developmental Biology,
Yale University, New Haven, Connecticut 06520-8103, USA

J. Hans Sietsma, Molecular Plant Biology Laboratory, Groningen Biomolecular
Sciences and Biotechnology Institute (GBB), University of Groningen, Kerklaan 30,
9751 NN Haren, The Netherlands

Michael Snyder (e-mail: Michael.Snyder@yale.edu, Tel.: +1-203-4326139,
Fax.: +1-203-4326161) Department of Molecular, Cellular and Developmental
Biology, Yale University, New Haven, Connecticut 06520-8103, USA

Nicholas J. Talbot (e-mail: N.J.Talbot@exeter.ac.uk, Tel.: +44-1392-264673,
Fax.: +44-1392-264668) Department of Biological Sciences, University of Exeter,
Exeter EX4 4QJ, UK

Herman Van Den Ende, Institute of Molecular Cell Biology,
University of Amsterdam, Faculty of Biology, Nieuwe Achtergracht 166, 1018 WV,
The Netherlands

Growth, Morphogenesis and Pathogenicity

1 Biomechanics of Invasive Hyphal Growth

Nicholas P. Money

CONTENTS

I. Introduction

The process of invasive hyphal growth allows fungi to acquire nutrients from diverse solid materials of biological and synthetic origin. It is a defining characteristic of the fungi and is at the root of their evolutionary origins (Bartnicki-Garcia 1987; Money 1999a). The essential cell biological processes that attend invasive growth are identical to those that operate during non-invasive hyphal extension. In both cases, polarized synthesis of new plasma membrane and cell wall advances the position of the hyphal tip, creating a cylindrical cell of increasing length. However, the mechanical challenges encountered by hyphae growing over surfaces and in broth culture are very different from those associated with invasive growth. Fungi secrete an abundance of enzymes to dissolve their surroundings (Walton 1994), and evidence suggests that turgor pressure provides hyphae with the necessary invasive force to overcome obstacles that are not liquefied ahead of their growing apices (Money 1998). In general, highly pressurized hyphae can penetrate tougher substrates than those with lower pressures, and fungi that naturally invade hard materials generate extraordinarily high pressures and exert concomitantly large invasive forces (e.g., penetration hyphae of certain plant pathogens; Sect. III.D). In striking contrast, it has been argued that turgor is almost irrelevant in understanding the fundamental mechanism of polarized hyphal growth, and is not even necessary for hyphal extension in some species of the Oomycota (Money 1997a; Johns et al. 1999). To examine the role of turgor in hyphal growth it is therefore crucial to specify the environmental context in which cell extension is analyzed (see Chap. 6, this Vol.). This chapter is concerned principally with the significance of mechanical factors when hyphae invade their food sources.

II. Experimental Subjects

Experiments on the large vegetative hyphae of species within the family Saprolegniaceae (Oomycota) and on appressoria formed by foliar pathogens are the source of most current wisdom on mechanical aspects of invasive growth. The appressorium has proven a powerful model for understanding the initial events in plant infection and these insights have informed parallel experiments on vegetative hyphae. The penetration hypha that develops from an appressorium is a specialized device that performs the exclusive task of host penetration, whereas vegetative hyphae also function in nutrient absorption while they burrow through their substrates. Vegetative hyphae differ from penetration hyphae (or pegs) in other ways: they are much larger, can elaborate thicker cell walls, generate lower turgor pressures,

Department of Botany, Miami University, Oxford, Ohio 45056, USA

The Mycota VIII
Biology of the Fungal Cell
Howard/Gow (Eds.)
© Springer-Verlag Berlin Heidelberg 2001

and are usually components of a more extensive mycelium.

Because many biomechanical experiments involve micromanipulation, a premium is placed upon hyphal size. Non-septate coenocytic hyphae produced by *Achlya* and other Saprolegniaceae can exceed diameters of 30 µm (Fig. 1A; though *Saprolegnia* is smaller); these beautiful cells scintillate under microscope illumination, with parallel streams of organelles running through their translucent cytoplasm. By contrast, hyphae of many basidiomycete and ascomycete fungi are 5 µm or less in diameter (Fig. 1B) and micromanipulation of these cells is difficult. The speed of hyphal extension is a further experimental asset of the Saprolegniaceae, and large diameters coupled with extension rates exceeding 10 µm min^{-1} support unparalleled rates of volumetric expansion. Hyphae of insect pathogens within the Entomophthorales (Zygomycota) can be even larger than those of the Saprolegniaceae (Fig. 1C), but the fact that they grow slowly in culture offsets the utility of their size. Hyphae of mucoralean fungi such as *Rhizopus* offer the best combination of size and speed outside the Saprolegniaceae, and while they are still considerably smaller than *Achlya* and its relatives, they may prove useful for future biomechanical research.

In adopting the Oomycota as experimental models for studying hyphal mechanics, an important question must be addressed: Can we apply our findings from studies on these organisms to other mycelial fungi? If we restrict our frame of reference to the mechanics of hyphal growth I believe we can be confident in an affirmative answer to this question. The reason is simple. The Oomycota are subject to the same constraints of eukaryote structure and physiology as other fungi, and the fact that they do not form a Spitzenkörper, for example, is of little significance from a biomechanical perspective, and nor is their unique cell wall composition. When any hypha pushes through solid material it must exercise enough force to overcome the mechanical resistance of its surroundings. Exoenzymes probably weaken the substrate, but unless the material surrounding the growing apex is liquefied, the cell must exert force to sustain penetration. For these reasons, what we learn about the mechanics of invasive growth from *Achlya* is very likely to inform us about the fundamental processes used by ascomycete pathogens to proliferate within leaves, and how fungi like *Candida albicans* penetrate solid tissues in immune-compromised human hosts. Nevertheless, the scientific value of the Saprolegniaceae is limited by the fact that they live in freshwater

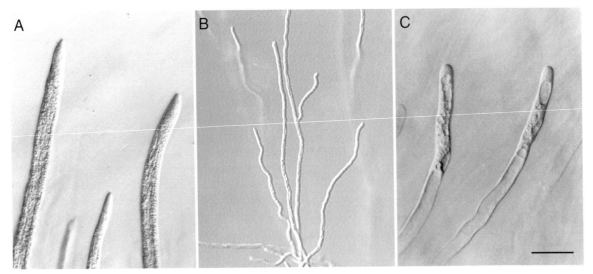

Fig. 1. Range of hyphal size among fungi. **A** The saprobe *Achlya bisexualis* (Saprolegniaceae; Oomycota) produces non-septate hyphae with acute apices (maximum diameter of the two large hyphae is 23 µm). **B** Hyphae of *Magnaporthe grisea* (Ascomycotina), the rice blast fungus, are less than 5 µm in diameter, typical of species of Ascomycotina and Basidiomycotina. **C** *Basidiobolus microsporus* is a member of the family Entomophthoraceae (Zygomycota) that includes many important insect pathogens. This strain was isolated from lizard dung. Its hyphae are sufficiently large for conventional micromanipulation (maximum hyphal diameter in this image is 18 µm), but grow very slowly in culture. *Bar* 50 µm. (**B** kindly supplied by Timothy M. Bourett and Richard J. Howard)

habitats, rebuff standard methods of genetic modification, and only in rare cases colonize living organisms. Progress in hyphal growth research would undoubtedly be accelerated if some pathogenic hyphal giant were discovered, better still if it could be crossed easily and transformed.

Experiments on the melanized appressoria of the pathogens *Magnaporthe grisea* and *Colletotrichum* species, including *C. graminicola*, have also greatly influenced our knowledge of fungal mechanics. Naturally, these cells form on and penetrate grass leaves, but they will also invade a variety of synthetic materials with hydrophobic properties. This is useful because materials impervious to enzymatic degradation can be studied, allowing the investigator to dissociate the effects of mechanical factors from enzymes in the invasive process. The fact that mutant strains of these fungi have been isolated that are defective in certain facets of appressorial development and function has proven a great bonus for experimental mycologists. Both vegetative hyphae and penetration hyphae will be examined in this chapter.

III. Invasive Force

A. Force Measurements from Vegetative Hyphae

Hyphal force is derived from turgor pressure (see Sect. V), but data on turgor only provide a measure of the force per unit area that presses the plasma membrane against the inner surface of the hyphal wall, not the actual force that the cell utilizes for invasive growth. Only when the cell wall of the fungus yields does some proportion of the internal force act upon the material in contact with the hyphal tip (Money 1995). However, although turgor does not equate directly with invasive force, accurate turgor measurements are important in understanding invasive growth because they provide an estimate of the maximum force that the hypha can exert. Such estimates (force = turgor × tip area; Table 1) reflect the invasive force available to a hypha whose apical wall is sufficiently loose that it absorbs none of the internal pressure, and allows all of it to push against the substrate. In this hypothetical situation, hyphal rupture would no longer be obviated by the cell wall, but only through the support offered by the

substrate. Predictions based on this scenario indicate that hyphae have the potential to exert forces between 10^{-5} and 10^{-4} N (Table 1).

These calculations can be compared with direct measurements of hyphal force made with an ultrasensitive silicon bridge strain gauge or force transducer. The resolution of this instrument is better than $1\,\mu N$, equivalent to the force produced by gravity acting on a mass of $100\,\mu g$. Measurements of hyphal force have been made from cells growing from an agar shelf into a liquid-filled well in a culture dish (Fig. 2; Johns et al. 1999). In these experiments the strain gauge was positioned with a micromanipulator so that the tip of its silicon beam was submerged in the liquid in a vertical orientation. As the hypha pushed against the flat beam, its electrical output changed in proportion to the applied force (Fig. 3). Hyphal force was then measured for a few minutes, and in successful recordings, the transducer output returned to the baseline voltage as soon as the beam was moved away from the hyphal tip. For calibration the instrument was clamped in a horizontal orientation and microgram and milligram weights were balanced on the end of the beam to provide a range of micronewton forces.

Force measurements using this system have been made from a number of fungi, and data obtained from *Achlya bisexualis* (Saprolegniaceae) have been published (Johns et al. 1999). There are no standing gradients of turgor within the coenocytic hyphae of the Oomycota (Money 1990), and because turgor is isotropic in nature, internal pressure exerts an identical force over every square micrometer of the plasma membrane. As argued above, a proportion of this force is transmitted to the substrate according to the degree of wall yielding at the hyphal tip. It follows that for the same level of turgor and wall compliance, larger hyphae will exert more force than smaller cells. This is indeed the case in *A. bisexualis*, in which forces of $100\,\mu N$ or greater were measured from unusually large hyphae, while more typical hyphae, with diameters of $15-25\,\mu m$, exerted a mean force of $12.0 \pm 5.8\,\mu N$ ($n = 17$) at $22\,°C$. By estimating the area of contact between the hyphal apex and beam, the force data can be corrected for hyphal size and expressed in terms of force per unit area (or pressure). These calculations show that hyphae of *A. bisexualis* exert forces between 0.02 and $0.07\,\mu N\,\mu m^{-2}$ (or MPa, the units are identical), at least when they push against a silicon beam. Reference to Table 1

Table 1. Peak hyphal forces predicted from turgor pressure measurements

Organism	Hyphal turgor pressure[a] (MPa or $\mu N\,\mu m^{-2}$)	Hyphal radius (μm)	Contact area between hyphal apex and substrate[b] (μm^2)	Total invasive force[c] (μN)	Source
Basidiomycota					
Uromyces appendiculatus	0.3	5	79	24	Terhune et al. (1993)
Ascomycota					
Morchella esculenta	0.2	5	79	16	Amir et al. (1995)
Neurospora crassa	0.2	7.5	177	35	Money (unpubl. data)
Zygomycota					
Basidiobolus microsporus	0.4	10	314	126	Money (unpubl. data)
Rhizopus oryzae	0.2	7.5	177	35	Money (unpubl. data)
Oomycota					
Achlya bisexualis	0.7	12.5	491	344	Money and Harold (1992)
Saprolegnia ferax	0.4	7.5	177	71	Money and Harold (1993)

[a] Table shows only those data obtained using the pressure probe and reflect maximum values reported for each species. Note: $1\,MPa = 10^6\,N\,m^{-2}$.
[b] Area based on circle with radius equal to subtending hypha, because hyphal apex flattens in contact with solids.
[c] Force calculated from product of turgor and tip area.

indicates that 10% or less of the force available from turgor is actually applied by the hyphal apex of *A. bisexualis* against its surroundings, consistent with the idea that the mechanical strength of the apical wall governs the magnitude of the invasive force. The implications of these measurements are discussed further in Section V.

B. Fluctuations in Hyphal Force

Strain-gauge recordings from hyphal apices of *A. bisexualis* reveal that force is not constant, but fluctuates around an average value with a maximum frequency of one or two cycles per minute ($= 10^{-2}\,Hz$; Johns et al. 1999). The recording shown in Fig. 4A was made from the tip of a large hypha exerting a peak force of more than $100\,\mu N$, with maximum excursions of $20\,\mu N$. This behavior correlates with dramatic swings in the rate of hyphal extension. Figure 4B shows that the extension rate oscillates as much as 50% in *A. bisexualis*, falling from a maximum rate of $0.075\,\mu m\,s^{-1}$ ($4.5\,\mu m\,min^{-1}$) to $0.050\,\mu m\,s^{-1}$ within 20–30 s, and then increasing again at a similar rate. The similarity in the frequency of the changes in force and extension rate suggests that fluctuations in force reflect underlying changes in turgor pressure as the cell circulates through periods of faster and slower growth. Similar changes in extension rate have been documented previously in other

Fig. 2. Hypha of *Achlya bisexualis* growing against the silicon beam of a miniature strain gauge. The beam is $100\,\mu m$ thick and the hypha projects $100\,\mu m$ from the agar shelf into liquid medium; recordings of force are made before hyphae reach this length, as the cell apex emerges from the agar. (Johns et al. 1999)

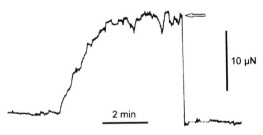

Fig. 3. Force recording from a single hypha of *Achlya bisexualis*. In this example, hyphal force builds to a maximum value of $19\,\mu N$ following initial contact between the growing apex and the silicon beam. When the beam is moved away from the hyphal apex (*arrow*), the strain-gauge output falls immediately to the baseline

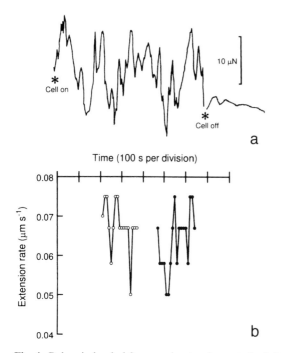

Fig. 4. Pulses in hyphal force and extension rate in *Achlya bisexualis*. **a** A 17-min recording of hyphal force from a single large hypha exerting a total force of 107 μN; time scale in Fig. 2B applies. **b** Fluctuations in the extension rate of two hyphae calculated from video recordings by measuring the position of hyphal tips at 10-s intervals. (Johns et al. 1999)

fungi (Castle 1940; López-Franco et al. 1994), and in pollen tubes (Pierson et al. 1996).

Pulses in pressure have not been measured with the pressure probe because they are beneath its resolution. To appreciate this technical limitation it is necessary to consider some numbers. An unbranched hypha of *A. bisexualis* with a diameter of 16 μm and length of 500 μm contains a volume of about 100,000 μm^3 (or 100 picoliters). At an extension rate of 0.08 μm s^{-1} its volume would increase by 1000 μm^3 every minute, or become 1% larger. In the absence of immediate and perfect osmotic adjustment, turgor would also decline by up to 1% (Boyle's law, pressure × volume = constant), or 0.007 MPa for a hypha pressurized to 0.69 MPa (the mean turgor for this species in dilute culture medium). Alter the value of turgor, cell volume, or growth rate and it is unlikely that the pressure probe, with a resolution of 0.02 MPa, could detect anything but constant turgor during growth. Changes in turgor are buffered by the entire volume of the cell, and are always likely to be small. Pressure pulses, however, are reflected in the force measurements from the hyphal apex. A pressure decrease of 0.007 MPa will reduce the

force at the hyphal tip by 1.4 μN, which is within the resolution of the strain gauge. Dynamic changes in the mechanical properties of the hyphal wall complicate the interpretation of the force measurements because wall hardening will result in a decrease in hyphal force, and loosening will increase force without any alterations in turgor (Sect. V).

C. Other Approaches to Measuring Hyphal Forces

It is difficult to devise methods to measure forces from cells that are buried in a solid substrate, and for this reason the strain-gauge technique is used to determine force as hyphae emerge from agar (Sect. III.A). A number of alternative approaches to measuring forces exerted by single cells have been developed, and it is worth reviewing these in the hope that mycological applications may be developed. Burton and Taylor (1997) published measurements of the forces exerted by mammalian cells during cell division. These authors determined traction forces of cultured fibroblasts attached to thin sheets of silicone polymer from wrinkles in the sheets induced by dividing cells (the technique was developed by Harris et al. 1980). The sheets were calibrated by measuring the distortion caused by microneedles; the microneedles were themselves calibrated by bending their tips with attached glass beads of known weight. Traction forces during cytokinesis were in the nanonewton range, three orders of magnitude smaller than the forces measured from hyphae. This method does not appear suitable for measuring hyphal force because (1) most vegetative hyphae do not adhere tightly to synthetic surfaces, and (2) hyphal force would only be reflected accurately if the cell apex pushed into the polymer (see Sect. III.D). Other researchers have measured the forces exerted by leukocytes as they crawl through glass pipettes by applying pressure to the interior of the pipettes to oppose cell motion (Usami et al. 1992). The leukocytes produced maximum forces of 30 nN, again much smaller than those measured from hyphae. This method is impractical for fungal hyphae that do not adhere tightly to glass (or plastics) and would be dislodged once pressure was applied to the pipette. The atomic force microscope (AFM) seems a useful instrument for hyphal force experiments, but it is difficult to grow and view hyphae at high magnification in the vertical orientation necessitated by contact with the cantilever tip of the AFM. If a suitable culture system can be developed, the AFM will not provide a measure of the total force exerted at the hyphal apex because the contact area between the cantilever tip and hypha is so small. Nevertheless, the phenomenal sensitivity of the AFM could provide useful information on dynamic changes in extension growth and hyphal force in relation to vesicular fusion at the hyphal apex (Johns et al. 1999).

D. Force Measurements from Penetration Hyphae

Unlike vegetative hyphae, penetration hyphae emerging from appressoria of the cereal pathogens *M. grisea* and *C. graminicola* are firmly anchored by an adhesive secreted by the base of

the appressorium, and the hypha drives straight down into the underlying surface (Fig. 5A,B). Following appressorial adhesion, a thin layer of dihydroxynaphthalene (DHN) melanin is deposited on the inner surface of its cell wall (next to the plasma membrane), greatly reducing appressorial permeability to small molecules (Howard 1997). This permeability barrier enables the appressorium to accumulate high levels of cytoplasmic osmolytes and generate enormous turgor pressure (Howard et al. 1991; Money and Howard 1996; De Jong et al. 1997; Money 1997b). Penetration of the host cuticle and epidermal cell wall is then accomplished by the thin penetration hypha that extends from the non-melanized base of the appressorium. Appressoria are effective symbols for the awesome mechanical capabilities of fungi: their turgor matches the hydrostatic pressure at an ocean depth of one mile, greatly exceeds that

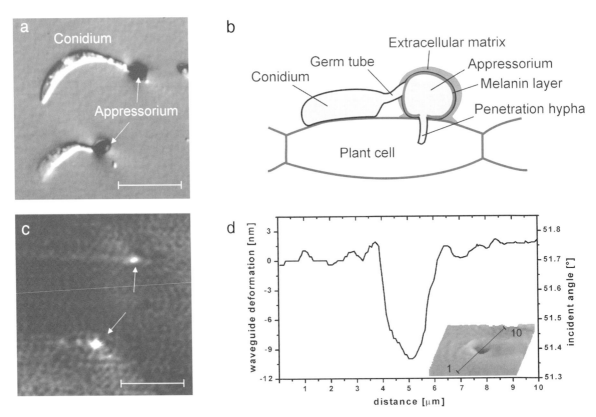

Fig. 5. a Light micrograph showing appressoria of *Colletotrichum graminicola* attached to the aluminum "skin" of a waveguide. *Bar* 10 μm. **b** Drawing of the infection apparatus. **c** Reflected light image of the waveguide area shown in (**a**); the bright spots correspond to the position of the appressoria and indicate indentation of the waveguide by penetration hyphae. **d** Plot of a cross section of the waveguide surface showing an indentation beneath a single appressorium. Measurements indicate that a 10-nm-deep impression was made by a 2-μm-diameter penetration hypha. *Inset* shows topography of the waveguide surface over the region of measurement. Raised ring around indentation is consistent with tight adhesion between the appressorial base and waveguide surface. Reprinted with permission from Bechinger et al. (1999), Copyright 1999 American Association for the Advancement of Science

produced by the toe of a 400-lb dancer engaged in a brief pirouette (2 or $3\,\mu N\,\mu m^{-2}$), and is sufficient to punch holes through Kevlar (Howard et al. 1991).

The developmental behavior of appressoria encouraged an intriguing experiment in which invasive force was measured using an optical method. Appressoria of *C. graminicola* were allowed to form on a film called a waveguide, in which two 13-nm-thick layers of aluminum sandwiched a 1-μm-thick layer of silicone gel (Bechinger et al. 1999; Money 1999b,c). Penetration hypha developing from the base of the appressoria depressed the aluminum surface, pushing the metal a few nanometers into the gel. The depth of the impression under each appressorium (proportional to the force applied by the hypha) was measured by scanning the upper silicone-Al interface with a laser and collecting data on the intensity of reflected light. Using this method, Bechinger and colleagues created three-dimensional images of the dented waveguide surface with vertical spatial resolution in the nanometer range (Fig. 5D). The waveguide was calibrated by compressing the surface with a glass capillary of known spring constant. According to these experiments, penetration hyphae of *C. graminicola* exert an average force of $17\,\mu N$, which is close to the estimate of $8\,\mu N$ published for *M. grisea* based on a turgor pressure of 8.0 MPa and tip area of $1\,\mu m^2$ (force = pressure × area; Money 1995).

Further interpretation of the force measurements requires data on turgor from *C. graminicola*. Penetration hyphae of this fungus are about $2\,\mu m$ in diameter, suggesting that the measured force could be supplied by a turgor pressure of 5.4 MPa, although resistance from the cell wall at the tip of the hypha would necessitate a higher cytoplasmic pressure. The force estimate of $8\,\mu N$ for *M. grisea* was based on the assumption that the wall of the penetration hypha offered no resistance to extension, so that all of the internal force was allowed to act on the underlying substrate. Indeed, it is possible that appressoria operate in this fashion, creating a very fluid wall at the tip of the penetration hypha and allowing the host tissue to buffer the cell against bursting. Clearly, the way that the pathogen controls the mechanical properties of its cell wall at the base of the appressorium before invasion, and then at the tip of the penetration hypha, is a vital element of the penetration mechanism. While appressoria are highly

pressurized structures, it is important to realize that because penetration hyphae are only 1–2 μm in diameter, the stress acting in the wall that covers their apices is no more than the stress produced at the tips of 10- or 20-μm-diameter vegetative hyphae with tenfold lower turgor (Money 1998). In terms of mechanical stability then, there is nothing extraordinary about penetration hyphae. What is more remarkable, and very poorly understood, is how all hyphae extend and weave new walls without rupturing their existing extracellular textiles.

Holland et al. (1999) suggested that mechanical penetration is also an important feature of the mechanism used by *Paecilomyces lilacinus* to invade nematode eggs. Hyphopodia of this fungus develop from an adhesive mycelium on the surface of the egg and produce penetration hyphae that breach the chitinous shell. Secreted proteinases and lipases have been characterized from *P. lilacinus*, but in common with other pathogens, the relative significance of enzymes and mechanics in the invasive process has not been established.

IV. Substrate Resistance and Tissue-Degrading Enzymes

To appreciate their significance in invasive growth, micronewton hyphal forces must be compared with the strength of the solid materials colonized by fungi. When filamentous fungi penetrate plant and animal tissues their tip-growing hyphae extend through complex mixtures of polymers, many of which are covalently linked and surely present formidable obstacles to motility. In addition to the purely mechanical aspects of invasive growth discussed in the preceding sections, it has been assumed for decades that enzyme-catalyzed degradation of specific macromolecules reduces the strength of these tissues and thereby aids invasion (see Pryce-Jones et al. 1999). However, experimental evidence for this thesis is limited. Gene-disruption experiments show that while the secretion of enzymes is a critical aspect of tissue invasion by certain human pathogens (e.g., secreted aspartyl proteinases in *Candida albicans*; Hube et al. 1997; Sanglard et al. 1997), disease progression is not blocked by the targeted prohibition of one or more enzymes from the host-pathogen interaction. Precisely the same results have been obtained from long-running studies on phytopathogens (Walton 1994; Scott-Craig et al. 1998). Gene-disruption experiments have probably been equivocal because fungi have evolved multiple enzyme systems to degrade biological materials, rendering the loss of one or more enzymes of limited consequence to the fungus. Furthermore, it has been suggested that rather than trickling out a careful mix of proteins in response to specific materials that the organism encounters, fungi may vomit a cocktail of enzymes as they attack tissue barriers (Knogge 1996). With this degree of physiological flexibility, underpinned by a high degree of genetic redundancy, pathogenic fungi are well equipped to thwart efforts to understand and inhibit their growth mechanism by mutating one or two genes that encode exoenzymes. Unfortunately, even if a sufficient congregation of enzyme-encoding genes were disrupted to limit invasive growth, this would do little to resolve the relative significance of force versus enzymes. This is because exoenzymes presumably play dual roles in tissue invasion by (1) reducing substrate resistance, and (2) drenching the fungus with low molecular weight nutrients. Therefore, it would be impossible to know whether growth ceased because the mutant encountered an insoluble barrier, or because its enzyme deficiency limited its food supply. For these reasons the significance of substrate-degrading enzymes in the invasive process remains an unresolved question. A potential solution could be sought by comparing substrate strengths with hyphal forces in the presence or absence of putative tissue-degrading enzymes. Such experiments have not been performed, but the necessary methods are in place.

An extensive biomechanical literature reports various aspects of the strength of biological materials (Vogel 1988), but there have been no measurements that accurately reflect the resistance of plant and animal tissues to hyphal penetration. However, data on the resistance of synthetic films and agar media that have been used in invasive growth research are available. Howard et al. (1991) quantified the strength of Mylar [poly(ethylene terephthalate)] films employed in appressorial studies by measuring the impressions made by metal probes at various loads (the method is referred to as Vicker's indentation technique). Experiments showed that appressoria could penetrate films with hardness values exceeding $2 \times 10^8\,\mathrm{N\,m^{-2}}$ or 200 MPa. This observation predicts an appressorial pressure of 200 MPa and an invasive force of $200\,\mu\mathrm{N}$ ($2 \times 10^8\,\mathrm{N\,m^{-2}} \times 10^{-12}\,\mathrm{m^2}$) at the tip of the penetration hypha. However, the turgor measurements and waveguide data discussed in the previous section indicate that penetration hyphae exert less than 10% of this force, presenting a paradox (Money 1999b,c). Either the macroscopic test for film strength does not reflect the penetrative force required on a microscopic scale, or the fungus utilizes some mechanism for softening the film. While thin Mylar films are punctured by appressoria, the compound is incredibly resistant to microbial decomposition and Mylar-degrading enzymes are unknown. It is therefore critical to measure the strength of Mylar films using a microprobe with a tip that matches the size and shape of a penetration hypha. Until this is done we cannot be sure that appressoria penetrate such tough materials by mechanics alone.

Unlike the development of penetration hyphae, sustained invasive growth by vegetative hyphae is dependent on an exogenous supply of nutrients. Therefore, biomechanical experiments on vegetative hyphae have employed nutrient media solidified with a range of concentrations of agar to intensify the resistance of the substrate to hyphal invasion (Money 1995; Brush and Money 1999). The strength of these agar gels has been measured using macroscopic and microscopic probes. Soil scientists use instruments called penetrometers to measure the resistance of soils to root penetration, and fruit growers use smaller versions of the same tools to determine crop ripeness. Hand-held penetrometers contain spring-loaded plungers with tips that range from 1–25 mm in diameter. The tip of the plunger is pushed into the sample and an analog scale (proportional to the spring constant of the instrument) provides a reading of the material's strength, or resistance to compression. The penetrometer readings show that medium saturated with agar (8% w/v) has a strength of about 0.1 MPa (Fig. 6). Although 8% agar offers minimal resistance compared to the Mylar film penetrated by appressoria (Sect. III.D), it is effective at slowing invasive growth by vegetative hyphae.

Measurements of substrate strength have also been obtained with a microprobe device that approaches the size and shape of the hyphal apex (Money, unpubl. data). In these experiments a glass micropipette (microprobe) with a smooth, closed tip was attached to a miniature strain gauge

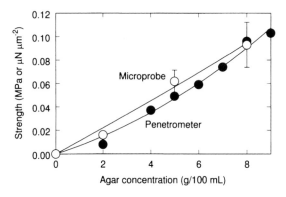

Fig. 6. Agar strength (resistance to penetration) measurements made with a hand-held penetrometer and microprobe device described in the text

(the same type used for the hyphal force measurements described in Sect. III.A), and pushed into agar gel to a depth of $40\,\mu$m. The change in electrical output from the strain gauge reflected the force required to insert the microprobe. There is surprisingly good correspondence between these measurements and those made with the hand-held penetrometer; in Fig. 6, data are shown for a 25-μm-diameter tip equal to the size of an *A. bisexualis* hypha. To penetrate 2% agar it was necessary to apply a force of $8\,\mu$N, which represents a force per unit area (or pressure) of $0.02\,\mu$N$\,\mu$m^{-2} at the tip of the microprobe (a 25-μm-diameter probe has a tip area of about $500\,\mu$m^2). Similarly, a force of $46\,\mu$N or $0.09\,\mu$N$\,\mu$m^{-2} was needed to overcome 8% agar. As we have seen, forces of these magnitudes and higher have been measured from hyphae of *A. bisexualis* (Sect. III), and these cells are capable of penetrating 8% agar (though their rate of invasive growth is slowed). Based on the available measurements and estimates of hyphal force and agar strength it is not surprising that experimental reduction of turgor handicaps invasive growth, and that the inhibitory effect of depressurization is amplified in high concentrations of agar (Money 1995). The finding that hyphae produce micronewton forces and that forces of the same magnitude are needed to probe solid media is gratifying. The combination of physical data from living hyphae and inert substrates offers convincing support for the hypothesis that mechanical penetration plays a central role in invasive hyphal growth. On a less positive note, it is unsettling that so few studies have tested the importance of hyphal mechanics in host invasion, particularly given the agricultural and clinical

significance of this fungal activity. It is also disturbing that we know almost nothing about the mechanical properties of plant and animal tissues from the perspective of microbial colonization.

V. A Simple Mathematical Model for Invasive Hyphal Growth

As we have seen in Sect. III.A, none of the force exerted on the inner surface of the hyphal wall pushes against the substrate unless the apical cell wall of the fungus yields (Money 1995). The force exerted during invasive growth, which allows the fungus to drive forward, is determined by the difference between the turgor within the hypha and the critical pressure required to induce wall yielding. This critical pressure has been termed the yield threshold (Ortega 1990), and was one component of the classical equations for cell elongation derived by Lockhart (1965). These equations provide a model for growth based on the precept that simultaneous water uptake and wall yielding drive extension. At full turgor *A. bisexualis* hyphae appear to exert 10% or less of the available internal force at their extending apices (Sect. III.A), suggesting that the wall absorbs 90% of the force exerted by turgor pressure. The rigidity of the cell wall may be affected both by its molecular architecture (including the density of molecules that cross-link microfibrils), and by any restraining forces produced by the cytoskeleton that resist cellular extension (e.g., through integrin-like connections that tether the actin cytoskeleton to the wall; Kaminskyj and Heath 1996; see also Chap. 10, this Vol.). These considerations indicate that the mechanical properties of the cell wall are subject to many levels of cellular and environmental control, including nutrient supply and demand, wall biosynthesis and enzymatic loosening, cytoskeletal dynamics, and countless other cytoplasmic processes. But although the determination of a cell's growth rate is far more complex than Lockhart's model allows (Money 1997a), the concept of the yield threshold is useful for grasping the fundamental mechanics that must operate during invasive growth. The central idea has been formalized in a mathematical model (Money 1998) whose most important features can be illustrated with a single equation:

$$\text{Invasive force }(\mu N) = F_{inv} = [(\Psi_p + P_{cs}) - Y] \cdot A$$

The equation predicts that the force exerted by the apex of a hypha pushing through its substrate (F_{inv}) is determined by turgor pressure (Ψ_p), any pressure (force per unit area) exerted by the cytoskeleton against the inner surface of the apical cell wall (P_{cs}), the yield threshold of the wall (Y), and the surface area of the hyphal tip (A). According to the model, hyphal force does not necessarily parallel changes in turgor, but can increase or decrease in response to changes in any of the other variables in the equation. For example, the model proposes that hyphal force can be boosted by the secretion of enzymes that loosen the apical hyphal wall and thereby decrease its yield threshold, without any change in turgor pressure (Fig. 7).

In addition to the theoretical considerations discussed in the preceding paragraphs, a substantial body of data supports the model:

– **Turgor powers invasive growth.** Experiments on saprobic Oomycota (citations in Money 1997a) and on the rice pathogen *Magnaporthe grisea* (Howard et al. 1991; Howard 1997) show that invasive growth is inhibited by turgor reduction. In vitro experiments on the human pathogen *Wangiella dermatitidis* are also consistent with a physical mechanism of penetration (Brush and Money 1999; Sect. VI), though it may be argued that tissue-degrading enzymes play a more important role when the fungus invades a living host.

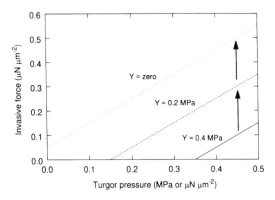

Fig. 7. Predictions of model positing that invasive force is elevated by increases in turgor pressure but limited by wall resistance. Plots show invasive forces ($\mu N\,\mu m^{-2}$) for a hypha generating turgor pressures between 0 and 0.5 MPa for three values of yield threshold (Y). Calculations based on constant low pressure exerted by the cytoskeleton of 0.05 MPa. *Arrows* indicate that invasive force is boosted when the yield threshold decreases (i.e., when the apical cell wall is loosened), without any change in turgor pressure

– **Importance of the cytoskeleton.** Calculations show that cytoskeletal expansion has the capacity to exert forces of a few hundredths of one megapascal against the apical plasma membrane of the hypha (Money 1997a). These estimates suggest that the cytoskeleton does little to power invasive growth, although at low levels of turgor, actin polymerization or myosin-based sliding of microfilaments may help sustain non-invasive hyphal extension (Gupta and Heath 1997; Heath and Steinberg 1999).

– **Significance of wall compliance.** Experiments showing that hyphal force can be much lower than the maximum levels predicted from turgor measurements indicate that the compliance of the cell wall controls the amount of force applied by the hyphal apex (Sect. III.A). Changes in the tensile strength of the hyphal wall (that are proportional to the yield threshold) in three species within the Saprolegniaceae are controlled by the secretion of 1,4-β-D-glucan glucanhydrolases (or endoglucanases; Money and Hill 1997). These experiments suggest that the hypha can tune its invasive force to the resistance of its surroundings by regulating the secretion of lytic enzymes that loosen its own wall.

– **Significance of hyphal size.** Force is equal to the product of pressure and area, and so for the same turgor pressure and level of wall compliance, larger hyphae tend to exert greater total force (Sect. III.A). It is likely that hyphae of certain sizes are better adapted to penetration than others, and that this optimum size varies with substrate identity.

The third law of motion asserts that an equal and opposite reaction will occur in response to any action. Applied to invasive hyphae we recognize that the cell must always be stabilized behind its tip if it is to avoid sliding back through the substrate when it applies force. Invasive mycelia are secured by the intimacy between the surface area of the whole mycelium and the enveloping substrate, so that numerous interconnected hyphae and lateral branches provide each hyphal tip with perfect physical stability. Penetration of the surface of a potential food source requires a different strategy because only a small part of the fungal wall is in contact with the substrate. For this reason, appressoria become firmly affixed to the plant cuticle by a ring of adhesive that supplies the requisite brace for extension of the penetration hypha. The domed part of the appressorial wall does not expand after melanization, but the pore region at the base of the cell in contact with the substrate is non-melanized, and it is from this translucent site that the hypha emerges. So, although turgor acts equally over every square micrometer of the plasma membrane inside the appressorium, focused wall synthesis and yielding in the pore region restrict application of invasive force to the tip of the penetration hypha.

VI. Mycelial Growth in Plants and Animals

Mycelial development within plant and animal tissues following the preliminary events of host penetration is very poorly understood, and its analysis represents a considerable challenge for fungal biologists. From the limited perspective of biomechanics, the model for invasive growth offered in the previous section is likely to be relevant to mycelial motility within host tissues, but the paucity of information on enzyme action is a significant problem. Tissue degradation by secreted enzymes is indicated by evidence of wall damage seen in histological studies of infected plants (Mendgen and Deising 1993), but as we have seen, the significance of enzymes remains unproven by molecular genetic analyses (Sect. IV).

Some of the most important fungal pathogens of humans form hyphae only outside the host and proliferate as yeast cells within infected tissues (Sect. VII), but there are a large number of opportunistic mycoses in which hyphae colonize solid tissues. If culture medium solidified with 8% agar is stab-inoculated with yeast cells of *Candida albicans* or *Wangiella dermatitidis*, for example, these fungi proliferate initially by budding along the track left by the inoculation needle. Within a few days, however, they switch to hyphal morphology and erupt into the medium, spawning dendritic colonies of branched hyphae around the central axis of yeast cells (Fig. 8; Brush and Money 1999). The genetic control of hyphal induction is quite well understood with the characterization of a growing number of *C. albicans* genes that encode the responsible transcription factors (Mitchell 1998). Much less is known about the genes that control the morphogenetic processes themselves, including the determinants of polarized

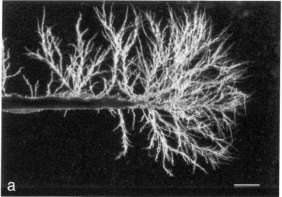

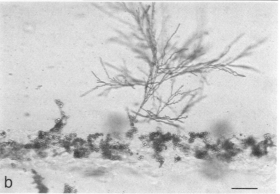

Fig. 8. Invasive hyphae of *Wangiella dermatitidis* penetrating 8% agar. **a** Low-magnification dark-field image showing dendritic morphology of colony developing from an inoculum of yeast cells. **b** Higher magnification view of branched hyphae extending into agar from yeast cells confined at the inoculation site. *Bars* (**a**) 250 µm, (**b**) 100 µm. (Brush and Money 1999)

growth and hyphal-specific patterns of wall assembly. Experiments on *W. dermatitidis* hint at the significance of biomechanical factors in disease progression.

Wangiella dermatitidis is one of the agents of phaeohyphomycosis, a retinue of diseases caused by melanotic fungi (Horré and de Hoog 1999). This fungus is described as neurotropic, meaning that it shows marked propensity for growth in the nervous systems of human patients and laboratory animals. Recent research involving albino mutants of *W. dermatitidis* has shown that its rate of invasive hyphal growth is dependent on melanization and that only melanized cells can swiftly penetrate growth medium solidified with high concentrations of agar (Brush and Money 1999). This is a tantalizing observation because the same strains of the fungus exhibit very similar behavior in mice brains, and in the absence of fast hyphal proliferation, animals survive infection by albino strains.

Unlike appressoria, cells of *W. dermatitidis* do not lose their permeability or accumulate glycerol after melanization, and the mode-of-action of hyphal melanin has not been solved. Nevertheless, these experiments are important because they signify that the mechanical behavior of cells is an important determinant of the outcome of the interaction between humans and our parasites. There is practical significance to this insight. By informing us about very general processes that control fungal growth, biomechanical experiments may help identify novel broad-spectrum targets for antifungal therapy.

VII. Invasive Yeast Cells

Biomechanical factors also play a significant role in the mechanism used by yeast cells to proliferate within solid substrates. If culture medium solidified with 8% agar is stab-inoculated with the melanotic pathogen *Cryptococcus neoformans*, the fungus proliferates by bud formation along the needle track. When it has filled the available space, and probably depleted its nutrient supply, it begins to penetrate the agar with the formation of vast lens-shaped colonies of yeast cells (Fig. 9; Brush and Money 1999). Each cell is pressurized, and so once a bud forms it cannot be pushed back into its mother, and slow motion is achieved by the accretion of millions of cells. In vivo, the process is complicated by enzyme action that may create pockets of softened tissue into which the daughter cells can slip, but the golden rule of "enough force to overcome the strength of the surrounding tissue" cannot be defeated. Unlike mycelia composed of hyphae, colonies of *C. neoformans* yeast cells expand without apparent direction. The same phe-

nomenon can be reproduced with yeast cells of *Saccharomyces cerevisiae*. By contrast, the pseudo-hyphal or filamentous growth form of *S. cerevisiae* first described by Gimeno et al. (1992) exhibits polarized invasive growth similar to the pattern of agar penetration shown by hyphae of *W. dermatitidis* (see Sect. VI).

VIII. Rock Invasion

Fungi also penetrate rocks, and are surprisingly prevalent within the surface layers of masonry (Jongmans et al. 1997; Sterflinger and Krumbein 1997). Many of these fungi are mutualists with algae (lichens) or vascular plants (mycorrhizal fungi), but a number of non-lichenized fungi also thrive on the products of rock dissolution and airborne sources of nutrients. Details of the process of rock penetration are sketchy and most information has been obtained from careful sectioning of rocks. Indeed, it is difficult to envisage experiments that could elucidate such an achingly slow process as rock penetration. Secretion probably plays a key role, though in this case fungi solubilize the surrounding substrate with organic acids such as citrate, oxalate, and succinate, rather than enzymes. Thin sections of feldspars show hyphae situated in smooth tubes that perfectly match their contours (Fig. 10), which is suggestive of chemical leaching during mycelial development. In some cases secreted acids may solubilize the substrate ahead of the hyphal apex reducing the significance of mechanical penetration, but in other examples mechanical penetration appears vital. Sterflinger and Krumbein (1997) suggest that dematiaceous fungi exploit microscopic fissures between crystals in marble and pry the crystals apart as they penetrate deeper and deeper into the rock. In common with the invasion of plant and animal tissues, the only logical source of power for this type of microbial mining is hyphal turgor pressure.

IX. Evasive Hyphae

The formation of aerial hyphae, conidiophores, sporangia, and fruiting bodies plays the invasive growth video in reverse, with the fungus emerging from its exhausted substrate. While truly aquatic fungi can form and release spores under water, terrestrial fungi growing in wet environments

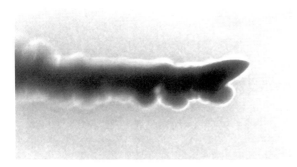

Fig. 9. Invasive yeast cells of *Cryptococcus neoformans* penetrating 8% agar 1 week after stab inoculation. Diameter of colony 0.5–1 mm

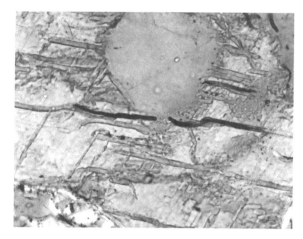

Fig. 10. Thin-sectioned sample of alkali feldspar permeated with 5-μm-diameter tubes formed by rock-invading fungus. Tubes with a black interior contain hyphae. Micrograph kindly supplied by Toine Jongmans

must breach the elastic interface between water and air before sporulating. The force necessary to break though the surface is given by the product of surface tension ($7.3 \times 10^{-2}\,\mathrm{J\,m^{-2}}$) and the length of the contact line between the hypha and the interface. For hyphae projecting perpendicular to the interface the contact line is a circumference. Use of the maximum circumference is likely to overestimate the necessary force because the pointed hypha first pushes against the interface with a much smaller part of its tip. However, applying the larger figure for the purpose of estimation, the predicted force for a 5-μm-diameter hypha is $1.1 \times 10^{-6}\,\mathrm{J\,m^{-1}}$ or $1.1\,\mu$N. The tip of a 5-μm-diameter cell producing the same force per unit area as *A. bisexualis* (between 0.02–$0.07\,\mu$N $\mu\mathrm{m}^{-2}$; Sect. III) would exert a total force of 0.4–$1.4\,\mu$N. The overlap between these force estimates (based on data from a single species), and the strength of the water-air interface, suggests that submerged hyphae face a significant barrier when they grow into air.

Small glycosylated proteins called hydrophobins secreted by many fungi create a water-repellent coat on the surface of aerial hyphae and fruiting bodies, and also self-assemble to form microscopic mosaics at water-air interfaces. In the latter configuration, polymeric complexes of hydrophobins may facilitate the formation of aerial hyphae by reducing the surface tension of the interface. A mutant strain of *Schizophyllum commune* (Basidiomycota), in which the gene that encodes a hydrophobin called SC3 was disrupted, formed few aerial hyphae (van Wetter et al. 1996).

Formation of aerial hyphae was revived in this strain, $\Delta SC3$, when purified SC3 protein was added to its culture medium (Wösten et al. 1999a). An explanation for these observations is found by examining the effect of hydrophobins on the surface tension of the interface. Surface tension was measured by comparing the contact angle (or degree of flattening) of water droplets on hydrophobic films in the presence and absence of the protein. The wild-type strain of *S. commune* secretes high concentrations of SC3 and can reduce the surface tension of its culture medium to $3.0 \times 10^{-2}\,\mathrm{J\,m^{-2}}$ (Wösten et al. 1999a,b). This corresponds to a 59% diminution in the "evasive force" necessary to breach the interface from $1.1\,\mu$N to just $0.5\,\mu$N ($= 3.0 \times 10^{-2}\,\mathrm{J\,m^{-2}} \times 1.5 \times 10^{-5}\,\mathrm{m}$). Even in the absence of SC3 the mutant reduced the surface tension of the water-air interface to $4.5 \times 10^{-2}\,\mathrm{J\,m^{-2}}$, presumably by secreting other surfactant molecules. According to the same formula, the surface tension of this interface requires an evasive force of $0.7\,\mu$N (for a 5-μm hypha), and the fact that aerial hyphae are only plentiful above SC3-coated water predicts that *S. commune* hyphae cannot exercise this much force.

The physical interaction between the hypha and the interface is of course more subtle than the preceding calculations suggest, but this treatment will provide a useful basis for further discussion if surface tension data can be compared with measurements of hyphal force from *S. commune*. Part of the intricacy lies in the degree to which the submerged mycelium is free to move, because unless the emerging hyphae are supported firmly they will be pushed away from the interface, back down into the fluid as they apply force (Sect. V). There is an interesting analogy here between this evasive or emergent growth, and the processes of tissue and rock invasion. In all cases a mechanical barrier appears to be weakened, whether by exoenzymes, metabolic acids, or surfactants, reducing the mechanical force required to overcome the obstacle to growth.

X. Conclusions

The mechanical attributes of a fungus are a physical expression of its genome, reflecting a much higher level of biological order than protein structure, cytoskeletal organization, or even the entire architecture of the cell. When inquiry shifts from the function of single enzymes to metabolic pathways, and then to a process like leaf penetration, the relationship between the genes and the behavior of the organism becomes increasingly complex. While genetic analysis has identified certain determinants of fungal mechanics (such as the link between melanin and appressorial function), the interaction between a fungus and its food source unfolds from many simultaneous and interdependent waves of gene expression.

Though choreographed in outline, genomic control of the mechanical behavior of a fungus does not seem specified in detail. Invasive hyphae are genetically programmed to exert force at their apices, but factors such as turgor, wall strength, and a farrago of other cellular and environmental variables determine exactly how much force. For this reason, many investigators have found that disruption of individual genes has offered few insights into hyphal behavior. Recognition of this labyrinthine quality of fungal growth, development, and pathogenesis places a premium upon the biomechanical research discussed in this chapter. Measurements of hyphal force have now been made and we are edging close to understanding how fungi control this physical attribute. With growing appreciation of the mechanical powers of fungi it has become absurd to discount the significance of cellular mechanics in any fungal activity. With great excitement we can proceed with this inquiry, because for the first time in the history of experimental mycology satisfactory tools are available to study the mechanical behavior of individual cells.

Acknowledgements. Research in the author's laboratory on *Wangiella* was financed by the National Institutes of Health, and the National Science Foundation provided support for research on hyphal biomechanics. The author is indebted to Paul Taylor (Canada) for suggesting use of the strain-gauge system to measure hyphal force. Parts of Section III.D were adapted from Money (1999c) with permission from Academic Press.

References

Amir R, Steudle E, Levanon D, Hadar Y, Chet I (1995) Turgor changes in *Morchella esculenta* during translocation and sclerotial formation. Exp Mycol 19:129–136

Bartnicki-Garcia S (1987) The cell wall: a crucial structure in fungal evolution. In: Rayner ADM, Brasier CM, Moore D (eds) Evolutionary biology of the fungi. Cambridge Univ Press, Cambridge, pp 389–403

Bechinger C, Giebel K-F, Schnell M, Leiderer P, Deising HB, Bastmeyer M (1999) Optical measurements of invasive forces exerted by appressoria of a plant pathogenic fungus. Science 285:1896–1899

Brush L, Money NP (1999) Invasive hyphal growth in *Wangiella dermatitidis* is induced by stab inoculation and shows dependence upon melanin biosynthesis. Fungal Genet Biol 28:190–200

Burton K, Taylor L (1997) Traction forces of cytokinesis measured with optically modified elastic substrata. Nature 385:450–454

Castle ES (1940) Discontinuous growth of single plant cells measured at short intervals, and the theory of intussusception. J Cell Comp Physiol 15:285–298

Davis DJ, Burlak C, Money NP (2000) Biochemical and biomechanical aspects of appressorial development in *Magnaporthe grisea*. In: Tharreau D, Lebrun MH, Talbot NJ, Notteghem JL (eds) Advances in rice blast research. Kluwer, Dordrecht, pp 248–256

De Jong JC, McCormack BJ, Smirnoff N, Talbot NJ (1997) Glycerol generates turgor in rice blast. Nature 389:244–245

Gimeno CJ, Ljungdahl CA, Styles CA, Fink GR (1992) Unipolar cell divisions in the yeast *S. cerevisiae* lead to filamentous growth: regulation by starvation and RAS. Cell 68:1077–1090

Gupta GD, Heath IB (1997) Actin disruption by latrunculin B causes turgor-related changes in tip growth of *Saprolegnia ferax* hyphae. Fungal Genet Biol 21:64–75

Harris AK, Wild P, Stopak D (1980) Silicone rubber substrata: a new wrinkle in the study of cell locomotion. Science 208:177–189

Heath IB, Steinberg G (1999) Mechanisms of hyphal tip growth: tube dwelling amebae revisited. Fungal Genet Biol 28:79–93

Holland RJ, Williams KL, Khan A (1999) Infection of *Meloidogyne javanica* by *Paecilomyces lilacinus*. Nematology 1:131–139

Horré R, de Hoog GS (1999) Primary cerebral infections by melanized fungi: a review. In: de Hoog GS (ed) Studies in mycology 43. Ecology and evolution of black yeasts and their relatives. CBS, Baarn, The Netherlands, pp 176–193

Howard RJ (1997) Breaching the outer barriers – cuticle and cell wall penetration. In: Carroll G, Tudzynski P (eds) The Mycota, vol 5, part A. Plant relationships. Springer, Berlin Heidelberg New York, pp 43–60

Howard RJ, Ferrari MA, Roach DH, Money NP (1991) Penetration of hard substrates by a fungus employing enormous turgor pressures. Proc Natl Acad Sci USA 88:11281–11284

Hube B, Sanglard D, Odds FC, Hess D, Monod M, Schafer W, Brown AJP, Gow NAR (1997) Disruption of each of the secreted aspartyl proteinase genes *SAP1*, *SAP2*, and *SAP3* of *Candida albicans* attenuates virulence. Infect Immun 65:3529–3538

Johns S, Davis CM, Money NP (1999) Pulses in turgor pressure and water potential: resolving the mechanics of hyphal growth. Microbiol Res 154:225–231

Jongmans AG, van Breemen N, Lundström U, van Hees PAW, Finlay RD, Srinivasan M, Unestam T, Giesler R,

Melkerud P-A, Olsson M (1997) Rock-eating fungi. Nature 389:682–683

Kaminskyj SGW, Heath IB (1996) Studies on *Saprolegnia ferax* suggest the general importance of the cytoplasm in determining hyphal morphology. Mycologia 88: 20–37

Knogge W (1996) Fungal infection of plants. Plant Cell 8:1711–1722

Lockhart JA (1965) An analysis of irreversible plant cell elongation. J Theor Biol 8:264–275

López-Franco R, Bartnicki-Garcia S, Bracker CE (1994) Pulsed growth of fungal hyphal tips. Proc Natl Acad Sci USA 91:12228–12232

Mendgen K, Deising H (1993) Infection structures of fungal plant pathogens – a cytological and physiological evaluation. New Phytol 124:193–213

Mitchell AP (1998) Dimorphism and virulence in *Candida albicans*. Curr Opin Microbiol 1:687–692

Money NP (1990) Measurement of hyphal turgor. Exp Mycol 14:416–425

Money NP (1995) Turgor pressure and the mechanics of fungal penetration. Can J Bot 73 [Suppl 1]: S96–S102

Money NP (1997a) Wishful thinking of turgor revisited: the mechanics of fungal growth. Fungal Genet Biol 21:173–187

Money NP (1997b) Mechanism linking cellular pigmentation and pathogenicity in rice blast disease: a commentary. Fungal Genet Biol 22:151–152

Money NP (1998) Mechanics of invasive fungal growth and the significance of turgor in plant infection. In: Kohmoto K Yoder OC (eds) Molecular genetics of host-specific toxins in plant disease. Kluwer, Dordrecht, pp 261–271

Money NP (1999a) On the origin and functions of hyphal walls and turgor pressure. Mycol Res 103:1360

Money NP (1999b) Fungus punches its way in. Nature 401:332–333

Money NP (1999c) To perforate a leaf of grass. Fungal Genet Biol 28:146–147

Money NP, Harold FM (1992) Extension growth in the water mold *Achlya*: interplay of turgor and wall strength. Proc Natl Acad Sci USA 89:4245–4249

Money NP, Harold FM (1993) Two water molds can grow in the absence of measurable turgor pressure. Planta 190:426–430

Money NP, Hill T (1997) Correlation between endoglucanase secretion and cell wall strength in oomycete fungi: implications for growth and morphogenesis. Mycologia 89:777–785

Money NP, Howard RJ (1996) Confirmation of a link between fungal pigmentation, turgor pressure, and pathogenicity using a new method of turgor measurement. Fungal Genet Biol 20:217–227

Ortega JKE (1990) Governing equations for plant cell growth. Physiol Plant 79:116–121

Pierson E, Miller DD, Callaham DA, van Aken J, Hackett G, Hepler PK (1996) Tip-localized calcium entry fluctuates during pollen tube growth. Dev Biol 174: 160–173

Pryce-Jones E, Carver T, Gurr S (1999) The roles of cellulase enzymes and mechanical force in host penetration by *Erysiphe graminis* f. sp. *hordei*. Physiol Mol Plant Pathol 55:175–182

Sanglard D, Hube B, Monod M, Odds FC, Gow NAR (1997) A triple deletion of the secreted aspartyl proteinase genes *SAP4*, *SAP5*, and *SAP6* of *Candida albicans* causes attenuated virulence. Infect Immun 65:3539–3546

Scott-Craig JS, Apel-Birkhold PC, Görlach JM, Nikolskaya A, Pitkin JW, Ransom RF, Sposato P, Ahn J-H, Tonukari NJ, Wegener S, Walton JD (1998) Cell wall degrading enzymes in HST-producing fungal pathogens. In: Kohmoto K, Yoder OC (eds) Molecular genetics of host-specific toxins in plant disease. Kluwer, Dordrecht, pp 245–252

Sterflinger K, Krumbein WE (1997) Dematiaceous fungi as a major agent for biopitting on Mediterranean marbles and limestone. Geomorphol J 14:219–230

Terhune BT, Bojko RJ, Hoch HC (1993) Deformation of stomatal guard cell lips and microfabricated artificial topographies during appressorium formation by *Uromyces*. Exp Mycol 17:70–78

Usami S, Wung S-L, Skierczynski BA, Skalak R, Chien S (1992) Locomotion forces generated by a polymorphonuclear leukocyte. Biophys J 63:1663–1666

van Wetter M-A, Schuren FHJ, Wessels JGH (1996) Targeted mutation of the Sc3 hydrophobin gene of *Schizophyllum commune* affects formation of aerial hyphae. FEMS Microbiol Lett 140:265–270

Vogel S (1988) Life's devices. The physical world of animals and plants. Princeton Univ Press, Princeton, NJ

Walton JD (1994) Deconstructing the cell wall. Plant Physiol 104:1113–1118

Wösten HAB, van Wetter M-A, Lugones LG, van der Mei HC, Busscher HJ, Wessels JGH (1999a) How a fungus escapes the water to grow into the air. Curr Biol 9:85–88

Wösten HAB, Richter M, Willey JM (1999b) Structural proteins involved in emergence of microbial aerial hyphae. Fungal Genet Biol 27:153–160

2 Control of Cell Polarity and Shape

YI-JUN SHEU and MICHAEL SNYDER

CONTENTS

Department of Molecular, Cellular and Developmental Biology, Yale University, PO Box 208103, New Haven, Connecticut 06520-8103, USA

I. Introduction

Cell polarity results from asymmetric cell growth (polarized growth) and cell division from a specific plane (polarized division), two phenomena fundamental for the development of organisms ranging from yeasts to humans. Polarized growth leads to the formation of unique cell shapes and subcellular structures required for many cell types to carry out specialized functions. Sensory transduction by the neurites of neurons, nutrient absorption by the microvilli of intestinal epithelial cells and plant fertilization by navigation of pollen tubes are examples of cellular processes that require polarized growth (Mooseker 1985; Bedinger et al. 1994; Eisen 1994). The ability to undergo morphological change is essential for the virulence of many pathogenic fungi (Shepherd 1988; see also Chap. 5, this Vol.). Appropriate division plane selection is important for the equal distribution of essential organelles, such as the nucleus and mitochondria, as well as the generation of differential cell fates. For example, in *Drosophila*, division planes specify the unequal partitioning of localized cell fate determinants which determine whether the daughters of dividing sensory organ precursor cells become either the precursor of hair and socket cells or the precursor of neuronal and sheath cells (Jan and Jan 1998). In *Caenorhabditis elegans*, the division plane directs proper cell-cell contact and partitioning of cytoplasmic components (Strome 1993). Positioning of division sites is presumably the basis for morphogenesis of many multicellular structures since it affects the higher order organization among cells.

Saccharomyces cerevisiae undergoes polarized cell growth that leads to distinct cell shapes and exhibits special division patterns. The shape and division pattern of a yeast cell are influenced by its genetic background and growth conditions. In a favorable environment, *S. cerevisiae* propagates by budding in which cell growth is directed

The Mycota VIII
Biology of the Fungal Cell
Howard/Gow (Eds.)
© Springer-Verlag Berlin Heidelberg 2001

into the bud; the place of bud formation is ultimately the site of cell division. When nutrient sources are limited, *S. cerevisiae* cells undergo "filamentous" differentiation and adopt a highly elongated cell shape and form chains of connected cells. This feature allows cells to forage and maximize access to nutrients (Scherr and Weaver 1953; Gimeno et al. 1992). Finally, haploid *S. cerevisiae* cells respond to mating pheromones and form projections toward their mating partners to facilitate cell fusion (Sprague and Thorner 1992).

The budding pattern (Fig. 1) is controlled by genetic and environmental factors (Freifelder 1960; Hicks et al. 1977; Snyder 1989; Madden et al. 1992; Chant and Pringle 1995; Costigan and Snyder 1998). Haploid cells (*MATa* or *MATα* cells) use an axial budding pattern in which cells form new buds adjacent to the previous site of cytokinesis. In the wild, the axial budding pattern may place cells of different mating types in close proximity and facilitate formation of diploids, which are genetically more stable and more resistant to environmental stresses (Nasmyth 1982; Costigan and Snyder 1998). *MATa/MATα* cells (usually diploids) use a bipolar budding pattern.

That is, the first bud of a virgin cell is almost exclusively at the distal pole, opposite to the birth site of this cell. Subsequent budding events use either pole, with younger mothers exhibiting a distal pole preference. The bipolar budding pattern is advantageous for spreading across solid surfaces (Gimeno et al. 1992; Madden et al. 1992). Thus, although proper bud site selection is not essential for the survival of *S. cerevisiae*, proper budding patterns are probably favorable for *S. cerevisiae* cells living in the real world (Gimeno et al. 1992; Madden et al. 1992).

In this chapter, we will discuss the processes and molecular components involved in bud growth, bud site selection, filamentous differentiation and mating in yeasts. Since many of the components important for controlling cell shape and polarity in *S. cerevisiae* have homologues in multicellular organisms, these cellular processes are likely to be highly conserved among eukaryotes.

II. Processes Involved in Cell Morphogenesis

Cell wall labeling experiments have revealed distinct phases of cell surface remodeling during bud formation in *S. cerevisiae* (Tkacz and Lampen 1972; Farkas et al. 1974; Lew and Reed 1993). Shortly after bud emergence at the end of the G1/S period of the cell cycle, new cell wall components are primarily deposited at bud tips (apical growth). Later, in the G2 phase of the cell cycle, the deposition of the new cell wall material occurs throughout the entire bud surface (isotropic growth), although still restricted to the bud. Finally, new cell wall material is deposited at the neck between the mother cell and the bud (repolarization) in preparation for cytokinesis. Figure 2 illustrates different phases of polarized growth during bud formation.

Apical growth allows bud elongation and thus is a key process controlling the shape of yeast cells. Prolonged apical growth during budding results in elongated cells whereas insufficient apical growth leads to rounder cells (Lew and Reed 1995a). Diploid cells, which appear substantially more elongated than haploid cells (Galitski et al. 1999), presumably have a longer apical growth phase. Less concentrated apical growth makes bud elongation less efficient and causes a rounder final shape (Sheu et al. 2000). Therefore, components

Haploid - axial budding

Diploid - bipolar budding

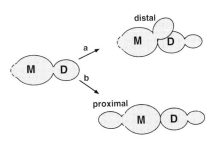

Fig. 1. *Saccharomyces cerevisiae* cells exhibit defined budding patterns. Haploid *MATa* and *MATα* cells use an axial budding pattern in which both mother (*M*) and daughter (*D*) cells bud adjacent to the latest site of cytokinesis. Diploid *MATa/MATα* cells use a bipolar budding pattern in which daughter cells bud almost exclusively at the distal pole, opposite to their birth sites, whereas mother cells bud at either the distal pole (*a*) or the proximal pole (*b*). The birth scar on the diploid mother cells is marked by a *dashed line*

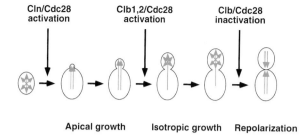

Apical growth   Isotropic growth   Repolarization

Fig. 2. Phases of polarized growth during bud formation. *Arrows in gray* indicate the direction of growth. The key cell cycle events that control the initiation and termination of each growth processes are indicated. See later section for discussion on cell cycle control

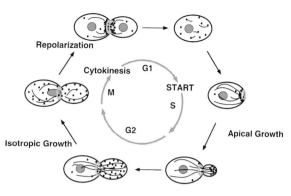

Fig. 3. Localization of the actin and septin cytoskeletons during cell cycle. Actin exists as cortical patches (*dots*) and cables (*wavy lines*). Actin patches form a ring at the incipient bud site soon after START. These patches remain concentrated in the small bud and become dispersed throughout the cortex within the bud as it grows larger. As the bud reaches its maximal size, actin patches depolarize briefly throughout the cell and repolarize toward the mother-bud neck at time of cytokinesis. Septins (the *gray ring*) assemble into a ring structure at the incipient bud site at about the same time as actin. The septin ring remains at the mother-bud neck throughout the rest of the cell cycle and appears as a double-ring structure during mitosis; the double ring splits after cytokinesis. The *shaded circle/oval* represents the nucleus

that (1) regulate the onset and termination of apical growth, (2) establish apical growth and (3) help concentrate apical growth are the key factors that influence cell shape.

III. Structural Components Important for Cell Shape and Polarity

The cytoskeleton and cell wall are structural components involved in the establishment and maintenance of cell shape and polarity. There are at least three cytoskeletal systems in budding yeast: the actin, microtubule and septin cytoskeletons. Actin and septins are crucial structural components for cellular morphogenesis and cell polarity. Microtubules are not essential for polarized growth; however, they are crucial for other cellular processes, such as spindle orientation, in response to cell polarity (see also Chap. 10, this Vol.).

A. The Actin Cytoskeleton

The actin cytoskeleton is highly polarized in *S. cerevisiae*. Staining with rhodamine-conjugated phalloidin, which binds to F-actin, reveals two major actin structures, actin patches and actin cables (Adams and Pringle 1984; Kilmartin and Adams 1984). Actin patches reside at the cell cortex and actin cables run parallel to the polarity axis (the mother-bud axis during budding and longitudinal to the projection during mating), often intersecting actin patches at their ends. These structures exhibit cell cycle dependent distribution (Fig. 3). Actin patches are highly motile

and concentrate at sites of active growth and cell wall deposition (Adams and Pringle 1984; Kilmartin and Adams 1984; Doyle and Botstein 1996; Waddle et al. 1996). Patches form a ring at the incipient bud site before bud emergence and remain concentrated at the surface of growing buds. As the bud grows larger, the patches become dispersed throughout the cortex. These patches repolarize to the mother-bud neck at cytokinesis. The distribution of actin structures also changes in response to environment. Actin patches are hyperpolarized during filamentous differentiation (Kron et al. 1994; Cali et al. 1998), and become briefly depolarized under heat shock and osmotic stress (Chowdhury et al. 1992; Palmer et al. 1992; Lew and Reed 1993; Lillie and Brown 1994).

In budding yeast, actin is encoded by the *ACT1* gene (Gallwitz and Sures 1980; Ng and Abelson 1980), which is essential for viability. Mutants with temperature-sensitive alleles of *ACT1* often form large mother cells with small buds at the restrictive temperature, suggestive of a defect in which growth occurs throughout the mother cell rather than directed into the bud (Novick and Botstein 1985). Consistent with this idea, chitin, an important component of the *S. cerevisiae* cell wall, is delocalized throughout the cell in actin mutants instead of accumulating at bud scars (see below). Mutations in the *ACT1* gene also cause defects in a variety of other cellular processes, including cytokinesis, mating projection formation, pseudohyphal growth, nuclear migration, organelle movement, and endocytosis (Shortle et al. 1982, 1984; Novick and Botstein 1985; Palmer et al. 1992; Drubin et al. 1993; Kubler and Riezman 1993; Munn et al. 1995; Yang et al. 1997; Cali et al. 1998).

The primary role for actin cables in polarized growth in *S. cerevisiae* is likely through the directional targeting of secretory vesicles to growth sites. The *act1-2* mutants and cells lacking Tpm1, the yeast tropomyosin, have almost no actin cables and accumulate vesicles (Novick and Botstein 1985; Liu and Bretscher 1989b, 1992). Tpm1 localizes to cables and overexpression of this protein can restores actin cables in the *act1-2* mutant (Liu and Bretscher 1989a). Thus, Tpm1 may function to stabilize actin cables. A conditional allele of *MYO2*, encoding a type-V myosin, causes phenotypes similar to the *tpm1Δ* mutant (Johnston et al. 1991; Liu and Bretscher 1992). Myo2 localizes to bud tips and bud necks and is required for vesicular transport (Johnston et al. 1991; Liu and Bretscher 1992; Lillie and Brown 1994; Govindan et al. 1995; Santos and Snyder 1997). Thus, Myo2 may be responsible for carrying secretory vesicles to polarized growth sites along the Tpm1 stabilized cables.

In spite of the essential role for actin in polarized growth, the identity and function of actin patches are mysterious. Electron microscopic studies indicate that cortical actin patches are sites of plasma membrane invagination, and it has been suggested that these patches represent sites of membrane and cell wall deposition (Mulholland et al. 1994). Alternatively, actin patches may correspond to endocytic sites since actin also plays a role in endocytosis (Kubler and Riezman 1993; Munn et al. 1995). Consistent with this possibility, many *S. cerevisiae* mutants defective in endocytosis also delocalize the actin cytoskeleton (Munn et al. 1995; Tang and Cai 1996; Tang et al. 1997). Pan1, a homologue of mammalian EH domain proteins, which associate with the plasma membrane clathrin adapter complex involved in endocytosis, localizes to actin patches (Tang and Cai 1996). Two homologous kinases, Ark1 and Prk1, are found in cortical patches and are involved in regulation of both Pan1 and actin (Cope et al. 1999; Zeng and Cai 1999). In addition, the Arp2/3 complex (Arp:

actin-related protein) controls actin assembly in many organisms (reviewed in Machesky and Gould 1999). This complex colocalizes with actin patches and controls patch movement and endocytosis in budding *S. cerevisiae* (Moreau et al. 1996, 1997). Whether all the patches are the same and whether their distribution directly influences cell polarity, or their polarization is merely the outcome of polarized growth, remain to be addressed.

Localization of actin patches at the mother-bud neck indicates that the actin cytoskeleton participates in cytokinesis. Many actin regulators also localize to the mother-bud neck and affect cytokinesis. Like many other organisms, budding *S. cerevisiae* has a contractile actomyosin ring containing actin and the type-II myosin encoded by the *MYO1* gene (Field et al. 1999). *S. cerevisiae* strains deleted for *MYO1* fail to form this ring and undergo septation with less efficiency (Bi et al. 1998).

A large number of actin-interacting proteins have been identified using biochemical and genetic methods (reviewed in Ayscough 1998; Costigan and Snyder 1998). Many of them, such as profilin, cofilin, fimbrin, capping proteins, several myosins, and tropomyosin, are highly conserved throughout all eukaryotes studied. Some of these proteins affect the general properties of actin filament assembly and mutations in some of these proteins have many defects similar to actin mutants.

B. Septins

The septin cytoskeleton in budding *S. cerevisiae* participates in cell morphogenesis, cytokinesis, bud site selection, polarized growth, a morphogenesis checkpoint, and spore wall formation (Longtine et al. 1996). The septins were first recognized in *S. cerevisiae* as a group of homologous proteins required for cytokinesis (Hartwell et al. 1974). Temperature-sensitive mutations in four septin genes, *CDC3*, *CDC10*, *CDC11* and *CDC12*, cause the formation of long chains of connected cells at the restrictive temperature (Hartwell 1971). Since their initial discovery, three additional septins have been found, two of which participate in sporulation (De Virgilio et al. 1996; Fares et al. 1996). Septins have now been found in other fungi, insects and mammals, and are probably ubiquitous (Neufeld and Rubin 1994; Fares et al. 1995; Longtine et al. 1996). As in budding yeast, septins in other organisms are important for cytokinesis.

Septins are likely to form filaments, perhaps in a fashion analogous to their distant relatives, the tubulins (Flescher et al. 1993). All septins contain a guanine nucleotide-binding motif (Longtine et al. 1996). Many, but not all, septins also contain a coiled-coil domain. By immunolocalization using fluorescence and electron microscopy, septins localize to the cell periphery of the neck where 10-nm electron-dense filaments reside closely associated with the plasma membrane (Byers 1981; Haarer and Pringle 1987; Ford and Pringle 1991; Kim et al. 1991). Mutations in *CDC3*, *CDC10*, *CDC11* and *CDC12* result in loss of the 10-nm filaments, suggesting that septins comprise this structure. Septin complexes purified from fly embryos form filaments; they bind and hydrolyze GTP (Field et al. 1996; Longtine et al. 1996). Cdc3, Cdc10, Cdc11 and Cdc12 also copurify with each other from yeast extracts and form long, paired filaments (Frazier et al. 1998). The long filaments were not found when the complexes were isolated from strains missing one of the non-essential septins (Cdc10 or Cdc11); instead, shorter filaments were found.

The localization of septins is cell-cycle-regulated (Fig. 3; Haarer and Pringle 1987; Ford and Pringle 1991; Kim et al. 1991). Cdc3, Cdc10, Cdc11 and Cdc12 form a ring structure at the incipient bud site at approximately the same time as the actin cytoskeleton, and they remain at the mother-bud neck after bud emergence and throughout the rest of the cell cycle. During mitosis, septins appear as a double-ring structure, which splits after cytokinesis. Localization of septins to the incipient bud site depends on the function of the polarity establishment genes *CDC42* and *CDC24*, but not actins (Longtine et al. 1996; Pringle et al. 1995; Richman et al. 1999). Septins still localize at the presumptive bud site in cells treated with Latrunculin A, an actin-depolymerizing drug, and actin still localizes to the incipient bud site in septin mutants (Adams and Pringle 1984; Ford and Pringle 1991; Ayscough et al. 1997). Thus, in spite of their colocalization to the presumptive bud site in a manner dependent on polarity establishment components, septins and actins localize to this site independently from each other.

Although proper localization and filament formation of septins are important for their function in cytokinesis, their exact role in this process is not clear. Since septins are the first components for cytokinesis to arrive at the bud neck, the septin cytoskeleton might provide a scaffold for the assembly of proteins participating in cytokinesis and new cell wall synthesis at the region. Consistent with this idea, the formation and maintenance of the actomyosin ring at the division site is septin-dependent (Bi et al. 1998). In addition, septin mutants display delocalized chitin deposition (Roberts et al. 1983; Flescher et al. 1993; DeMarini et al. 1997). Thus, septins may comprise a polarity determinant at sites of cytokinesis.

In addition to their role in cytokinesis, septins are also important for maintaining the asymmetric bud growth during the isotropic phase of bud formation (Barral and Snyder, 2000). Temperature-sensitive septin mutants undergoing isotropic growth depolarize several polarity proteins and actin stability factors at semi-permissive temperatures. Perhaps, the septin cytoskeleton forms a physical barrier at the mother-bud neck to confine components crucial for growth within the bud during isotropic growth.

C. The Cell Wall

The cell wall is the exoskeleton that maintains the shape and structural integrity of the cell (see Chap. 9, this Vol.). The *S. cerevisiae* cell wall accounts for 20–30% of the dry weight of the cell (reviewed in Klis 1994; Cid et al. 1995; Smits et al. 1999). It consists mainly of β-glucans, mannoproteins and chitin. β-Glucans form the main structural framework to which mannoproteins are attached. Chitin is present in small amounts but is essential for cell growth (Shaw et al. 1991). The deposition of chitin in the cell wall is spatially and temporally regulated (Bulawa 1993; Cid et al. 1995). During vegetative growth, chitin is mainly present at the mother-bud neck and the bud scar (the chitin-rich remnants on the mother surface after cell separation) and is present in small amounts in the lateral wall (Cabib and Bowers 1971; Molano et al. 1980). During mating, chitin is synthesized and deposited at the base of the mating projection (Lipke et al. 1976; Schekman and Brawley 1979).

The yeast cell wall is a dynamic structure rather than merely an inert hard outer covering. Constant remodeling and modification of the composition and structure of the cell wall occur throughout bud emergence, bud growth, septation and cell separation, mating projection formation, cell fusion between mating partners and sporulation. These changes must be properly controlled according to cell cycle stages and environmental signals. In addition, the yeast cell wall is also involved in controlling cell polarity, such as bud site selection (Brown and Bussey 1993; Yabe et al. 1996). Moreover, the asymmetric nature of the cell wall architecture (e.g., chitin deposition and localization of some cell wall proteins) may contribute to cell polarity in yeast.

Three chitin synthase activities (CHSI, CHSII and CHSIII) have been identified; each has a distinct function. CHSI is responsible for repairing and replenishing chitin hydrolyzed by chitinase during cell separation (Cabib et al. 1989, 1992). CHSII functions to synthesize the primary septum (Silverman et al. 1988; Shaw et al. 1991). CHSIII activity is required for the formation of chitin rings at bud necks and bud scars as well as for chitin synthesis during mating (Shaw et al. 1991; Valdivieso et al. 1991). Localization of CHSIII activity is regulated during polarized growth. One component of the catalytic subunit, Chs3, is concentrated at sites of polarized growth (Santos and Snyder 1997). Proper localization of Chs3 to polarized growth sites in yeast is through complex assembly involving the septins, a septin-interacting protein, Bni4, and Chs4, a putative regulatory subunit of CHSIII (DeMarini et al. 1997). Localization of Chs3 also requires the function of Chs5, a trans-Golgi network protein, and Myo2 (Santos et al. 1997; Santos and Snyder 1997). Interestingly, Chs5 is also involved in control of other cell polarity processes, such as proper bipolar bud site selection, mating and cell fusion (Santos et al. 1997), suggesting a role for chitin in these processes.

The expression of many cell wall proteins is cell cycle regulated and affected by nutrient availability and mating pheromone (summarized and reviewed by Smits et al. 1999). Some of the cell wall proteins, such as the flocculin Flo11, which is involved in cell-cell adhesion, are up-regulated during filamentous differentiation (see below), suggesting a role for cell wall proteins in morphogenesis (Lo and Dranginis 1998; Rupp et al. 1999). Another putative cell wall protein Fig. 2, induced by mating pheromone, affects morphogenesis during cell fusion (Erdman et al. 1998).

IV. Polarized Secretion

Cell growth and morphogenesis require the expansion of the cell surface presumably through the targeted transport of secretory vesicles. A detailed review of this process is presented elsewhere (Kaiser et al. 1996; see also Chap. 11, Sect. II, this Vol.). Like the actin cytoskeleton, membrane vesicles are polarized to a region near the cell surface before bud emergence and in the nascent bud; both Golgi and endoplasmic reticulum markers are present in these vesicles (Pruess et al. 1991, 1992). The density of vesicles appears to decrease as the bud enlarges, presumably due to their rapid fusion with the plasma membrane.

Consistent with the expectation that secretion is polarized in yeast, cytological staining of several secreted proteins indicates that they accumulate near the bud surface. Furthermore, several proteins involved in the late steps of secretion are polarized to growth sites. Sec4, a small GTPase of the rab family that plays a central role in regulating the post-Golgi stage of secretion in *S. cerevisiae* (Novick and Brennwald 1993), and components of exocyst (required for exocytosis), such as Sec6, Sec8 and Sec15, concentrate at the bud tip during apical growth and at the mother-bud neck during cytokinesis (Goud et al. 1988; TerBush and Novick 1995). Growth site localization of Sec4 depends on its regulator Sec2 but not exocyst components (Walch-Solimena et al. 1997). Localization of Sec4 also depends on actin assembly (Ayscough et al. 1997), consistent with the role for actin cables in transporting secretory vesicles (see above). Another important component involved in polarized secretion is Sec3, which forms a tight patch at the incipient bud site and bud tip at early stages of bud formation and relocalizes to the bud neck during cytokinesis when secretion is polarized towards that region (Finger and Novick 1997; Finger et al. 1998). Sec3 may be the spatial landmark for polarized secretion since Sec3 localization does not rely on vesicle trafficking, actin assembly and components crucial for early stages of bud growth (Finger et al. 1998).

Different types of secretory vesicles exist in *S. cerevisiae* cells (Harsay and Bretscher 1995). It has been shown that *sec4* mutants accumulate two types of vesicles: one contains endoglucanase and the major plasma membrane ATPase, the other contains invertase and acid phosphatase. In addition, deletion of *TPM1* or *MYO2* causes accumulation of vesicles, but cells still secret invertase at normal kinetics (Liu and Bretscher 1989a,b; Johnston et al. 1991; Amatruda et al. 1992; Govindan et al. 1995). Moreover, the Golgi protein Chs5 is required for localization of Chs3 but not other secretory proteins (Santos and Snyder 1997; Santos et al. 1997). Thus, different types of vesicles may be responsible for transporting different cellular components to the cell surface.

V. Regulation of Structural Components

A. Establishment of Polarized Growth Site

The Rho-type small GTPase Cdc42 module is a key component for the establishment of cell polarity, mainly by directing polarized assembly of components important for budding, such as actin, septins, Spa2, and chitin (Sloat and Pringle 1978; Adams et al. 1990; Johnson and Pringle 1990; Snyder et al. 1991; Ziman et al. 1991, 1993; Zheng et al. 1994; Li et al. 1995). A detailed review of Cdc42 is available (Johnson 1999). The Cdc42 GTPase module consists of Cdc42, the guanine nucleotide exchange factor (GEF), Cdc24 (Sloat and Pringle 1978; Sloat et al. 1981), the guanine nucleotide dissociation factor (GDI), Rdi1 (Masuda et al. 1994), and the GTPase-activating proteins (GAPs) Bem3, Rga1(Dbm1) and Rga2 (Stevenson et al. 1992, 1995; Zheng et al. 1994; Chen et al. 1996). Cdc42, Cdc24 and Bem1 are critical for bud emergence and are commonly referred to as polarity-establishment proteins. Temperature-sensitive mutants defective in genes encoding these proteins fail to form buds and give rise to large, round, unbudded cells with multiple nuclei (Sloat and Pringle 1978; Sloat et al. 1981; Bender and Pringle 1989, 1991; Adams et al. 1990). In addition, dominant active *CDC42* alleles cause cells to form multiple buds (Ziman et al. 1991).

Cdc42 is highly conserved throughout organisms. Cdc42 is prenylated in all organisms examined so far, and this prenylation is essential for association with membrane (Johnson 1999). In budding *S. cerevisiae*, Cdc42 localizes to sites of polarized growth and cytokinesis (Ziman et al. 1991, 1993; Johnson 1999), indicating that it functions in multiple stages of the life cycle. Polarized localization of Cdc42 does not rely on a functional actin cytoskeleton (Ayscough et al. 1997). Rather, Cdc42 functions to promote actin assembly at

growth sites. *cdc42* mutants exhibit depolarized actin patches (Adams et al. 1990), and Cdc42 can stimulate actin polymerization in permeabilized yeast cells (Li et al. 1995). Cdc42 also interacts with the *S. cerevisiae* formin Bni1, which is implicated in modulating actin functions (see below). In other organisms, Cdc42 also regulates actin assembly (Hall 1994).

Cdc24 is the Cdc42 GEF that facilitates formation of the active GTP-bound form of Cdc42. Cdc24 contains a Dbl homology (DH) domain with high similarity to the Dbl family of GEFs in mammals, a pleckstrin homology (PH) domain and two potential Ca^{2+}-binding domains (Miyamoto et al. 1987). PH domains interact with phospholipids and the $G\beta\gamma$ subunit of the heterotrimeric G-protein (Harlan et al. 1994; Touhara et al. 1994). A functional GFP-Cdc24 fusion localizes to the incipient bud sites, mother-bud necks and mating projection tips (Johnson 1999; Nern and Arkowitz 1999), consistent with its function as a Cdc42 regulator.

Bem1, a protein containing two SH3 domains at its N-terminus (Chenevert et al. 1992), was isolated as a multicopy suppressor of both *cdc24-4* and *cdc42-1* temperature-sensitive mutants (Bender and Pringle 1989), suggesting its role in facilitating the Cdc42 module. Consistent with this function, Bem1 interacts with Cdc24 (Peterson et al. 1994; Zheng et al. 1995). In addition, the *bem1* mutation is lethal in combination with a mutant allele of *BUD5*, involved in general bud site selection function (see below), and mutations in the Mpk1 (Slt2) MAP kinase pathway (see below), suggesting its position as a link between bud site selection and bud growth. Bem1 was proposed to be a scaffold protein that links the Cdc42 module to the mating signaling pathway (Leeuw et al. 1995; Lyons et al. 1996; see below). Whether Bem1 plays a similar role during vegetative growth is not known.

Cdc42 also regulates the actin cytoskeleton during cytokinesis. Its function in cytokinesis is probably connected to the IQGAP Iqg1(Cyk1) and the p21-activated kinase (PAK) Cla4 (Cvrckova et al. 1995; Benton et al. 1997; Epp and Chant 1997; Osman and Cerione 1998). IQGAPs in higher organisms are capable of interacting with F-actin, calmodulin, and Cdc42 (Machesky 1998). Iqg1 is thought to be a scaffold protein at the site of cytokinesis that mediates the formation of the actomyosin ring (Epp and Chant 1997; Lippincott and Li 1998). Cla4 interacts with the Cdc42 and its kinase activity peaks at G2/M (Benton et al. 1997), suggesting the function of both proteins in cytokinesis.

B. Components That Ensure the Quality of Polarized Growth

The non-essential polarity proteins Spa2, Pea2, Bud6 and Bni1 also participate in multiple aspects of *S. cerevisiae* polarized growth and establishment of cell polarity. Mutants defective in any of these proteins have many common phenotypes. These cells appear rounder than wild-type cells and have defects in pseudohypha formation and cytokinesis (Mosch and Fink 1997; Roemer et al. 1998; Sheu et al. 2000). They are also defective in mating projection formation and have decreased mating efficiency (Gehrung and Snyder 1990; Valtz and Herskowitz 1996; Amberg et al. 1997; Evangelista et al. 1997). In addition, these mutants display a diploid-specific random budding pattern (Snyder 1989; Valtz and Herskowitz 1996; Zahner et al. 1996; Amberg et al. 1997). All these phenotypes suggest that these proteins function in a common cellular process that affects both morphogenesis and bipolar budding.

In addition to the similar mutation phenotypes, Spa2, Pea2, Bud6 and Bni1 also display very similar localization patterns. They localize to sites of polarized growth during vegetative growth and during mating (Snyder 1989; Gehrung and Snyder 1990; Snyder et al. 1991; Jansen et al. 1996; Valtz and Herskowitz 1996; Amberg et al. 1997; Evangelista et al. 1997). At the onset of apical growth, these proteins concentrate at the incipient bud site and remain at the bud tip during the apical growth phase. Their bud tip localization becomes more dispersed or not detectable in large budded cells undergoing isotropic growth. At the repolarization phase prior to cytokinesis, these proteins concentrate at the mother-bud neck.

Spa2 is a large protein containing several domains, including a predicted coiled-coil region, a stretch of 25 nine-amino-acid repeats, and five regions that are conserved with those of a related *S. cerevisiae* protein, Sph1 (Gehrung and Snyder 1990; Arkowitz and Lowe 1997; Roemer et al. 1998). These five regions are termed Spa2 homology domain (SHD)-I–V. The N-terminal SHD-I, which interacts with MEKs (MAP kinase kinases, see below and Fig. 4), is also present in proteins from a wide variety of eukaryotes (Roemer et al. 1998; Sheu et al. 1998). SHD-II of Spa2 overlaps with part of the coiled-coil region and interacts with Pea2, which also contains a putative coiled-coil domain (Valtz and Herskowitz 1996; Sheu et al. 1998). SHD-II is required for proper localization of both Spa2 and Pea2. Spa2 and Pea2 form a tight complex in vivo, as demonstrated by coimmunoprecipition experiments. Spa2 also displays a two-hybrid interaction with Bud6(Aip3; Aip: actin-interacting protein) (Amberg et al. 1997; Sheu et al. 1998). Association of Spa2 with Bud6 is not as strong as that with Pea2 since Spa2 and Bud6 do not coimmunoprecipitate. Nevertheless, Spa2, Pea2 and Bud6 cofractionate in a sucrose gradient as part of a 12S complex, termed the polarisome. Spa2 and Bud6 also interact with Bni1 (Evangelista et al. 1997; Fujiwara et al. 1998),

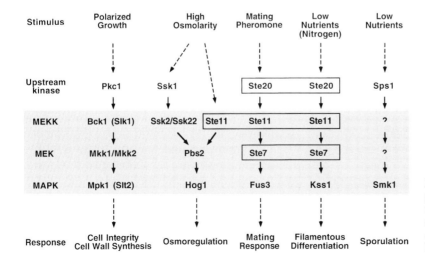

Fig. 4. Summary of the mitogenactivated protein kinase (MAPK) pathways in budding *S. cerevisiae*. *MAPK* MAP kinase; *MEK* MAPK kinase; *MEKK* MEK kinase

a member of the formin homology (FH) protein family (reviewed by Wasserman 1998). Formin is present in a wide variety of organisms and participates in actin-mediated functions, such as the formation and maintenance of the contractile ring during cytokinesis and other processes involved in control of cell shape and polarity. Like many proteins of the FH family, Bni1 interacts with Rho-type GTPases and profilin (Kohno et al. 1996; Evangelista et al. 1997); all of them function in regulating actin dynamics.

The interactions between Spa2, Pea2, Bud6 and Bni1, their identical localization patterns, and the common phenotypes of mutants lacking these proteins suggest that these proteins function as part of a multiprotein complex which influences polarized morphogenesis, perhaps via regulation of actin-mediated cellular processes. Interestingly, many components of this complex also interact with proteins of signaling pathways (Sheu et al. 1998). Spa2 interacts with the MEKs Mkk1 and Mkk2 of the Mpk1 MAPK pathway, and Ste7 and Ste11, the MEK and MEKK, respectively, of the Fus3/Kss1 MAPK pathway (see below and Fig. 4). Bud6 interacts with Ste11. In addition to interaction with MAPK pathways, one of the complex components, Bni1, interacts with the effector domain of Cdc42 and Rho1 (Kohno et al. 1996; Evangelista et al. 1997), both of which function upstream of MAPK pathways (see below). Cdc42 helps activate Ste20, which, in turn, activates Ste11 (Leberer et al. 1992; Ramer and Davis 1993). One of the Rho1 effectors is the protein kinase C, Pkc1, which functions upstream of the Mpk1 signaling pathway (Nonaka et al. 1995; Kamada et al. 1996). The biological meaning of these complex interaction networks remains elusive. Yet clearly the processes involving all these components are interconnected. Interestingly, the Spa2-homologous protein, Sph1, interacts with the same signaling components as Spa2 in two-hybrid assays (Roemer et al. 1998). Sph1 also interacts with Bud6, but not Pea2. Although deletion of *SPH1* results in relatively mild defects in many aspects of morphogenesis, Sph1 appears to function in the same process as Spa2. The circumstances under which Sph1 is important remain unknown.

What processes does the complex containing Spa2, Pea2, Bud6 and Bni1 participate in? Mutational analyses of Spa2 with deletions and small insertions have failed to associate any specific region with a single phenotype of the *spa2* mutant (Sheu et al. 1998); that is, all defective mutant alleles display all the phenotypes of *spa2Δ* cells. This suggests that all the common phenotypes observed in *spa2Δ*, *pea2Δ*, *bud6Δ* and *bni1Δ* mutants result from a single cellular defect. Recent evidence has shown that Spa2, Pea2, Bud6 and Bni1 are important for the apical growth (Sheu et al. 2000). These proteins are required for the elongation of cells and buds in wild-type and *cdc34*, *cdc12* and *clb2* mutants. Since the elongation of buds and cells depend on apical growth, these components may control cell shape by influencing apical growth. This idea helps explain why mutants defective in *SPA2*, *PEA2* and *BNI1* fail to form elongated cells during filamentous differentiation (Mosch and Fink 1997; Roemer et al. 1998).

What is the role of this complex in apical growth and how does the failure in fulfilling this function lead to the multiple defects in morphogenesis and cell polarity? These proteins are not required for the initiation of apical growth, since, unlike mutations in polarity establishment genes, such as *cdc42* and *cdc24* (Pringle et al. 1995; Drubin and Nelson 1996), *spa2Δ*, *pea2Δ*, *bud6Δ* and *bni1Δ*, cells are able to form buds and exhibit polarized actin distribution. In addition, direct observation of cell surface expansion using FITC-Con A labeling and return-to-growth experiments have shown that these mutants (except *bni1Δ*) and wild-type cells have a very similar proportion of buds undergoing apical growth (Sheu et al. 2000). However, these mutants fail to concentrate apical growth in a small region at the bud tip. The staining pattern of a polarized secretion marker, Sec4, is more diffuse at bud tips in *spa2Δ*, *pea2Δ*, *bud6Δ* and *bni1Δ* mutants, compared with the staining pattern in wild-type cells. Sec4 staining is also more diffuse at the mother-bud neck in these mutants, suggesting their defect in cytokinesis is caused by the failure to concentrate secretion and cell wall deposition during the repolarization phase. Interestingly, the Sec4 staining pattern is most diffuse and aberrant in the *bni1Δ* mutant, which displays the most severe defects in apical growth and in all aspects of morphogenesis and polarity functions (Zahner et al. 1996; Evangelista et al. 1997; Long et al. 1997; Sheu et al. 2000). Thus, the primary function of the complex containing Spa2, Pea2, Bud6 and Bni1 is to help concentrate polarized growth to a small region and thereby governs the quality of polarized growth. Elimination of this function does not abolish polarized growth completely but leads to formation of aberrant mating projections with wider tips, cell elongation defects during vegetative growth and filamentous differentiation, and inefficient cell separation after cytokinesis.

C. Rho-Type Guanosine Triphosphatases (GTPases)

In addition to Cdc42, there are a number of other Rho-type GTPases, including Rho1, Rho2, Rho3 and Rho4, that are involved in bud growth (Madaule et al. 1987; Matsui and Toh-e 1992). Rho1 is essential for viability and is most related to the mammalian RhoA, which is required for assembly of actin-based stress fiber. Rho3 is required for viability at high temperature, whereas rho2Δ or rho4Δ alone do not confer any growth defects. Like Cdc42, Rho1, Rho3 and Rho4 have been shown to interact with the *S. cerevisiae* formin Bni1 (Kohno et al. 1996; Imamura et al. 1997), suggesting that part of their functions in cell growth is through regulating the actin cytoskeleton. Recent studies have revealed some common and unique functions for these *S. cerevisiae* Rho proteins.

Rho1 localizes to sites of polarized growth, including the incipient bud site, the bud tip and the site of cytokinesis (Yamochi et al. 1994), suggesting that it is involved in multiple steps of bud growth. However, Rho1 is not required for bud emergence. A temperature-sensitive mutant of *RHO1*, *rho1-104*, arrests at the restrictive temperature with small buds (Yamochi et al. 1994). The GEFs, Rom1 and Rom2, and the GAP, Bem2 (Kim et al. 1994; Peterson et al. 1994; Ozaki et al. 1996), regulate Rho1. Deletion of both *ROM1* and *ROM2* causes nonviability and *rom1 rom2* double mutants arrest with small buds like *rho1* mutants. Rom2 can be activated by cell wall alterations (Bickle et al. 1998). The *S. cerevisiae* phosphatidylinositol kinase Tor2 can also activate Rho1 and Rho2, partly via Rom2 (Schmidt et al. 1997). Like Rho1, Tor2 is also implicated in regulation of actin organization.

Rho1 is involved in maintaining cell wall integrity during polarized growth. This function is mediated partly through the Pkc1-Mpk1 cell integrity pathway (see below and Fig. 4), which is responsible for activating the transcription of a number of genes involved in cell wall functions (Igual et al. 1996). Rho1 binds Pkc1 and is

required for Pkc1 activation in vivo (Nonaka et al. 1995; Kamada et al. 1996). In addition, overexpression of *PKC1* can rescue the *rho1* mutant. Overexpression of *PKC1* also suppresses the temperature-sensitive phenotype of the *tor2* mutant (Helliwell et al. 1998).

Rho1 is also required for β-1,3-glucan synthesis (Qadota et al. 1996). β-1,3-Glucan is synthesized at the cell surface by the glucan synthases, Fks1 and Fks2 (Masur et al. 1995; Ram et al. 1995; Qadota et al. 1996). Rho1 copurifies with the β-1,3-glucan synthase activity and associates with Fks1, which colocalizes with Rho1 to growth sites. Thus, Rho1 functions as a regulatory subunit of β-1,3-glucan synthase.

Rho2 is highly homologous to Rho1, and may have redundant functions in cell growth and be regulated by the same GAPs and GEFs (Madaule et al. 1987; Wang and Bretscher 1995; Manning et al. 1997). *rho2Δ*, *rom2Δ* and *bem2Δ* are supersensitive to a microtubule depolymerization drug, benomyl (Wang and Bretscher 1995; Manning et al. 1997). This phenotype suggests a role for Rho2 in microtubule function and raises the possibility that Rho2 may coordinate polarized growth with microtubule-dependent cellular processes.

The functions of Rho3 and Rho4 partially overlap but are distinct from the functions of Rho1 and Rho2. Disruption of *RHO3* results in slow growth and cell lysis at high temperature, which is exacerbated by deletion of *RHO4* altogether; this defect can be alleviated by overexpression of *RHO4* and the polarity establishment genes *BEM1* and *CDC42*, but not *RHO1* and *RHO2* (Matsui and Toh-e 1992). In addition, expression of an activated allele of *RHO3* causes cold-sensitivity and elongated cells that are bent (Imai et al. 1996). Rho3 localizes to sites of polarized growth and interacts with Myo2, and Exo70, a component of the exocyst complex, in the two-hybrid system and in vitro binding with purified proteins (TerBush et al. 1996; Robinson et al. 1999). Thus, Rho3 appears to influence cell growth by regulating polarized secretion and actin cytoskeleton.

D. p21-Activated Kinases (PAKs)

The p21-activated kinase (PAK) interacts with the effector domain of Cdc42/Rac small GTPases and is generally thought to be regulated by Cdc42/Rac

(Sells and Chernoff 1997). In budding *S. cerevisiae*, there are four PAKs, at least two of which, Ste20 and Cla4, are involved in polarized morphogenesis. Both Ste20 and Cla4 interact with the activated form of Cdc42 (Cvrckova et al. 1995; Simon et al. 1995; Zhao et al. 1995; Peter et al. 1996; Benton et al. 1997; Leberer et al. 1997).

1. Ste20

Ste20 functions upstream of MAPK pathways involved in mating and pseudohypha formation and is required for these two developmental events in *S. cerevisiae* (Fig. 4; Chapter 3 Leberer et al. 1992; Liu et al. 1993; Neiman and Herskowitz 1994; Roberts and Fink 1994). Ste20 is concentrated at the tips of both emerging and small buds and later disperses as the buds enlarge (Peter et al. 1996; Leberer et al. 1997); Ste20 also localizes to mating projection tips. Interaction with Cdc42 is required for localizing Ste20 to the growth site but not for activating the mating pathway since *ste20* mutants lacking the Cdc42-binding site show nearly wild-type levels of pheromone-induced transcription (Peter et al. 1996; Leberer et al. 1997). However, the same mutants are defective in signaling during pseudohyphal growth (Peter et al. 1996; Leberer et al. 1997).

Ste20 may also have functions independent of the MAPK cascade (see below). Genetic analysis has shown that Ste20 is involved in regulating the duration of apical growth (Sheu et al. 2000); this function does not require the downstream components Ste11 and Ste7 of the MAPK cascade. Since Ste20 can be phosphorylated in a Cln2-Cdc28 dependent manner (Oehlen and Cross 1998; Wu et al. 1998), and Cln2-Cdc28 is involved in activation of apical growth (see below), it is likely that Ste20 is activated by the G1 cyclin-dependent kinases (CDKs) to regulate the timing of apical growth through polarization of the actin cytoskeleton. The role of Ste20 as a regulator of actin assembly is further supported by the finding that overexpression of *STE20* suppresses *cdc42-1* growth defects and cytoskeletal defects and that Ste20 corrects actin-assembly defects of permeabilized *cdc42-1* cells (Eby et al. 1998). Thus, Ste20 may function to mediate the role for Cdc42 in regulation of the actin cytoskeleton.

Ste20 is required for budding at the distal pole in bipolar budding cell types (Sheu et al. 2000). This function is also independent of Ste11 and Ste7, and may, at least in part, be related to its function in apical growth. Alternatively, Ste20 may directly regulate distal bud site tags (see below).

2. Cla4

Another *S. cerevisiae* PAK, Cla4, was originally identified in genetic screens for mutants that were nonviable in the absence of the G1 cyclins, Cln1 and Cln2 (Benton et al. 1993; Cvrckova and Nasmyth 1993). The terminal phenotypes include elongated buds, wide mother-bud necks and multiple nuclei, suggesting roles for Cla4 in bud morphogenesis and cytokinesis. Another mutant identified in the same screen is allelic to the septin gene *CDC12* (see above); this mutant gives similar phenotypes as above (Cvrckova and Nasmyth 1993), suggesting that Cla4 may have septin-related functions.

Cla4 has significant sequence similarity to Ste20 within the kinase domain and the Cdc42-binding domain. In addition, Cla4 has a PH domain that is absent in Ste20. Ste20 and Cla4 are not essential for viability. However, cells lacking both PAKs cannot survive (Cvrckova et al. 1995). The co-lethality of *cla4Δ ste20Δ* double mutants suggests that Ste20 and Cla4 overlap in some vegetative functions. This function may involve polarization of the actin cytoskeleton since inactivation of both Ste20 and Cla4 at either G1, S/G2 or M phases leads to depolarization of the cortical actin cytoskeleton (Holly and Blumer 1999). How Ste20 and Cla4 regulate the actin cytoskeleton is not clear. Both kinases have been shown to phosphorylate and activate type-I myosins (Wu et al. 1996, 1997) that are involved in actin regulation and endocytosis (Geli and Riezman 1996; Goodson et al. 1996). Thus, Ste20 and Cla4 may affect assembly of actin structures through regulation of type-I myosins.

The functions of Cla4 and Ste20 overlap but are not completely identical. Deletion of *CLA4* causes elongated buds (Cvrckova et al. 1995), suggesting a role for Cla4 different from that of Ste20 in controlling cell shape. In addition, unlike Ste20, Cla4 is not required for bud site selection (Sheu et al. 2000).

The kinase activity of Cla4 is stimulated by the GTP-bound form of Cdc42 in vivo and peaks at mitosis (Benton et al. 1997). However, the Cdc42-binding domain of Cla4 is not required for its kinase activity in vitro. Perhaps, in living cells, Cdc42 recruits Cla4 to sites of polarized growth and spatially regulates Cla4 activation. Indeed,

2322

Cla4 localizes to sites of polarized growth independent of intact actin structures (Holly and Blumer 1999). In addition, Cla4 is hyperphosphorylated during mitosis (Tjandra et al. 1998); this hyperphosphorylation depends on Clb2-Cdc28, the GTP-bound form Cdc42 and the Elm1 kinase, which is also involved in morphogenesis (Tjandra et al. 1998; Sreenivasan and Kellog 1999). Hyperphosphorylation of Cla4 is responsible for relaying the signal to activate the Gin4 kinase (Tjandra et al. 1998). Interestingly, septins are also required for activation of Gin4 during mitosis (Carroll et al. 1998). Gin4 and its homologues, Hsl1 and Kcc4, are involved in the regulation of septin organization and checkpoint control (Longtine et al. 1998; Barral et al. 1999; see below).

E. Mitogen-Activated Protein (MAP) Kinase Pathways

Mitogen-activated protein (MAP) kinase pathways are used by all eukaryotes to transduce signals from the environment to elicit cellular responses. MAP kinase (MAPK) pathways are composed of a cascade of protein kinases acting sequentially, presumably to amplify the signal. The basic module of the kinase cascade includes a MAP kinase, its activator, the MAP kinase kinase (MEK or MAPKK), and the MAP kinase kinase kinase (MEKK or MAPKKK) (Marshall 1994). There are six MAPKs identified in budding S. cerevisiae (Hunter and Plowman 1997) and, of these, five have known associated biological functions (Fig. 4; Chapter 3 Herskowitz 1995; Gustin et al. 1998). Mpk1 (Slt2) functions to maintain cell integrity; Fus3 mediates mating response; Kss1 regulates filamentous differentiation; Hog1 is required for response to high external osmolarity and Smk1 is needed for sporulation. These MAPK pathways form complex signaling networks through shared components and upstream activators and cross activations. For example, Ste20, Ste11 and Ste7 function in both mating and filamentous differentiation (Liu et al. 1993; Roberts and Fink 1994); Ste11 also functions in the high osmolarity response (Posas and Saito 1997). In addition, Kss1 is known to function in the mating signaling pathway in the absence of Fus3 (Madhani et al. 1997). While sharing components may facilitate coordination among pathways during complex biological processes, it also raises the problem of specificity. Thus, mechanisms that

help coordinate and confer specificity for these MAPK pathways are important for cell growth.

In budding S. cerevisiae, the Mpk1, Fus3 and Kss1 MAPK pathways participate in controlling cell polarity in response to external cues and internal status of cell-cycle progression. The Mpk1 pathway participates in polarized growth during vegetative growth, filamentous growth and mating. The Mpk1 signaling pathway plays a central role in monitoring cell wall integrity during polarized growth and coordinating cell wall synthesis with different stages of cell growth. Components of this pathway include Mpk1 (Torres et al. 1991; Mazzoni et al. 1993), two homologous and redundant MEKs, Mkk1 and Mkk2 (Irie et al. 1993), and the MEKK, Bck1 (Slk1) (Lee and Levin 1992; Costigan et al. 1992). Mutations in this pathway lead to cell lysis defects at high temperature. In addition, cells lacking Mpk1 accumulate secretory vesicles in the bud (Mazzoni et al. 1993), suggesting that this pathway positively regulates secretion to the cell surface.

Activation of the Mpk1 pathway by external stimuli depends on Pkc1 and transmembrane protein sensors. Pkc1 activates Bck1 by phosphorylating it (Levin et al. 1994; Levin and Errede 1995). The Pkc1-MAPK pathway is induced drastically by heat stress in a Pkc1- and Bck1-dependent manner (Kamada et al. 1995). High temperature induces tyrosine phosphorylation and kinase activity of Mpk1 (Kamada et al. 1995; Zarzov et al. 1996). A drug that induces inward membrane stretch, chlorpromazine, also elicits a similar response, raising the possibility that heat stress induces membrane disturbance, which then transduces the signal to Pkc1. A group of homologous transmembrane proteins, Wsc1 (Hcs77), Wsc2 and Wsc3, are the candidate sensors for cell wall stress (Gray et al. 1997; Verna et al. 1997). Deletion of the genes encoding these proteins diminishes heat-induced activation of Mpk1. Deletion of all three WSC genes along with PKC1 does not yield more severe defects than deletion of PKC1 alone, suggesting that Wsc proteins and Pkc1 function in a linear pathway (Verna et al. 1997). Another transmembrane protein, Mid2, isolated as a dosage suppressor of the cell lysis defect of wsc1Δ, shares with Wsc1 an essential function as the cell surface sensors for cell integrity signaling during vegetative growth (Ketela et al. 1999; Rajavel et al. 1999). Both Mid2 and Wsc1 are uniformly distributed throughout the plasma membrane. Mid2 is also required for Mpk1 signaling in response to mating pheromone. Thus, Mid2 and Wsc1 function to relay the signal of cell wall stress to the Pkc1-Mpk1 pathway during polarized growth.

The Pkc1-Mpk1 pathway is intimately linked to cell cycle control. Mutations in this pathway are co-lethal with *cdc28* mutations (Mazzoni et al. 1993; Marini et al. 1996). Increasing or decreasing the activity of Cdc28 results in a corresponding increase or decrease of the tyrosine phosphorylation and kinase activity of Mpk1 (Marini et al. 1996; Zarzov et al. 1996), suggesting that Mpk1 activity is cell cycle regulated. Moreover, tyrosine phosphorylation of Mpk1 peaks at the G1/S transition (Zarzov et al. 1996), the time of the onset of apical growth (Lew and Reed 1993). The transcription of a group of genes involved in cell wall functions also peaks at the G1/S transition and is dependent on the activation of the Pkc1-Mpk1 MAPK pathway (Igual et al. 1996). Thus, Cdc28 may regulate cell wall function through the cell integrity pathway during polarized growth.

There is reciprocal control over cell cycle machinery by the Mpk1 pathway. Genetic and biochemical studies indicate that components of the cell cycle regulator SBF are targets of Mpk1 (Igual et al. 1996; Madden et al. 1997). SBF is a heterodimer composed of Swi4 and Swi6 and regulates the G1/S transition by activating transcription of the G1 cyclins (Andrews and Herskowitz 1989; Nasmyth and Dirick 1991; Ogas et al. 1991; Espinoza et al. 1994; Measday et al. 1994). SBF plays a role in polarized growth and phosphorylation of Swi6 depends on the activity of Mpk1 (Madden et al. 1997). Thus, the Mpk1 pathway may function to modulate and coordinate cell cycle with polarized growth through regulation of SBF.

Other downstream targets of the Mpk1 pathway include Rlm1 (Watanabe et al. 1995, 1997; Dodou and Treisman 1997), Rom2 (S. Vidan and M. Snyder, unpubl.), Nhp6A and Nhp6B (Costigan et al. 1994). Rlm1 is a transcription factor, whose activity depends on Mpk1. Rlm1 mediates transcription of a group of cell wall proteins and enzymes involved in cell wall biosynthesis in response to Mpk1 activation (Jung and Levin 1999). As discussed earlier, Rom2 is a positive regulator of Rho1, which functions upstream of the cell integrity pathway. The discovery of Rom2 as a target of Mpk1 raises the possibility of a positive feedback loop that may boost the activity of the pathway during polarized growth. The roles of Nhp6A and Nhp6B in polarized growth remain undetermined.

The Mpk1 MAPK pathway is involved in mating. Mutants lacking genes in this pathway are defective in mating projection morphogenesis (Costigan et al. 1992; Madden et al. 1997). Mating pheromone (Errede et al. 1995; Zarzov et al. 1996) also induces the tyrosine phosphorylation and kinase activity of Mpk1. Ste20, but not Ste11 or Ste12, is required for activation of Mpk1 by mating pheromone, suggesting that Ste20 regulates the cell integrity pathway during mating by a mechanism independent of the regular pheromone response pathway; Ste20 is not required for the heat-induced Mpk1 phosphorylation (Zarzov et al. 1996). More studies are needed to understand the relationship between pheromone response and activation of the Mpk1 pathway.

The MAPK pathway required for the mating response is composed of Ste11 (the MEKK), Ste7 (the MEK) and Fus3. This pathway shares with the Kss1 filamentation response pathway the same MEKK and MEK as well as the upstream activator, Ste20, and the downstream transcription factor, Ste12. Several possible mechanisms are responsible for the specificity of the responses elicited by these two pathways (Madhani and Fink 1998). First, the mating response is prevented from occurring in diploid cells because they do not express some of the components (e.g. pheromone receptors) required for this response. Second, the Ste12 transcription factor cooperates with different partners during different signaling response. Third, Ste11, Ste7 and Fus3 complex with the scaffold protein Ste5 during mating response (Choi et al. 1994; Marcus et al. 1994) but not during filamentous growth response (Liu et al. 1993; Roberts and Fink 1994). Thus, Ste5 may tether these kinases and insulate them from unwanted cross-talk (Yashar et al. 1995). Finally, the MAP kinase for each pathway is unique. Different kinases may have different substrate specificity. Furthermore, both Fus3 and Kss1 have a unique kinase-independent inhibitory function toward the other pathway (Madhani et al. 1997).

The activation of the MAPK pathways involved in mating and filamentous differentiation will be discussed in later sections.

F. Cell Cycle Control of Morphogenesis

In *S. cerevisiae*, the progression of cellular events, such as bud growth and nuclear division, are controlled temporally by cyclin-dependent kinases (CDKs). Two CDKs, Cdc28 and Pho85, are involved in bud morphogenesis. Early in the cell cycle, Cdc28 forms complexes with and is activated by the G1 cyclins, Cln1, Cln2 and Cln3 (Cross 1995; Nasmyth 1996). The G1 cyclins for Pho85 are Pcl1 and Pcl2 (Espinoza et al. 1994; Measday et al. 1994). Activation of Cdc28 by G1 cyclins triggers passage through START and assembly of actin structures at the incipient bud site (Fig. 3; Lew and Reed 1993). Other polarity components, such as Cdc42, Bem1, Spa2, and septins are also assembled at the incipient bud site at this time. The concerted functions of actin and these polarity components promote apical growth. It is thus proposed that G1-Cdc28 triggers apical growth (Fig. 2; Lew and Reed 1993). Consistent with the role of G1 cyclins in activating apical growth, overexpression of Cln1 or Cln2 causes elongated bud shape and hyperpolarization of the actin cytoskeleton. Moreover, deletion of *GRR1* stabilizes Cln1 and Cln2 and leads to cell elongation (Barral et al. 1995). Overexpression of Cln3 does not lead to bud elongation, suggesting that Cln1,2-Cdc28 phosphorylates specific substrates during bud morphogenesis. It is thought that the primary role of Cln3 is to promote expression of SBF-dependent genes (Breeden 1996), such as *CLN1* and *CLN2*, and thus is the key to controlling the timing of START; the timing of START affects cell size. Thus, Cln3 may be the critical protein involved in cell size control. Consistent with this idea, translation of the *CLN3* mRNA is very sensitive to the level/activity of ribosomes (Polymenis and Schmidt 1997; Hall et al. 1998), which reflects the overall energy status of the cell.

Activation of Cdc28 complexed with the G2 cyclins Clb1–6 triggers the process of nuclear division and the switch from apical to isotropic bud growth (Fig. 2; Lew and Reed 1993). The main G2 cyclin involved in the apical-to-isotropic transition appears to be Clb2. *clb2Δ* cells are elongated (Surana et al. 1991; Richardson et al. 1992), an indication of prolonged apical growth and a delay in activation of the G2 cyclins. Indeed, analysis of cell surface expansion reveals that deletion of *CLB2* greatly delays the apical-to-isotropic switch, and deletion of both *CLB1* and *CLB2* prevents the switch (Lew and Reed 1993). In contrast, overexpression of *CLB1* or *CLB2* accelerates the switch. Clb3 and Clb4 do not appear to affect morphogenesis since cells lacking these two cyclins do not exhibit alterations in cell shape. Inactivation of the G2 form of Cdc28 is required for exiting mitosis and the initiation of processes in preparation for cytokinesis, such as the repolarization of the actin cytoskeleton (and perhaps some other polarity proteins) to the mother-bud neck (Fig. 2).

Pho85 is a non-essential CDK that is involved in a wide spectrum of cellular processes (Lenburg and O'Shea 1996). The pleiotropic nature of Pho85 may be attributed to its association with multiple cyclin partners; ten potential cyclins for Pho85 exist (Andrews and Measday 1998). Pho85 and three of the more related Pho85 cyclins, Pcl1, Pcl2 and Pcl9, are involved in cell morphogenesis and control of cell polarity in *S. cerevisiae* (Madden et al. 1997; Lee et al. 1998; Tennyson et al. 1998). Transcription of *PCL9* is controlled by the transcription factor Swi5, and peaks at M/G1, suggestive of a function in early G1 (Tennyson et al. 1998). Transcription of *PCL1* and *PCL2* is controlled by SBF, and peaks at START (Andrews and Measday 1998). Suppression of the temperature sensitivity and the lysis defect of *mpk1Δ* cells by overexpression of *PCL1* and *PCL2* further suggests that SBF mediates the function of Mpk1 in coordinating cell cycle with polarized growth through activation of Pcl1 and Pcl2 (Madden et al. 1997). Consistent with the role of these cyclins in polarized growth, Pcl2 interacts in vivo with Rvs167, which is involved in regulation of the actin cytoskeleton and bipolar budding (Bauer et al. 1993; Sivadon et al. 1995; Lee et al. 1998). In addition, Pcl2-Pho85 phosphorylates Rvs167 in vitro. Deletion of *PHO85* or all three of *PCL1*, *PCL2* and *PCL9* also causes random budding in diploid cells (Tennyson et al. 1998). Thus, Pho85 and its cyclins *PCL1*, *PCL2* and *PCL9* contribute to polarized growth through regulation of the actin cytoskeleton early in the cell cycle.

G. Budding Checkpoint

Checkpoint controls are regulatory pathways that act to delay cell cycle progression to ensure that one series of cellular events must be completed before the next set of events can be initiated (Hartwell and Weinert 1989). The checkpoints that monitor nuclear events, such as DNA replication and DNA damage, and mitotic spindle assembly have been quite well studied. Much less is known about the coordination of nuclear division with cytoplasmic events. In budding *S. cerevisiae*, perturbations that disrupt the septin or the actin cytoskeleton cause a delay in nuclear division (Lew and Reed 1995b; McMillan et al. 1998; Barral et al. 1999). The delay is mediated through the function of the tyrosine kinase Swe1, the *S. pombe* Wee1 homologue (Russell and Nurse 1987a), and through the delayed accumulation of the G2 cyclins, Clb1 and Clb2 (Booher et al. 1993; Lew and Reed 1995b; McMillan et al. 1998; Barral et al. 1999). Swe1 inhibits the CDK by specifically phosphorylating its tyrosine 19 residue (Sia et al. 1996), and by an as-yet-unclear mechanism, which does not require the Swe1 kinase activity (McMillan et al. 1999).

Septin mutants incubated at the restrictive temperature have a prolonged apical growth phase and exhibit a nuclear division delay in a Swe1-dependent manner (Fig. 5; Barral et al. 1999). This mitotic delay is necessary to maintain cell viability in septin mutants. Proper assembly of Hsl1 onto the septin cytoskeleton (Barral et al. 1999) monitors the integrity of septin organization. Hsl1 is a homologue of the *S. pombe* Nim1 kinase that phosphorylates and inhibits Wee1 (Russell and Nurse 1987b). Hsl1 interacts with septins in vivo and its kinase activity depends on proper septin organization. Hsl1 and its homologous kinases in *S. cerevisiae*, Gin4 and Kcc4, localize to the mother-bud neck in a septin-dependent manner (Longtine et al. 1998; Barral et al. 1999). Moreover, deletion of all three Nim1-related

kinases causes a cellular morphology very similar to that of septin mutants (Barral et al. 1999). Therefore, association with septins activates the Hsl1 kinase, which negatively regulates Swe1 and allows normal cell progression. When the septin cytoskeleton is disrupted, the Hsl1 kinase remains inactive, and Swe1 can thus inhibit the G2 CDK and cause a cell cycle delay.

During normal cell cycle progression, phosphorylation of Swe1 by the Hsl1 kinase is thought to promote its recognition and destruction by the ubiquitination complex at G2/M (Kaiser et al. 1998; Sia et al. 1998). Hsl7 also negatively regulates Swe1 (Ma et al. 1996); both Hsl1 and Hsl7 are required for rapid degradation of Swe1. Hsl7 is a novel protein that interacts with both Hsl1 and Swe1 (McMillan et al. 1999; Shulewitz et al. 1999); localization of Hsl7 to the neck region requires Hsl1 (Shulewitz et al. 1999). It is hypothesized that the interaction of Hsl1 with the septin ring recruits Hsl7, which then presents Swe1 to Hsl1 for the destructive phosphorylation. Thus, activation of the morphogenesis checkpoint in response to septin defects may stabilize Swe1 by negatively regulating Hsl1 and Hsl7.

The manner in which actin organization is scrutinized by the checkpoint pathway is not clear. Actin might affect the targeting of Hsl1 or Swe1 to the septin ring (Barral and Snyder, unpubl.). Alternatively, since the G2 CDK is the ultimate target of the checkpoint kinase cascade, actin might affect the shuttling of signaling

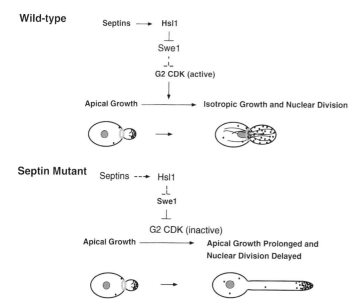

Fig. 5. Cell cycle checkpoint monitors the integrity of septins. The G2 cyclin-dependent kinase (CDK) triggers the switch from apical growth to isotropic growth and the onset of nuclear division. The activity of the G2 CDK is controlled by its inhibitory kinase, Swe1, which is negatively regulated by the septin-associated kinase Hsl1. Proper organization of the septin cytoskeleton is essential for activating Hsl1

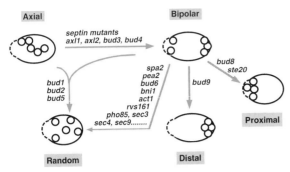

Fig. 6. Budding patterns in wild-type and mutant *S. cerevisiae* cells. Some mutations affect both axial and bipolar budding whereas others specifically affect one or the other. Mutations in general bud site selection components, such as Bud1, Bud2 and Bud5, lead to random budding for all cell types. Mutations specifically affecting axial budding cause cells to use the bipolar budding pattern. Some mutations affect only bipolar budding. Most of these mutations lead to random budding whereas a few cause budding exclusively at the proximal pole (such as *bud8* and *ste20*) or distal pole (such as *bud9*). The birth scar region (corresponding to the proximal pole) is marked by a *dashed line*. The *small circles* represent bud scars. The patterns of bud site selection are indicated in the *shaded box*

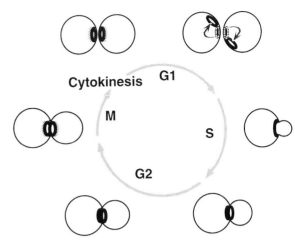

Fig. 7. The cytokinesis tag model for axial budding. The septin rings (*dark rings*) at the mother-bud neck serve as a template for assembly of axial budding components (*hatched rings*), such as Bud3, Bud4 and a portion of Axl2, at about G2/M of the cell cycle. These components at sites of cytokinesis provide the positional signal recognized by the general bud site selection proteins, which then direct polarity establishment (including the assembly of a new septin ring) at the axial sites in the subsequent cell cycle (*arrows inside the cell*)

components to and from the nucleus. Additional experiments are required to address these possibilities.

VI. Control of Division Site Selection

As noted above, the mating locus and cellular environments influence bud site selection. Haploid cells normally use an axial budding pattern whereas diploid cells use a bipolar pattern. These patterns are easily envisioned by staining cells with Calcofluor, a fluorescent dye that binds to chitin and stains the chitin-rich bud scars left on the mother surface after cell separation. The axial budding pattern results in a continuous chain of bud scars on older mother cells, whereas bipolar budding cells have bud scars at both poles. It is thought that, for both of these patterns, cortical tags help mark bud sites and that additional cellular components are important for establishing and recognizing the tags. Many mutations influence bud site selection. Some of these mutations and their resulting budding patterns are summarized in Fig. 6.

A. Establishment of Bud Site Tags for Axial Budding Pattern

In the axial budding pattern, a bud always forms adjacent to the latest previous bud site; hence, a cytokinesis tag model has been proposed to explain this pattern (Fig. 7; Chant and Herskowitz 1991; Snyder et al. 1991; Madden et al. 1992; Flescher et al. 1993). The septin cytoskeleton is important for the axial budding pattern; septin mutations lead to bipolar budding and its localization to the site of cytokinesis suggests that the septin cytoskeleton may be part of the tag (Flescher et al. 1993; Chant et al. 1995).

In addition to septins, Bud3, Bud4, Axl1 and Axl2(Bud10) are also important for axial budding (Chant and Herskowitz 1991; Fujita et al. 1994; Chant et al. 1995; Roemer et al. 1996a; Sanders and Herskowitz 1996). Mutations in any genes encoding these proteins cause axial budding cell types to use a bipolar pattern. *BUD3* and *BUD4* were identified in a screen for mutants defective in bud site selection (Chant and Herskowitz 1991). The sequence of Bud3 is not homologous to other known proteins (Chant et al. 1995), whereas Bud4 contains a putative GTP-binding motif near its carboxyl terminus (Sanders and Herskowitz 1996); the sequences around this motif are homologous to the integrin-like protein Int1 of *Candida albicans* (Gale et al. 1996). Both Bud3 and Bud4 are present at the mother-bud neck as a ring at around the time of mitosis (Chant et al. 1995; Sanders and Herskowitz 1996). A double-ring structure forms and splits after cytokinesis, resulting in a ring that persists in both the mother and daughter cells during most of the G1 phase of the next cell cycle. Formation of the double-ring structure containing Bud3 and Bud4 depends on the septins, and efficient assembly of the Bud4 ring requires the function of Bud3 (Chant et al. 1995; Sanders and Herskowitz 1996; Frazier et al. 1998). Therefore, Bud3 and Bud4 proteins may assemble onto the septin cytoskeleton during mitosis to mark the site for axial budding in the next cell cycle.

AXL2 was first identified as a multicopy suppressor of co-lethality in *spa2Δ cdc10-10* double mutant (Roemer et al. 1996a). Axl2 is a type-I transmembrane protein (with single transmembrane domain and carboxyl terminus facing cytoplasm) with N- and O-linked glycosylations (Roemer et al. 1996a; Sanders et al. 1999). Axl2 can be detected as a patch at the incipient bud site, at the periphery of a small bud, and as a ring at the mother-bud neck in a medium or large budded cell. The bud neck localization persists through cytokinesis until the G1 stage of the next cell cycle (Halme et al. 1996; Roemer et al. 1996a). Like Bud3 and Bud4, Axl2 depends on the septins for localization to the mother-bud neck (Roemer et al. 1996a), whereas it does not require septins for localization to the incipient bud site and the small bud, suggesting a septin-independent function of Axl2 in early bud growth. In addition, localization of the Cdc3 septin and Bud4 does not require Axl2, and Axl2 does not depend on Bud3 for localization to the neck region. Axl2 may function together with Bud3 and Bud4 as part of the cytokinesis tag, or Axl2 may recognize and direct polarized growth to the marked site (Halme et al. 1996; Roemer et al. 1996); the latter possibility is also consistent with a role for Axl2 in early bud growth.

Axl1 may be the cell-type-specific determinant for an axial budding pattern because it is the only haploid-specific gene identified so far to be essential for this purpose (Fujita et al. 1994; Adames et al. 1995). Loss of *AXL1* leads to bipolar budding in haploid cells and ectopic expression of *AXL1* in diploid a/α cells results in an axial budding pattern (Fujita et al. 1994). Axl1 also functions as a protease for processing of a-factor mating pheromone, but the protease activity is not essential for axial bud site selection (Adames et al. 1995). Thus, Axl1 is likely the key factor for making the axial site preferred over the bipolar sites, either by activating axial bud site selection or by inactivating bipolar bud site selection. The localization pattern of Axl1 has not been determined, and this information should help understand how Axl1 functions to specify the preference for axial budding in haploid cells.

The cytokinesis tag in haploid cells is transient because cells that reach the stationary phase bud at distal sites upon returning to fresh medium (Madden and Snyder 1992; Chant and Pringle 1995). The transient nature of the tag is due, at least in part, to the loss of Bud4, which disappears as cells enter the stationary phase (Sanders and Herskowitz 1996). Furthermore, the fact that cells recovering from the stationary phase bud at the distal pole indicates that the bipolar tag(s) is more persistent.

B. Establishment of Bud Site Tags for Bipolar Budding Pattern

The characteristics of bipolar bud site tags and the establishment of these tags are clearly different from those of axial tags since many genes are specifically required for one budding pattern but not the other. Since bipolar budding requires marking both the distal pole and the proximal pole (next to the birth scar, where the daughter cell separate from the mother), the establishment of bipolar bud site tags requires a mechanism that is able to build multiple positional signals.

A large number of genes are found to be important for bipolar bud site selection (Fig. 6). Mutations in any of these genes cause diploid cells to use a non-bipolar budding pattern, but have no effect on axial budding in haploid cells. Surprisingly, most mutations lead to a random budding pattern. All of these genes function in bud growth, suggesting that processes involved in bud growth are also responsible for setting up bipolar bud site signals. Several lines of evidence suggest that apical growth and repolarization to the neck are processes important for establishing bipolar bud site tags (Fig. 8; Sheu et al. 2000). First, polarity proteins, such as Spa2, Pea2, Bud6 and Bni1, which are required for concentrating polarized growth at both poles, are also required for the bipolar budding pattern (Snyder 1989;

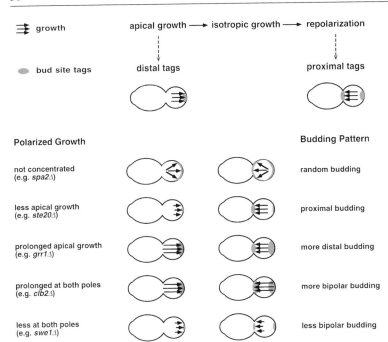

Fig. 8. Model for the control of bipolar bud site selection by processes of polarized growth. Apical growth at the bud tip is essential for proper establishment of distal tags and repolarization at the mother-bud neck helps build proximal tags. Thus, components involved in these polarized growth processes not only influence the cell shape, but also affect distribution and strength of bud site tags required for bipolar budding. Examples of mutants that affect both polarized growth and bipolar budding are depicted below according to the model. The relative length of each growth period corresponds to the length of the arrows. Growth direction is represented by the direction of *arrowheads*. *Parallel arrows* indicate a concentrated growth pattern while *dispersed arrows* indicate diffuse growth. *Gray-shaded areas* indicate the distribution of bipolar bud site tags

Valtz and Herskowitz 1996; Zahner et al. 1996; Amberg et al. 1997). *spa2Δ*, *pea2Δ*, *bud6Δ* and *bni1Δ* mutants can not concentrate apical growth and may therefore have diffuse bud site tags at the distal pole. Likewise, the failure of these mutants to concentrate repolarization to the mother-bud neck may lead to diffuse proximal bud site tags. As a result, these cells choose the budding sites at the two poles with less precision and display a "random-like" budding pattern. Many mutants that affect the actin cytoskeleton (e.g. *act1*, *rvs161*, *rvs167*, etc.) and the secretory pathway (e.g. *sec3*, *sec4* and *sec9*) display a diploid-specific random budding phenotype similar to *spa2Δ* (Bauer et al. 1993; Durrens et al. 1995; Sivadon et al. 1995; Haarer et al. 1996; Finger and Novick 1997; Yang et al. 1997). Their phenotype probably also results from defects in apical growth and repolarization at the mother-bud neck. Detailed budding pattern analyses show that most of these mutants, with the exception of *bni1Δ*, are able to place their first buds at the distal pole (Valtz and Herskowitz 1996; Zahner et al. 1996; Yang et al. 1997), suggesting that a functional bud site selection signal is present at the distal pole. Moreover, in the *ste20Δ* background, which buds proximally, deletion of *SPA2* does not lead to random budding, but causes budding exclusively in the proximal hemisphere of the cell, albeit in a more scattered fashion. This indicates that the proximal pole is also marked in the absence of the Spa2 function.

Diploid *ste20Δ* mutants rarely bud at the distal pole, suggesting that the distal tag is not properly deposited. As mentioned earlier, Ste20 is involved in regulation of apical growth: deletion of *STE20* leads to a shorter apical growth phase (Sheu et al. 2000). It is thus likely that insufficient apical growth in *ste20Δ* mutants leads to the defect in bud site selection. Ste20 does not seem to be essential for cytokinesis or other functions at the bud neck when another homologous kinase, Cla4, is present (Cvrckova et al. 1995), and thus it is likely that repolarization and hence the proximal tag are not affected in the *ste20Δ* mutant.

Additional evidence that apical growth is important for building the distal bud site tags comes from analysis of the cell cycle mutants that affect the timing of the switches from apical growth to isotropic growth and the timing of repolarization. These mutants also influence bipolar budding. In diploid *grr1Δ* mutants, which stabilize G1 cyclins (Barral et al. 1995), apical growth is enhanced and buds are preferentially formed at distal poles (Sheu et al. 2000). Clb2-Cdc28 kinase functions to trigger the switch from apical growth to isotropic growth, and inactivation of this kinase is essential for repolarization and cytokinesis (Lew and Reed 1993). Deletion of *CLB2* prolongs the apical growth and repolarization phases and improves accuracy of bipolar budding in wild-type and *spa2Δ* and *ste20Δ* mutants. In contrast, deletion of *SWE1*, which inhibits Clb2-CDK, decreases the fidelity of bipolar bud site selection. Therefore, the length of apical growth and repolarization affects not only cell shape, but also the fidelity of bipolar budding.

The model described above suggests that most mutants affect bipolar budding through their effects on polarized growth, a process important for establishment of the bipolar tags. What then constitutes the persistent bipolar bud site tags? There are a few mutants that can uncouple budding at two poles from each other. For example, Bud8 is specifically required for budding at the distal pole and Bud9 is specifically required for budding at the proximal pole (Zahner et al. 1996), raising the possibility that these two proteins may constitute the distal and proximal tags, respectively. Consistent with this possibility, both proteins are transmembrane glycoproteins that localize to the two poles of the cell (J. Pringle, pers. comm.). However, their localization patterns do not appear to be persistent for more than one cell cycle but are dynamic at different stages of bud growth. Alternatively, Bud8 and Bud9 may function to position or respond to the persistent positional signal (Zahner et al. 1996).

The budding pattern of the *bud8Δ* mutant is the same as that of the *ste20Δ* mutant (Sheu et al. 2000), suggesting that Bud8 and Ste20 function in the same pathway for bud site selection. Thus, Bud8 may function in apical growth like Ste20 or, alternatively, this kinase may have a direct role in activating the distal tag in addition to its role in apical growth. Whether Ste20 can phosphorylate and regulate Bud8, or Bud8 can regulate the kinase activity of Ste20, remains to be determined.

C. Recognition of Bud Site Tags

Both axial and bipolar budding patterns require a GTPase module for proper bud site selection. This module contains a Ras-related small G-protein Rsr1(Bud1) (Bender and Pringle 1989), its GEF, Bud5, and its GAP, Bud2 (Chant et al. 1991; Chant and Herskowitz 1991; Powers et al. 1991; Bender 1993; Park et al. 1993). Deletion of the gene encoding any of these proteins leads to random budding in all cell types. The requirement for both the GEF and GAP of Rsr1 in

proper bud site selection suggests that cycling of Rsr1 between the GTP- and GDP-bound forms is essential for its function in bud site selection. Indeed, *rsr1* mutants locked in either GTP- or GDP-bound states also cause a random budding phenotype (Ruggieri et al. 1992). Rsr1 interacts with the polarity establishment proteins Cdc24, Cdc42 and Bem1 in a guanine nucleotide-dependent manner: the GTP-bound form interacts with Cdc24 and the GDP-bound Cdc42 whereas the GDP-bound Rsr1 interacts with Bem1 (Zheng et al. 1995; Michelitch and Chant 1996; Park et al. 1996). The interaction between Rsr1 and the Cdc42 GTPase module suggests that the Rsr1 GTPase module functions to recruit polarity establishment components to the selected bud site. Consistent with this idea, overexpression of *CDC42* and mutation alleles of *CDC24* displays random budding phenotypes in both haploid and diploid cells (Sloat et al. 1981; Johnson and Pringle 1990; Zheng et al. 1994). However, Rsr1 does not localize specifically to the incipient bud site but localizes throughout the cortex (Michelitch and Chant 1996). Rather, the regulators of Rsr1, Bud2 and Bud5 localize to the bud site (Park et al. 1999). Localization of Bud2 to the incipient bud site is independent of Rsr1 and changes over the cell cycle. Thus, the function of Rsr1 in directing cell polarity may be assisted by localized activators and/or localized effectors.

The localization patterns of Bud2 and Bud5 suggest that these two proteins may be responsible for a direct link to the bud site tags. Interestingly, some alleles of *BUD2* and *BUD5* specifically cause random budding in diploid cells but do not affect axial budding (Zahner et al. 1996), suggesting that these mutant alleles encode proteins that retain the ability to recognize the axial bud site signal but lose the ability to recognize the bipolar signal. It is possible that Bud2 and Bud5 contain distinct tag-interacting domains that are differentially regulated in different cell types, or the relative activities of these proteins differ in different cell types so that one type of the tag is preferred to the other.

BUD2 is essential in cells lacking both Cln1 and Cln2 (Cvrckova and Nasmyth 1993; Benton et al. 1997). The lethality is likely due to the accumulation of the GTP-bound form of Rsr1 because additional deletion of *RSR1* suppresses the lethality. The nature of this genetic interaction is not clear, but nonetheless suggests a role for these proteins in bud formation.

D. Regulation of Bud Site Selection by the Environment

The environment also influences bud site selection. Haploid cells exiting stationary phase tend to bud at the distal pole (Madden et al. 1992; Chant and Pringle 1995); this feature may maximize the exposure of the colony to nutrients. Haploid cells exposed to low levels of mating pheromone also display a bipolar-like budding pattern reminiscent of filamentous differentiation (Erdman and Snyder 2000). Such budding pattern alteration maybe result from cross-talk between signaling pathways of mating response and filamentous differentiation since these two signaling pathways share many common components (see below). This response presumably enhances the ability of haploid cells to find mating partners.

Diploid cells under stress, such as low nitrogen or osmotic shock, also bud with increased frequency at the distal pole (Gimeno et al. 1992; Brewster and Gustin 1994). Again, this is believed to be an adaptive response that allows cells to grow away from an adverse environment.

VII. Partitioning of Cellular Components in Response to Cell Polarity

After bud formation, special mechanisms are needed to ensure the proper segregation of the nucleus, mitochondria, vacuole and cell fate determinants into the bud. We will review the segregation of the nucleus, mitochondria and a cell fate determinant, Ash1. Much less is known about vacuolar segregation and it will not be discussed here.

A. Segregation of the Nucleus

Segregation of the nucleus depends on the function of the mitotic spindle. In animal cells, the position of the spindle determines the division plane (Rappaport 1986). However, in budding *S. cerevisiae*, the division plane is established before the formation of the mitotic spindle. Therefore, orienting the spindle relative to the division plane is pivotal for proper chromosome segregation in *S. cerevisiae*. Studies in both fixed and living cells have revealed the processes of spindle positioning

in response to cell polarity in *S. cerevisiae* (Fig. 9; Snyder et al. 1991; Carminati and Stearns 1997; Shaw et al. 1997). Prior to bud emergence, the spindle pole body (SBP) moves toward the incipient bud site and the cytoplasmic microtubule (cMT) emanating from the SBP intersects this region; the initial movement of the SPB toward the emerging bud requires the function of Bni1 (Lee et al. 1999). At about the time that SPBs separate and the bipolar spindle starts to form, astral microtubules from one spindle pole penetrate into the bud and reach the cortex of the bud tip and the spindle becomes aligned with the mother-bud axis.

Alignment of the mitotic spindle with the axis of cell division is thought to be mediated by the interaction of the cMT with the capture site at the cell cortex (Snyder et al. 1991; Hyman and Stearns 1992). Kar9 localizes to the bud tip and is required for proper cMT orientation and is thus a good candidate for cMT capturing in the cortex of the bud (Miller and Rose 1998). Another cortical protein, Num1, localizes to the cortex of the mother cell, and is also required for spindle orientation. Therefore, Num1 may function to capture cMT in the mother cortex during the S/G2 period, when the

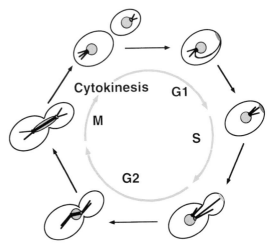

Fig. 9. Spindle positioning in response to cell polarity in budding *S. cerevisiae*. In the G1 phase of the cell cycle, the cytoplasmic microtubule (cMT) emanating from the spindle pole body (SPB) interacts with the cMT capture site (the *shaded patch*) at the incipient bud site. As a result, the SPB orients toward this region. The SPB starts duplicating at about this time. At about the time of SPB separation, the cMT from one spindle pole penetrates into the bud and reaches the cortex of the bud tip. At about G2, the bipolar spindle forms, the nucleus migrates toward the mother-bud neck region and then the spindle becomes aligned with the mother-bud axis

protein is transiently expressed (Farkasovsky and Kuntzel 1995). Despite their cortical localization and involvement in spindle orientation, there is no evidence that Num1 and Kar9 associate with any microtubule-binding activity. In contrast, the *S. cerevisiae* coronin Crn1, which colocalizes with cortical actin patches, has been shown to have microtubule-binding activity (Goode et al. 1999), although *crn1* mutants have relatively mild defects in microtubule polymerization and spindle orientation (Heil-Chapdelaine et al. 1998; Goode et al. 1999). It is likely that multiple factors function together to help capture the cMT.

Interestingly, many components involved in bipolar bud site selection are also required for proper spindle orientation (Lee et al. 1999; Miller et al. 1999). Both *bni1* and *bud6* cells exhibit defects in spindle orientation; Bni1 is also required for proper localization of Kar9 (Miller et al. 1999). In addition, the actin cytoskeleton is also important for spindle orientation (Palmer et al. 1992; Yang et al. 1997). It is likely that, like the bipolar bud site tag, the cMT capture components are also established during apical growth.

B. Segregation of Mitochondria

Mitochondria are essential organelles responsible for energy generation within the cell. Since mitochondria cannot be synthesized de novo, mitochondria must grow, divide and be properly distributed to daughter cells prior to cell division. In budding *S. cerevisiae*, a portion of the mitochondrial structure is delivered to the growing bud (Stevens 1981). Unlike segregation of nuclei, mitochondrion partitioning does not require microtubules (Huffaker et al. 1988; Jacobs et al. 1988; Weisman and Wickner 1988; cf. Chap. 11, Sect. IIA, this Vol.). The actin cytoskeleton is involved in organization and morphogenesis of mitochondria (Drubin et al. 1993). A mutation in the *MDM20* gene disrupts both actin cables and mitochondria segregation (Hermann et al. 1997). The defect can be suppressed by an extra copy of *TPM1* that encodes tropomyosin (Liu and Bretscher 1989a), suggesting that actin cables are important for mitochondria segregation. In addition to the actin cytoskeleton, an intermediate filament-like protein encoded by the *MDM1* gene is required for segregating mitochondria into the bud (McConnell et al. 1990; McConnell and Yaffe 1992, 1993).

C. Asymmetric Segregation of Cell Fate Determinants

In the wild, yeast cells are believed to be homothalic (Herskowitz 1988; Nasmyth 1993). Mother cells switch their mating type after each cell division, but daughter cells do not switch. Mating type switch requires the HO endonuclease to initiate gene conversion at the mating type locus. Transcription of *HO* occurs only in the mother cell but not in the daughter cell due to asymmetric accumulation of Ash1, a transcription repressor of *HO*, within the daughter nucleus (Bobola et al. 1996; Jansen et al. 1996; Sil and Herskowitz 1996). FISH (fluorescent in situ hybridization) analysis has revealed that the *ASH1* mRNA localizes to the tip of large budded cells (Long et al. 1997; Takizawa et al. 1997). Its 3′ untranslated region is sufficient to act in a *cis* fashion to localize a chimeric RNA to the bud. Proper localization of the *ASH1* mRNA to the bud and accumulation of the Ash1 protein to the daughter nucleus require the functions of actin, Myo4 (a type-V myosin, also isolated as She1), Bni1 (also isolated as She5), She2, She3 and She4 (Bobola et al. 1996; Long et al. 1997; Takizawa et al. 1997). Mutations in actin, tropomyosin and profilin disrupt actin cables and proper localization of the *ASH1* mRNA. These results raise the possibility that Myo4 may carry the mRNA along actin cables to the bud tip and other She proteins help anchor the mRNA in the bud. It is not clear whether other polarity proteins are involved in the asymmetric localization of mRNAs. It is unknown whether the mRNAs of other asymmetrically distributed proteins display a similar asymmetric localization.

VIII. Filamentous Differentiation

Depending on growth conditions, budding *S. cerevisiae* cells undergo a dimorphic switch between the yeast form and a "filamentous" form in which cells still bud, but the buds are highly elongated, and do not separate from mother cells (Roemer et al. 1998; Shepherd 1988; Gimeno et al. 1992; Kron et al. 1994; Roberts and Fink 1994; Erdman and Snyder 2000). Daughter cells bud from the distal site, resulting in the formation of long chains of connected cells. These features presumably allow cells in the wild to move away from the colony. In pathogenic fungi, the dimorphic switch to the filamentous form is thought to be critical for pathogenesis (see Chapters 1 and 3).

Both haploid and diploid cells can undergo filamentous differentiation. Diploid yeast cells undergo pseudohyphal growth when grown in solid medium deprived of nitrogen source and other nutrients (Gimeno et al. 1992). Haploid cells undergo invasive growth when grown for long periods in rich agar medium, which presumably leads to a depletion of glucose (Roberts and Fink 1994). Haploid cells also exhibit filamentous growth in the presence of low levels of mating pheromone (Erdman and Snyder 2001). Filamentous differentiation in budding *S. cerevisiae* is reversible. When returned to favorable growth conditions, filamentous growing cells switch back to the yeast form. Most laboratory *S. cerevisiae* strains cannot form filaments, presumably due to an accumulation of mutations that prevent this process from occurring. For example, the S288C strain has a mutation in *FLO8*, which is required for filamentous growth (Liu et al. 1996), and the W303 strain has a mutation in Sph1, which participates in cell morphogenesis (Roemer et al. 1998).

The similarity between the budding process in yeast-form cells and filamentous cells suggests that many of the mechanisms that control the cell shape and polarity are common for both cell types. For example, polarity establishment proteins, the actin cytoskeleton and its regulators are all important for this developmental event (Mosch et al. 1996; Mosch and Fink 1997; Cali et al. 1998; Roemer et al. 1998). Of particular importance are the genes involved in apical growth (e.g. *SPA2*, *PEA2* and *BNI1*) and bipolar budding (*BUD1*, *BUD2*, *BUD5* and *BUD8*); mutations in these genes cause defects in pseudohyphal and haploid invasive growth (Roberts and Fink 1994; Mosch and Fink 1997; Roemer et al. 1998). The fact that filamentous cells are connected suggests that components involved in regulation of septation are important for this process. Interestingly, mutations in *CLB2*, *CLA4* and *ELM1* affect cell separation and also cause cell elongation (Surana et al. 1991; Richardson et al. 1992; Benton et al. 1993, 1997; Cvrckova and Nasmyth 1993), presumably through delaying cell cycle progression in G2. These observations suggest that regulators of septation components (e.g. the septins) influence both cytokinesis and cell elongation.

In addition to morphological changes, cell cycle control is also altered in filamentous cells.

Cln1 and Cln2 are partially stabilized in pseudohyphal cells and are required for filamentous growth (Barral and Mann 1995; Oehlen and Cross 1998). Moreover, deletion of *CLB2* makes cells more sensitive to pseudohyphal induction (Kron and Gow 1995). In addition, it has been proposed that Elm1, Hsl1 and Swe1 form a kinase cascade that modulates Cdc28 activity during filamentous differentiation (Edgington et al. 1999). All these cell cycle regulators are involved in regulation of apical growth. Thus, it is likely that cell cycle machinery, with the help of polarity proteins, functions to promote filamentous differentiation by prolonging apical growth, which leads to cell elongation and more distal budding (Sheu et al. 2000), two prominent morphological features of filamentous cells.

Many components participate in signaling for filamentous differentiation. The small G-protein Ras2 stimulates filamentous differentiation through two downstream signaling pathways, the Kss1 MAPK cascade and the cAMP-protein kinase A (PKA) pathway (Gimeno et al. 1992; Liu et al. 1993; Mosch et al. 1996; Madhani et al. 1997). The 14-3-3 proteins, Bmh1 and Bmh2 (Roberts et al. 1997), facilitate signaling from Ras2. Cdc42 functions downstream of Ras2 and acts through its effector Ste20, which, in turn, activates the Kss1 MAPK cascade (Mosch et al. 1996; Peter et al. 1996; Leberer et al. 1997). During haploid invasive growth, the two parallel Ras2-activated signaling pathways can stimulate expression of reporter genes with the filamentation response element (FRE) in a Ste12- and Tec1-dependent manner (Liu et al. 1993; Gavrias et al. 1996; Mosch et al. 1996, 1999). However, cAMP branch-induction of the cell surface flocculin Flo11 – which is required for filamentous differentiation (Lo and Dranginis 1998) – is mediated by another transcription factor, Flo8, rather than Ste12/Tec1 (Liu et al. 1996; Rupp et al. 1999). Other transcription regulators, such as Phd1, Sok2 and Ash1, have been found to be involved in filamentous growth (Gimeno and Fink 1994; Ward et al. 1995; Chandarlapaty and Errede 1998).

IX. Mating Cells

Haploid budding yeast cells (*MATa* cells or *MAT*α cells) undergo physiological and morphological changes in response to the mating pheromone from cells of the opposite mating type (i.e. α-factor from *MATα* cells or **a**-factor from *MATa* cells, respectively). The mating responses include cell cycle arrest in G1, polarized growth to form mating projections, expression of proteins required for cell adhesion and cell fusion, and ultimately nuclear fusion (Fig. 10).

The mating responses require activation of the mating signaling pathway. The activation is initiated by the binding of mating pheromones to the corresponding receptor (Ste2 for α-factor and Ste3 for **a**-factor), a seven-transmembrane-spanning cell surface protein coupled to a heterotrimeric G-protein complex (Sprague and Thorner 1992). This binding induces the dissociation of Gα from the G$\beta\gamma$ (Ste4/Ste18) dimer. The

release of G$\beta\gamma$ allows activation of the kinase cascade scaffold protein Ste5 and the Ste20 PAK, which, in turn, activate the Ste11→Ste7→Fus3 MAPK cascade. Ste4 (Gβ) may activate Ste5 by recruiting it to the plasma membrane; addition of a membrane targeting domain to Ste5 causes constitutive activation of the signaling pathway in a manner bypassing the requirement for Ste4 but still dependent on Ste20 (Pryciak and Huntress 1998). Mating pheromone also induces interaction of Ste4 with Ste20 (Leeuw et al. 1998), which functions as an upstream kinase of the MEKK Ste11 (Leberer et al. 1992; Neiman and Herskowitz 1994). How this interaction leads to Ste20 activation is not clear. Another link between the G-protein complex and the MAPK cascade is Ste50,

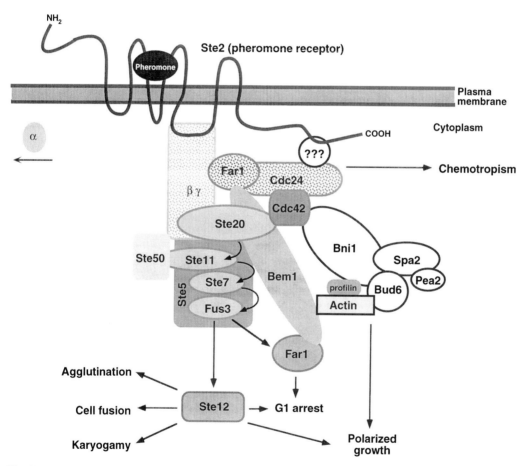

Fig. 10. Model for activation of cellular events in response to perception of the mating pheromone. Binding of mating pheromone to its receptor (Ste2) induces the dissociation of Gα (with an *arrow* pointing away) from the G$\beta\gamma$ dimer, which can then activate Ste5 and Ste20, and hence the Fus3 MAPK cascade. Phosphorylation of Far1 by Fus3 leads to cell cycle arrest. Activation of the Ste12 transcription factor by Fus3 turns on genes involved in polarized growth, agglu-

tination, cell fusion and karyogamy. Polarity establishment proteins and polarisome components may function to promote polarized growth to form the mating projection and help concentrate growth at the projection tip via regulation of the actin cytoskeleton. Ste2, Cdc24, Far1 and G$\beta\gamma$ are also involved in chemotropism that orients the mating projection toward the pheromone source which interacts strongly with Ste11 and weakly with Ste4 in the two-hybrid system (Xu et al. 1996). How Ste50 fits into the signaling scheme remains elusive.

Fus3 activates Ste12 to turn on a group of pheromone-induced genes involved in polarized morphogenesis, agglutination, cell fusion, karyogamy and adaptation/recovery from mating response (Sprague and Thorner 1992). In addition, phosphorylation of the G1-CDK inhibitor (CKI) Far1 by Fus3 leads to cell cycle arrest in G1 (Chang and Herskowitz 1990, 1992; Peter and Herskowitz 1994). Apart from its function as a CKI, Far1 is required for chemotropism (Segall 1993; Dorer et al. 1995; Valtz et al. 1995; see below). Two distinct domains (Valtz et al. 1995) mediate these two functions of Far1. A C-terminal region is required for chemotropism while an N-terminal one is responsible for the CKI activity.

One prominent event during the mating response is the formation of a mating projection from one edge of the cell. This process accompanies considerable cell wall remodeling. Chitin is deposited at the base of the mating projection and a unique set of cell surface proteins is produced (Lipke et al. 1976; Schekman and Brawley 1979; Tkacz and MacKay 1979). Septins are also present at the base of the projection (Ford and Pringle 1991; Kim et al. 1991). Similar to bud emergence, the actin cytoskeleton and membrane vesicles concentrate at a region of the cell where the mating projection emerges (Hasek et al. 1987; Baba et al. 1989). Growth continues at the tip of the projection in a manner very much like in the apical phase of bud growth (Tkacz and MacKay 1979). Later, cortical actin patches are concentrated at the projection tip whereas actin cables are aligned along the projection axis. Cdc42, Cdc24, Bem1, Spa2, Pea2, Bud6 and Bni1 and many other proteins that localize to bud tips are present at the mating projection tip (Gehrung and Snyder 1990; Chenevert et al. 1992; Ziman et al. 1993; Valtz and Herskowitz 1996; Amberg et al. 1997; Evangelista et al. 1997; Nern and Arkowitz 1999).

Cdc42, Cdc24 and Bem1 are likely involved in both signaling and polarity establishment to initiate mating projection formation (Chenevert 1994; Chenevert et al. 1992). Mutation alleles of *CDC42* and *CDC24* that affect pheromone signaling have been isolated, and overexpression of the active form of Cdc42 induces the pathway (Simon et al. 1995; Zhao et al. 1995). The mating

pheromone signaling mediated by Cdc42 and Cdc24 is probably Ste20-independent since overexpression of Cdc42 improves mating efficiency of *ste20Δ* mutants and the Cdc42-Ste20 interaction is not essential for normal pheromone-induced signal transduction (Leberer et al. 1996). However, Ste20 may mediate the signaling-independent role for Cdc42 in mating projection morphogenesis. Deletion of the Cdc42-binding domain from Ste20 does not affect signaling but results in defects in bilateral mating and cell fusion (Peter et al. 1996; Leberer et al. 1997); these phenotypes are commonly observed in mutants defective in mating projection morphogenesis like *spa2* (Gehrung and Snyder 1990; Gammie et al. 1998). The broader mating projection tip found in *spa2* and *pea2* mutants are consistent with the role of these proteins in concentrating growth at a small region at the mating projection tip similar to their function during apical bud growth (Gehrung and Snyder 1990; Valtz and Herskowitz 1996; Sheu et al. 2000). Bem1 interacts with actin and Ste20 during the mating response and the Bem1-Ste20 interaction is required for association of Ste20 with actin (Leeuw et al. 1995). The Bem1-bound Ste5 is associated with the Fus3 MAPK cascade and is required for efficient signal transduction (Lyons et al. 1996). Bem1 also interacts with the Far1 and overexpression of *BEM1* suppresses the G1 arrest defect of the *fus3-2* mutant. Thus, Bem1 may function to coordinate signaling, morphogenesis and the cell cycle arrest during mating.

Mating projections from two partners grow toward each other by tracking pheromone gradients (Jackson and Hartwell 1990; Segall 1993; Dorer et al. 1995). This chemotropic behavior is important for efficient mating. Pheromone receptors accumulate at the tips of growing projections and may contribute to the spatial regulation of signaling (Jackson et al. 1991). Consistent with this idea, Ste2 truncation mutants lacking part of the C-terminal cytoplasmic domain exhibit defects in forming and orienting projections along pheromone gradients but have normal signaling activity (Vallier et al., unpubl., cited in Roemer et al. 1996b). Far1, Cdc24 and Gβγ are required for chemotropism during mating (Valtz et al. 1995; Schrick et al. 1997; Nern and Arkowitz 1998). In response to pheromone, Far1 shuttles from the nucleus to the cytoplasm and interacts with Gβγ, Bem1, Cdc24 and Cdc42 (Butty et al. 1998; Nern and Arkowitz 1999). Thus, these components may form a complex to orient the cytoskeleton during chemotropic growth of the mating projection.

In the presence of a very high concentration of the mating pheromone such that the pheromone gradient no longer exists (e.g. addition of exogenous pheromone to the mating mixture), cells form mating projections adjacent to the previous bud site (Madden and Snyder 1992; Dorer et al. 1995). This observation suggests that bud site selection signals direct projection formation when cells fail to sense the pheromone gradient. Consistent with this idea, in a *cdc24-m1* mutant which is defective in projection orientation, deletion of *BUD1* causes defects in projection formation and a further decrease in mating efficiency (Nern and Arkowitz 1999); in *CDC24* cells, Bud1 is not essential for projection morphogenesis and efficient mating. Thus, it is likely that, in the absence of chemotropism, Bud1 directs polarized growth during mating response in a way similar to axial bud site selection.

After mating partners find each other, the two haploid cells fuse to form a diploid zygote. Cell surface proteins involved in cell adhesion (agglutinins), cell fusion (e.g. Fus1 and Fus2) and karyogamy are upregulated in response to mating pheromone and become concentrated at tips of mating projections (reviewed in Lipke and Kurjan 1992; Sprague and Thorner 1992; Marsh and Rose 1997). Interestingly, proteins involved in cell polarity control, such as Spa2, Pea2, Bni1 and Rvs161, are also involved in cell fusion (Dorer et al. 1997; Gammie et al. 1998). These polarity proteins may function to concentrate secretory vesicles toward mating projection tips and sites of cell fusion, similar to their role in apical growth. During cell fusion, cell wall is removed from the center region of cell-cell contact so that the plasma membrane of the two cells can fuse (Gammie et al. 1998). Since removal of the cell wall can be detrimental to cell integrity, it is crucial that cell wall removal is precisely regulated both spatially and temporally (Philips and Herskowitz 1997). Two homologous proteins, Kel1 and Kel2, localize to the site where cell fusion occurs and may function to regulate localized cell wall degradation by antagonizing the Pkc1-Mpk1 cell integrity pathway at fusion sites (Philips and Herskowitz 1998). After the cell membrane fuses, the nuclei from the two cells can then fuse through a process called karyogamy, which involves microtubules, a group of motor proteins and their regulator.

The process of karyogamy is reviewed elsewhere (Marsh and Rose 1997).

X. Conclusions

The combination of genetics, cell biology and biochemical techniques has elucidated a wealth of knowledge about fundamental processes and molecular mechanisms that control cell polarity and cell shape in budding *S. cerevisiae*. Detailed phenotypic analyses of changes in cell shape, division patterns and subcellular structures in various mutants reveal the cellular processes that certain genes are participating in. Determining the localization pattern of a protein tells the possible site of its action. For example, many structural and regulatory components involved in cell polarity localize to sites of polarized growth.

It has been proposed that bud growth follows a hierarchy of successive actions: bud site determination, bud initiation and bud enlargement (Drubin 1991; Madden et al. 1992). This concept provides a general framework; however, processes of cell growth are not just linear pathways but complex networks in which one functional module may simultaneously influence several cellular processes, and processes in one cell cycle may affect those in subsequent generations. For example, some components involved in apical growth affect not only cell shape, but also bud site selection, nuclear positioning and cell fusion. This may be important for spatial and temporal coordination of bud growth and nuclear division. The morphological similarities (i.e. incomplete cell separation, cell elongation and more distal budding) between filamentous cells and cells with defective septin organization raise the interesting possibility that regulation of septin organization may be used to achieve morphological changes during filamentous growth. That is, by downregulating septin assembly (which leads to incomplete cell separation), cells can activate the signaling pathway used in the septin-integrity checkpoint to increase apical growth (which leads to cell elongation), which, in turn, promotes budding at distal poles. The existence of such a regulation mechanism remains to be addressed.

Components involved in controlling polarized growth in *S. cerevisiae* often have equivalents in other organisms and thus studies in *S. cerevisiae* provide a theoretical framework for understanding the molecular mechanisms of morphogenesis and cell polarity in other eukaryotes.

Acknowledgements. We thank many of our colleagues for communication of unpublished results. We also thank S. Bidlingmaier and G. Michaud for critical comments on the manuscript and members of the Snyder laboratory for helpful discussions.

References

Adames N, Blundell K, Ashby M, Boone C (1995) Role of yeast insulin-degrading enzyme homologs in pro-pheromone processing and bud site selection. Science 270:464–467

Adams A, Pringle J (1984) Relationship of actin and tubulin distribution to bud growth in wild-type and morphogenetic-mutant *Saccharomyces cerevisiae*. J Cell Biol 98:934–945

Adams AEM, Johnson DI, Longnecker RM, Sloat BF, Pringle JR (1990) *CDC42* and *CDC43*, two additional genes involved in budding and the establishment of cell polarity in the yeast *Saccharomyces cerevisiae*. J Cell Biol 111:131–142

Amatruda JF, Gattermeir DJ, Karpova TS, Cooper JA (1992) Effects of null mutations and overexpression of capping protein on morphogenesis, actin distribution, and polarized secretion in yeast. J Cell Biol 119:1151–1162

Amberg DC, Zahner JE, Mulholland JW, Pringle JR, Botstein D (1997) Aip3p/Bud6p, a yeast actin-interacting protein that is involved in morphogenesis and the selection of bipolar budding sites. Mol Biol Cell 8:729–753

Andrews B, Measday V (1998) The cyclin family of budding yeast: abundant use of a good idea. Trends Genet 14:66–72

Andrews BJ, Herskowitz I (1989) The yeast SWI4 protein contains a motif present in developmental regulators and is part of a complex involved in cell-cycle-dependent transcription. Nature 342:830–833

Arkowitz RA, Lowe N (1997) A small conserved domain in the yeast Spa2p is necessary and sufficient for its polarized localization. J Cell Biol 138:17–36

Ayscough KR (1998) In vivo functions of actin-binding proteins. Curr Opin Cell Biol 10:102–111

Ayscough KR, Stryker J, Pokala N, Sanders M, Crews P, Drubin DG (1997) High rates of actin filament turnover in budding yeast and roles for actin in the establishment and maintenance of cell polarity revealed using the actin inhibitor latrunculin A. J Cell Biol 137:399–416

Baba M, Baba N, Ohsumi Y, Kanaya K, Osumi M (1989) Three-dimensional analysis of morphogenesis induced by mating pheromone α factor in *Saccharomyces cerevisiae*. J Cell Sci 94:207–216

Barral Y, Mann C (1995) G1 cyclin degradation and cellular differentiation of *Saccharomyces cerevisiae*. CR Acad Sci III 318:43–50

Barral Y, Jentsch S, Mann C (1995) G1 cyclin turnover and nutrient uptake are controlled by a common pathway in yeast. Genes Dev 9:399–409

Barral Y, Parra M, Bidlingmaier S, Snyder M (1999) Nim1-related kinases coordinate cell cycle progression with organization of the peripheral cytoskeleton. Genes Dev 13:176–187

Bauer F, Urdaci M, Aigle M, Crouzet M (1993) Alteration of a yeast SH3 protein leads to conditional viability with defects in cytoskeletal and budding patterns. Mol Cell Biol 13:5070–5084

Bedinger PA, Hardeman KJ, Loukides CA (1994) Travelling in style: the cell biology of pollen. Trends Cell Biol 4:132–138

Bender A (1993) Genetic evidence for the roles of the bud-site-selection genes *BUD5* and *BUD2* in control of the Rsr1p (Bud1p) GTPase in yeast. Proc Natl Acad Sci USA 90:9926–9929

Bender A, Pringle JR (1989) Multicopy suppression of the *cdc24* budding defect in yeast by *CDC42* and three newly identified genes including the *ras*-related gene *RSR1*. Proc Natl Acad Sci USA 86:9976–9980

Bender A, Pringle JR (1991) Use of a screen for synthetic lethal and multicopy suppressee mutants to identify two new genes involved in morphogenesis in *Saccharomyces cerevisiae*. Mol Cell Biol 11:1295–1305

Benton BK, Tinkelenberg AH, Jean D, Plump SD, Cross FR (1993) Genetic analysis of Cln/Cdc28 regulation of cell morphogenesis in budding yeast. EMBO J 12:5267–5275

Benton BK, Tinkelenberg A, Gonzalez I, Cross F (1997) Cla4p, a *Saccharomyces cerevisiae* Cdc42p-activated kinase involved in cytokinesis, is activated at mitosis. Mol Cell Biol 17:5067–5076

Bi E, Maddox P, Lew DJ, Salman ED, MacMillan JN, Yeh E, Pringle JR (1998) Involvement of an actomyosin contractile ring in *Saccharomyces cerevisiae* cytokinesis. J Cell Biol 142:1301–1312

Bickle M, Delly P-A, Schmidt A, Hall MN (1998) Cell wall integrity modulates Rho1 activity via the exchange factor Rom2. EMBO J 17:2235–2245

Bobola N, Jansen R-P, Shin TH, Nasmyth K (1996) Asymmetric accumulation of Ash1p in postanaphase nuclei depends on a myosin and restricts yeast mating-type switching to mother cells. Cell 84:699–709

Booher RN, Deshaies RJ, Kirschner MW (1993) Properties of *Saccharomyces cerevisiae* wee1 and its differential regulation of p34*CDC28* in response to G1 and G2 cyclins. EMBO J 12:3417–3426

Breeden L (1996) Start-specific transcription in yeast. In: Farnham P (ed) Transcriptional control of cell growth: the E2F gene family. Current Topics in Microbiology and Immunology, vol 208. Springer, Berlin Heidelberg New York, pp 95–127

Brewster JL, Gustin MC (1994) Positioning of cell growth and division after osmotic stress requires a MAP kinase pathway. Yeast 10:425–439

Brown J, Bussey H (1993) The yeast *KRE9* gene encodes an O-glycoprotein involved in cell surface β-glucan assembly. Mol Cell Biol 13:6346–6356

Bulawa CE (1993) Genetics and molecular biology of chitin synthesis in fungi. Annu Rev Microbiol 47:505–534

Butty A-C, Pryciak PM, Huang LS, Herskowitz I, Peter M (1998) The role of Far1p in linking the heterotrimeric

G protein to polarity establishment proteins during yeast mating. Science 282:1511–1516

Byers B (1981) Cytology of the yeast life cycle. In: Strathern JN, Jones E, Broach J (eds) The molecular biology of the yeast *Saccharomyces*: life cycle and inheritance, vol 1. Cold Spring Harbor Laboratory, New York, pp 59–96

Cabib E, Bowers B (1971) Chitin and yeast budding: localization of chitin in yeast bud scars. J Biol Chem 246: 152–159

Cabib E, Sburlati A, Bowers B, Silverman SJ (1989) Chitin synthase 1, an auxiliary enzyme for chitin synthesis in *Saccharomyces cerevisiae*. J Cell Biol 108:1665–1672

Cabib E, Silverman SJ, Shaw JA (1992) Chitinase and chitin synthase 1: counterbalancing in cell separation of *Saccharomyces cerevisiae*. J Gen Microbiol 138:97–102

Cali BM, Doyle TC, Botstein D, Fink GR (1998) Multiple functions for actin during filamentous growth of *Saccharomyces cerevisiae*. Mol Biol Cell 9:1873–1889

Carminati JL, Stearns T (1997) Microtubules orient the mitotic spindle in yeast through dynein-dependent interactions with the cell cortex. J Cell Biol 138:629–641

Carroll CW, Altman R, Schieltz D, Yates JR, Kellogg D (1998) The septins are required for the mitosis-specific activation of the Gin4 kinase. J Cell Biol 143:709–718

Chandarlapaty S, Errede B (1998) Ash1, a daughter cell-specific protein, is required for pseudohyphal growth of *Saccharomyces cerevisiae*. Mol Cell Biol 18:2884–2891

Chang F, Herskowitz I (1990) Identification of a gene necessary for cell cycle arrest by a negative growth factor of yeast: FAR1 is an inhibitor of a G1 cyclin, CLN2. Cell 63:999–1011

Chang F, Herskowitz I (1992) Phosphorylation of *FAR1* in response to α-factor: a possible requirement for cell-cycle arrest. Mol Biol Cell 3:445–450

Chant J, Herskowitz I (1991) Genetic control of bud-site selection in yeast by a set of gene products that comprise a morphogenetic pathway. Cell 65:1203–1212

Chant J, Pringle JR (1995) Patterns of bud-site selection in the yeast *Saccharomyces cerevisiae*. J Cell Biol 129:751–765

Chant J, Corrado K, Pringle JR, Herskowitz I (1991) Yeast BUD5, encoding a putative GDP-GTP exchange factor, is necessary for bud site selection and interacts with bud formation gene BEM1. Cell 65:1213–1224

Chant J, Mischke M, Mitchell E, Herskowitz I, Pringle JR (1995) Role of Bud3p in producing the axial budding pattern of yeast. J Cell Biol 129:767–778

Chen GC, Zheng L, Chan CS (1996) The LIM domain-containing Dbm1 GTPase-activating protein is required for normal cellular morphogenesis in *Saccharomyces cerevisiae*. Mol Cell Biol 16:1376–1390

Chenevert J (1994) Cell polarization directed by extracellular cues in yeast. Mol Biol Cell 5:1169–1175

Chenevert J, Corrado K, Bender A, Pringle J, Herskowitz I (1992) A yeast gene (*BEM1*) necessary for cell polarization whose product contains two SH3 domains. Nature 356:77–79

Choi K-Y, Satterberg B, Lyons DM, Elion EA (1994) Ste5 tethers multiple protein kinases in the MAP kinase cascade required for mating in *S. cerevisiae*. Cell 78:499–512

Chowdhury S, Smith KW, Gustin MC (1992) Osmotic stress and the yeast cytoskeleton: phenotype-specific suppression of an actin mutation. J Cell Biol 118:561–571

Cid V, Duran A, del Rey F, Snyder M, Nombela C, Sanchez M (1995) Molecular basis of cell integrity and morphogenesis in *Saccharomyces cerevisiae*. Microbiol Rev 59:345–386

Cope MJTV, Yang S, Shang C, Drubin D (1999) Novel protein kinases Ark1 and Prk1 associate with and regulate the cortical actin cytoskeleton in budding yeast. J Cell Biol 144:1203–1218

Costigan C, Gehrung S, Snyder M (1994) A synthetic lethal screen identifies *SLK1*, a novel protein kinase homolog important in yeast cell morphogenesis and cell growth. Mol Cell Biol 12:1162–1178

Costigan C, Snyder M (1998) Cell polarity in the budding yeast, *Saccharomyces cerevisiae*. Adv Mol Cell Biol 26:1–66

Costigan C, Kolodrubetz D, Snyder M (1994) *NHP6A* and *NHP6B*, which encode HMG1-like proteins, function downstream in the yeast SLT2 MAPK pathway. Mol Cell Biol 14:2391–2403

Cross FR (1995) Starting the cell cycle: what's the point? Curr Opin Cell Biol 7:790–797

Cvrckova F, Nasmyth K (1993) Yeast G1 cyclins *CLN1* and *CLN2* and a GAP-like protein have a role in bud formation. EMBO J 12:5277–5286

Cvrckova F, Virgilio CD, Manser E, Pringle J, Nasmyth K (1995) Ste20-like protein kinases are required for normal localization of cell growth and for cytokinesis in budding yeast. Genes Dev 9:1817–1830

DeMarini DJ, Adams AEM, Faras H, Virgilio CD, Valle G, Chuang JS, Pringle JR (1997) A septin-based hierarchy of proteins required for localized deposition of chitin in the *Saccharomyces cerevisiae* cell wall. J Cell Biol 139:75–93

De Virgilio C, DeMarini DJ, Pringle JR (1996) SPR28, a sixth member of the septin gene family in *Saccharomyces cerevisiae* that is expressed specifically in sporulating cells. Microbiology 142:2897–2905

Dodou E, Treisman R (1997) The *Saccharomyces cerevisiae* MADS-box transcription factor RLM1 is a target for the MPK1 mitogen-activated protein kinase pathway. Mol Cell Biol 17:1848–1859

Dorer R, Pryciak PM, Hartwell LH (1995) *Saccharomyces cerevisiae* cells execute a default pathway to select a mate in the absence of pheromone gradients. J Cell Biol 131:845–861

Dorer R, Boone C, Kimbrough T, Kim J, Hartwell LH (1997) Genetic analysis of default mating behavior in *Saccharomyces cerevisiae*. Genetics 146:39–55

Doyle T, Botstein D (1996) Movement of yeast cortical actin cytoskeleton visualized in vivo. Proc Natl Acad Sci USA 93:3886–3891

Drubin D (1991) Development of cell polarity in budding yeast. Cell 65:1093–1096

Drubin DG, Nelson WJ (1996) Origins of cell polarity. Cell 84:335–344

Drubin DG, Jones HD, Wertman KF (1993) Actin structure and function: roles in mitochondrial organization and morphogenesis in budding yeast and identification of the phalloidin-binding site. Mol Biol Cell 4:1277–1294

Durrens P, Revardel E, Bonneu M, Aigle M (1995) Evidence for a branched pathway in the polarized cell division of *Saccharomyces cerevisiae*. Curr Genet 27: 213–216

Eby JJ, Holly SP, van Drogen F, Grishin AV, Peter M, Drubin DG, Blumer KJ (1998) Actin cytoskeleton organization regulated by the PAK family of protein kinases. Curr Biol 8:967–970

Edgington NP, Blacketer MJ, Bierwagen TA, Myers AM (1999) Control of *Saccharomyces cerevisiae* filamentous growth by cyclin-dependent kinase Cdc28. Mol Cell Biol 19:1369–1380

Eisen JS (1994) Development of motoneuronal phenotype. Annu Rev Neurosci 17:1–30

Epp JA, Chant J (1997) An IQGAP-related protein controls actin-ring formation and cytokinesis in yeast. Curr Biol 7:921–929

Erdman S, Snyder M (2001) Yeast mating differentiation involves a searching response Genetics (in press)

Erdman S, Lin L, Malczynski M, Snyder M (1998) Pheromone-regulated genes required for yeast mating differentiation. J Cell Biol 140:461–483

Errede B, Cade RM, Yasar BM, Kamada Y, Levin DE, Irie K, Matsumoto K (1995) Dynamics and organization of MAP kinase signal pathways. Mol Reprod Dev 42: 477–485

Espinoza FH, Orgas J, Herskowitz I, Morgan DO (1994) Cell cycle control by a complex of the cyclin HCS26 (PCL1) and the kinase PHO85. Science 266:1388–1391

Evangelista M, Blundell K, Longtine MS, Chow CJ, Adames N, Pringle JR, Peter M, Boone C (1997) Bni1p, a yeast formin linking Cdc42p and the actin cytoskeleton during polarized morphogenesis. Science 276:118–122

Fares H, Peifer M, Pringle JR (1995) Localization and possible functions of *Drosophila* septins. Mol Biol Cell 12:1843–1859

Fares H, Goetsch L, Pringle JR (1996) Identification of a developmentally regulated septin and involvement of the septins in spore formation in *Saccharomyces cerevisiae*. J Cell Biol 132:399–411

Farkas SK, Kovarik J, Kosinova A, Bauer S (1974) Autoradiographic study of mannan incorporation into the growing cell walls of *Saccharomyces cerevisiae*. J Bacteriol 117:265–269

Farkasovsky I, Kuntzel H (1995) Yeast Num1p associates with the mother cell cortex during S/G2 phase and affects microtubular functions. J Cell Biol 131:1003–1014

Field C, Al-Awar O, Rosenblatt J, Wong M, Alberts B, Mitchison TJ (1996) A purified *Drosophila* septin complex forms filaments and exhibits GTPase activity. J Cell Biol 133:605–616

Field C, Li R, Oegema K (1999) Cytokinesis in eukaryotes: a mechanistic comparison. Curr Opin Cell Biol 11: 68–80

Finger FP, Novick P (1997) Sec3p is involved in secretion and morphogenesis in *Saccharomyces cerevisiae*. Mol Biol Cell 8:647–662

Finger FP, Hughes TE, Novick P (1998) Sec3p is a spatial landmark for polarized secretion in budding yeast. Cell 92:559–571

Flescher EG, Madden K, Snyder M (1993) Components required for cytokinesis are important for bud site selection in yeast. J Cell Biol 122:373–386

Ford S, Pringle J (1991) Cellular morphogenesis in the *Saccharomyces cerevisiae* cell cycle: localization of the *CDC11* gene product and the timing of events at the budding site. Dev Genet 12:281–292

Frazier JA, Wong ML, Longtine MS, Pringle JR, Mann M, Mitchison TJ, Field C (1998) Polymerization of purified yeast septins: evidence that organized filament arrays may not be required for septin function. J Cell Biol 143:737–749

Freifelder D (1960) Bud position in *Saccharomyces cerevisiae*. J Bacteriol 124:511–523

Fujita A, Oka C, Arikawa Y, Katagal T, Tonouchi A, Kuhara S, Misumi Y (1994) A yeast gene necessary for budsite selection encodes a protein similar to insulin-degrading enzymes. Nature 372:567–569

Fujiwara T, Tanaka K, Mino A, Kikyo M, Takahashi K, Shimizu K, Takai Y (1998) Rho1p-Bni1p-Spa2p interactions: implication in localization of Bni1p at the bud site and regulation of the actin cytoskeleton in *Saccharomyces cerevisiae*. Mol Biol Cell 9:1221–1233

Gale C, Finkel D, Tao N, Meinke M, McClellan M, Olson J, Kendrick K, Hostetter M (1996) Cloning and expression of a gene encoding an integrin-like protein in *Candida albicans*. Proc Natl Acad Sci USA 93:357–361

Galitski T, Saldanha AJ, Styles CA, Lander ES, Fink GR (1999) Ploidy regulation of gene expression. Science 285:251–254

Gallwitz D, Sures I (1980) Structure of a split gene: complete nucleotide sequence of the actin gene in *Saccharomyces cerevisiae*. Proc Natl Acad Sci USA 77:2546–2550

Gammie AE, Brizzio V, Rose MD (1998) Distinct morphological phenotypes of cell fusion mutants. Mol Biol Cell 9:1395–1410

Gavrias V, Andrianopoulos A, Gimeno CJ, Timberlake WE (1996) *Saccharomyces cerevisiae TEC1* is required for pseudohyphal growth. Mol Microbiol 19:1255–1263

Gehrung S, Snyder M (1990) The *SPA2* gene of *Saccharomyces cerevisiae* is important for pheromone-induced morphogenesis and efficient mating. J Cell Biol 111:1451–1464

Geli MI, Riezman H (1996) Role of type I myosins in receptor-mediated endocytosis in yeast. Science 272: 533–535

Gimeno CJ, Fink GR (1994) Induction of pseudohyphal growth by overexpression of *PHD1*, a *Saccharomyces cerevisiae* gene related to transcriptional regulators of fungal development. Mol Cell Biol 14:2100–2112

Gimeno CJ, Ljungdahl PO, Styles CA, Fink GR (1992) Unipolar cell divisions in the yeast *S. cerevisiae* lead to filamentous growth: regulation by starvation and *RAS*. Cell 68:1077–1090

Goode BL, Wang JJ, Butty A-C, Peter M, McCormack AL, Yates JR, Drubin DG, Barnes G (1999) Coronin promotes the rapid assembly and cross-linking of actin filaments and may link the actin and microtubule cytoskeletons in yeast. J Cell Biol 144:83–98

Goodson HV, Anderson BL, Warrick HM, Pon LA, Spudich JA (1996) Synthetic lethality screen identifies a novel yeast myosin I gene (MYO5): myosin I proteins are required for polarization of the actin cytoskeleton. J Cell Biol 133:1277–1291

Goud B, Salminen A, Walworth NC, Novick PJ (1988) A GTP-binding protein required for secretion rapidly associates with secretory vesicles and the plasma membrane in yeast. Cell 53:753–768

Govindan B, Bowser R, Novick P (1995) The role of Myo2, a yeast class V myosin in vesicular transport. J Cell Biol 128:1055–1068

Gray JV, Ogas JP, Kamada YMS, Levin DE, Herskowitz I (1997) A role for the Pkc1 MAP kinase pathway of *Saccharomyces cerevisiae* in bud emergence and identification of a putative upstream regulator. EMBO J 16:4929–4937

Gustin MC, Albertyn J, Alexander M, Davenport K (1998) MAP kinase pathways in the yeast *Saccharomyces cerevisiae*. Microbiol Mol Biol Rev 62:1264–1300

Haarer B, Pringle JR (1987) Immunofluorescence localization of the *Saccharomyces cerevisiae* CDC12 gene product to the vicinity of the 10 nm filaments in the mother-bud neck. Mol Cell Biol 7:3678–3687

Haarer BK, Corbett A, Kweon Y, Petzold AS, Silver P, Brown SS (1996) *SEC3* mutations are synthetically lethal with profilin mutations and cause defects in diploid-specific bud-site selection. Genetics 144:495–510

Hall A (1994) Small GTP-binding proteins and the regulation of the actin cytoskeleton. Annu Rev Cell Biol 10:31–54

Hall DD, Markwardt DD, Parviz F, Heideman W (1998) Regulation of Cln3-Cdc28 kinase by cAMP in *Saccharomyces cerevisiae*. EMBO J 17:4370–4378

Halme A, Michelitch M, Mitchell EL, Chant J (1996) Bud10p directs axial cell polarization in budding yeast and resembles a transmembrane receptor. Curr Biol 6:570–579

Harlan JE, Hajduk PJ, Yoon HS, Fesik SW (1994) Pleckstrin homology domains bind to phosphatidylinositol-4,5-bisphosphate. Nature 371:168–170

Harsay E, Bretscher A (1995) Parallel secretory pathways to the cell surface in yeast. J Cell Biol 131:297–310

Hartwell LH (1971) Genetic control of the cell division cycle in yeast. IV. Genes controlling bud emergence and cytokinesis. Exp Cell Res 69:265–276

Hartwell LH, Weinert TA (1989) Checkpoints: controls that ensure the order of cell cycle events. Science 246:629–634

Hartwell LH, Culotti J, Pringle JR, Reid BJ (1974) Genetic control of the cell division cycle in yeast. Science 183:46–51

Hasek J, Rupes I, Svobodova J, Streiblova E (1987) Tubulin and actin topology during zygote formation of *Saccharomyces cerevisiae*. J Gen Microbiol 133:3355–3363

Heil-Chapdelaine RA, Tran NK, Cooper JA (1998) The role of *Saccharomyces cerevisiae* coronin in the actin and microtubule cytoskeletons. Curr Biol 8:1281–1284

Helliwell SB, Howald I, Barbet N, Hall MN (1998) *TOR2* is part of two related signaling pathways coordinating cell growth in *Saccharomyces cerevisiae*. Genetics 148:99–112

Hermann GJ, King EJ, Shaw JM (1997) The yeast gene, *MDM20*, is necessary for mitochondrial inheritance and organization of the actin cytoskeleton. J Cell Biol 137:141–153

Herskowitz I (1988) Life cycle of the budding yeast *Saccharomyces cerevisiae*. Microbiol Rev 52:536–553

Herskowitz I (1995) MAP kinase pathways in yeast: for mating and more. Cell 80:187–197

Hicks JB, Strathern JN, Herskowitz I (1977) Interconversion of yeast mating types. III. Action of the homothallism (HO) gene in cells homozygous for the mating type locus. Genetics 85:395–405

Holly SP, Blumer KJ (1999) PAK-family kinases regulate cell and actin polarization throughout the cell cycle of *Saccharomyces cerevisiae*. J Cell Biol 147:845–856

Huffaker TC, Thomas JH, Botstein D (1988) Diverse effects of β-tubulin mutations on microtubule formation and function. J Cell Biol 106:1997–2010

Hunter T, Plowman GD (1997) The protein kinases of budding yeast: six score and more. Trends Biochem Sci 22:18–22

Hyman AA, Stearns T (1992) Spindle positioning and cell polarity. Curr Biol 2:469–471

Igual JC, Johnson AL, Johnston LH (1996) Coordinated regulation of gene expression by the cell cycle transcription factor SWI4 and the protein kinase C MAP kinase pathway for yeast cell integrity. EMBO J 15:5001–5013

Imai J, Toh-e A, Matsui Y (1996) Genetic analysis of the *Saccharomyces cerevisiae RHO3* gene, encoding a Rho-type small GTPase provides evidence for a role in bud formation. Genetics 142:359–369

Imamura H, Tanaka K, Hihara T, Umikawa M, Kamei T, Takahashi K, Sasaki T, Takai Y (1997) Bni1p and Bnr1p: downstream targets of the Rho family small G-proteins which interact with profilin and regulate actin cytoskeleton in *Saccharomyces cerevisiae*. EMBO J 16:2745–2755

Irie K, Takase M, Lee K, Levin D, Araki H, Matsumoto K, Oshima Y (1993) *MKK1* and *MKK2*, which encode *Saccharomyces cerevisiae* mitogen-activated protein kinase-kinase homologs, function in the pathway mediated by protein kinase C. Mol Cell Biol 13:3076–3083

Jackson CL, Hartwell LH (1990) Courtship in *S. cerevisiae*: both cell types choose mating partners by responding to the strongest pheromone signal. Cell 63:1039–1051

Jackson CL, Konopka JB, Hartwell LH (1991) *S. cerevisiae* α-pheromone receptors activate a novel signal transduction pathway for mating partner discrimination. Cell 67:389–402

Jacobs CW, Adams AEM, Szaniszlo PJ, Pringle JR (1988) Functions of microtubules in the *Saccharomyces cerevisiae* cell cycle. J Cell Biol 107:1409–1426

Jan YN, Jan LY (1998) Asymmetric cell division. Science 392:775–778

Jansen RP, Dowzer C, Michaelis C, Galova M, Nasmyth K (1996) Mother cell-specific HO expression in budding yeast depends on the unconventional myosin myo4p and other cytoplasmic proteins. Cell 84:687–697

Johnson DI (1999) Cdc42: an essential Rho-type GTPase controlling eukaryotic cell polarity. Microbiol Mol Biol Rev 63:54–105

Johnson DI, Pringle JR (1990) Molecular characterization of *CDC42*, a *Saccharomyces cerevisiae* gene involved in the development of cell polarity. J Cell Biol 111:143–152

Johnston GC, Prendergast JA, Singer RA (1991) The *Saccharomyces cerevisiae MYO2* gene encodes an essential myosin for vectorial transport of vesicles. J Cell Biol 113:539–551

Jung US, Levin D (1999) Genome-wide analysis of gene expression regulated by the yeast cell wall integrity signalling pathway. Mol Microbiol 34:1049–1057

Kaiser CA, Gimeno RE, Shaywitz DA (1996) Protein secretion, membrane biogenesis, and endocytosis. In: Pringle JR, Broach JR, Jones EW (eds) The Molecular and Cellular Biology of the Yeast *Saccharomyces*. Cold Spring Harbor Laboratory Press, Cold Spring Harbor, pp 91–228

Kaiser P, Sia RAL, Bardes ESG, Lew DJ, Reed SI (1998) Cdc34 and the F-box protein Met30 are required for degradation of the Cdk-inhibitory kinase Swe1. Genes Dev 12:2587–2597

Kamada Y, Jung US, Piotrowski J, Levin DE (1995) The protein kinase C-activated MAP kinase pathway of *Saccharomyces cerevisiae* mediates a novel aspect of the heat shock response. Genes Dev 9:1559–1571

Kamada Y, Qadota H, Python CP, Anraku Y, Ohya Y, Levin DE (1996) Activation of yeast protein kinase C by Rho1 GTPase. J Biol Chem 271:9193–9196

Ketela T, Green R, Bussey H (1999) *Saccharomyces cerevisiae* Mid2p is a potential cell wall stress sensor and upstream activator of the PKC1-MPK1 cell integrity pathway. J Bacteriol 181:3330–3340

Kilmartin JV, Adams AEM (1984) Structural rearrangements of tubulin and actin during the cell cycle of the yeast *Saccharomyces*. J Cell Biol 98:922–933

Kim HB, Haarer BK, Pringle JR (1991) Cellular morphogenesis in the *Saccharomyces cerevisiae* cell cycle: localization of the CDC3 gene product and the timing of events at the budding site. J Cell Biol 112:535–544

Kim YJ, Francisco L, Chen GC, Marcotte E, Chan CSM (1994) Control of cellular morphogenesis by the Ipl2/Bem2 GTPase-activating protein: possible role of protein phosphorylation. J Cell Biol 127:1381–1394

Klis FM (1994) Cell wall assembly in yeast. Yeast 10:851–869

Kohno H, Tanaka K, Mino A, Umikawa M, Imamura H, Fujiwara T, Fujita Y, Hotta K, Qadota H, Watanabe T, Ohya Y, Takai Y (1996) Bni1p implicated in cytoskeletal control is a putative target of Rho1p small GTP binding protein in *Saccharomyces cerevisiae*. EMBO J 15:6060–6068

Kron SJ, Gow NAR (1995) Budding yeast morphogenesis: signalling, cytoskeleton and cell cycle. Curr Opin Cell Biol 7:845–855

Kron SJ, Styles CA, Fink GR (1994) Symmetric cell division in pseudohyphae of the yeast *Saccharomyces cerevisiae*. Mol Biol Cell 5:1003–1022

Kubler E, Riezman H (1993) Actin and fimbrin are required for the internalization step of endocytosis in yeast. EMBO J 12:2855–2862

Leberer E, Dignard D, Harcus D, Thomas DY, Whiteway M (1992) The protein kinase homologue Ste20p is required to link the yeast pheromone response G-protein beta gamma subunits to downstream signalling components. EMBO J 11:4815–4824

Leberer E, Chenevert J, Leeuw T, Harcus D, Herskowitz I, Thomas DY (1996) Genetic interactions indicate a role for Mdg1p and the SH3 domain protein Bem1p in linking the G-protein mediated yeast pheromone signalling pathway to regulators of cell polarity. Mol Gen Genet 252:608–621

Leberer E, Wu C, Leeuw T, Fourest-Lieuvin A, Segall JE, Thomas DY (1997) Functional characterization of the Cdc42p binding domain of yeast Ste20p protein kinase. EMBO J 16:83–97

Lee K, Levin D (1992) Dominant mutations in a gene encoding a putative protein kinase (*BCK1*) bypass the requirement for a *Saccharomyces cerevisiae* protein kinase C homolog. Mol Cell Biol 12:172–182

Lee J, Colwill K, Aneliunas V, Tennyson C, Moore L, Ho Y, Andrews B (1998) Interaction of yeast Rvs167 and Pho85 cyclin-dependent kinase complexes may link

the cell cycle to the actin cytoskeleton. Curr Biol 8:1310–1321

Lee L, Klee SK, Evangelista M, Boone C, Pellman D (1999) Control of mitotic spindle position by the *Saccharomyces cerevisiae* formin Bni1p. J Cell Biol 144:947–961

Leeuw T, Fourest-Lieuvin A, Wu C, Chenevert J, Clark K, Whiteway M, Thomas DY, Leberer E (1995) Pheromone response in yeast: association of Bem1p with proteins of the MAP kinase cascade and actin. Science 270:1210–1213

Leeuw T, Wu C, Schrag J, Whiteway M, Thomas DY, Leberer E (1998) Interaction of a G-protein β-subunit with a conserved sequence in Ste20/PAK family protein kinases. Nature 391:191–195

Lenburg ME, O'Shea EK (1996) Signaling phosphate starvation. Trends Biochem Sci 21:383–387

Levin DE, Errede B (1995) The proliferation of MAP kinase signaling pathways in yeast. Curr Opin Cell Biol 7:197–202

Levin DE, Bowers B, Chen C-Y, Kamada Y, Watanabe M (1994) Dissecting the protein kinase C/MAP kinase signalling pathway of *Saccharomyces cerevisiae*. Cell Mol Biol Res 40:229–239

Lew DJ, Reed SI (1993) Morphogenesis in the yeast cell cycle: regulation by Cdc28 and cyclins. J Cell Biol 120:1305–1320

Lew D, Reed S (1995a) Cell cycle control of morphogenesis in budding yeast. Curr Opin Genet Dev 5:17–23

Lew DJ, Reed SI (1995b) A cell cycle checkpoint monitors cell morphogenesis in budding yeast. J Cell Biol 129:739–749

Li R, Zheng Y, Drubin DG (1995) Regulation of cortical actin cytoskeleton assembly during polarized cell growth in budding yeast. J Cell Biol 128:599–615

Lillie SH, Brown SS (1994) Immunofluorescence localization of the unconventional myosin, Myo2p, and the putative kinesin-related protein, Smy1p, to the same regions of polarized growth in *Saccharomyces cerevisiae*. J Cell Biol 125:825–842

Lipke PN, Kurjan J (1992) Sexual agglutination in budding yeasts: structure, function, and regulation of adhesion glycoproteins. Microbiol Rev 56:180–194

Lipke PN, Taylor A, Ballou CE (1976) Morphogenic effects of α-factor on *Saccharomyces cerevisiae* **a** cells. J Bacteriol 127:610–618

Lippincott J, Li R (1998) Sequential assembly of myosin II, an IQGAP-like protein and filamentous actin to a ring structure involved in budding yeast cytokinesis. J Cell Biol 140:355–366

Liu H, Bretscher A (1989a) Disruption of the single tropomyosin gene in yeast results in the disappearance of actin cables from the cytoskeleton. Cell 57:233–242

Liu H, Bretscher A (1989b) Purification of tropomyosin from *Saccharomyces cerevisiae* and identification of related proteins in *Schizosaccharomyces* and *Physarum*. Proc Natl Acad Sci USA 86:90–93

Liu H, Bretscher A (1992) Characterization of *TPM1* disrupted yeast cells indicates an involvement of tropomyosin in directed vesicular transport. J Cell Biol 118:285–299

Liu H, Styles CA, Fink GR (1993) Elements of the yeast pheromone response pathway required for filamentous growth of diploids. Science 262:1741–1744

Liu H, Styles CA, Fink GR (1996) *Saccharomyces cerevisiae* S288C has a mutation in *FLO8*, a gene required for filamentous growth. Genetics 144:967–978

Lo WS, Dranginis AM (1998) The cell surface flocculin Flo11 is required for pseudohyphae formation and invasion by *Saccharomyces cerevisiae*. Mol Biol Cell 9:161–171

Long RM, Singer RH, Meng X, Gonzalez I, Nasmyth K, Jansen R-P (1997) Mating type switching in yeast controlled by asymmetric localization of *ASH1* mRNA. Science 277:383–387

Longtine MS, DeMarini DJ, Valencik ML, Al AO, Fares H, De VC, Pringle JR (1996) The septins: roles in cytokinesis and other processes. Curr Opin Cell Biol 8:106–119

Longtine MS, Fares H, Pringle JR (1998) Role of the yeast Gin4p protein kinase in septin assembly and the relationship between septin assembly and septin function. J Cell Biol 143:719–736

Lyons DM, Mahanty SK, Choi K-Y, Manandhar M, Elion EA (1996) The SH3-domain protein Bem1 coordinates mitogen-activated protein kinase cascade activation with cell cycle control in *Saccharomyces cerevisiae*. Mol Cell Biol 16:4095–4106

Ma XJ, Lu Q, Grunstein M (1996) A search for proteins that interact genetically with histone H3 and H4 amino termini uncovers novel regulators of the Swe1 kinase in *Saccharomyces cerevisiae*. Genes Dev 10:1327–1340

Machesky LM (1998) Cytokinesis: IQGAPs find a function. Curr Biol 8:R202–R205

Machesky LM, Gould KL (1999) The Arp2/3 complex: a multifunctional actin organizer. Curr Opin Cell Biol 11:117–121

Madaule P, Axel R, Myers AM (1987) Characterization of two members of the *rho* gene family from the yeast *Saccharomyces cerevisiae*. Proc Natl Acad Sci USA 84:779–783

Madden K, Snyder M (1992) Specification of sites of polarized growth in *Saccharomyces cerevisiae* and the influence of external factors on site selection. Mol Biol Cell 3:1025–1035

Madden K, Costigan C, Snyder M (1992) Cell polarity and morphogenesis in *Saccharomyces cerevisiae*. Trends Cell Biol 2:22–29

Madden K, Sheu Y-J, Baetz K, Andrews B, Snyder M (1997) SBF cell cycle regulator as a target of the yeast SLT2 MAP kinase pathway. Science 275:1781–1784

Madhani HD, Fink GR (1998) The riddle of MAP kinase signaling specificity. Trends Genet 14:151–155

Madhani HD, Styles CA, Fink GR (1997) MAP kinases with distinct inhibitory functions impart signaling specificity during yeast differentiation. Cell 91:673–684

Manning BD, Padmanabha R, Snyder M (1997) The Rho-GEF Rom2p localizes to sites of polarized cell growth and participates in cytoskeletal functions in *Saccharomyces cerevisiae*. Mol Biol Cell 8:1829–1844

Marcus S, Polverino A, Barr M, Wigler M (1994) Complexes between STE5 and components of the pheromone-responsive mitogen-activated protein kinase module. Proc Natl Acad Sci USA 91:7762–7766

Marini NJ, Meldrum E, Buehrer B, Hubberstey AV, Stone DE, Traynor-Kaplan A, Reed SI (1996) A pathway in the yeast cell division cycle linking protein kinase C (Pkc1) to activation of Cdc28 at START. EMBO J 15:3040–3052

Marsh L, Rose MD (1997) The pathway of cell and nuclear fusion during mating in *S. cerevisiae*. In: Pringle JR, Broach JR, Jones EW (eds) The molecular biology of the yeast *Saccharomyces*: cell cycle and cell biology. Cold Spring Harbor Laboratory Press, Cold Spring Harbor, pp 657–744

Marshall CJ (1994) MAP kinase kinase kinase, MAP kinase kinase, and MAP kinase. Curr Opin Genet Dev 4:82–89

Masuda TKT, Nonaka H, Yamochi W, Maeda A, Takai Y (1994) Molecular cloning and characterization of yeast rho GDP dissociation inhibitor. J Biol Chem 269:19713–19718

Masur P, Morin N, Baginsky W, El-Sherbeini M, Clemas JA, Nielsen JB, Foor F (1995) Differential expression and function of two homologous subunits of yeast β-1,3-D-glucan synthase. Mol Cell Biol 15:5671–5681

Matsui Y, Toh-e A (1992) Isolation and characterization of two novel *ras* superfamily genes in *Saccharomyces cerevisiae*. Gene 114:43–49

Mazzoni C, Zarzov P, Rambourg A, Mann C (1993) The *SLT2(MPK1)* MAP kinase homolog is involved in polarized cell growth in *Saccharomyces cerevisiae*. J Cell Biol 123:1821–1833

McConnell SJ, Yaffe MP (1992) Nuclear and mitochondrial inheritance in yeast depends on novel cytoplasmic structures defined by the MDM1 protein. J Cell Biol 118:385–395

McConnell SJ, Yaffe MP (1993) Intermediate filament formation by a yeast protein essential for organelle inheritance. Science 260:687–689

McConnell SJ, Stewart LC, Talin A, Yaffe MP (1990) Temperature-sensitive yeast mutants defective in mitochondrial inheritance. J Cell Biol 111:967–976

McMillan JN, Sia RAL, Lew DJ (1998) A morphogenesis checkpoint monitors the actin cytoskeleton in yeast. J Cell Biol 142:1487–1499

McMillan JN, Sia RAL, Bardes ESG, Lew DJ (1999) Phosphorylation-independent inhibition of Cdc28p by the tyrosine kinase Swe1p in the morphogenesis checkpoint. Mol Cell Biol 19:5981–5990

Measday V, Moore L, Ogas J, Tyers M, Andrews B (1994) The PCL2 (ORFD)-PHO85 cyclin dependent kinase complex: a cell cycle regulator in yeast. Science 266:1391–1395

Michelitch M, Chant J (1996) A mechanism of Bud1p GTPase action suggested by mutational analysis and immunolocalization. Curr Biol 6:446–454

Miller RK, Rose MD (1998) Kar9 is a novel cortical protein required for cytoplasmic microtubule orientation in yeast. J Cell Biol 140:377–390

Miller RK, Metheos D, Rose MD (1999) The cortical localization of the microtubule orientation protein, Kar9, is dependent upon actin and proteins required for polarization. J Cell Biol 144:963–975

Miyamoto S, Ohya Y, Ohsumi Y, Anraku Y (1987) Nucleotide sequence of the *CLS4* (*CDC24*) gene of *Saccharomyces cerevisiae*. Gene 54:125–132

Molano J, Bowers B, Cabib E (1980) Distribution of chitin in the yeast cell wall: a structural and biochemical study. J Cell Biol 85:199–212

Mooseker MS (1985) Organization, chemistry, and assembly of the cytoskeletal apparatus of the intestinal brush border. Annu Rev Cell Biol 1:209–241

Moreau V, Madania R, Martin P, Winsor P (1996) The *Saccharomyces cerevisiae* actin-related protein Arp2 is involved in the actin cytoskeleton. J Cell Biol 134:117–132

Moreau V, Galan JM, Devilliers G, Haguenauer-Tsapis R, Winsor P (1997) The yeast actin-related protein Arp2 is required for the internalization step of endocytosis. Mol Biol Cell 8:1368–1375

Mosch H-U, Roberts R, Fink GR (1996) Ras2 signals via the Cdc42/Ste20/mitogen-activated kinase module to induce filamentous growth in *Saccharomyces cerevisiae*. Proc Natl Acad Sci USA 93:5352–5356

Mosch H-U, Fink GR (1997) Dissection of filamentous growth by transposon mutagenesis in *Saccharomyces cerevisiae*. Genetics 145:671–684

Mosch H-U, Kubler E, Krappmann S, Fink GR, Braus GH (1999) Crosstalk between the Ras2-controlled mitogen-activated protein kinase and cAMP pathways during invasive growth of *Saccharomyces cerevisiae*. Mol Biol Cell 10:1325–1335

Mulholland J, Preuss D, Moon A, Wong A, Drubin D, Botstein D (1994) Ultrastructure of the yeast actin cytoskeleton and its association with the plasma membrane. J Cell Biol 125:381–391

Munn A, Stevenson B, Geli M, Riezman H (1995) *end5*, *end6*, and *end7*: mutations that cause actin delocalization and block the internalization step of endocytosis in *Saccharomyces cerevisiae*. Mol Biol Cell 6:1721–1742

Nasmyth KA (1982) Molecular genetics of yeast mating type. Annu Rev Genet 16:439–500

Nasmyth KA (1993) Control of the yeast cell cycle by the Cdc28 protein kinase. Curr Opin Cell Biol 5:166–179

Nasmyth KA (1996) At the heart of the budding yeast cell cycle. Trends Genet 12:405–412

Nasmyth KA, Dirick L (1991) The role of *SWI4* and *SWI6* in the activity of G1 cyclins in yeast. Cell 66:995–1013

Neiman AM, Herskowitz I (1994) Reconstitution of a yeast protein kinase cascade in vitro: activation of the yeast MEK homologue STE7 by STE11. Proc Natl Acad Sci USA 91:3398–3402

Nern A, Arkowitz RA (1998) A GTP-exchange factor required for cell orientation. Nature 391:195–198

Nern A, Arkowitz RA (1999) A Cdc24p-Far1p-Gβγ protein complex required for yeast orientation during mating. J Cell Biol 144:1187–1202

Neufeld TP, Rubin GM (1994) The Drosophila *peanut* gene is required for cytokinesis and encodes a protein similar to yeast putative bud neck filament proteins. Cell 77:371–379

Ng R, Abelson J (1980) Isolation of the gene for actin in *Saccharomyces cerevisiae*. Proc Natl Acad Sci USA 77:3912–3916

Nonaka H, Tanaka K, Hirano H, Fujiwara T, Kohno H, Umikawa M, Mino A, Takai Y (1995) A downstream target of RHO1 small GTP binding protein is PKC1, a homolog of protein kinase C, which leads to activation of the MPK1 kinase cascade in *Saccharomyces cerevisiae*. EMBO J 14:5931–5938

Novick P, Botstein D (1985) Phenotypic analysis of temperature-sensitive yeast actin mutants. Cell 40:405–416

Novick P, Brennwald P (1993) Friends and family: the role of the Rab GTPases in vesicular traffic. Cell 75:597–601

Oehlen LJWM, Cross FR (1998) Potential regulation of Ste20 function by the Cln1-Cdc28 and Cln2-Cdc28 cyclin-dependent protein kinases. J Biol Chem 273:25089–25097

Ogas J, Andrews BJ, Herskowitz I (1991) Transcriptional activation of *CLN1*, *CLN2*, and a new G1 cyclin (*HCS26*) by *SWI4*, a positive regulator of G1-specific transcription. Cell 66:1015–1026

Osman MA, Cerione RA (1998) Iqg1p, a yeast homologue of the mammalian IQGAPs, mediates Cdc42p effects on the actin cytoskeleton. J Cell Biol 142:443–455

Ozaki K, Tanaka K, Imamura H, Hihara T, Kameyama T, Nonaka H, Hirano H, Matsuura Y, Takai Y (1996) Rom1p and Rom2p are GDP/GTP exchange proteins (GEPs) for the Rho1p small GTP binding protein in *Saccharomyces cerevisiae*. EMBO J 15:2196–2207

Palmer RE, Sullivan DS, Huffaker T, Koshland D (1992) Role of astral microtubules and actin in spindle orientation and migration in the budding yeast, *Saccharomyces cerevisiae*. J Cell Biol 119:583–593

Park HO, Chant J, Herskowitz I (1993) *BUD2* encodes a GTPase-activating protein for Bud1/Rsr1 necessary for proper bud-site selection in yeast. Nature 365:269–274

Park HO, Bi E, Pringle JR, Herskowitz I (1996) Two active states of the Ras-related Bud1/Rsr1 protein bind to different effectors to determine yeast cell polarity. Proc Natl Acad Sci USA 94:4463–4468

Park HO, Sanson A, Herskowitz I (1999) Localization of Bud2p, a GTPase-activating protein necessary for programming cell polarity in yeast to the presumptive bud site. Genes Dev 13:1912–1917

Peter M, Herskowitz I (1994) Direct inhibition of the yeast cyclin-dependent kinase Cdc28-Cln by Far1. Science 265:1228–1231

Peter M, Neiman AM, Park H-O, van Lohuizen M, Herskowitz I (1996) Functional analysis of the interaction between the small GTP binding protein Cdc42 and the Ste20 protein kinase in yeast. EMBO J 15:7046–7059

Peterson J, Zheng Y, Bender L, Myers A, Cerione R, Bender A (1994) Interactions between the bud emergence proteins Bem1p and Bem2p and Rho-type GTPases in yeast. J Cell Biol 127:1395–1406

Philips J, Herskowitz I (1997) Osmotic balance regulates cell fusion during mating in *Saccharomyces cerevisiae*. J Cell Biol 138:961–974

Philips J, Herskowitz I (1998) Identification of Kel1, a Kelch domain-containing protein involved in cell fusion and morphology in *Saccharomyces cerevisiae*. J Cell Biol 143:375–389

Polymenis M, Schmidt EV (1997) Coupling of cell division to cell growth by translational control of the G1 cyclin Cln3 in yeast. Genes Dev 11:2522–2531

Posas F, Saito H (1997) Osmotic activation of the HOG MAPK pathway via Ste11p MAPKKK: scaffold role of Pbs2p MAPKK. Science 276:1702–1705

Powers S, Gonzales E, Christensen T, Cubert J, Broek D (1991) Functional cloning of *BUD5*, a *CDC25*-related gene from S. cerevisiae that can suppress a dominant-negative *RAS2* mutant. Cell 65:1225–1231

Pringle J, Bi E, Harkins H, Zahner J, Devirgilio C, Chant J, Corado K, Fares H (1995) Establishment of cell polarity in yeast. Cold Spring Harbor Symp Quant Biol 60:729–744

Pruess D, Mulholland J, Kaiser CA, Orlean P, Albright C (1991) Structure of the yeast endoplasmic reticulum localization of ER proteins using immunofluorescence and immunolocalization microscopy. Yeast 7: 891–911

Pruess D, Mulholland J, Franzusoff A, Segev N, Botstein D (1992) Characterization of the *Saccharomyces* Golgi complex through the cell cycle by immunoelectron microscopy. Mol Cell Biol 3:789–803

Pryciak PM, Huntress FA (1998) Membrane recruitment of the kinase cascade scaffold protein Ste5 by the Gβγ complex underlies activation of the yeast pheromone response pathway. Genes Dev 12:2684–2697

Qadota H, Python CP, Inoue SB, Arisawa M, Anraku Y, Zheng Y, Watanabe T, Levin DE, Ohya Y (1996) Identification of yeast Rho1p GTPase as a regulatory subunit of 1,3-β-glucan synthase. Science 272:279–281

Rajavel M, Philip B, Buehrer BM, Errede B, Levin DE (1999) Mid2 is a putative sensor for cell integrity signaling in *Saccharomyces cerevisiae*. Mol Cell Biol 19:3969–3976

Ram AFJ, Brekelmans SSC, Oehlen LJWM, Klis FM (1995) Identification of two cell cycle regulated genes affecting the β-1,3-glucan content of cell wall in *Saccharomyces cerevisiae*. FEBS Lett 358:165–170

Ramer SW, Davis RW (1993) A dominant truncation allele identifies a gene, *STE20*, that encodes a putative protein kinase necessary for mating in *Saccharomyces cerevisiae*. Proc Natl Acad Sci USA 90:452–456

Rappaport R (1986) Establishment of the mechanism of cytokinesis in animal cells. Int Rev Cytol 105:245–281

Richardson H, Lew DJ, Henze M, Sugimoto K, Reed SI (1992) Cyclin-B homologs in *Saccharomyces cerevisiae* function in S phase and in G2. Gene Dev 6: 2021–2034

Richman TJ, Sawyer MM, Johnson DI (1999) The Cdc42p GTPase is involved in a G2/M checkpoint regulating the apical-isotropic switch and nuclear division in yeast. J Biol Chem 274:16861–16870

Roberts R, Fink GR (1994) Elements of a single MAP kinase cascade in *Saccharomyces cerevisiae* mediate two developmental programs in the same cell type: mating and invasive growth. Genes Dev 8:2974–2985

Roberts RL, Bowers B, Slater ML, Cabib E (1983) Chitin synthesis and localization in cell division cycle mutants of *Saccharomyces cerevisiae*. Mol Cell Biol 3:922–930

Roberts RL, Mosch H-U, Fink GR (1997) 14-3-3 proteins are essential for RAS/MAPK cascade signalling during pseudohyphal development in *S. cerevisiae*. Cell 89:1055–1065

Robinson NGG, Guo L, Imai J, Toh-e A, Matsui Y, Tamanoi F (1999) Rho3 of *Saccharomyces cerevisiae*, which regulates the actin cytoskeleton and exocytosis, is a GTPase which interacts with Myo2 and Exo70. Mol Cell Biol 19:3580–3587

Roemer T, Madden K, Chang J, Snyder M (1996a) Selection of axial growth sites in yeast requires Axl2p, a novel plasma membrane glycoprotein. Genes Dev 10:777–793

Roemer T, Vallier L, Snyder M (1996b) Selection of polarized growth sites in yeast. Trends Cell Biol 6:434–441

Roemer T, Vallier L, Sheu Y-J, Snyder M (1998) The Spa2p-related protein, Sph1p, is important for polarized growth in yeast. J Cell Sci 111:479–494

Ruggieri R, Bender A, Matsui Y, Powers S, Takai Y, Pringle JR, Matsumoto K (1992) *RSR1*, a *ras*-like gene homologous to *Krev-1* (*smg21A/rap1A*): role in the development of cell polarity and interactions with the Ras pathway in *Saccharomyces cerevisiae*. Mol Cell Biol 12:758–766

Rupp S, Summers E, Lo HJ, Madhani H, Fink GR (1999) MAP kinase and cAMP filamentation signaling pathways converge on the usually large promoter of the yeast *FLO11* gene. EMBO J 18:1257–1269

Russell P, Nurse P (1987a) Negative regulation of mitosis by wee1+, a gene encoding a protein kinase homolog. Cell 49:559–567

Russell P, Nurse P (1987b) The mitotic inducer nim1+ functions in a regulatory network of protein kinase homologs controlling the initiation of mitosis. Cell 49:569–576

Sanders SL, Herskowitz I (1996) The Bud4 protein of yeast, required for axial budding, is localized to the mother-bud neck in a cell cycle-dependent manner. J Cell Biol 134:413–427

Sanders SL, Gentzsch M, Tanner W, Herskowitz I (1999) O-Glycosylation of Axl2/Bud10p by Pmt4p is required for its stability, localization, and function in daughter cells. J Cell Biol 145:1177–1188

Santos B, Snyder M (1997) Targeting of chitin synthase 3 to polarized growth sites in yeast requires Chs5p and Myo2p. J Cell Biol 136:95–110

Santos B, Duran A, Valdivieso MH (1997) CHS5, a gene involved in chitin synthesis and mating in *Saccharomyces cerevisiae*. Mol Cell Biol 17:2485–2496

Schekman R, Brawley V (1979) Localized deposition of chitin on the yeast cell surface in response to mating pheromone. Proc Natl Acad Sci USA 76:645–649

Scherr GH, Weaver RH (1953) The dimorphism phenomenon in yeasts. Bacteriol Rev 17:51–92

Schmidt A, Bickle M, Beck T, Hall MN (1997) The yeast phosphatidylinositol kinase homolog TOR2 activates RHO1 and RHO2 via the exchange factor ROM2. Cell 88:531–542

Schrick K, Garvik B, Hartwell LH (1997) Mating in *Saccharomyces cerevisiae*: the role of the pheromone signal transduction pathway in the chemotropic response to pheromone. Genetics 147:19–32

Segall JE (1993) Polarization of yeast cells in spatial gradients of α-factor. Proc Natl Acad Sci USA 90:8332–8336

Sells MA, Chernoff J (1997) Emerging from the Pak: the p21-activated protein kinase family. Trends Cell Biol 7:162–167

Shaw JA, Mol PC, Bowers B, Silverman SJ, Valdivieso MH, Duran A, Cabib E (1991) The function of chitin synthases 2 and 3 in the *Saccharomyces cerevisiae* cell cycle. J Cell Biol 114:111–123

Shaw SL, Yeh E, Maddox P, Salmon ED, Bloom K (1997) Astral microtubule dynamics in yeast: a microtubule-based searching mechanism for spindle orientation and nuclear migration into the bud. J Cell Biol 139:985–994

Shepherd MG (1988) Morphogenetic transformation of fungi. Curr Top Med Mycol 2:278–304

Sheu Y-J, Santos B, Fortin N, Costigan C, Snyder M (1998) Spa2p interacts with cell polarity proteins and signaling components involved in yeast cell morphogenesis. Mol Cell Biol 18:4053–4069

Sheu Y-J, Barral Y, Snyder M (2000) Polarized growth controls cell shape and bipolar bud site selection in *Saccharomyces cerevisiae*. Mol Cell Biol (in press)

Shortle D, Haber JE, Botstein D (1982) Lethal disruption of the yeast actin gene by integrative DNA transformation. Science 217:371–373

Shortle D, Novick P, Botstein D (1984) Construction and genetic characterization of temperature-sensitive mutant alleles of the yeast actin gene. Proc Natl Acad Sci USA 81:4889–4893

Shulewitz MJ, Inouye CJ, Thorner J (1999) Hsl7 localizes to a septin ring and serves as an adapter in a regulatory pathway that relieves tyrosine phosphorylation of Cdc28 protein kinase in *Saccharomyces cerevisiae*. Mol Cell Biol 19:7123–7137

Sia RA, Herald HA, Lew DJ (1996) Cdc28 tyrosine phosphorylation and the morphogenesis checkpoint in budding yeast. Mol. Biol Cell 7:1657–1666

Sia RAL, Bardes ESG, Lew DJ (1998) Control of Swe1p degradation by the morphogenesis checkpoint. EMBO J 17:6678–6688

Sil A, Herskowitz I (1996) Identification of an asymmetrically localized determinant, Ash1p, required for lineage-specific transcription of the yeast HO gene. Cell 84:711–722

Silverman SJ, Sburlati A, Slater ML, Cabib E (1988) Chitin synthase 2 is essential for septum formation and cell separation in *Saccharomyces cerevisiae*. Proc Natl Acad Sci USA 85:4735–4739

Simon M, Virgilio CD, Souza B, Pringle JR, Abo A, Reed SI (1995) Role for the Rho-family GTPase Cdc42 in yeast mating-pheromone signal pathway. Nature 376: 702–705

Sivadon P, Bauer F, Aigle M, Crouzet M (1995) Actin cytoskeleton and budding pattern are altered in the yeast *rvs161* mutant: the Rvs161 protein shares common domains with the brain protein amphiphysin. Mol Gen Genet 246:485–495

Sloat B, Pringle J (1978) A mutant of yeast defective in cellular morphogenesis. Science 200:1171–1173

Sloat BF, Adams A, Pringle JR (1981) Roles of the *CDC24* gene product in cellular morphogenesis during the *Saccharomyces cerevisiae* cell cycle. J Cell Biol 89:395–405

Smits GJ, Kapteyn JC, van den Ende H, Klis FM (1999) Cell wall dynamics in yeast. Curr Opin Microbiol 2:348–352

Snyder M (1989) The SPA2 protein of yeast localizes to sites of cell growth. J Cell Biol 108:1419–1429

Snyder M, Gehrung S, Page BD (1991) Studies concerning the temporal and genetic control of cell polarity in *Saccharomyces cerevisiae*. J Cell Biol 114:515–532

Sprague GF, Thorner J (1992) Pheromone response and signal transduction during the mating process of *Saccharomyces cerevisiae*. In: Broach JR, Pringle JR, Jones E (eds) The molecular biology of the yeast *Saccharomyces*. Cold Spring Harbor Laboratory Press, Cold Spring Harbor, pp 657–744

Sreenivasan A, Kellog D (1999) The Elm1 kinase functions in a mitotic signaling network in budding yeast. Mol Cell Biol 19:7983–7994

Stevens B (1981) Mitochondrial structure. In: Strathern JN, Jones EW, Broach JR (eds) The molecular biology of the yeast *Saccharomyces cerevisiae*: life cycle and inheritance. Cold Spring Harbor Laboratory Press, Cold Spring Harbor, pp 471–504

Stevenson BJ, Ferguson B, DeVirgilio C, Bi E, Pringle JR, Ammerer GGF, Sprague J (1995) Mutation of *RGA1*, which encodes a putative GAP for the polarity-establishment protein Cdc42p, activates the pheromone response pathway in the yeast *Saccharomyces cerevisiae*. Genes Dev 9:2949–2963

Stevenson BJ, Rhodes N, Errede B, Sprague GF (1992) Constitutive mutants of the protein kinase STE11 activate the yeast pheromone response pathway in the absence of G protein. Genes Dev 6:1293–1304

Strome S (1993) Determination of cleavage planes. Cell 72: 3–6

Surana U, Robitsch H, Price C, Schuster T, Fitch I, Futcher A, Nasmyth K (1991) The role of CDC28 and cyclins during mitosis in the budding yeast *S. cerevisiae*. Cell 65:145–161

Takizawa PA, Sil A, Swedlow JR, Herskowitz I, Vale RD (1997) Actin-dependent localization of an RNA encoding a cell fate determinant in yeast. Nature 389:90–93

Tang HY, Cai M (1996) The EH-domain-containing protein Pan1 is required for normal organization of the actin cytoskeleton in *Saccharomyces cerevisiae*. Mol Cell Biol 16:4897–4914

Tang HY, Munn A, Cai M (1997) EH domain proteins Pan1 and End3 are components of a complex that plays a dual role in organization of the cortical actin cytoskeleton and endocytosis in *Saccharomyces cerevisiae*. Mol Cell Biol 17:4294–4303

Tennyson CN, Lee J, Andrews BJ (1998) A role for the Pcl9-Pho85 cyclin-cdk complex at the M/G1 boundary in *Saccharomyces cerevisiae*. Mol Microbiol 28:69–79

TerBush DR, Novick P (1995) Sec6, Sec8, and Sec15 are components of a multisubunit complex which localizes to small bud tips in *Saccharomyces cerevisiae*. J Cell Biol 130:299–312

TerBush DR, Maurice T, Roth D, Novick P (1996) The exocyst is a multiprotein complex required for exocytosis in *Saccharomyces cerevisiae*. EMBO J 15:6483–6494

Tjandra H, Compton J, Kellogg D (1998) Control of mitotic events by the Cdc42 GTPase, the Clb2 cyclin and a member of the PAK kinase family. Curr Biol 8:991–1000

Tkacz JS, Lampen JO (1972) Wall replication in *Saccharomyces* species: use of fluorescein-conjugated concanavalin A to reveal the site of mannan insertion. J Gen Microbiol 72:243–247

Tkacz JS, MacKay VL (1979) Sexual conjugation in yeast: cell surface changes in response to the action of mating hormones. J Cell Biol 80:326–333

Torres L, Martin H, Garcia-Saez MI, Arroyo J, Molina M, Sanchez M, Nombela C (1991) A protein kinase gene complements the lytic phenotype of *Saccharomyces cerevisiae lyt2* mutants. Mol Microbiol 5:2845–2854

Touhara K, Inglese J, Pitcher JA, Shaw G, Lefkowitz RJ (1994) Binding of G protein $\beta\gamma$-subunits to pleckstrin homology domains. J Biol Chem 269:10217–10220

Valdivieso MH, Mol PC, Shaw JA, Cabib E, Duran A (1991) *CAL1*, a gene required for activity of chitin synthase 3 in *Saccharomyces cerevisiae*. J Cell Biol 114:101–109

Valtz N, Herskowitz I (1996) Pea2 protein of yeast is localized to sites of polarized growth and is required for

efficient mating and bipolar budding. J Cell Biol 135:
725–739

Valtz N, Peter M, Herskowitz I (1995) FAR1 is required for
oriented polarization of yeast cells in response to
mating pheromones. J Cell Biol 131:863–873

Verna J, Lodder A, Lee K, Vagts A, Ballester R (1997) A
family of genes required for maintenance of cell wall
integrity and for the stress response in Saccharomyces
cerevisiae. Proc Natl Acad Sci USA 94:13804–13809

Waddle JA, Karpova TS, Waterston RH, Cooper JA (1996)
Movement of cortical actin patches in yeast. J Cell
Biol 132:861–870

Walch-Solimena C, Collins RN, Novick PJ (1997) Sec2p
mediates nucleotide exchange on Sec4p and is
involved in polarized delivery of post-Golgi vesicles. J
Cell Biol 137:1495–1509

Wang T, Bretscher A (1995) The rho-GAP encoded by
BEM2 regulates cytoskeletal structure in budding
yeast. Mol Biol Cell 6:1011–1024

Ward MP, Gimeno CJ, Fink GR, Garrett S (1995) SOK2
may regulate cyclic AMP-dependent protein kinase-
stimulated growth and pseudohyphal development by
repressing transcription. Mol Cell Biol 15:6854–6863

Wasserman S (1998) FH proteins as cytoskeletal organiz-
ers. Trends Cell Biol 8:111–115

Watanabe Y, Irie K, Matsumoto K (1995) Yeast RLM1
encodes a serum response factor-like protein that
may function downstream of the Mpk1 (Slt2)
mitogen-activated protein kinase pathway. Mol Cell
Biol 15:5740–5749

Watanabe Y, Takaesu G, Hagiwara M, Irie K, Matsumoto
K (1997) Characterization of a serum response factor-
like protein in Saccharomyces cerevisiae, RLM1,
which has transcriptional activity regulated by the
MPK1(SLT2) mitogen-activated protein kinase
pathway. Mol Cell Biol 17:2615–2623

Weisman LS, Wickner W (1988) Intervacuole exchange in
the yeast zygote: a new pathway in organelle communication. Science 241:589–591

Wu C, Lee S-F, Furmaniak-Kazmierczak E, Cote GP,
Thomas DY, Leberer E (1996) Activation of myosin I
by members of the Ste20 protein kinase family. J Biol
Chem 271:31787–31790

Wu C, Lytvyn V, Thomas DY, Leberer E (1997) The phosphorylation site for Ste20-like protein kinases is essential for the function of myosin-I in yeast. J Biol Chem
272:30623–30626

Wu C, Leeuw T, Leberer E, Thomas DY, Whiteway M
(1998) Cell cycle- and Cln2p-Cdc28p-dependent phosphorylation of the yeast Ste20p protein kinase. J Biol
Chem 273:28107–28115

Xu G, Jansen G, Thomas DY, Ramesani Rad M (1996)
Ste50p sustains mating pheromone-induced signal

transduction in the yeast Saccharomyces cerevisiae.
Mol Microbiol 20:773–783

Yabe T, Yamada-Okabe T, Kasahara S, Furuichi Y,
Nakajima T, Ichishima E, Arisawa M, Yamada-Okabe
H (1996) HKR1 encodes a cell surface protein that
regulates both cell wall B-glucan synthesis and
budding pattern in the yeast Saccharomyces cerevisiae.
J Bacteriol 178:477–483

Yamochi W, Tanaka K, Nonaka H, Maeda A, Musha T,
Takai Y (1994) Growth site localization of Rho1 small
GTP-binding protein and its involvement in bud formation in Saccharomyces cerevisiae. J Cell Biol
125:1077–1093

Yang S, Ayscough KR, Drubin DG (1997) A role for the
actin cytoskeleton of Saccharomyces cerevisiae in
bipolar bud-site selection. J Cell Biol 136:111–123

Yashar B, Irie K, Printen JA, Stevenson BJ, Sprague
GFJ, Matsumoto K, Errede B (1995) Yeast MEK-
dependent signal transduction: response thresholds
and parameters affecting fidelity. Mol Cell Biol
15:6545–6553

Zahner JE, Harkins HA, Pringle JR (1996) Genetic analysis of the bipolar pattern of bud site selection in the
yeast Saccharomyces cerevisiae. Mol Cell Biol 16:
1857–1870

Zarzov P, Mazzoni C, Mann C (1996) The SLT2(MPK1)
MAP kinase is activated during periods of polarized
cell growth in yeast. EMBO J 15:83–91

Zeng G, Cai M (1999) Regulation of the actin cytoskeleton
organization in yeast by a novel serine/threonine
kinase Prk1p. J Cell Biol 144:71–82

Zhao Z, Leung T, Manser E, Lim L (1995) Pheromone sig-
nalling in Saccharomyces cerevisiae requires the small
GTP-binding protein Cdc42p and its activator
CDC24. Mol Cell Biol 15:5246–5257

Zheng Y, Cerione R, Bender A (1994) Control of the yeast
bud-site assembly GTPase Cdc42. Catalysis of guanine
nucleotide exchange by Cdc24 and stimulation of
GTPase activity by Bem3. J Biol Chem 269:2369–2372

Zheng Y, Cerione R, Bender A (1995) Interactions among
proteins involved in bud-site selection and bud-site
assembly in Saccharomyces cerevisiae. J Biol Chem
270:626–630

Ziman M, O'Brien JM, Ouellette LA, Church WR, Johnson
DI (1991) Mutational analysis of CDC42Sc, a Saccharomyces cerevisiae gene that encodes a putative GTP-
binding protein involved in the control of cell polarity.
Mol Cell Biol 11:3537–3544

Ziman M, Preuss D, Mulholland J, O'Brien JM, Botstein D,
Johnson DI (1993) Subcellular localization of Cdc42p,
a Saccharomyces cerevisiae GTP-binding protein
involved in the control of cell polarity. Mol Biol Cell
4:1307–1316

3 Signal Transduction and Morphogenesis in *Candida albicans*

ALISTAIR J.P. BROWN and NEIL A.R. GOW

CONTENTS

I. Introduction

There are several well-studied model systems of fungal growth and morphogenesis. These include *Saccharomyces cerevisiae*, *Schizosaccharomyces pombe*, *Aspergillus nidulans* and *Neurospora crassa*, all of which are tractable to both classical and molecular genetics. In these systems developmental and signalling pathways have been identified, often via the analysis of mutants that exhibit developmental arrest. Subsequently, the cognate genes have been defined and examined at the molecular level. Many of these pathways are conserved in fungi and in other organisms although the signals and outputs from the pathways may differ significantly. Identification of conserved components of such pathways has provided one experimental route to establishing their role in other less genetically tractable organisms. The major human pathogen, *Candida albicans*, on which this review is focused, is a case in point. The major tool for understanding key aspects of the morphogenesis and pathogenesis of this fungus

has undoubtedly been the resource and insights provided by analysis of *S. cerevisiae*. This is now paying handsome dividends not only in furthering our understanding of conserved aspects of the biology of *C. albicans*, but by also enabling the unique and contrasting properties of this organism, which are not features of model fungi, to be investigated.

This review describes progress that has been made in understanding a central aspect of the biology of most fungi – their ability to grow either by isodiametric expansion of spheres (yeast cells, spores, spherules, vesicular bodies etc.) or by apical expansion of hyphal tips. In *C. albicans* and many other fungi this feature is highly developed and vegetative growth can occur either in a budding yeast form or as branching hyphae. These fungi are said to be "**dimorphic**". *C. albicans* can also generate a range of intermediate morphologies called pseudohyphae and so it has been argued that it is more correctly "**pleiomorphic**".

The regulation of morphogenesis in *C. albicans*, and other fungi, can be reduced to a question of how growth (more specifically expansion of the cell surface) is polarized. Cell polarity is a highly regulated and much studied aspect of cellular physiology (see Chap. 12, this Vol.). We will focus on signal transduction leading to polarized growth rather than the mechanism of polarized growth per se.

Part of the impetus behind the considerable efforts made in recent years to establish the molecular mechanisms underlying dimorphic regulation in *C. albicans* is the notion that hyphal growth is an attribute of pathogenesis, or even a *bona fide* virulence factor (Ryley and Ryley 1990; Cutler 1991; Gow 1997; Brown and Gow 1999). Hyphae are commonly observed in biopsies of *C. albicans*-infected organs (Odds 1988) and tip growth of hyphae seems well suited to a role for penetrating host tissues. Studies of more or less filamentous natural isolates or heavily mutagenized strains that are debilitated in hyphal development have shown

Department of Molecular and Cell Biology, Institute of Medical Sciences, University of Aberdeen, Foresterhill, Aberdeen AB25 2ZD, UK

The Mycota VIII
Biology of the Fungal Cell
Howard/Gow (Eds.)
© Springer-Verlag Berlin Heidelberg 2001

that such strains are poorer in establishing infections in animal models (Martin et al. 1984; Sobel et al. 1984; Shepherd 1985; Ryley and Ryley 1990). Evidence from earlier literature that hypha development is a virulence factor remained equivocal because clinical lesions with remarkably few hyphal forms have also been reported and because random methods of mutagenesis that existed at that time almost certainly generated mutations at a wide range of other loci that encoded potential virulence factors. Also, there is no absolute relationship between hyphal growth and pathogenesis in clinically relevant fungi because many important human pathogens, including *Histoplasma capsulatum*, *Paracoccidioides brasiliensis* and *Coccidioides immitis*, are pathogenic in the yeast form and saprophytic in the hyphal form (Kwon-Chung and Bennett 1992). A range of *Candida* species, including those such as *C. glabrata*, which is not able to generate true filamentous forms, can cause infections and so again no absolute correlation between hyphal development and *Candida* disease is observed. That being the case, the most recent research, described below, favors the view that hyphal development is indeed a relevant and important aspect of the pathogenesis of *C. albicans*.

C. *albicans* now represents the third or fourth most common agent of microbial septicemia in US and western hospitals and has outstripped many bacterial infections in terms of incidence and morbidity (Beck-Sagué and Jarvis 1993; Jamal et al. 1999). *C. albicans* is a classical opportunistic agent of infection and septicemia is almost always associated with some underlying immune deficiency such as treatment for cancer or immunosuppression following bone marrow or organ transplantation. It is much more commonly seen as a mucosal pathogen of the oral or vaginal cavity where the disease is called thrush or a "yeast infection".

Prior to the era of molecular genetic analysis of hypha morphogenesis in *C. albicans* an enormous effort was expended on attempting to characterize the environmental conditions that were conducive and stimulatory to dimorphic switching. Unfortunately, strain variability and the lack of systems to perform targeted mutagenesis or mutant screens obscured the data that emerged. In the new era, the construction of isogenic mutants by systems such as the "ura-blaster" protocol (Fonzi and Irwin 1993), first devised for analysing diploid *S. cerevisiae* strains (Alani et al. 1987), has allowed the functional dissection of morpho-genetic signalling pathways and hence the mechanisms that are essential for yeast-hypha conversion.

II. Cell Morphology

Most studies of *C. albicans* morphogenesis have focused on the transition from the budding, yeast-like form to the hyphal growth form. Although this transition is reversible, little work has addressed hypha-yeast morphogenesis. Furthermore, the yeast and hyphal forms represent the morphological extremes that *C. albicans* is able to manifest. This fungus can grow in various filamentous forms, collectively termed pseudohyphae (Merson-Davies and Odds 1989). Pseudohyphae display constrictions at their septa, whereas true hyphae do not. Merson-Davies and Odds (1989) described the morphology index (Mi; based on the relative length, and the maximum and septal diameters of cellular compartments) that provides a definitive guide to cell shape in *C. albicans*. The Mi of budding and hyphal cells are approximately 1.5 and 4.0, respectively, while pseudohyphal cells display a range of Mi values from about 2.0 to 3.5.

It is attractive to view pseudohyphal cells as intermediates between budding and hyphal cells, and indeed this might be the case. For example, pseudohyphal cells may form in response to moderate morphogenetic signals, whereas hyphal development might be triggered by more extreme morphogenetic conditions. However, several observations suggest that pseudohyphae may represent a developmental form that is both morphologically and genetically distinct from the budding and hyphal growth forms (Fig. 1). Hyphal cells are generally narrower than pseudohyphal, having maximum diameters of about 2 and 5 μm respectively (Sevilla and Odds 1986). Specific media promote the growth of pseudohyphal cells as opposed to hyphae (Merson-Davies and Odds 1989). Pseudohyphal cells divide synchronously, whereas hyphal cells grow asynchronously, indicating that there are distinct differences in the cell cycle of these two growth forms (Kron and Gow 1995; Gow 1997). Also, hypha-defective and yeast-defective *C. albicans* mutants can grow as pseudohyphae (Braun and Johnson 1997; Stoldt et al. 1997), suggesting that these growth forms are genetically distinct.

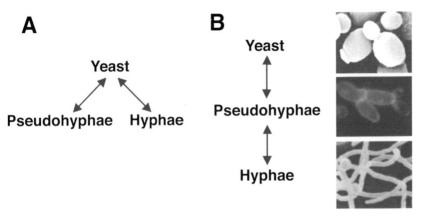

Fig. 1. Morphogenesis in *Candida albicans*. Does the formation of pseudohyphae represent a separate developmental pathway (*model A*), or do they represent a developmental intermediate between budding and hyphal growth forms (*model B*)?

Understanding morphogenesis in *C. albicans* will not be complete until the relationship between the hyphal and pseudohyphal growth forms has been clarified. Two contrasting models predict that pseudohyphal cells represent an intermediate growth form, or a distinct developmental form (Fig. 1). None of the above data distinguish between these models unambiguously, although they do favor model A. A critical test will be to examine whether any well-defined *C. albicans* mutants that are unable to undergo the yeast-pseudohypha transition are still able to form hyphae. In contrast to model B, model A predicts that such mutants could exist (Fig. 1). Greater clarity in the description of morphological phenotypes in the literature will also help to establish the relationship between hyphal and pseudohyphal growth forms. The remainder of this review focuses on the molecular mechanisms that trigger the development of true hyphae.

III. Morphogenetic Signals

Yeast-hypha morphogenesis in *C. albicans* can be triggered by numerous different treatments in vitro (Odds 1988). These in vitro conditions presumably reflect the various signals that promote hyphal growth in vivo. Many signals that promote hyphal development impose a stress on the *C. albicans* cell. For example, changes in temperature, rapid pH shifts, or starvation, all promote hyphal development, and all impose stresses (Brown and Gow 1999). It is conceivable, therefore, that a common signalling pathway(s) mediates the morphogenetic responses to these various stresses. Hence, morphogenetic signals may be divisible into distinct functional categories including stresses or nutrient starvation. However, this model must take account of the fact that not all stresses stimulate morphogenesis – for example, hyperosmotic stresses inhibit hyphal development (Monge et al. 1999). Also, a number of recent studies now suggest that several morphogenetic signalling pathways exist in *C. albicans*, and that these pathways mediate responses to distinct sets of environmental triggers (Fig. 2; Sect. IV).

One class of stimulatory signals is the putative **morphogens**, the nature of which has not been established. Serum is the most powerful inducer of hyphal development. When serum addition is combined with an increase in temperature, from 25 to 37 °C, more that 95% of *C. albicans* cells form germ tubes – the progenitors of true hyphae – within 60 min. It has been argued that serum is composed mainly of proteins that represent an inaccessible nutrient source until they are hydrolyzed and, hence, that serum addition imposes nitrogen starvation (Brown and Gow 1999). However, serum addition still stimulates hyphal growth, even when added in a rich source of nutrients, for example YPD consisting of 2% glucose, 2% bacteriological peptone, and 1% yeast extract (Swoboda et al. 1994), suggesting that something other than nitrogen starvation is responsible for serum-stimulated morphogenesis. However, the relief of nutrient stress may be offset by the hypha-stimulating effect of inoculations into fresh media at elevated pH.

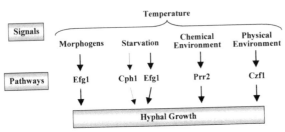

Fig. 2. Distinct environmental signals promote morphogenesis via specific signalling pathways in *Candida albicans*. Signals include poorly defined morphogens (e.g. in serum), nutrient starvation (particularly nitrogen and carbon starvation), chemical parameters (particularly ambient pH), and physical parameters (especially embedding in an agar matrix). The efficacy of most, if not all, of these signals is increased by temperature elevation to 37 °C. Specific signalling pathways appear to mediate the responses to these signals. Each signalling pathway can be defined operationally by a transcription factor that is required for the functionality of that pathway (*Efg1*, *Cph1*, *Prr2* and *Czf1*; see text for explanation). There appears to be some functional overlap between some of these pathways. Also, with the exception of ambient pH, the exact nature of the signal(s) that activates each pathway remains to be clarified

Fink and coworkers have shown recently that a serum filtrate can stimulate hyphal development efficiently (Feng et al. 1999), suggesting that serum might contain a low molecular weight morphogen(s). Such morphogens might include exogenous human hormones such as oestrodiol and progesterone, since they appear to enhance hyphal development (Kinsman et al. 1988; Bramley et al. 1991; Zhao et al. 1995), and *C. albicans* expresses hormone-binding proteins (Williams et al. 1990; Malloy et al. 1993). Other morphogens may exist, however, because serum stimulates hyphal development much more effectively than when these hormones are added in isolation. The medium M199 is commonly used to promote the yeast-hypha transition (Lo et al. 1997; Sharkey et al. 1999; Staab et al. 1999). M199 and serum may stimulate morphogenesis by similar mechanisms. This view is supported by the observation that numerous *C. albicans* morphogenetic mutants display similar responses to M199 and serum (Lo et al. 1997; Davis et al. 2000).

Prior starvation of *C. albicans* cells – for example, for several hours in H_2O – increases the inducibility of yeast cells to form hyphae upon exposure to morphogenetic conditions, but starved cells form hyphae less quickly than metabolically active yeast cells (Buchan and Gow 1991). The use of growth media containing only low concentrations of serum, which therefore may expose cells to both starvation and morphogenetic signals, efficiently promotes hyphal development (Gow 1997). Media containing only a poor carbon and nitrogen source, such as *N*-acetylglucosamine or proline, induce the yeast-hypha transition (Mattia et al. 1982; Torosantucci et al. 1984). Supplementing serum or Glc-*N*Ac media with amino acids and glucose causes hyphae to revert to yeast cells, suggesting that starvation contributes significantly to the hyphal stimulus.

Specific aspects of the chemical environment promote morphogenesis, most notably the H^+ concentration (pH). [Some of the above stimuli (morphogens, the lack of efficient nutrients, for example) could be viewed as components of the chemical environment, but for the purposes of this discussion they are considered separately.] A shift from acidic (pH 4.5) to more alkaline conditions (pH 6.5 or higher) promotes hyphal growth when combined with a shift in temperature from 25 to 37 °C (Lee et al. 1975; Buffo et al. 1984). These conditions also promote a rise in internal pH that has been shown to be a strong correlate of yeast-hypha morphogenesis and may act as a signal transduction system in its own right (Stewart et al. 1988, 1989). A distinct signalling pathway appears to be responsible for mediating the effects of pH (Porta et al. 1999; Davis et al. 2000), apparently setting this stimulus apart from other hypha-inducing signals. These pH changes may contribute to the efficacy of other hypha-inducing conditions. For example, the addition of 10% serum, or serum filtrate, can increase the pH of YPD-based media from about 5 to 6.5. Clearly, this might complicate the interpretation of much published data.

Recently, the physical environment of *C. albicans* cells has been shown to influence yeast-hypha morphogenesis. Cells embedded in agar form hyphae, irrespective of chemical composition of the growth medium (Brown et al. 1999), suggesting that *C. albicans* can respond to physical cues to regulate its morphology. This is an attractive idea because previous work has demonstrated that *C. albicans* responds to physical cues perhaps via stretch-activated channels to control the direction of hyphal growth (Sherwood et al. 1992; Watts et al. 1998).

It is clear, therefore, that *C. albicans* responds to a variety of environmental cues to regulate yeast-hypha morphogenesis (Fig. 2). However, one factor that appears to overlay most of these conditions is temperature: an increase in temperature to about 37 °C is required before most of the above conditions can affect morphogenesis. Is this because temperature elevation imposes a mild stress, or can other environmental stimuli only activate morphogenesis once a minimum temperature is attained? Physical triggers may be a possible exception to this "rule", since embedded colonies form hyphae when grown at 25 °C (Brown et al. 1999). However, cells undergo a heat shock during the embedding process, and this might activate a putative temperature switch.

IV. Morphogenetic Pathways

Several morphogenetic signalling pathways have been identified in *C. albicans*. Molecular analysis of these pathways has been hampered by various features of this fungus: (1) it is diploid (Magee 1998), (2) it displays an alternative genetic code (Santos et al. 1993), and (3) although it carries a mating-type-like locus (*MLT*; Hull and Johnson 1999) and strains can be engineered to mate (Gow et al. 2000; Hull et al. 2000; Magee and Magee 2000), no full sexual cycle or meiotic behaviour has yet been identified. For these reasons, most experimental approaches toward the identification of signalling components in *C. albicans* have exploited the genetically tractable model, *S. cerevisiae*. Screens have exploited the ability of *C. albicans* genes to complement or suppress mutations in the corresponding *S. cerevisiae* genes (Liu et al. 1994; Clark et al. 1995), or to interfere with or constitutively activate the corresponding signalling pathways in *S. cerevisiae* (Whiteway et al. 1992; Stoldt et al. 1997). More recently, *C. albicans* genome sequencing has facilitated the direct isolation of signal transduction genes from *C. albicans* on the basis of sequence similarities (see list of sequenced *C. albicans* genes at http://alces.med.umn.edu:80/bin/genelist?seqs). Nevertheless, the functional analysis of these genes still relies heavily upon *S. cerevisiae* because the corresponding regulatory hierarchies are better characterized in this organism.

Both positive and negative signalling pathways regulate the yeast-hypha transition in *C. albicans*. The Cph1, Efg1, Prr2 and Czf1 pathways are

thought to stimulate, whereas Tup1 represses hyphal development.

A. Chp1 Pathway

The first morphogenetic signalling components to be identified in *C. albicans* were members of the Cph1 pathway. Several screens were employed initially (Whiteway et al. 1992; Liu et al. 1994; Clark et al. 1995), each exploiting the detailed understanding of the mating pathway in *S. cerevisiae* (see, for example, Chap. 2, this Vol.). *C. albicans* was not thought to have a sexual cycle, but this pathway was known to share components with the pseudohyphal-signalling pathway (Liu et al. 1993; Roberts and Fink 1994; see also Chap. 2, this Vol.). Therefore, components of the *C. albicans* Cph1 pathway were initially isolated in *S. cerevisiae*, and then their function investigated by examining the phenotype of *C. albicans* null mutants and overexpressing strains. It remains the best-defined signalling pathway in *C. albicans*, although some components remain to be identified.

Cph1 pathway mutants display characteristic phenotypes. Null mutants, in which this pathway is blocked, are unable to form hyphae on solid Spider medium – which contains the poor carbon source, mannitol – but they still undergo morphogenesis in response to other stimuli (serum, pH) and in liquid media, including Spider medium (e.g. Liu et al. 1994; Leberer et al. 1996). Constitutive activation of the Cph1 pathway, for example by inactivation of the negative regulator Cpp1 (Csank et al. 1997) or overexpression of components of the mitogen-activated protein (MAP) kinase module (Leberer et al. 1996), promotes constitutive hyphal development. Taken together, these data suggest that the Cph1 pathway might activate morphogenesis in response to specific starvation conditions (Fig. 2). Parallel pathways promote hyphal development in response to other stimuli, and there is probably some functional overlap between these pathways and the Cph1 pathway.

Consistent with these in vitro observations, the Cph1 pathway displays functional specialization in vivo. Inactivation of the Cph1 pathway attenuates virulence in the mammary glands of lactating mice, preventing tissue colonization (Guhad et al. 1998). However, Cph1 pathway mutants retain their virulence in the mouse model of systemic candidosis (Leberer et al. 1996; Lo et al. 1997). Hence, the Cph1 pathway is only required for infection in certain specific environments in vivo.

As yet, no receptors have been identified that are specific for the Cph1 pathway. By analogy with *S. cerevisiae* (Lorenz and Heitman 1997), a Ras protein is thought to activate both the Cph1 and Efg1 pathways (Fig. 3), and the phenotypes of a *C. albicans ras1* null mutant are consistent with this model. Like a *cph1, efg1* double mutant (Lo et al. 1997), this *ras1* null mutant is unable to develop hyphae in response to serum stimulation, whereas single *cph1* and *efg1* mutants are able to develop some hyphae under these conditions (Feng et al. 1999).

By analogy with the *S. cerevisiae* pseudohyphal pathway, one would expect the signal to be passed from Ras2 to Ste20 homologues via Cdc42 and the 14-3-3 proteins Bmh1 and Bmh2 (Mosch et al. 1996; Roberts et al. 1997). Although a Cdc42 homologue has been identified in *C. albicans*, its

role in morphogenesis has not been established (Mirbod et al. 1997). In contrast, Cst20, a member of the Ste20/p65[pak] family of protein kinases, has been shown to lie on the Cph1 pathway (Köhler and Fink 1996; Leberer et al. 1996). Unlike other Cph1 pathway mutants, a *C. albicans cst20* null mutant displays attenuated virulence in systemic infections (Leberer et al. 1996), suggesting that Cst20 probably executes additional functions on top of its role in the Cph1 pathway. By analogy with *S. cerevisiae* Ste20, Cst20 might integrate morphogenetic signalling with the control of cell polarity.

At the core of the Cph1 pathway lies a MAP kinase module that is activated by Cst20 (Fig. 3). As yet, the MAP kinase kinase kinase in this module, presumably a *S. cerevisiae* Ste11 homologue (Fig. 3: Chap. 2, this Vol.), has not been described. However, the MAP kinase kinase (Hst7: Clark et al. 1995; Köhler and Fink 1996;

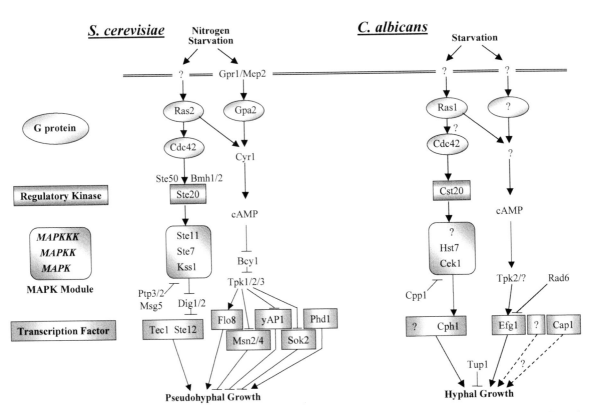

Fig. 3. Conserved mitogen-activated protein (MAP) kinase and Ras-cAMP pathways promote morphogenesis in *Saccharomyces cerevisiae* and *Candida albicans*. In *C. albicans*, the MAP kinase and Ras-cAMP pathways correspond to the Cph1 and Efg1 pathways respectively (Fig. 2)

Leberer et al. 1996) and MAP kinase (Cek1; Whiteway et al. 1992; Csank et al. 1998) have been characterized, and their relative positions on the pathway have been established by epistasis analyses (Leberer et al. 1996; Csank et al. 1998). The activity of the MAP kinase appears to be negatively regulated by the protein phosphatase, Cpp1, since the constitutive hyphal development observed in a *cpp1* null mutant is inhibited by inactivation of Cek1 (Csank et al. 1998). However, Cpp1 might also regulate other signalling pathways in *C. albicans* because other protein phosphatases are known to have multiple substrates.

The function of the MAP kinase module is dependent on the transcription factor Cph1, a functional homologue of *S. cerevisiae* Ste12 (Liu et al. 1994). In *S. cerevisiae*, Ste12 acts in concert with Tec1 to stimulate the transcription of genes containing the filamentation response element (FRE: Madhani and Fink 1997). Activation of Ste12-Tec1 by the MAP kinase module is mediated by the inhibition of the repressors Dig1 and Dig2 (Fig. 3: Cook et al. 1996). Given the high degree of functional conservation between components of these MAP kinase pathways in *S. cerevisiae* and *C. albicans*, the identification of Tec1, Dig1 or Dig2 homologues in *C. albicans* is likely in the near future.

The gene targets of the Cph1 pathway, and hence the specific role(s) of this pathway in morphogenesis and virulence, remain obscure. No hypha-specific *C. albicans* genes have been identified, whose expression is dependent on this pathway (Brown et al. 2000; Sharkey et al. 1999). However, this might reflect the integration of multiple pathways at hypha-specific promoters (Brown et al. 2000). Hence, the gene targets of the Cph1 pathway might be better defined by identifying those genes that are derepressed following constitutive activation of this pathway, e.g. in *cpp1* null strains (Csank et al. 1997).

B. Efg1 Pathway

Ernst and coworkers were the first to identify a component of a second morphogenetic signalling pathway in *C. albicans* (Stoldt et al. 1997). They showed that **Efg1** expression *Enhances Filamentous Growth* in *S. cerevisiae*, and that Efg1 depletion blocks hyphal development in *C. albicans*. Significantly, the morphogenetic phenotypes of *efg1* mutants are stronger than those for *cph1* null

mutants and their phenotypes appear additive (Lo et al. 1997; Stoldt et al. 1997). This suggests that Efg1 defines a distinct and major morphogenetic pathway in *C. albicans* (Fig. 3). An *efg1* null mutant displays hyphal defects on solid and liquid serum-containing media and a range of media that might be expected to impose starvation (Lo et al. 1997; Sonneborn et al. 2000). This indicates that serum stimulation and activation on solid surfaces depend specifically upon the Efg1 pathway (Fig. 2). This pathway also appears to mediate morphogenetic signals in response to starvation signals. The exact demarcation between those starvation signals that specifically stimulate the Cph1 or Efg1 pathways remains to be defined.

Efg1 depletion or inactivation generates elongated pseudohyphal-like cells under growth conditions that favor growth in the yeast form (Lo et al. 1997; Stoldt et al. 1997). This has led Ernst and coworkers to suggest that Efg1 might act as a repressor as well as an activator of morphogenesis (Stoldt et al. 1997). The situation is complicated further by the observation that Efg1 plays roles in other processes such as phenotypic switching (Sonneborn et al. 1999). Therefore, the role(s) of Efg1 appears to be complex, and this must be considered when studying the phenotypic effects of disturbing Efg1 structure or expression.

Consistent with the strong morphological phenotypes observed in vitro, *efg1* null mutants display significantly attenuated virulence and hyphal development in systemic infections (Lo et al. 1997), suggesting that the Efg1 pathway plays an important role(s) in pathogenesis. However, Efg1 is not required for the development of candidosis in all microenvironments in vivo. *Efg1* mutants can form hyphae and retain their virulence in infections of the tongues of immunosuppressed gnotobiotic pigs (Riggle et al. 1999), an infection model that is thought to reflect stimulation via physical cues in the environment (Sect. IV.D).

Efg1 is a bHLH transcription factor that displays sequence similarity to morphogenetic regulators in other fungi, including Sok2 and Phd1 (APSES proteins: Stoldt et al. 1997; Sonneborn et al. 2000). Sok2 is thought to operate on a Ras-cAMP signalling pathway to regulate pseudohyphal development in *S. cerevisiae* (Ward et al. 1995). Also, the filamentous phenotypes of Efg1-expressing *S. cerevisiae* cells can be modulated by changes that modulate Ras-cAMP pathway activity (Rademacher et al. 1998). Therefore, Efg1 might lie on a Ras-cAMP pathway in *C. albicans* (Brown and Gow 1999).

The downstream elements of the morphogenetic Ras-cAMP pathway are better characterized in *S. cerevisiae*. Ras2 is thought to stimulate pseudohyphal development by activating adenyl cyclase. However, adenyl cyclase can also be activated by Gpa2 (Fig. 3; Kubler et al. 1997; Lorenz and Heitman 1997). Activation of adenyl cyclase generates a transient increase in cAMP levels that, in turn, leads to the activation of protein kinase A – mainly encoded by *TPK2* (Pan and Heitman 1999). Protein kinase A then downregulates negative regulators of pseudohyphal growth, such as Sok2, Msn2/4 and yAP-1, and upregulates Flo8, which activates pseudohyphal development (Fig. 3; Liu et al. 1996; Pan and Heitman 1999; Stanhill et al. 1999). The Efg1-like factor Phd1 is a weak activator of pseudohyphal development that acts independently of this Ras-cAMP pathway and the Cph1-like MAP kinase pathway (Fig. 3; Gimeno and Fink 1994; Chandarlapaty and Errede 1998; Pan and Heitman 1999).

Many components of this pathway are conserved in *C. albicans*. *C. albicans ras1* null mutants display similar phenotypes to *efg1* mutants in that they are defective in hyphal growth following serum stimulation. Also, a dominant active RAS^{V13} allele enhances hyphal development, and a dominant negative RAS^{V13} allele reduces hyphal growth (Feng et al. 1999). These data provide convincing evidence that *CaRAS1*, whose closest homologue in *S. cerevisiae* is *RAS2*, plays a key role in morphogenetic signalling (Fig. 3). G proteins have been identified in *C. albicans* (Paveto et al. 1992; Sadhu et al. 1992), but as yet none of these proteins has been shown to play a role in yeast-hypha morphogenesis. Similarly, the involvement of a *C. albicans* adenyl cyclase gene in morphogenetic signalling remains to be confirmed. However, the stimulatory activity of cAMP upon hyphal development is well documented (Egidy et al. 1990; Sabie and Gadd 1992).

Recently, a *C. albicans* protein kinase A gene (*TPK2*) has been reported. A *tpk2* mutant displays hyphal defects that are suppressed by Efg1 overexpression, but *TPK2* overexpression does not suppress an *efg1* mutation, suggesting that Tpk2 lies on the Efg1 pathway (Sonneborn et al. 2000). Significantly, *TPK2* overexpression can suppress a *cph1* mutation, indicating that the Cph1 and Tpk2-Efg1 pathways are distinct (Fig. 3). However, the phenotypes of the *tpk2* null mutant are not as

strong as the *efg1* mutant, suggesting that additional functionally overlapping *TPK* genes might exist (Sonneborn et al. 2000). Rad6 negatively regulates the activity of this putative Ras1-Tpk2-Efg1 pathway (Leng et al. 2000). Rad6 depletion enhances pseudohyphal development, whereas Rad6p overexpression inhibits hyphal development. Rad6 is an ubiquitin-conjugating enzyme that might target a component of the Efg1p pathway for ubiquitination-mediated protein degradation (Leng et al. 2000).

How does activation of the Efg1 pathway mediate changes in cell shape? In *S. cerevisiae*, the Ras-cAMP pathway contributes to the regulation of *FLO11* (*MUC1*), which encodes a cell surface glycoprotein protein, the expression of which appears necessary and sufficient for pseudohyphal development and invasive growth (Lambrechts et al. 1996; Lo and Dranginis 1998; Rupp et al. 1999). In *C. albicans*, some Efg1 gene targets have been identified. Inactivation of Efg1 blocks the expression of the hypha-specific genes *HWP1*, *HYR1*, *ECE1* and *ALS8* (formally known as *ALS7*), but none of these genes are essential for hyphal development (Sharkey et al. 1999; Brown et al. 2000; Brown et al., unpubl.). Recent data indicate that Efg1 interacts directly with hypha-specific promoters to regulate their expression (Brown et al., unpubl.). Most of these Efg1 gene targets encode hypha-specific cell wall components and adhesion factors (Birse et al. 1993; Bailey et al. 1996; Staab et al. 1999; Brown et al., unpubl.). However, it is not clear how these (or other) Efg1 targets mediate the changes in cell shape required for hyphal development.

C. Prr2 Pathway

It has been known for some time that increases in pH, combined with elevated temperatures, promote yeast-hypha morphogenesis (Buffo et al. 1984; Soll 1986). Screens for *C. albicans* genes that are regulated during pH-stimulated morphogenesis led to the identification of *PHR1* and then *PHR2*, which encode pH-regulated glucosidases involved in maintaining cell wall structure and cell shape (Saporito-Irwin et al. 1995; Muhlschlegel and Fonzi 1997; Fonzi 1999). Inactivation of the alkaline-expressed gene *PHR1* causes pH-conditional defects in the polarity of growth and reduced virulence in systemic infection models, whereas null mutations in the acid-expressed gene

PHR2 block growth at pH 4 and attenuate virulence in vaginal infections (Saporito-Irwin et al. 1995; Ghannoum et al. 1995; Muhlschlegel and Fonzi 1997; De Bernardis et al. 1998). Constitutive expression of Phr1 from a *TEF-PHR1* fusion rescues the *prr2* mutant phenotype, indicating that Phr1 and Phr2 are functionally equivalent.

The pH-signalling pathways have been characterized in the greatest detail in *Aspergillus nidulans* (Denison et al. 1998; Mingot et al. 1999), and this pathway is conserved in other fungi including *S. cerevisiae* (Fig. 4: see Ramon et al. 1999, Davis et al. 2000 and references therein). In *A. nidulans*, the integral membrane proteins PalH and PalI, whose functions have not been established, probably detect an alkaline ambient pH. A signalling cascade involving PalA, PalB, PalC and PalF then transduces this stimulus, but their biochemical activity and their hierarchy in the pathway have not been established. They lead to the activation of the zinc finger transcription factor, PacC. PacC activation involves a specific cleavage to remove the inhibitory C-terminal

region, allowing the N-terminal region to activate the expression of alkaline-expressed genes, and repress the transcription of acid-expressed genes (Fig. 4; Mingot et al. 1999). The cleavage of PacC involves a cysteine protease, and *palB* encodes a protein with similarity to cysteine proteases; however, PalB does not appear to catalyze PacC cleavage (Denison et al. 1995).

C. albicans homologues of these pH regulators have been identified through their sequence similarities to their fungal counterparts (Porta et al. 1999; Ramon et al. 1999; Wilson et al. 1999; Davis et al. 2000). At present, the gene nomenclature is somewhat confusing because the Fonzi group has named these genes according to their functions (*PRR*), whereas the Mitchell group has used the names of the *S. cerevisiae* homologues that were originally identified on the basis of their meiotic phenotypes (*RIM*) (Fig. 4).

Prr2/rim101 and *prr1/rim8* null mutants display similar phenotypes. Alkaline-expressed genes (e.g. *PHR1*) are no longer induced, and acid-expressed genes (e.g. *PHR2*) are no longer

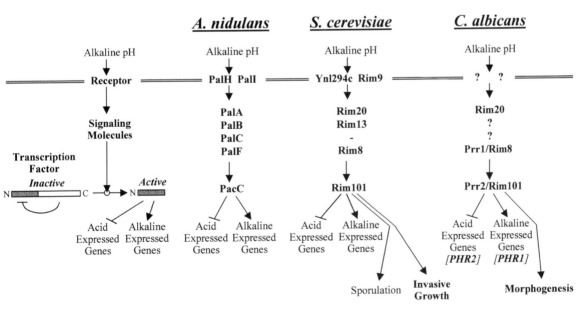

Fig. 4. A conserved signal transduction pathway mediates responses to ambient pH in *Aspergillus nidulans*, *Saccharomyces cerevisiae* and *Candida albicans*. The pH response pathway, which involves C-terminal truncation to activate a critical transcription factor, is illustrated on the left. In *C. albicans*, this transcription factor is called Prr2 (Fig. 2) or Rim101 (Sect. IV). The epistatic relationships between the signalling molecules have not been clarified. Hence, they have been ordered alphabetically in the *A. nidulans* pathway, and those in *S. cerevisiae* and *C. albicans* have then been placed in the equivalent positions to their *A. nidulans* homologue

repressed in these mutants (Porta et al. 1999; Davis et al. 2000). These phenotypes are consistent with the role of this pH-response pathway in pH-regulated gene expression in other fungi (Fig. 4). Interestingly, *prr2/rim101* and *prr1/rim8* mutants also displayed morphogenetic defects. On solid media at neutral/alkaline pHs (pH 7.5–8.0), hyphal growth and invasion are blocked on serum and M199 medium (M199), and hyphal development and invasion are reduced on Spider and Lee's medium. In liquid media at neutral/alkaline pHs, these null mutants can form hyphae in serum but not in M199, Spider or Lee's media (Porta et al. 1999; Ramon et al. 1999; Wilson et al. 1999; Davis et al. 2000). Similarly, a *rim20* mutant displays a hyphal defect in M199, but not upon serum stimulation in liquid media (Davis et al. 2000). The expression of a C-terminally truncated version of Prr2/Rim101 – equivalent to the removal of the inhibitory C-terminal region of PacC, as shown in Fig. 4 (Mingot et al. 1999) – suppressed the filamentous defects of *prr1/rim8* and *rim20* mutants at alkaline pH, and induced weak filamentation at acid pH (Davis et al. 2000). These data confirmed the epistatic relationship between these genes on the Prr2 pathway, as well as the role of this pathway in controlling pH-regulated morphogenesis.

Therefore, both morphogenesis and pH-regulated gene expression are controlled by the Prr2 pathway. Interestingly, two observations indicate that these phenotypes are separable. Firstly, the expression of the hypha-specific gene *HWP1* is not affected by inactivation of *PRR1/RIM8*, indicating that different control mechanisms mediate hypha-specific and pH-regulated gene expression (Porta et al. 1999). Secondly, constitutive expression of Phr1 does not rescue the hyphal defects of *prr2/rim101* or *prr1/rim8* null mutants, indicating that it is not the lack of Phr1 that prevents hyphal development in these strains (Porta et al. 1999). Hence, in *C. albicans*, there appears to be bifurcation of the pH-response pathway at Prr2/Rim101, one fork involved in the control of pH-regulated gene expression, and the other fork activating yeast-hypha morphogenesis (Fig. 4).

D. Czf1 Pathway

The recent identification and characterization of the *CZF1* gene has defined an additional mor-

phogenetic pathway in *C. albicans* (Brown et al. 1999). This gene was identified on the basis that ectopic expression of Czf1 promotes hyphal development when *C. albicans* cells are embedded in YP agar containing sucrose or galactose. Czf1 contains a zinc finger and is thought to be a transcription factor. Inactivation of Czf1 attenuates hyphal growth under these conditions, but not following stimulation with Spider medium, serum or *N*-acetylglucosamine (Brown et al. 1999). In contrast, an *efg1,cph1* double mutant retains the ability to form hyphae upon embedding in agar (Riggle et al. 1999). Thus the phenotype of a *czf1* mutant is distinct from those of Cph1 and Efg1 pathway mutants, indicating that Czf1 defines a separate pathway.

A unique aspect of the Czf1 pathway is that it appears to respond to physical cues in the environment, rather than chemical cues such as pH, morphogens or starvation (Brown et al. 1999). However, the Czf1 pathway does appear to interact in some way with the Chp1 pathway, since a *czf1,cph1* double mutant displays a more dramatic hyphal defect than a *czf1* single mutant (Brown et al. 1999). The definition of physical cues that trigger this pathway, the identification of other components of the Czf1 pathway, and the characterization of interactions with other morphogenetic signalling pathways represent major challenges for the future.

E. Other Factors

Some other factors have been implicated in the control of yeast-hypha morphogenesis in *C. albicans*, but their relationships to the above morphogenetic signalling pathways have not been established. These include Tup1p, a repressor of hyphal development (Braun and Johnson 1997). *C. albicans tup1* null mutants form hyphae under conditions that normally promote growth in the yeast form. In addition, the expression of hypha-specific genes is derepressed in these strains (Brown et al. 2000). In *S. cerevisiae*, Tup1 is a key component of transcription complexes that repress specific subsets of yeast genes involved in both developmental and metabolic processes (Keleher et al. 1992; Gancedo 1998). Hence, in *C. albicans*, Tup1 probably mediates its effects on hyphal development via transcriptional repression, possibly acting downstream of the Efg1p and Cph1p pathways. Given the broad role of

Tup1 in *S. cerevisiae*, *C. albicans* Tup1 might regulate additional processes that are unrelated to morphogenesis.

The analysis of a *tup1* null mutant first highlighted the importance of negative regulation in the control of yeast-hypha morphogenesis in *C. albicans* (Braun and Johnson 1997). Since then, other observations have reinforced this view. Inactivation of Sir2, a factor that is thought to maintain gene silencing in *C. albicans*, enhances filamentous growth under conditions that normally promote growth in the yeast form (Perez-Martin et al. 1999). Similarly, depletion of Rad6 increases the growth of filamentous cells (Leng et al. 2000). Rad6 is known to play important roles in gene silencing in *S. cerevisiae*. Also, overexpression of Nhp6, an HMG-like nonhistone protein involved in chromatin modelling, inhibits yeast-hypha morphogenesis in *C. albicans* (Brown et al., unpubl.). Therefore, several lines of evidence suggest that gene silencing plays a role in repressing hyphal development.

Rbf1 was isolated on the basis that it interacts with the RPG box (Ishii et al. 1997a). This short DNA sequence plays an important role in the regulation of several classes of *S. cerevisiae* genes, including ribosomal protein, transcription, translation and glycolytic genes. Rbf1 exerts a negative effect upon hyphal development in *C. albicans*, since *rbf1* mutants form hyphae constitutively (Ishii et al. 1997b). The mechanism by which Rbf1p mediates this regulatory effect is not clear, but it might act indirectly to inhibit morphogenesis, for example by affecting metabolism and/or imposing stresses upon the *C. albicans* cell.

As described above (Sect. III), various stresses, such as temperature elevation and starvation, promote yeast-hypha morphogenesis. This suggests a correlation between increased stress and hyphal development in *C. albicans* under some conditions. In contrast, in *S. cerevisiae*, the activation of yAP-1 and Msn2/4-mediated stress responses appears to inhibit invasive growth. Mutations that suppress these stress responses also promote invasive growth, and Ras2 mutations that stimulate invasive growth inhibit these stress responses (Fig. 3) (Stanhill et al. 1999). Presumably, this effect is mediated through negative regulation of key transcription factors such as Msn2 by protein kinase A mediated phosphorylation (Gorner et al. 1998). Therefore, *S. cerevisiae* and *C. albicans* might differ significantly with respect to the integration of their stress and morphogenetic responses (Figs. 2–5). As yet, *C. albicans* homologues of Msn2 and Msn4 have not been described. However, a *C. albicans* yAP-1 homologue, Cap1, has been reported. Inactivation of Cap1 reduces the resistance of *C. albicans* to oxidative stress responses and multidrug resistance (Alarco and Raymond 1999), but its effect on yeast-hypha morphogenesis is not known. The *C. albicans* CAT1 gene, which encodes catalase, is a putative target of the Cap1 bZip transcription factor. A *cat1* null mutant displays normal hyphal development in cell culture medium, and increased susceptibility to killing by neutrophils (Wysong et al. 1998). However, the critical experiments, involving measurements of hyphal development in response to stress-inducing stimuli, have not been reported. Therefore, possible interactions between Cap1 and Msn2/4 stress response pathways remain to be elucidated (Fig. 3).

Two further *C. albicans* proteins might play roles in the integration of morphogenetic signalling with changes in cell polarity. Cla4 is a Ste20-like protein that is required for hyphal growth under all conditions tested (Leberer et al. 1997). *S. cerevisiae* Cla4p is a protein kinase that activates myosin 1 and interacts with the G protein Cdc42p. Hence, Cla4p exerts its effects on *C. albicans* morphogenesis through roles in cytoskeletal organization. Inactivation of *C. albicans* Int1 inhibits hyphal development and adhesion (Gale et al. 1998). Also, *S. cerevisiae* cells expressing Int1 generate extended parallel-sided filaments. Int1 is located at the cell surface and has a cytoplasmic tail. Hence, like Cla4, Int1 might coordinate morphogenetic signalling pathways with changes in cell polarity.

F. Calcium and Inositol Phosphates

Changes in the concentration of free cytoplasmic calcium ions act as a signal that regulates many aspects of eukaryotic cell biology, including fungal morphogenesis (see Chap. 4, this Vol.). Calcium ions bind calcium-binding proteins such as calmodulin that, in turn, modulates the activity of a number of calmodulin-dependent protein kinases resulting in new cellular responses. Calcium can also bind and affect the activity of some protein kinases directly, e.g. protein kinase C. However, the *C. albicans* PKC1 gene has been disrupted and this does not seem to affect yeast-hypha morphogenesis (Paravicini et al. 1996). The

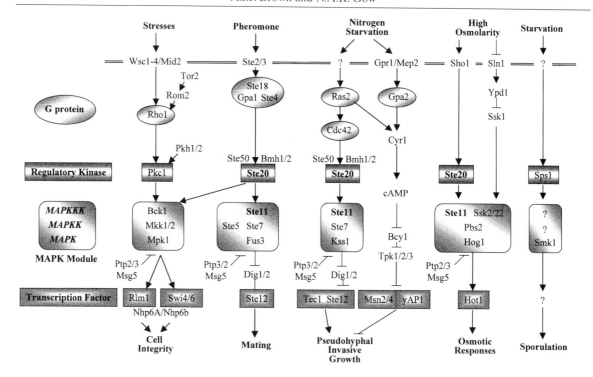

Fig. 5. Current models of mitogen-activated protein kinase (MAPK) signalling pathways in *S accharomyces cerevisiae*. Known signals, G proteins, regulatory kinases, MAPK module components, transcription factors and responses are shown

release of calcium ions from stores such as the vacuole and mitochondria is influenced by other signalling molecules, such as inositol 4,5 triphosphate, which is a product of phospholipase C action on phosphatidylinositol in the plasma membrane. The *C. albicans* phospholipase C has been cloned but lacks the normal calcium-binding EF hand at the N-terminus of the predicted protein (Bennett et al. 1998). The effect of the disruption of this gene has not yet been reported. The importance of calcium, calmodulin and inositol phosphate signalling in yeast-hypha morphogenesis has not yet been fully explored but there is some evidence for a role for calcium signalling in yeast-hypha regulation. For example, calmodulin inhibitors reportedly block the yeast-to-hypha transition (Eilam et al. 1987; Roy and Datta 1987; Sabie and Gadd 1989; Buchan et al. 1993) and calcium ion uptake has been suggested to have a repressing effect on mycelium formation (Holmes et al. 1991). Calmodulin levels have also been reported as elevated in mycelial forms (Paranjape et al. 1990), although this was not reflected in differential levels of expression of the calmodulin

gene *CMD1* (Saporito and Sypherd 1991). Therefore, the roles of calcium ions and inositol phosphates in dimorphic transition of *C. albicans* are not yet clear but their potential as important regulators of morphogenesis must be considered.

V. Other Outputs of Morphogenetic Pathways

Downstream substrates of the morphogenetic signalling pathways listed above are much less clear. In order to achieve a change in cell shape a number of physiological functions must be modulated. These include cell polarity, the cell cycle and, ultimately, cell wall biosynthesis. Mechanistically, changes in cell shape are mediated via alterations in the vectorial synthesis of the cell wall. This, in turn, is controlled by actin microfilaments and the secretory pathway that delivers the membrane vesicles required for expansion of the plasma membrane and the enzymes for elaboration of the new cell wall (see Gow and Gadd 1995 for reviews in this area, and Chaps. 10 and 11, this Vol.). Hyphal cells exhibit polarized cell growth in which cell expansion is confined to the apex of the hypha (Staebell and Soll 1985; Anderson and Soll 1986). In yeast cell growth the new bud grows initially by apical expansion, and finally by isotropic growth (Staebell and Soll 1985). In *S. cerevisiae*, the G2/M interface of the cell cycle regulates the transition from apical to isotropic growth of buds (Lew and Reed 1995). Activation of the Cdc28 cyclin-dependent protein kinase by specific cyclins plays a key role in cellular polarity responsible for both triggering polarized bud development in the G1 phase and then regulating the transition from polarized to isotropic bud development at G2/M (Lew and Reed 1993, 1995). The G1 cyclins Cln1, Cln2 and Cln3 are required for activation of polarized growth and the mitotic cyclins Clb1 and Clb 2 are required for the depolarization of growth (Lew and Reed 1995). In *C. albicans* the Cln1 G1 cyclin exhibits cell cycle dependent expression and is required for the maintenance of filamentous growth, since *cln1/cln1* mutants revert more quickly to the yeast form (Loeb et al. 1999). This mutant was deficient in hypha formation on several solid media but was still competent to form hyphae in liquid media. This suggests that CaCln1 regulation of hyphal growth may be coordinated with specific signal transduction pathways that operate under specific environmental conditions (Loeb et al. 1999). In *S. cerevisiae* a protein kinase cascade involving Elm1, Hsl1 and Swe1 modulates the regulation of filamentous growth by Cdc28 (Edgington et al. 1999). This suggests that a complex phosphorylation cascade lies upstream of the events that ultimately regulate cell shape by modulating the organization of the actin cytoskeleton and the secretory apparatus (see Chaps. 2 and 10, this Vol.).

In *S. cerevisiae* the MAP kinase pathway that activates Ste12 is involved in the regulation of polarized shmoo formation during mating (Fig. 5). Until recently, *C. albicans* was not thought to mate and was apparently believed to be constitutively diploid. However, recently, a mating-type locus reminiscent of that in *S. cerevisiae* has been identified (Hull and Johnson 1999). The *C. albicans* mating locus is heterozygous at the MTL (mating type locus) and carries *MTLa1* on one of the chromosome 5 pairs and *MTLα1* and *MTLα2* on the other. Therefore, most *C. albicans* strains may be unable to mate because they encode the two essential components of the a1/α2

heterodimeric protein that normally represses mating in diploid strains of yeast fungi like *S. cerevisiae*. Elimination of either single genes, or whole copies of chromosome 5 – to construct strains that were *MTLa* or *MTLα* – generated strains that could be mated to form stable tetraploids (Gow et al. 2000; Hull et al. 2000; Magee and Magee 2000). It is not yet known what the mating interaction looks like and whether it involves elongated projections akin to shmoo formation. However, it is very likely that the regulation of mating in *C. albicans* involves transcriptional activation of genes by Cph1p, the homologue of the *S. cerevisiae* transcription factor Ste 12 that resides at the bottom of the STE MAP kinase pathway.

Finally, it is possible, or even likely, that virulence attributes other than filamentous growth are regulated by these signalling pathways. The *EFG1/efg1* heterozygous mutant forms normal hyphae but is less virulent than the *EFG1/EFG1* parent (Lo et al. 1997). This may indicate that Efg1p may also regulate other virulence factors in addition to playing a role in yeast-hypha morphogenesis (Brown and Gow 1999). Therefore, the outputs of the morphogenetic pathways described here are complex and involve regulation of a wide range of factors important for pathogenesis in *C. albicans*.

VI. Conclusions

Twenty years ago organisms such as *C. albicans* were considered insufficiently tractable for their research to make a significant contribution to our understanding of basic mechanisms in cell biology and pathogenesis. However, new widely applicable molecular tools and methodologies for the functional analysis of fungal genes and fungal genomes has launched an exciting era in fungal cell biology for genetically recalcitrant species and is facilitating important research into how fungal morphogenesis influences fungal pathogenesis. A number of signalling pathways have been identified that impinge on transcriptional regulators that influence yeast-hypha morphogenesis and, most likely, other virulence factors. It seems unlikely that these pathways function in mutual isolation from one another but this possibility has barely been addressed experimentally. It remains to be established exactly what signals are fed into these pathways and what outputs are regulated by them.

Answers to these questions will provide important insights into how *C. albicans* interacts and perceives different microenvironments in the human host.

Acknowledgements. AJPB and NARG are supported by the UK Biotechnology and Biological Sciences Research Council, the Wellcome Trust, and the European Commission.

References

Alani E, Cao L, Kleckner N (1987) A method for gene disruption that allows repeated use of *URA3* selection in the construction of multiply disrupted yeast strains. Genetics 116:541–545

Alarco A-M, Raymond M (1999) The bZip transcription factor Cap1p is involved in multidrug resistance and oxidative stress response in *Candida albicans*. J Bacteriol 181:700–708

Anderson J, Soll DR (1986) Differences in actin localisation in bud and hypha formation in the yeast *Candida albicans*. J Gen Microbiol 132:2035–2047

Beck-Sagué CM, Jarvis WR (1993) National nosocomial infections surveillance system. Secular trends in the epidemiology of nosocomial fungal infections in the United States, 1980–1990. J Infect Dis 167:1247–1251

Bennett DE, McCreary CE, Coleman DC (1998) Genetic characterization of phospholipase C gene from *Candida albicans*: presence of homologous species in *Candida* species other than *Candida albicans*. Microbiology 144:55–72

Bailey DA, Feldmann PJF, Bovey M, Gow NAR, Brown AJP (1996) The *Candida albicans HYR1* gene, which is activated in response to hyphal development, belongs to a gene family encoding yeast cell wall proteins. J Bacteriol 178:5353–5360

Birse CE, Irwin MY, Fonzi WA, Sypherd PS (1993) Cloning and characterization of *ECE1*, a gene expressed in association with cell elongation of the dimorphic pathogen *Candida albicans*. Infect Immun 61:3648–3655

Bramley TA, Menzies GS, Williams RJ, Kinsman OS, Adams DJ (1991) Binding sites for LH in *Candida albicans*: comparison with the mammalian corpus luteum LH receptor. J Endocrinol 130:177–190

Braun BR, Johnson AD (1997) Control of filament formation in *Candida albicans* by the transcriptional repressor *TUP1*. Science 277:105–109

Brown AJP, Gow NAR (1999) Regulatory networks controlling *Candida albicans* morphogenesis. Trends Microbiol 7:333–338

Brown AJP, Barelle CJ, Budge S, Duncan J, Harris S, Lee PR, Leng P, Macaskill S, Abdul Murad AM, Ramsdale M, Wiltshire C, Wishart JA, Gow NAR (2000) Gene regulation during morphogenesis in *Candida albicans*. In: Ernst JF, Schmidt A (eds) Contributions to microbiology: dimorphism in human pathogenic and apathogenic yeasts, vol 5. Karger, Basel, pp 112–125

Brown DH, Giusani AD, Chen X, Kumamoto CA (1999) Filamentous growth of *Candida albicans* in response to physical environmental cues and its regulation by the unique *CZF1* gene. Mol Microbiol 34:651–662

Buchan ADB, Gow NAR (1991) Rates of germ tube formation from growing and non-growing yeast cells of *Candida albicans*. FEMS Microbiol Lett 81:15–18

Buchan ADB, Kelly VA, Kinsman OS, Gooday GW, Gow NAR (1993) Effect of trifluoperazine on growth, morphogenesis and pathogenicity of *Candida albicans*. J Med Vet Mycol 31:427–433

Buffo J, Herman MA, Soll DR (1984) A characterization of pH regulated dimorphism in *Candida albicans*. Mycopathologia 86:21–30

Chandarlapaty S, Errede B (1998) Ash1, a daughter cell-specific protein, is required for pseudohyphal growth of *Saccharomyces cerevisiae*. Mol Cell Biol 18:2884–2891

Clark KL, Feldmann PJF, Dignard D, Larocque R, Brown AJP, Lee MG, Thomas DY, Whiteway M (1995) Constitutive activation of the *Saccharomyces cerevisiae* mating response pathway by a MAP kinase kinase from *Candida albicans*. Mol Gen Genet 249:609–621

Cook JG, Bardwell L, Kron SJ, Thorner J (1996) Two novel targets of the MAP kinase Kss1 are negative regulators of invasive growth in the yeast *Saccharomyces cerevisiae*. Genes Dev 10:2831–2848

Csank C, Makris C, Meloche S, Schroppel K, Rollinghoff M, Dignard D, Thomas DY, Whiteway M (1997) Derepressed hyphal growth and reduced virulence in a VH1 family-related protein phosphatase mutant of the human pathogen *Candida albicans*. Mol Biol Cell 8:2539–2551

Csank C, Schroppel K, Leberer E, Harcus D, Mohamed O, Meloche S, Thomas DY, Whiteway M (1998) Roles of the *Candida albicans* mitogen-activated protein kinase homolog, Cek1p, in hyphal development and systemic candidosis. Infect Immun 66:2713–2721

Cutler JE (1991) Putative virulence factors of *Candida albicans*. Annu Rev Microbiol 45:187–218

Davis D, Wilson RB, Mitchell AP (2000) *RIM101*-dependent and -independent pathways govern pH responses in *Candida albicans*. J Bacteriol 20:971–978

De Bernardis F, Muhlschlegel FA, Cassone A, Fonzi WA (1998) The pH of the host niche controls gene expression in and virulence of *Candida albicans*. Infect Immun 66:3317–3325

Denison SH, Orejas M, Arst HN Jr (1995) Signalling of ambient pH in *Aspergillus* involves a cysteine protease. J Biol Chem 270:28519–28522

Denison SH, Negrete-Urtasun S, Mingot JM, Tilburn J, Mayer WA, Goel A, Espeso EA, Penalva MA, Arst HN Jr (1998) Putative membrane components of signal transduction pathways for ambient pH regulation in *Aspergillus* and meiosis in *Saccharomyces* are homologous. Mol Microbiol 30:259–264

Edgington NP, Blacketter MJ, Bierwagen TA, Myers AM (1999) Control of *Saccharomyces cerevisiae* filamentous growth by cyclin-dependent kinase Cdc28. Mol Cell Biol 19:1369–1380

Egidy G, Paveto C, Passeron S, Galvagno MA (1990) cAMP levels and *in situ* measurement of cAMP related enzymes during yeast-to-hyphae transition in *Candida albicans*. Cell Biol Int Rep 14:59–68

Eliam Y, Polacheck I, Ben-Gigi G, Chernichovsky D (1987) Activity of phenothiazines against medically important yeasts. Antimicrob Agents Chemother 31:834–836

Feng Q, Summers E, Guo B, Fink G (1999) Ras signalling is required for serum-induced hyphal differentiation in *Candida albicans*. J Bacteriol 181:6339–6346

Fonzi WA (1999) *PHR1* and *PHR2* of *Candida albicans* encode putative glycosidases required for proper cross-linking of bb-1,3- and aa-1,6-glucans. J Bacteriol 181:7070–7079

Fonzi WA, Irwin MY (1993) Isogenic strain construction and gene mapping in *Candida albicans*. Genetics 134:717–728

Gale CA, Bendel CM, McClellan M, Hauser M, Becker JM, Berman J, Hostetter MK (1998) Linkage of adhesion, filamentous growth, and virulence in *Candida albicans* to a single gene, *INT1*. Science 279:1355–1358

Gancedo JM (1998) Yeast carbon catabolite repression. Microbiol Mol Biol Rev 62:334–361

Ghannoum MA, Spellberg B, Saporito-Irwin SM, Fonzi WA (1995) Reduced virulence of *Candida albicans PHR1* mutants. Infect Immun 63:4528–4530

Gimeno CJ, Fink GR (1994) Induction of pseudohyphal growth by overexpression of *PHD1*, a *Saccharomyces cerevisiae* gene related to transcriptional regulators of fungal development. Mol Cell Biol 14:2100–2112

Gorner W, Durchschlag E, Martinez-Pastor MT, Estruch F, Ammerer G, Hamilton B, Ruis H, Schuller C (1998) Nuclear localization of the C_2H_2 zinc finger protein Msn2p is regulated by stress and protein kinase A activity. Genes Dev 12:586–597

Gow NAR (1997) Germ tube growth in *Candida albicans*. Curr Top Med Mycol 8:43–55

Gow NAR, Brown AJP, Odds FC (2000) *Candida*'s arranged marriage. Science 289:256–257

Gow NAR, Gadd GM (eds) (1995) The growing fungus. Chapman and Hall, London

Guhad FA, Jensen HE, Aalbaek B, Csank C, Mohamed O, Harcus D, Thomas DY, Whiteway M, Hau J (1998) Mitogen-activated protein kinase-defective *Candida albicans* is avirulent in a novel model of localized murine candidosis. FEMS Microbiol Lett 166:135–139

Holmes AR, Cannon RD, Shepherd MG (1991) Effect of calcium ion uptake on *Candida albicans* morphology. FEMS Microbiol Lett 77:187–194

Hull CM, Johnson AD (1999) Identification of a mating type-like locus in the asexual pathogenic yeast *Candida albicans*. Science 285:1271–1275

Hull CM, Raiser RM, Johnson AD (2000) Evidence for mating of the "asexual" yeast *Candida albicans* in a mammalian host. Science 289:307–310

Ishii N, Yamamoto M, Lahm HW, Iizumi S, Yoshihara F, Nakayama H, Arisawa M, Aoki Y (1997a) A DNA binding protein from *Candida albicans* that binds to the RPG box of *Saccharomyces cerevisiae* and the telomeric repeat sequence of *Candida albicans*. Microbiology 143:417–427

Ishii N, Yamamoto M, Yoshihara F, Arisawa M, Aoki Y (1997b) Biochemical and genetic characterisation of Rbf1p, a putative transcription factor of *Candida albicans*. Microbiology 143:429–435

Jamal WY, El-Din K, Rotimi VO, Chugh TD (1999) An analysis of hospital-acquired bacteraemia in intensive care unit patients in a university hospital in Kuwait. J Hosp Infect 43:49–56

Keleher CA, Redd MJ, Schultz J, Carlson M, Johnson AD (1992) Ssn6-Tup1 is a general repressor of transcription in yeast. Cell 68:709–719

Kinsman OS, Pitblado K, Coulson CJ (1988) Effect of mammalian steroid hormones and luteinizing hormone on the germination of *Candida albicans* and implications for vaginal candidosis. Mycoses 31:617–626

Köhler JR, Fink GR (1996) *Candida albicans* strains heterozygous and homozygous for mutations in mitogen-activated protein kinase signalling components have defects in hyphal development. Proc Natl Acad Sci USA 93:13223–13228

Kron SJ, Gow NAR (1995) Budding yeast morphogenesis: signalling, cytoskeleton and cell cycle. Curr Opin Cell Biol 7:845–855

Kubler E, Mosch HU, Rupp S, Lisanti MP (1997) Gpa2p, a G-protein alpha-subunit, regulates growth and pseudohyphal development in *Saccharomyces cerevisiae* via a cAMP-dependent mechanism. J Biol Chem 272:20321–20323

Kwon-Chung KJ, Bennett JE (1992) Medical mycology. Lea and Febiger, Philadelphia

Lambrechts MG, Bauer FB, Marmur J, Pretorius IS (1996) Muc1, a mucin-like protein that is regulated by Mss10, is critical for pseudohyphal differentiation in yeast. Proc Natl Acad Sci USA 93:8419–8424

Leberer E, Harcus D, Broadbent ID, Clark KL, Dignard D, Ziegelbauer K, Schmit A, Gow NAR, Brown AJP, Thomas DY (1996) Homologs of the Ste20p and Ste7p protein kinases are involved in hyphal formation of *Candida albicans*. Proc Natl Acad Sci USA 93: 13217–13222

Leberer E, Ziegelbauer K, Schmidt A, Harcus D, Dignard D, Ash J, Johnson L, Thomas DY (1997) Virulence and hyphal formation of *Candida albicans* require the Ste20p-like protein kinase CaCla4p. Curr Biol 7:539–546

Lee KL, Buckler HR, Campbell CC (1975) An amino acid liquid synthetic medium for the development of mycelial and yeast forms of *Candida albicans*. Sabouraudia 13:148–153

Leng P, Sudbery PE, Brown AJP (2000) Rad6p represses yeast-hypha morphogenesis in the human fungal pathogen, *Candida albicans*. Mol Microbiol 35: 1264–1275

Lew DJ, Reed SI (1993) Morphogenesis in yeast cell cycle: regulation by Cdc28 and cyclins. J Cell Biol 120: 1305–1320

Lew DJ, Reed SI (1995) A cell cycle checkpoint monitors cell morphogenesis in budding yeast. J Cell Biol 129:739–749

Liu H, Styles CA, Fink GR (1993) Elements of the yeast pheromone response pathway required for filamentous growth of diploids. Science 262:1741–1744

Liu H, Köhler JR, Fink GR (1994) Suppression of hyphal formation in *Candida albicans* by mutation of a *STE12* homolog. Science 266:1723–1726

Liu H, Styles CA, Fink GR (1996) *Saccharomyces cerevisiae* S288C has a mutation in FLO8, a gene required for filamentous growth. Genetics 144:967–978

Lo HJ, Köhler JR, DiDomenico B, Loebenberg D, Cacciapuoti A, Fink GR (1997) Nonfilamentous *C. albicans* mutants are avirulent. Cell 90:939–949

Lo WS, Dranginis AM (1998) The cell surface flocculin Flo11 is required for pseudohyphae formation and invasion by *Saccharomyces cerevisiae*. Mol Cell Biol 9:161–171

Loeb JDJ, Sepulveda-Becerra A, Hazan I, Liu H (1999) A G₁ cyclin is necessary for maintenance of filamentous growth in *Candida albicans*. Mol Cell Biol 19: 4019–4027

Lorenz MC, Heitman J (1997) Yeast pseudohyphal growth is regulated by *GPA2*, a G protein alpha homolog. EMBO J 16:7008–7018

Madhani HD, Fink GR (1997) Combinatorial control required for the specificity of yeast MAPK signaling. Science 275:1314–137

Magee BB, Magee PT (2000) Induction of mating in *Candida albicans* by construction of MTLa and MTLα strains. Science 289:310–313

Magee PT (1998) Analysis of the *Candida albicans* genome. In: Brown AJP, Tuite MF (eds) Yeast gene analysis. Methods in Microbiology, vol 26. Academic Press, New York, pp 395–415

Malloy PJ, Zhao X, Madani ND, Feldman D (1993) Cloning and expession of the gene from *Candida albicans* that encodes a high-affinity corticosteroid-binding protein. Proc Natl Acad Sci USA 90:1902–1906

Martin MV, Craig GT, Lamb DJ (1984) An investigation of the role of true hypha production in the pathogenesis of experimental candidosis. J Med Vet Mycol 22: 471–476

Mattia E, Carruba G, Angiolella L, Cassone A (1982) Induction of germ tube formation by *N*-acetyl-D-glucosamine in *Candida albicans*: uptake of inducer and germinative response. J Bacteriol 152:555–562

Merson-Davies LA, Odds FC (1989) A morphology index for characterization of cell shape in *Candida albicans*. J Gen Microbiol 135:3143–3152

Mingot JM, Tilburn J, Diez E, Bignell E, Orejas M, Widick DA, Sarkar S, Brown CV, Caddick MX, Espeso EA, Arst HN Jr, Penalva MA (1999) Specificity determinants of proteolytic processing of *Aspergillus* PacC transcription factor are remote from the processing site, and processing occurs in yeast if pH signalling is bypassed. Mol Cell Biol 19:1390–1400

Mirbod F, Nakashima S, Kitajima Y, Cannon RD, Nozawa Y (1997) Molecular cloning of a Rho family, *CDC42Ca* gene from *Candida albicans* and its mRNA expression changes during morphogenesis. J Med Vet Mycol 35:173–179

Monge RA, Navarro-Garcia F, Molero G, Diez-Orejas R, Gustin M, Pla J, Sanchez M, Nombela C (1999) Role of the mitogen-activated protein kinase Hog1p in morphogenesis and virulence of *Candida albicans*. J Bacteriol 181:3058–3068

Mosch HU, Roberts RL, Fink GR (1996) Ras2 signals via the Cdc42/Ste20/mitogen-activated protein kinase module to induce filamentous growth in *Saccharomyces cerevisiae*. Proc Natl Acad Sci USA 93: 5352–5356

Muhlschlegel FA, Fonzi WA (1997) *PHR2* of *Candida albicans* encodes a functional homolog of the pH-regulated gene *PHR1* with an inverted pattern of expression. Mol Cell Biol 17:5960–5967

Odds FC (1988) *Candida* and candidosis, 2nd edn. Bailliere Tindall, London

Pan X, Heitman J (1999) Cyclic AMP-dependent protein kinase regulates pseudohyphal differentiation in *Saccharomyces cerevisiae*. Mol Cell Biol 19:4874–4887

Paranjape V, Roy BG, Datta A (1990) Involvement of calcium, calmodulin and protein phosphorylation in morphogenesis of *Candida albicans*. J Gen Microbiol 136:2119–2154

Paravicini G, Medoza A, Antonsson B, Cooper M, Losberger C, Payton MA (1996) The *Candida albicans PKC1* gene encodes a protein kinase C homolog necessary for cellular integrity but not dimorphism. Yeast 30:741–756

Paveto C, Montero L, Passeron S (1992) Enzymatic and immunological detection of G protein alpha-subunits in the pathogenic fungus *Candida albicans*. FEBS Lett 311:51–54

Perez-Martin J, Uria JA, Johnson AD (1999) Phenotypic switching in *Candida albicans* is controlled by a *SIR2* gene. EMBO J 18:2580–2592

Porta A, Ramon AM, Fonzi WA (1999) *PRR1*, a homolog of *Aspergillus nidulans palF*, controls pH-dependent gene expression and filamentation in *Candida albicans*. J Bacteriol 181:7516–7523

Rademacher F, Kehren V, Stoldt VR, Ernst JF (1998) A *Candida albicans* chaperonin subunit (CaCct8) as a suppressor of morphogenesis and Ras phenotypes in *C. albicans* and *Saccharomyces cerevisiae*. Microbiology 144:2951–2960

Ramon AM, Porta A, Fonzi WA (1999) Effect of environmental pH on morphological development of *Candida albicans* is mediated via the PacC-regulated transcription factor encoded by *PRR2*. J Bacteriol 181:7524–7530

Riggle PJ, Andrutis KA, Chen X, Tzipori SR, Kumamoto CA (1999) Invasion lesions containing filamentous forms produced by a *Candida albicans* mutant that is defective in filamentous growth in culture. Infect Immun 67:3649–3652

Roberts RL, Fink GR (1994) Elements of a single MAP kinase cascade in *Saccharomyces cerevisiae* mediate two developmental programs in the same cell type: mating and invasive growth. Genes Dev 8:2974–2985

Roberts RL, Mosch HU, Fink GR (1997) 14-3-3 proteins are essential for RAS/MAPK cascade signalling during pseudohyphal development in *S. cerevisiae*. Cell 89:1055–1065

Roy BG, Datta A (1987) A calmodulin inhibitor blocks morphogenesis in *Candida albicans*. FEMS Microbiol Lett 41:327–329

Rupp S, Summers E, Lo HJ, Madhani H, Fink GR (1999) MAP kinase and cAMP filamentous signalling pathways converge on the unusually large promoter of the yeast *FLO11* gene. EMBO J 18:1257–1269

Ryley JF, Ryley NG (1990) *Candida albicans* – do mycelia matter? J Med Vet Mycol 28:225–239

Sabie FT, Gadd GM (1989) Involvement of a Ca²⁺-calmodulin interaction in the yeast-mycelial (Y-M) transition of *Candida albicans*. Mycopathologia 198: 47–54

Sabie FT, Gadd GM (1992) Effect of nucleosides and nucleotides and the relationship between cellular adenosine 3':5'-cyclic monophosphate (cyclic AMP) and germ tube formation in *Candida albicans*. Mycopathologia 119:147–156

Sadhu C, Hoekstra D, McEachern MJ, Reed SI, Hicks JB (1992) A G-protein alpha subunit from asexual *Candida albicans* functions in the mating signal transduction pathway of *Saccharomyces cerevisiae* and is regulated by the a1-alpha 2 repressor. Mol Cell Biol 12:1977–1985

Santos MAS, Keith G, Tuite MF (1993) Non-standard translational events in *Candida albicans* mediated by an unusual seryl-tRNA with a 5′-CAG-3′ (leucine) anticodon. EMBO J 12:607–616

Saporito SM, Sypherd PS (1991) The isolation and characterization of a calmodulin-encoding gene (*CMD1*) from the dimorphic fungus *Candida albicans*. Gene 106:43–49

Saporito-Irwin SM, Birse CE, Sypherd PS, Fonzi WA (1995) *PHR1*, a pH-regulated gene of *Candida albicans*, is required for morphogenesis. Mol Cell Biol 15:601–613

Sevilla M-J, Odds FC (1986) Development of *Candida albicans* hyphae in different growth media: variation in growth rates, cell dimensions and timing of morphological events. J Gen Microbiol 132:3083–3088

Sharkey LL, McNemar MD, Saporito-Irwin SM, Sypherd PS, Fonzi WA (1999) *HWP1* functions in the morphological development of *Candida albicans* downstream of *EFG1*, *TUP1* and *RBF1*. J Bacteriol 181:5273–5279

Shepherd MG (1985) Pathogenicity of morphological and auxotrophic mutants of *Candida albicans* in experimental infections. Infect Immun 50:541–544

Sherwood J, Gow NAR, Gooday GW, Gregory DW, Marshall D (1992) Contact sensing in *Candida albicans*: a possible aid to epithelial penetration. J Med Vet Mycol 30:461–469

Sobel JD, Muller G, Buckley HR (1984) Critical role of germ tube formation in the pathogenesis of *Candida* vaginitis. Infect Immun 44:576–580

Soll DR (1986) The regulation of cellular differentiation in the dimorphic yeast *Candida albicans*. BioEssays 5:5–11

Sonneborn A, Tebarth B, Ernst JF (1999) Control of white-opaque phenotypic switching in *Candida albicans* by the Efg1p morphogenetic regulator. Infect Immun 67:4655–4660

Sonneborn A, Bockmuhl DP, Gerads M, Kurpanek K, Sanglard D, Ernst JF (2000) Protein kinase A encoded by *TPK2* regulates dimorphism of *Candida albicans*. Mol Microbiol 35:386–396

Staab JF, Bradway SD, Fidel PL, Sundstrom P (1999) Adhesive and mammalian transglutaminase substrate properties of *Candida albicans* Hwp1. Science 283:1535–1538

Staebell M, Soll DR (1985) Temporal and spatial differences in cell wall expansion during bud and mycelium formation in *Candida albicans*. J Gen Microbiol 131:1079–1087

Stanhill A, Schick N, Engelberg D (1999) The yeast Ras/cyclic AMP pathway induces invasive growth by suppressing the cellular stress response. Mol Cell Biol 19:7529–7538

Stewart E, Gow NAR, Bowen DV (1988) Cytoplasmic alkalinization during germ tube formation in *Candida albicans*. J Gen Microbiol 134:1079–1087

Stewart ES, Hawser S, Gow NAR (1989) Changes in internal and external pH accompanying growth of *Candida albicans*: studies of non-dimorphic variants. Arch Microbiol 151:149–153

Stoldt VR, Sonneborn A, Leuker C, Ernst JF (1997) Efg1p, an essential regulator of morphogenesis of the human pathogen *Candida albicans*, is a member of a conserved class of bHLH proteins regulating morphogenetic processes in fungi. EMBO J 16:1982–1997

Swoboda RK, Bertram G, Delbruck S, Ernst JF, Gow NAR, Gooday GW, Brown AJP (1994) Fluctuations in glycolytic mRNA levels during the yeast-to-hyphal transition in *Candida albicans* reflect underlying changes in growth rather than a response to cellular dimorphism. Mol Microbiol 13:663–672

Torosantucci A, Angiolella L, Caccone A (1984) Antimorphogenetic effcts of 2-deoxy-D-glucose in *Candida albicans*. FEMS Microbiol Lett 24:335–339

Ward MP, Gimeno CJ, Fink GR, Garrett S (1995) *SOK2* may regulate cyclic AMP-dependent protein kinase-stimulated growth and pseudohyphal development by repressing transcription. Mol Cell Biol 15:6854–6863

Watts HJ, Very AA, Perera THS, Davies JM, Gow NAR (1998) Thigmotropism and stretch activated channels in the pathogenic fungus *Candida albicans*. Microbiology 144:689–695

Whiteway M, Dignard D, Thomas DY (1992) Dominant negative selection of heterologous genes: isolation of *Candida albicans* genes that interfere with *Saccharomyces cerevisiae* mating factor-induced cell cycle arrest. Proc Natl Acad Sci USA 89:9410–9414

Williams RJ, Dickinson K, Kinsman OS, Bramley TA, Menzies GS, Adams DJ (1990) Receptor-mediated elevation of adenylate cyclase by luteinizing hormone in *Candida albicans*. J Gen Microbiol 136:2143–2148

Wilson RB, Davis D, Mitchell AP (1999) Rapid hypothesis testing with *Candida albicans* through gene disruption with short homology regions. J Bacteriol 181:1868–1874

Wysong DR, Christin L, Sugar AM, Robbins PW, Diamond RD (1998) Cloning and sequencing of a *Candida albicans* catalase gene and effects of disruption of this gene. Infect Immun 66:1953–1961

Zhao X, Malloy PJ, Ardies CM, Feldman D (1995) Oestrogen-binding protein in *Candida albicans*: antibody development and cellular localization by electron immunocytochemistry. Microbiology 141:2685–2692

4 Ions as Regulators of Growth and Development

BRIAN D. SHAW and HARVEY C. HOCH

CONTENTS

I. Introduction

Fungal cells respond to a multitude of environmental conditions for reproduction, dispersal, attachment to a substratum, growth, and in the case of pathogenic species penetration into host tissues. For phytopathogenic fungi, a set of prerequisite conditions must be met before a successful parasitic relationship can be established. The fungal propagule, usually a spore, is dispersed to the host surface and is secured to that surface in a way that optimizes the likelihood of remaining there until germination and growth occur. Optimally, growth is toward the most appropriate site for entry into the host. The environmental conditions that surround the fungal cell and promote these cell responses vary, but include minimally the availability of moisture, light, nutrients, and physical characteristics of the substratum such as wettability and topography. Salts, and more specifically ions, are also important to attachment, germination, growth, and morphological development of the fungal cell. While many studies have addressed, in part, the influence and importance of ions on fungal cell function, relatively few studies have been directed with these effects as the primary focus. Here, we discuss the importance and role of ions in fungal cell biology. Emphasis, although not exclusively so, is placed on fungal cells derived from spores. In addition, our interest centers on the direct consequence of the ions on the cell rather than indirect influences such as generation of ionic gradients within cells or on means of assessing intracellular ionic levels. These topics have been discussed elsewhere (Jackson and Heath 1993; Youatt 1993; Heath 1995; Hyde 1998).

Within this chapter ions will be discussed from the standpoint of their influence on

1. spore attachment to the substratum,
2. initiation of spore germination and subsequent growth, and
3. development of specialized structures such as appressoria and fruit body formation.

A summation of these effects on a number of fungal species is presented in Table 1. Most investigations have emphasized cations with the anionic component receiving little attention as an influencing factor.

II. Attachment

A. Adhesion of Spores

Adhesion of fungal spores is a result of an interaction between the physiochemical nature of both

Department of Plant Pathology, Cornell University, New York State Agricultural Experiment Station, Geneva, New York 14456, USA
Current address: B.D. Shaw, Department of Botany, 2502 Miller Plant Sciences, University of Georgia, Athens, Georgia 30602, USA

The Mycota VIII
Biology of the Fungal Cell
Howard/Gow (Eds.)
© Springer-Verlag Berlin Heidelberg 2001

Table 1. Influence of exogenous ions on attachment, growth and development of fungal cells

Organism	Developmental process	Reference
Alternaria solani	Sporulation	Moretto and Barreto (1995)
Aspergillus sp.	Sporulation	Pitt and Ugalde (1984)
Blastocladiella emersonii	Zoospore encystment and attachment	Soll and Sonneborn (1972)
	Cyst germination	Soll and Sonneborn (1972); Van Brunt and Harold (1980)
Candida albicans	Attachment	O'Shea (1991); Jones and O'Shea (1994)
Colletotrichum gloeosporioides	Spore germination	Kim et al. (1998)
	Appressorium formation	Kim et al. (1998)
Colletotrichum lindemuthianum	Spore attachment	Young and Kauss (1984)
Colletotrichum trifollii	Appressorium formation	Dickman et al. (1995); Warwar and Dickman (1996)
Fusarium graminearum	Hyphal extension and branching, and morphology	Robson et al. (1991a,b)
Metarhizium anisopliae	Spore germination	St. Leger et al. (1989, 1990)
	Appressorium formation	St. Leger et al. (1990, 1991)
Neurospora crassa	Hyphal extension and branching, and morphology	Schmid and Harold (1988); Dicker and Turian (1990)
Penicillium sp.	Sporulation	Pitt and Poole (1981); Pitt and Ugalde (1984); Pitt and Barnes (1993); Roncal et al. (1993); Pascual et al. (1997)
Phyllosticta ampelicida	Spore attachment	Kuo and Hoch (1996); Shaw et al. (1998)
	Germination	Shaw et al. (1998); Shaw and Hoch (2000)
	Appressorium formation	Shaw and Hoch (2000)
Phytophthora cinnamomi	Zoospore encystment and attachment	Gubler et al. (1989); Irving and Grant (1984)
	Cyst germination	Byrt et al. (1982); Irving and Grant (1984)
Phytophthora infestans	Germination and zoosporulation	Hill et al. (1998)
Phytophthora palmivora	Zoospore encystment and attachment	Irving et al. (1984)
	Cyst germination	Grant et al. (1986)
	Appressorium formation	Bircher and Hohl (1999)
Phytophthora parasitica	Zoospore encystment and attachment	Warburton and Deacon (1998)
	Cyst germination	von Broembsen and Deacon (1996); Warburton and Deacon (1998)
Phytophthora sojae	Cyst germination and sporulation	Xu and Morris (1998)
Pythium aphanidermatum	Zoospore encystment and attachment	Donaldson and Deacon (1992)
	Cyst germination	Donaldson and Deacon (1992)
Saccharomyces cerevisiae	Sporulation	Suizu et al. (1994, 1995)
Saprolegnia ferax	Hyphal extension and branching, and morphology	Jackson and Heath (1989)
Sporothrix schenckii	Spore germination	Rivera and Rodriguez (1992)
Trichoderma viride	Sporulation	Krystofova et al. (1995, 1996)
Uromyces appendiculatus	Spore germination	Baker et al. (1987)
	Appressorium formation	Staples et al. (1983); Kaminskyj and Day (1984); Hoch et al. (1987a); Stumpf et al. (1991)
Zoophthora radicans	Appressorium formation	Magalhaes et al. (1991)

the cell surface and that of the substratum. Processes involved in adhesion of spores serve as a basis for understanding how ions might be influential in spore attachment to surfaces, and many excellent reviews have been written on the topic of spore and germling adhesion (Nicholson and Epstein 1991; Hardham 1992; Braun and Howard 1994b; Jones 1994; Epstein and Nicholson 1997; see also Chap. 5, this Vol.) and thus the subject will be discussed briefly only to serve as a background for the processes involved.

Adhesion of microbial propagules to various substrata involves either specific or nonspecific mechanisms. **Specific attachment** occurs when the interaction involves molecular 'lock-and-key' recognitions, e.g., between a ligand and its receptor molecule. Typically, most fungal spores attach to host and artificial substrata *nonspecifically*.

Spores of *Cochliobolus heterostrophus* (Braun and Howard 1994a), *Nectria haematococca* (Kwon and Epstein 1993), and *Magnaporthe grisea* (Hamer et al. 1988), for example, exhibit little preference for the surface on which they will attach – host, non-host, hydrophobic, hydrophilic, etc. Spores of some fungi, however, do attach preferentially to surfaces with specific degrees of wettability, e.g., urediospores of rust fungi (Terhune and Hoch 1993), or with specific surface charge potentials, as discussed below. Furthermore, attachment may be either active or passive (Jones 1994). **Active mechanisms** for attachment occur when the cell is stimulated to produce adhesive polymers such as when zoospores sense and respond to a host surface by encysting (Hardham et al. 1991) or as in spores of *N. haematococca* (Kwon and Epstein 1993), adhesive material is produced in zucchini fruit extract but not in water. For the latter case, the adhesive material is produced within 20 min of exposure to the extract, well before germ-tube emergence (3–7 h). Active production of such adhesives is inhibited by various metabolic factors from living or killed spores. Also, active adhesion occurs as the cell becomes more active metabolically and develops a pronounced extracellular matrix (ECM) as it initiates development of a germ tube and begins growth on a substratum. **Passive attachment** occurs when preformed molecules on the spore surface affix the cell to a substratum. Packaging of such 'glues' may be in specially positioned compartments as in the "spore tip mucilage" of *M. grisea* (Hamer et al. 1988), in 'sheaths' that surround the spore as on pycnidiospores of *Phyllosticta ampelicida* (Kuo and Hoch 1995; Shaw et al. 1998), or on specialized appendages (Jones 1994). Passive attachment is often rapid and is not metabolically dependent. As noted below, cations have in some instances pronounced temporal effects on spore attachment.

Hydrophobicity and hydrophilicity, viz. **wettability**, of the spore surface and the substratum frequently dictate whether or not spores adhere (passively), much less whether attachment is weak or firm. Wettability of spore surfaces is influenced, in part, by the chemical nature of the spore surface, typically including a complex of proteins, e.g., the cysteine-rich hydrophobin polypeptides (Wessels 1996), and carbohydrates, usually as glycoproteins. Ions influence the hydrophobic domain surrounding these molecules. Addition of ions (increase in ionic strength of the surrounding

solution) increases hydrophobic adhesion not only because electrostatic interactions are suppressed (Ochoa 1978), but also because the surfaces become less polar as the ordered layer of water molecules is released into the 'bulk' aqueous phase (Rutter and Vincent 1984; Rosenberg and Kjelleberg 1986) leading to an increase in entropy. It is not surprising then that the addition of cations (e.g., Ca^{2+}, H^+) to spore suspensions of *P. ampelicida* enhances both the rate and tenacity of attachment to hydrophobic substrata (Kuo and Hoch 1996; Shaw and Hoch 1999), as discussed later. Assessment of fungal spore attachment to substrata of varying wettabilities has been reported for numerous species. On the other hand, where a colloid, bacterium, spore, etc. and the substratum are hydrophilic, short-range repulsive forces prevent attachment even with the addition of electrolytes (Harding 1971; Rutter and Vincent 1984). Many fungal spores attach well to both hydrophobic and hydrophilic surfaces, e.g., *M. grisea* (Hamer et al. 1988) and *Phytophthora cinnamoni* (Gubler et al. 1989), and some only or preferably to hydrophobic surfaces, e.g., *Uromyces appendiculatus* (Terhune and Hoch 1993), *Botrytis cinerea* (Doss et al. 1993), *Colletotrichum graminicola* (Mercure et al. 1994), and *Colletotrichum musae* (Sela Burrlage et al. 1991). Where the role of ionic influence on wettability has been investigated, attachment of spores was greatly enhanced. However, a simple concept dictating an increase in entropy upon ionic elevation was more often not the case as other factors must also be taken into account, e.g., how ions affect surface potential and counter repulsive effects of like-charged bodies.

B. Surface Potential and the Mechanisms Influencing Attachment

Most fungal propagules possess net negative charges (Kennedy 1991; O'Shea 1991; Pendland and Boucias 1991; Jones and O'Shea 1994; Kuo and Hoch 1995) due to the ionization of surface groups, e.g., carboxyls, glycolipids, acidic amino acids. Attachment of such propagules likely involves, in part, the physiochemical forces of repulsion/attraction derived from electrostatic and van der Waals forces between the surface charges of the two bodies (spores and the substratum). Together, the attraction-repulsion forces follow the lyophobic colloid theory referred to as the DLVO theory (Dejaguin and Landau 1941; Verwey and Overbeek 1948). It is, in part, the additive effects of the electrostatic repulsive forces and the attraction of the van der Waals forces that determine adhesion of colloids (or spores). Such forces are operative at distances of 1–10 nm, while the hydrophobic forces discussed above influence adhesion at greater distances (Gristina 1987; Israelachvili and McGuiggan 1988). According to DLVO theory, initial 'attachment' of a colloid, or in this case a spore, is weak, tenuous, and time-dependent (Rutter and Vincent 1984). The interaction between a charged spore and the substratum is such that at low electrolyte concentration a significant free energy barrier needs to be overcome before the spore is able to contact the surface. However, if the electrolyte level is increased the energy barrier is effectively depleted and a strong net attraction occurs between the spore and the substratum. Such enhanced contact and adhesion of fungal propagules with various charged substrata has been noted in a number of studies (e.g., Young and Kauss 1984; Jones and O'Shea 1994; Kuo and Hoch 1996; Shaw and Hoch 1999, 2000). Once the proximity of the spore to the substratum is determined by the physiochemical nature of the two interacting surfaces, biological glues of the propagule will assure attachment to the substratum.

Clearly, many factors mediate attachment of spores to substrata: wettability and surface potentials of both the substratum and of the spore, electrolyte species and concentration, adhesive nature of the spore 'glue' as well as possibly spore size. All are important in determining the proximity to which the spore reaches the substratum.

C. Ions as Regulators of Spore Attachment

Ionic mediation of fungal spore attachment to substrata has been reported for a number of genera. Acidification or elevation of the ionic concentration of media enhanced the rate at which conidia of *Phyllosticta ampelicida* attached to hydrophobic substrata (Kuo and Hoch 1996; Fig. 1), as well as the number of conidia that attached (Shaw and Hoch 1999). In the latter study, it was also noted that the addition of cations to a suspension of conidia induced attachment, albeit weakly, to hydrophilic surfaces. Normally, spores of *P. ampelicida* do not adhere to

hydrophilic surfaces when in ddH$_2$O or solutions of low ionic strength. This effect was valence-dependent, with La^{3+} inducing higher levels of attachment than Ca^{2+} or Mg^{2+}, which in turn mediated higher levels of attachment than K$^+$ or Na$^+$ (Fig. 2). In that study Cl$^-$ was the anion common to all cations tested with the conclusion that it was not directly effecting attachment. Similarly, attachment of *Colletotrichum lindemuthianum* conidia is also enhanced in solutions with elevated levels of various cations including, Na$^+$, K$^+$, Mg^{2+}, and Ca^{2+}, although no difference was noted between monovalent and divalent cations (Young and Kauss 1984). *Discula umbrinella* conidium attachment is pH-dependent. A low pH (4–5), viz. high H$^+$ concentrations, promoted maximum attachment while increasing pH disrupted attachment (Toti et al. 1992). Similarly, adhesion of *Candida albicans* yeast cells is also mediated by the addition of cations to the solution. In these studies divalent cations were 10–100 times more effective than monovalent cations in inducing attachment (O'Shea 1991). This effect was due to a reduction in the electrostatic potential of the *C. albicans* cell (Jones and O'Shea 1994). Acidification of the media also efficiently increased cell attachment. Although anions are frequently overlooked, it is significant that, by varying the anion valency using Cl$^-$ and SO$_4^{2-}$, no difference was noted (Jones and O'Shea 1994).

Cationic enhancement of attachment as described above is likely due, in part, to the attenuation of negatively charged groups on the surface of both spore and substratum. Together these systems exhibit attachment properties that are influenced as much by the physiochemical nature of the spore and of its substratum than by a more specific receptor ligand binding. An additional mechanism of cationic-enhanced attachment is suggested by adhesion of *Arthrobotrys oligospora* to its nematode host (Tunlid et al. 1992), and by zoospore cyst attachment in *Phytophthora cinnamomi* (Gubler et al. 1989). Attachment in each of these cases may involve a Ca^{2+}-mediated reorganization of carbohydrate-containing polymers on the spore surface. A discussion of this and other possible mechanisms for cation-mediated attachment is detailed below.

In the Oomycota adhesion of encysted zoospores is clearly enhanced by cationic amendment, particularly Ca^{2+}. In *Blastocladiella emersonii*, K$^+$, Na$^+$, Rb$^+$, Mg^{2+}, and Ca^{2+} stimulate

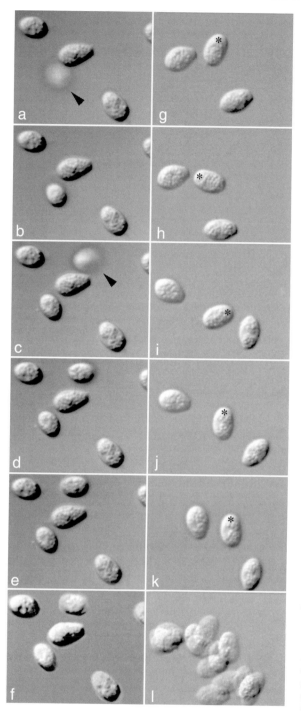

encystment of zoospores (Soll and Sonneborn 1972). Encystment of *B. emersonii* zoospores is preceded by a K^+-mediated depolarization of the plasma membrane (Jen and Haug 1981). The effect of many cation species on encystment of *Phytophthora cinnamomi* zoospores has been tested (Byrt et al. 1982; Irving and Grant 1984), with Ca^{2+} being most effective in inducing encystment at millimolar levels. Sr^{2+} also induced encystment. The monovalent cations Na^+ and Cs^+ had negligible effects on encystment; however, while K^+ did induce encystment, it became increasingly inhibitory to germination as its concentration increased (Byrt et al. 1982). In the latter case, it was noted that Ca^{2+} could override this 'germination' toxicity of K^+ (Irving and Grant 1984). The K^+ ionophore, valincomycin, and the Ca^{2+} ionophore, A23187, both induced encystment indicating that one or both cations are likely involved in the encystment process in *P. cinnamomi*. In one of the few studies that also analyzed anionic effects, ions Cl^-, NO^{3-}, SO_4^{2-}, and PO_4^{3-} were all tested as Na^+ salts and no difference in encystment induction was found in *P. cinnamomi* (Byrt et al. 1982). In another study, increasing the concentration of Ca^{2+} from 2 to 20 mM enhanced attachment of *P. cinnamomi* cysts, while chelation of Ca^{2+} with EGTA completely inhibited cyst adhesion (Gubler et al. 1989). Similarly, adhesion of *Pythium aphanidermatum* cysts was found to be dependent on Ca^{2+} availability, although Mg^{2+} and Sr^{2+} were also able to induce encystment. All other tested cations were either ineffective or toxic (Donaldson and Deacon 1992). This observation is

Fig. 1. Behavior of *Phyllosticta ampelicida* conidia settling onto hydrophobic (**a–f**) and hydrophilic (**g–l**) substrata as observed using an inverted light microscope. Spores in acidified (pH 4) water attach immediately (<0.03 s) upon contact with a hydrophobic surface. **a** Four spores are already attached and immobile with one "out-of-focus" spore (*arrowhead*) about to make contact. In **b** and **c** it makes contact and attaches immediately. A second spore (*arrowhead*) is about to contact the surface in frame **c**. Contact and attachment of the spore is noted in **d–e**. **f** A composite overlay of frames **a–e** illustrates the non-motile characteristic of attached conidia. Times for frames **a–e** are 03:23, 03:23, 04:08, 04:13, 04:15 (s:tenths s), respectively. Conidia in frames **g–k** on highly hydrophilic heat-treated glass do not attach (*asterisk* denotes the same spore). **l** A composite overlay of frames **g–k** illustrates that the conidia did not attach. Frames **g–k** taken at 10-min intervals. (Adapted from Kuo and Hoch 1996)

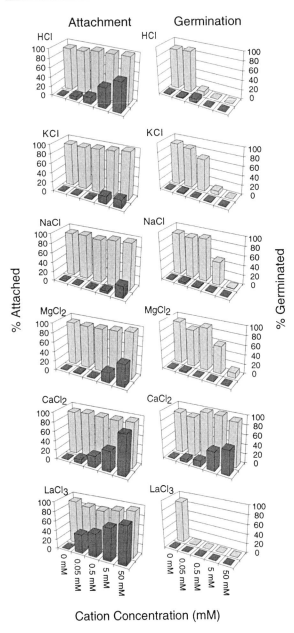

Fig. 2. Attachment and germination of pycnidiospores of *Phyllosticta ampelicida* as influenced by various cations of chloride salts. In distilled H_2O neither attachment nor germination occurred on hydrophilic glass substrata, but both occurred at high levels on hydrophobic polystyrene. Efficiency of attachment to hydrophilic substrata was a direct function of cation concentration and valence. All tested cations were inhibitory to germination, however, except Ca^{2+}. (Adapted from Shaw and Hoch 1999)

adhesion and they supported the earlier assertion of Irving and Grant (1984) that the monovalent cation Na^+ had negligible effects on cyst attachment. Additionally, transmembrane Ca^{2+} fluxes were associated with zoospore encystment of *Phytophthora parasitica*. Ca^{2+} from both internal stores and the external environment was essential for zoospore encystment, as determined by the use of channel blockers (lanthanum, verapamil, and TMB-8), the fluorescence indicator Fura-2, and by assessing influx and efflux of $^{45}Ca^{2+}$ (Warburton and Deacon 1998).

D. Mechanisms of Ionic Attachment Induction

The mechanism by which cations in the extracellular environment influence spore attachment is not well understood, in part because there are a multitude of interactions possible (viz. electrostatic, van der Waals forces, etc.) in addition to compounding associations of surface wettability and ligands. It is much more difficult to separate these effects in biological systems where surfaces are complex, compared to simplified inert colloidal systems, and arrive at simplified explanations. It is clear, however, that the presence of cations can induce attachment of many microorganisms to the substratum. Possibilities for the role of ions in attachment include (1) attenuation of the electronegative surface groups on the propagules and substratum (Fletcher 1980; Jones and O'Shea 1994); (2) divalent or trivalent ions bridging two negatively charged groups on the spore and the substratum; in the case of pH-dependent effects, (3) the inherent electronegativity of surface components may be attenuated by protonation of negatively charged groups, thus reducing electrostatic repulsive forces and enhancing attachment (O'Shea 1991; Jones and O'Shea 1994); (4) free cations may also chemically alter the adhesive material which mediates adhesion of many propagules, as suggested in the nematode trapping fungus *A. oligospora* (Tunlid et al. 1991) or in marine algae where cations have been shown to increase the adhesiveness of the spore glue (Cooksey 1981); (5) free ions may effect the thickness of the electrical double layer between adhering surfaces (Young and Kauss 1984; Jones 1994); and finally, (6) the hydrophobicity of the surface may also be attenuated by free ions in the media (Jones 1994).

most interesting considering that *Phytophthora palmivora* zoospores were found to release Ca^{2+} just prior to encystment (Irving et al. 1984). Gubler and colleagues also noted that Mg^{2+} was capable of inducing an eightfold increase in cyst

III. Germination and Apical Growth

A. Germination and Apical Extension

Once the fungal spore has been deposited in a new locale away from 'sister spores', sori and conidiogenous hyphae that may harbor self-germination-inhibiting substances, it has the opportunity to initiate restoration of metabolic and physiological activity, a process normally considered a prelude to germination. In the strictest sense germination begins when the resting stage (dormancy) ends and not when a germ tube is first visible. After all, at what level of resolution or detection should we consider a germ tube visible? Certainly other prerequisite conditions (or lack of) such as light, water, nutrients, aeration, etc. need to be satisfied as well. As will become evident, exogenous ions are frequently equally important in the germination process and in continued extension of the cell.

B. Influence of Ions on Germination

Unlike the studies of ionic influence on adhesion, much of the evidence concerning ions and germination focuses on Ca^{2+}. This is due primarily to the ubiquitous role that Ca^{2+} plays in signaling pathways in eukaryotes (Pitt and Ugalde 1984; Clapham 1995; Rudd and Franklin-Tong 1999; Sanders et al. 1999). Despite the ubiquity of the role of Ca^{2+} in a number of developmental processes, some fungal systems tested thus far require Ca^{2+} availability for germination, while others seem to germinate in its absence, at least to the extent of it not being added experimentally.

Many cations enhanced attachment of *P. ampelicida* conidia to a substratum; however, all except Ca^{2+} were inhibitory of germination (Shaw and Hoch 1999) (Fig. 2). Availability of external Ca^{2+} was essential for normal germination of *P. ampelicida* conidia. Using a Ca^{2+}-EGTA buffering system (Wayne 1985) to control the amount of free Ca^{2+} (0.1 nM–1 mM) in the external medium, at $10\,\mu M$ or higher the cation was found to be essential for normal germination to ensue (Shaw and Hoch 2000; Fig. 3). The presence of Ca^{2+} was required for the first 25–60 min following spore attachment to hydrophobic substrata. Conidia of *P. ampelicida* germinate only after they become attached to a hydrophobic surface. Attachment

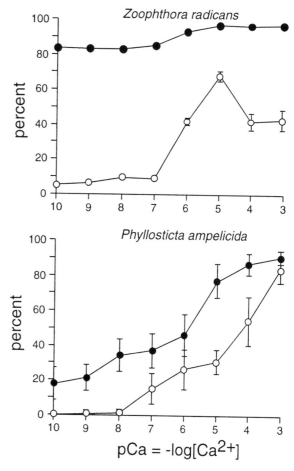

Fig. 3. Comparison of germination (●) and appressorium formation (○) in two fungal systems, *Zoophthora radicans*, adapted from Magalhaes et al. (1991) and *Phyllosticta ampelicida*, adapted from Shaw and Hoch (2000). Both studies used the Ca^{2+} buffering systems developed by Wayne (1985). Germination in *Z. radicans* was unaffected by external free Ca^{2+} while *P. ampelicida* germination required external free Ca^{2+}. Appressorium formation was dependent on availability of about pCa 7 in both systems

was a prerequisite for initiation of germination while availability of nutrients, normally considered adequate for germination of most fungal spores, had no influence on this fungus (Kuo and Hoch 1996). Some pharmacological agents known to block membrane resident ion channels, e.g., nicardipine and Gd^{3+}, effectively abolished germination of conidia. Additionally, calmodulin antagonists, compound 48/80 and compound R-24571 completely inhibited germination. Taken together, these data suggest that Ca^{2+} and Ca^{2+}-signaling pathways are essential for *P. ampelicida* pycnidiospore germination.

Ca^{2+}-calmodulin signaling pathways have been implicated in germination of *Metarhizium anisopliae* (St. Leger et al. 1990). Germination occurred in media containing 2 mM EGTA, viz., low levels of Ca^{2+}; however, when the ionophore A23187 was added, germination did not occur. This suggests that internal stores of Ca^{2+} are sufficient to allow for germination. Unfortunately, St. Leger and colleagues did not test internal Ca^{2+} channel blockers such as TMB-8 to ascertain this hypothesis. The plasma membrane Ca^{2+} channel blocker La^{3+}, as well as calmodulin antagonists, disrupted germination by interrupting normal protein phosphorylation (St. Leger et al. 1989). Similarly, in *Colletotrichum gloeosporioides* conidia, inhibition of calmodulin and calmodulin kinase associated processes with compound 48/80 or KN93, respectively, reduced germination (Kim et al. 1998). In the same study, chelation of Ca^{2+} with EGTA also reduced germination, as did U73122 inhibition of phospholipase C. Germination of *Sporothrix schenckii* conidia is also dependent on external Ca^{2+} availability or pharmacological stimulation of the calcium signaling pathway component protein kinase C (Rivera and Rodriguez 1992). Additionally, germination of *Uromyces* urediospores is enhanced with either Ca^{2+} or Mg^{2+}, but neither K^+ nor Na^+ was effective in stimulating germination in water purified with an ion-exchange resin (Baker et al. 1987). It seems likely that, at least for some fungal systems, Ca^{2+} dynamics are important regulators of spore germination.

In contrast to the previous discussion, external Ca^{2+} appears to have little or no role in germination in *Colletotrichum trifolli* (Dickman et al. 1995; Warwar and Dickman 1996). Neither Ca^{2+} chelation with EGTA, nor the use of the ionophore A23187, affected germination. This is surprising since both compounds are effectively toxic at high levels in most cell systems studied. It is possible that a more carefully constructed Ca^{2+} buffering system may effectively influence germination, possibly even at low Ca^{2+} concentrations. Additionally, various channel blockers had little effect on germination of *C. trifollii* conidia (Warwar and Dickman 1996) or the rate of germination in *Zoophthora radicans* (Magalhaes et al. 1991). In *Z. radicans* an EGTA buffering system was used to reduce Ca^{2+} to nM levels. Neither this treatment nor the use of channel blockers reduced germination levels appreciably, although multipolar germination was reported when Nd^{3+} was used to inhibit Ca^{2+} entry into the spores. Appressorium formation, however, was greatly effected in both

of these systems (see below). It should be noted that calmodulin inhibitors disrupted germination in both systems. Thus, Ca^{2+} regulation of germination in these systems remains unclear.

In the Oomycota, direct and indirect germination have been found to be influenced by ions, particularly Ca^{2+}. Germination of *Phytophthora infestans* sporangia is associated with Ca^{2+} availability (Hill et al. 1998). Chelation of divalent ions with either 1,2-bis(2-aminophenoxyethane-*N,N,N',N'*-tetraacetic acid (BAPTA) or ethylene glycol bis(2-aminoethyl ether)-*N,N,N',N'*-tetraacetic acid (EGTA) suppressed germination of *P. infestans* sporangia, as did treatment with the Ca^{2+} channel blockers lanthanum and gadolinium. Ca^{2+} was generally more effective than Mg^{2+} in overcoming these treatments, but levels of either cation above 1 mM inhibited germination. The developmental fate of germinating *Phytophthora sojae* cysts was dependent on Ca^{2+} availability (Xu and Morris 1998). If >10 mM Ca^{2+} was supplied to cysts, direct germination, i.e., emergence of a germ tube, took place. If, however, <10 mM Ca^{2+} was provided to cysts, indirect germination (i.e., zoospore release) occurred. No other tested ion, including Mg^{2+}, Mn^{2+}, K^+, and Na^+, had this effect on the cysts. Application of the Ca^{2+} channel blocker, verapamil, to *P. sojae* cysts also decreased direct germination and increased secondary zoospore release (Xu and Morris 1998). *Phytophthora parasitica* encysted zoospores lose their ability to germinate if they are diluted in H_2O. However, if 50 mM $CaCl_2$ is added to the cysts, germination returns to normal levels (von Broembsen and Deacon 1996). Additionally, transmembrane Ca^{2+} fluxes were associated with cyst germination of *Phytophthora parasitica* (Warburton and Deacon 1998). The cations Na^+, Ca^{2+}, and Sr^{2+} induced both encystment and germination of *Phytophthora palmivora* (Grant et al. 1986). However, in a later report, Grant and colleagues confirmed a requirement for Ca^{2+} in germination but could not confirm the requirement for Na^+ (Iser et al. 1989). Ca^{2+} from both internal stores and the external environment was essential for germination, as determined by the use of channel blockers (lanthanum, verapamil, and TMB-8), by the fluorescence indicator Fura-2, as well as influx and efflux dynamics using $^{45}Ca^{2+}$. Warburton and Deacon demonstrated a net Ca^{2+} influx during zoospore encystment followed by a net efflux, from internal stores, during germination. While any tested cation was effective in inducing encystment of *Phytophthora cinnamomi* zoospores, only Ca^{2+} induced germination (Byrt et al. 1982). None of the anions Cl^-, NO^{3-}, SO_4^{2-}, and PO_4^{3-} tested as Na^+ salts showed a detectable difference in induction of germination. In a later study, Ca^{2+} was shown to be essential for germination of *P. cinnamomi* cysts (Irving and Grant 1984).

In *Pythium aphanidermatum*, germination was enhanced by Ca^{2+}, as well as Mg^{2+} and Sr^{2+}, while all other tested ions, including K^+, Li^+, Na^+, Ba^{2+}, Mn^{2+}, Fe^{3+}, and Cu^{3+}, either had no effect on germination or were toxic (Donaldson and Deacon 1992). Ca^{2+} flux is essential for both encystment and germination in this system, as Ca^{2+} is released upon encystment and taken up during germination. The results of these studies led Donaldson and Deacon (1992) to propose an "auto-signaling" hypothesis for a role of Ca^{2+} in encystment and germination of zoospores, and to explain why cysts that maintain contact with a substratum germinate at much higher levels than those kept in suspension. Ca^{2+} secreted by encysting zoospores (Irving et al. 1984; Iser et al. 1989) enhances attachment, possibly through ionic bridging or polymerization of a surface molecule. More recent data suggests that Ca^{2+} is not secreted until the early stages of signaling for germination (Warburton and Deacon 1998). At that time, Ca^{2+} accumulates between the cyst and the substratum and is available for re-absorption by the cyst, possibly enhancing germination. It is clear that a flux of Ca^{2+} is essential for both encystment and germination of zoospores in many of the Oomycota. Knowledge of the role of calcium in encystment and germination in the Oomycota has led to at least one disease control recommendation (von Broembsen and Deacon 1997). Addition of 10 mM Ca^{2+} to an irrigation system interfered with the normal development of *Phytophthora parasitica* by inducing encystment and germination, without a corresponding shift to indirect germination and zoospore production.

Exogenously supplied K^+, Na^+, Rb^+, Mg^{2+}, and Ca^{2+} were demonstrated to stimulate germination of encysted *Blastocladiella emersonii* zoospores (Soll and Sonneborn 1972; Van Brunt and Harold 1980). The germination stimulus by K^+ may be the result of the depolarization of the membrane by its rapid uptake as determined by reversal of quenching of the fluorophore DiO-C$_6$ as well as the accumulation of $^{42}K^+$ (Van Brunt and Harold 1980). Using $^{45}Ca^{2+}$, it was shown that Ca^{2+} is not taken up by the spores, but likely binds to exter-

nal sites. In the same study it was shown that *B. emersonii* zoospores required Ca^{2+} and K^+ to germinate, but were inhibited from germinating (10% germination) if Na^+ was present. Germination of both *Aphanomyces astaci* and *A. euteiches* cysts was also induced by the addition of various cations, including K^+ and Ca^{2+} (Svensson and Unestam 1975; Deacon and Saxena 1998).

C. Ion Regulation of Growth Direction, Rate, and Morphology

Like studies of germination, work on the role of exogenously applied cations in growth and branching of hyphae has focused primarily on Ca^{2+}, and include studies with *Fusarium graminearum* (Robson et al. 1991a,b), *Neurospora crassa* (Schmid and Harold 1988), and *Saprolegnia ferax* (Jackson and Heath 1989). In *F. graminearum* the rate of hyphal extension and the amount of branching were both effected directly by the external Ca^{2+} concentration (Robson et al. 1991a,b). Hyphal growth rate increased as Ca^{2+} concentration increased from 48 nM to 10 μM, but concentrations >10 μM had no further effect. In addition, those mycelia that grew in <10 μM Ca^{2+} media had an altered growth habit with increased branching and bulbous-like hyphae. Growth rate decreased and branching increased by application of various Ca^{2+} channel blockers (Robson et al. 1991a). In addition, apical extension in *Neurospora crassa* (Schmid and Harold 1988) and in *Saprolegnia ferax* (Jackson and Heath 1989) was impaired by a decrease in external Ca^{2+} concentration. In both cases growth was accompanied by abnormal bulbous-like growth. Branching was also enhanced in *N. crassa* when grown in Ca^{2+}-depleted media or when exposed to Ca^{2+} channel blockers (Dicker and Turian 1990). In *N. crassa* the rate of apical extension was affected more strongly by a reduction in external, wall-bound $^{45}Ca^{2+}$ than by cytoplasmic $^{45}Ca^{2+}$ (Schmid and Harold 1988).

The mechanism(s) of Ca^{2+} influence on growth rate and branching is not known but may be due to intercellular influence on a cell 'tip-high' gradient or effects on apically localized cytoskeleton, vesicle distribution, or wall assembly (Heath 1995; Robson et al. 1991b). As models have been developed for the role(s) that these extracellular ions play in growth and branching, and methods of monitoring intercellular Ca^{2+} have improved, much work has focused on the possibility of a tip-high Ca^{2+} gradient directing such tip growth as well as initiating branches (Jackson and Heath 1993; Heath 1995; Grinberg and Heath 1997). Similarly, indirect effects of Ca^{2+} may influence cytosolic pH. However, controversy remains as to the role for both pH and Ca^{2+} tip-high gradients in fungal growth and branching, much less whether the apparent gradients are functional, or even exist, in physiologically normal cells (Dean 1997; Parton et al. 1997). Resolution to this quandary will rely in part on establishing certainty of the cytosolic Ca^{2+} status, viz., that means of detecting and measuring Ca^{2+} are truly doing so for cytosolic Ca^{2+} and not for Ca^{2+} bound within organelles. For more thorough treatments of internal ionic gradients influencing apical growth, as well as electrical dimensions of hyphal growth, several excellent reviews are available (Harold 1994; Heath 1995; Jackson and Heath 1993).

IV. Morphological Specialization

As in cell germination and growth, exogenous ions affect fungal cell differentiation. A variety of specialized structures including, but not limited to, production of asexual and sexual reproductive structures and their associated propagules (conidia, zoospores, etc.), morphologically distinct structures aiding ingress of the fungus (pathogen) into its host, or in the adsorption of nutrients (appressoria, hyphopodia, haustoria, etc.), are triggered by the presence of certain ions. In some situations the absence of ions are similarly stimulatory. Again, in the literature there appears to be an emphasis on calcium as the dominantion affecting these events, perhaps because more methods have been developed for its detection and monitoring, or, more likely, because calcium is one of the most important molecules in ionic signaling pathways.

A. Conidiation

A number of filamentous fungi reportedly initiate sporulation upon exposure to elevated levels of ions, notably calcium. Often, conidiogenous fungi sporulate under conditions of duress such as nutrient depletion and mechanical perturbation, or under certain light regimes. Conidiation in sub-

merged culture is rare among fungi; however, many species of *Penicillium* and a few *Aspergillus* species sporulate in submerged culture when nutrients are depleted (Pitt and Ugalde 1984). When nutrients are plentiful conidiation does not occur. Species including *P. notatum*, *P. griseofulvum*, *P. urticae*, *P. oxalicum*, and *P. cyclopium* have been observed to sporulate when supplied elevated levels of calcium (1–10 mM), even in nutritionally adequate media (Hadley and Harrold 1958a,b; Pitt and Poole 1981; Roncal et al. 1993; Pascual et al. 1997). The signaling process is not clearly understood, although it was demonstrated that under these conditions relatively short exposures to calcium (0.5–2 min) were all that were required for the process to be initiated, that very little calcium was taken up into the cytosol, and that considerable binding of calcium to the plasma membrane occurred (Ugalde and Pitt 1986; Ugalde et al. 1990). In addition, exposing cells to the chelating agent BAPTA within 2 h (Ugalde et al. 1990) can reverse calcium-induced conidiation. This suggests that calcium is bound primarily to the extracellular domains of the cell, where it possibly signals the conidiation process. Roncal and coworkers (1993) provide compelling evidence that calcium-induced alkalinization of cell apices triggers conidiation in *P. cyclopium* (Roncal et al. 1993). Alternatively, Pitt and Barnes associated calcium-induced conidiation with signaling via calmodulin and protein phosphorylation pathways (Pitt and Barnes 1993). The topic of signal transduction pathways and conidiation has been reviewed (Pitt and Kaile 1990).

Calcium-induced sporulation has also been reported in other fungi. In *Trichoderma viride* calcium supplied in millimolar quantities promoted conidiation as well as growth (Krystofova et al. 1995, 1996). When calcium was sequestered with EGTA the rate of conidiation was reduced significantly, but so was the rate of growth. This may indicate that cell health may have been compromised by the chelation treatment. Enhanced sporulation of *Alternaria solani* has been reported in culture media supplemented with calcium carbonate (Moretto and Barreto 1995); however, since several other variables were introduced into the experimental design of this study, it is difficult to assign a direct effect of calcium on sporulation. In *Fusarium graminearum* formation of conidia was reduced by 98% in 14 nM Ca^{2+}, compared to those formed in 50 μM Ca^{2+} (Robson et al. 1991b). In a diploid strain of *Saccharomyces cerevisiae*, meiotic sporulation (ascospore production) was induced when free external Ca^{2+} was reduced and internal cell levels were increased to a 3000× differential (Suizu et al. 1995). It was further determined that an influx of external Ca^{2+}, rather than a release of Ca^{2+} from internal stores, initiated meiosis and sporulation in this yeast. In contrast, no marked changes were found in the external or internal levels for other ions, including, Mg^{2+}, Na^+, or K^+. Additionally, sporulation was completely repressed when media containing no detectable Ca^{2+} or Mg^{2+} (determined by HPLC) were used (Suizu et al. 1994).

Sodium has been implicated in the promotion of conidiation in strains of *Aspergillus nidulans* bearing abnormal distributions of nuclei in most cells including hyphae, phialides, conidia, etc. (Queiroz and Azevedo 1998). When high levels of NaCl (0.5 M) are incorporated into the growth medium, conidiation is significantly increased, although no effect was noted on nuclear distribution.

B. Zoosporulation and Oosporulation

Sporulation, both asexual and sexual, in the Oomycota is also enhanced by, and in some case is dependent on, available Ca^{2+}. In *Saprolegnia* (Fletcher 1979), *Achlya* (Griffith et al. 1988), *Pythium* (Lenny and Klemmer 1966) and *Phytophthora* (Elliot 1972), Ca^{2+} has been implicated in formation of reproductive structures. *Saprolegnia terrestris* oogonial initials have an increased rate of abortive oogonia under Ca^{2+}-deficient conditions. Ultrastructural analysis showed a decreased density of lipid bodies and an increased density of mitochondria under these conditions (Fletcher 1983). In *Phytophthora palmivora* (Elliott 1986) and *P. cactorum* (Elliott 1988), trifluoperazine, an inhibitor of calmodulin, inhibited oosporogenesis and zoosporogenesis. *Aphanomyces euteiches* oospore formation is also thought to be dependent on Ca^{2+} availability (Yokosawa et al. 1995). In contrast to these examples, increasing Ca^{2+} availability (Fletcher 1988) reportedly reduces oogonium development in *Saprolegnia diclina*.

Micromolar levels of Ca^{2+} are required for sporangium development in *Achlya* sp. (Griffin 1966). Chelation of external Ca^{2+} with EGTA to levels below $0.1\,\mu M$ blocked sporulation of *Achlya bisexualis* (Thiel et al. 1988). Additionally, the Ca^{2+}

channel blockers lanthanum and gadolinium, at 20 and $50\,\mu M$, respectively, blocked sporangium development or sporulation in *A. bisexualis* depending on the developmental maturity at the time of application. Sporulation in sporangia of *Phytophthora infestans* is associated with Ca^{2+} availability (Hill et al. 1998). Either BAPTA or EGTA chelation suppressed zoospore development in *P. infestans* sporangia. Lanthanum and gadolinium also suppressed zoospore formation that could be overcome by addition of Ca^{2+}, and in some cases Mg^{2+}, but levels of either cation above 1 mM inhibited development. In another report, zoospore release from *P. infestans* cysts was stimulated by 0.3 mM Ca^{2+}, whereas Mg^{2+}, K^+, and Na^+ were each less effective at stimulating zoospore release (Sato 1994).

The release of secondary zoospores from an encysted zoospore, i.e. indirect germination, may also be influenced by Ca^{2+} dynamics. In at least six species of *Aphanomyces* secondary zoospore emergence was monitored in relation to Ca^{2+} dynamics (Cerenius and Soderhall 1985). If zoospores were exposed to 50 mM Ca^{2+} during the first 15 min of encystment, the cysts developed a germ tube and differentiated by direct germination. If, however, the zoospores were induced to encyst by physical agitation, and were denied Ca^{2+}, they subsequently germinated indirectly. It seems that, at least in *Aphanomyces*, indirect germination is a default pathway that can be disrupted by Ca^{2+}. Similarly, in *Phytophthora sojae*, <10 mM free Ca^{2+} stimulated secondary zoospore release, while >10 mM Ca^{2+} stimulated direct germination (Xu and Morris 1998).

C. Infection-Related Structures

Phytopathogenic fungi often gain access to their plant host through appressoria formed at germ-tube or hyphal apices (Hoch and Staples 1987; Emmett and Parbey 1975; Dean 1997). For most of these fungi, it is not clear what on the host surface serves to signal appressorium development; however, there are many 'all encompassing' vague features that have been identified including contact with a 'hard surface', a hydrophobic surface and in some instances a hydrophilic surface, etc. (Lee and Dean 1994; Xiao et al. 1994; Nicholson and Kunoh 1995; Beckerman and Ebbole 1996; Gilbert et al. 1996; Perfect et al. 1999). In rust fungi, urediospore germlings sense sub-micron topographical features that trigger appressorium development (Hoch et al. 1987b; Allen et al. 1991). Even so, the next level of signal perception beyond these surface features remains elusive, although there is some evidence for specific signaling pathway involvement, e.g. cAMP (Lee and Dean 1993; Yang and Dickman 1997; Choi et al. 1998). Similarly, ionic signaling mechanisms may be operative in appressorium induction of some fungal systems. Again, as already discussed, the ion that has received the most attention is Ca^{2+}.

Concentrations of Ca^{2+} >1 μM, but not K^+ or Cl^-, are required for appressorium formation in *Zoophthora radicans* (Magalhaes et al. 1991). Similarly, appressoria did not form on germlings of *Phyllosticta ampelicida* in a buffer containing <1 μM Ca^{2+} (Fig. 3). Appressorium formation approached 100% when 1 mM Ca^{2+} was available (Shaw and Hoch 2000). In both fungal systems, Ca^{2+} channel blockers indicated that both external and internal sources of Ca^{2+} were required for normal appressorium development. Additionally, calmodulin antagonists completely abolished appressorium formation, suggesting that Ca^{2+} signaling pathways may be involved in initiating differentiation in each system. Similarly, La^{3+} and Gd^{3+} disrupted appressorium formation in *Metarhizium anisopliae* (St. Leger et al. 1991), although earlier work by the same group showed that external Ca^{2+} was not required for normal appressorium formation (St. Leger et al. 1990). St. Leger and colleagues believe that it is the tip-high Ca^{2+} gradient that maintains polarity, and that disruption of this gradient is intimately involved in initiation of appressorium formation (St. Leger et al. 1991).

The addition of EGTA to the medium surrounding *Colletotrichum trifolii* reduced appressorium formation in a concentration-dependent manner (Dickman et al. 1995; Warwar and Dickman 1996). Additionally, the ionophore A23187 reduced appressorium formation. These data indicate that Ca^{2+} is likely involved in appressorium initiation in *C. trifolii*. Furthermore, the Ca^{2+} channel blockers nifedipine, neodymium, and TMB-8 had pronounced effects on appressorium formation, suggesting that more than one type of Ca^{2+} channel may be involved in regulation of appressorium development. Interestingly, when the flux of internal stores of Ca^{2+} was inhibited by TMB-8, normal melanization of appressoria was disrupted. In *C. gloeosporioides*, inhibition of

calmodulin or calmodulin kinase associated processes with compound 48/80 or KN93, respectively, reduced appressorium formation (Kim et al. 1998) and those appressoria that did form in the presence of KN93 were greatly reduced in melanin content. Chelation of Ca^{2+} with EGTA also reduced germination and appressorium formation in *C. gloeosporioides*, as did inhibition of phospholipase C with U73122.

In urediospore germlings of the rust fungi, surface topographical features generally stimulate appressorium formation (Hoch et al. 1987b). Mechanosensitive Ca^{2+} channels have been implicated in this response (Zhou et al. 1991). The presence of Ca^{2+}, Mg^{2+}, or K^+ can, however, stimulate appressorium formation on *Uromyces appendiculatus* urediospores germinating on agar or liquid media (Staples et al. 1983; Kaminskyj and Day 1984); however, these appressoria form aerially apart from the surface, a condition much different than in vivo growth on its host (Hoch et al. 1987a). On normally inductive topographies, however, the presence of mM levels of Ca^{2+}, K^+, or Na^+ inhibited appressorium formation (Stumpf et al. 1991). Sodium, as previously reported (Staples et al. 1983), was especially inhibitory toward appressorium development.

Ca^{2+} may also influence appressorium formation by germlings in the Oomycota. Reduction of external, free Ca^{2+} with EGTA inhibits appressorium formation in *Phytophthora palmivora* (Bircher and Hohl 1999) and, additionally, inhibited appressorium formation in these germlings. Antagonism of calmodulin, in this case with TFP, also inhibited appressorium formation. These data suggest that, as with the many examples above, Ca^{2+} is an integral component of initiating appressorium formation.

V. Conclusions and Future Directions

It has become clear that a dichotomy exists between studies concerned with the effect of exogenously supplied ions on attachment and the effects on other developmental processes in fungi. In the case of attachment, a variety of ions have been tested, and various theories have been developed to describe their role in that process. In studies of germination, growth, and differentiation emphasis usually has been on physiological events in the cytosol and their relationship to signaling cascades. In particular, Ca^{2+} has been the ion most implicated for cellular function(s) and thus has received the most attention. Conclusions regarding a specific role for Ca^{2+} in developmental processes perhaps should be made with a degree of caution. For example, Youatt and McKinnon (1993) have pointed out, and rightly so, that EGTA is capable of chelating other divalent ions in addition to Ca^{2+}. Thus, one cannot be certain which ion effects are responsible for experimental observations. In studies by Youatt, Zn^{2+}, Fe^{2+}, and Mn^{2+} all served as Ca^{2+} substitutes for growth of several fungi (Youatt and McKinnon 1993; Youatt 1994).

Assessment of ionic effects on fungal cell function can be advanced significantly using newer technological approaches as well as reconsidering how older methods might be better applied. For the latter, one must use more precise methods for ascertaining that the concentration of the ion in question is well established. This means using appropriate buffering systems such as Ca^{2+}-EGTA or -BAPTA where the free calcium ion is precisely controlled (Magalhaes et al. 1991; Bers et al. 1994; Shaw and Hoch 2000), rather than simply adding Ca^{2+} as a salt. Studying effects of ions on cells might be accomplished if their delivery could be more precisely targeted rather than bathing the entire cell or thallus in a solution of ions of interest. Such approaches might include the localized release of ions at selected extracellular regions on a cell using a micropipette delivery system or through caged-ions and their photoactivation and release. More technologically advanced approaches might take advantage of micro- and nanofabrication methods (Hoch et al. 1996). There are many opportunities for applying 'nanobiotechnology' to the study of fungal cell biology in which the substratum can be modified in numerous ways to accommodate nano-size analytical sampling devices, molecular supplying devices, fluidics (pumps, valves, etc.), topographical features, patterning of specific domains (hydrophobic, ligands, caged-ions, etc.), and micro-electronics. Currently, there are excellent opportunities to assess ion concentrations and species temporally and spatially using a variety of methods such as the vibrating ion-specific electrode (Smith 1995; Smith et al. 1994), fluorescent probes with affinities for specific ions, aequorin which emits light upon binding with calcium (Shimomura et al. 1989; Read et al. 1993; Kendall and Badminton 1998), and ion microscopy which detects ion species using a mass spectrometry based imaging technique (Chandra et al. 1999). Other methods will surely become available. Any one of these approaches provides tremendous opportunities for gaining a better understanding of fungal cell biology and how it is influenced by the surrounding ionic environment.

References

Allen EA, Hazen BE, Hoch HC, Kwon Y, Leinhos GME, Staples RC, Stumpf MA, Terhune BT (1991) Appressorium formation in response to topographical signals by 27 rust species. Phytopathology 81:323–331

Baker CJ, Tomerlin JR, Mock N, Davidson L, Melhuish J (1987) Effects of cations on germination of urediniospores of *Uromyces phaseoli*. Phytopathology 77:1556–1560

Beckerman JL, Ebbole DJ (1996) MPG1, a gene encoding a fungal hydrophobin of *Magnaporthe grisea*, is involved in surface recognition. Mol Plant Microbe Interact 9:450–456

Bers DM, Patton CW, Nuccitelli R (1994) A practical guide to the preparation of Ca^{2+} buffers. Methods Cell Biol 40:3–29

Bircher U, Hohl HR (1999) A role for calcium in appressorium induction in *Phytophthora palmivora*. Bot Helv 109:55–65

Braun EJ, Howard RJ (1994a) Adhesion of *Cochliobolus heterostrophus* conidia and germlings to leaves and artificial surfaces. Exp Mycol 18:211–220

Braun EJ, Howard RJ (1994b) Adhesion of fungal spores and germlings to host plant surfaces. Protoplasma 181:202–212

Byrt PN, Irving HR, Grant BR (1982) The effect of cations on zoospores of the fungus *Phytophthora cinnamomi*. J Gen Microbiol 128:1189–1198

Cerenius L, Soderhall K (1985) Repeated zoospore emergence as a possible adaptation to parasitism in *Aphanomyces*. Exp Mycol 9:259–263

Chandra S, Leinhos GME, Morrison GH, Hoch HC (1999) Imaging of total calcium in urediospore germlings of *Uromyces* by ion microscopy. Fungal Genet Biol 27:77–87

Choi WB, Kang SH, Lee YW, Lee YH (1998) Cyclic AMP restores appressorium formation inhibited by polyamines in *Magnaporthe grisea*. Phytopathology 88:58–62

Clapham DE (1995) Calcium signaling. Cell 80:259–268

Cooksey KE (1981) Requirement for calcium in adhesion of a fouling diatom to glass. Appl Environ Microbiol 41:1378–1382

Deacon JW, Saxena G (1998) Germination triggers of zoospore cysts of *Aphanomyces euteiches* and *Phytophthora parasitica*. Mycol Res 102:33–41

Dean RA (1997) Signal pathways and appressorium morphogenesis. Annu Rev Phytopathol 35:211–234

Dejaguin BV, Landau L (1941) Theory of the stability of strongly charged lyphobic sols and of the adhesion of strongly charged particles in solutions of electrolytes. Act Physiochim USSR 14:633–662

Dicker JW, Turian G (1990) Calcium deficiencies and apical hyperbranching in wild type and the frost and spray morphological mutants of *Neurospora crassa*. J Gen Microbiol 136:1413–1420

Dickman MB, Buhr TL, Warwar V, Truesdell GM, Huang CX (1995) Molecular signals during the early stages of alfalfa anthracnose. Can J Bot 73: S1169–S1177

Donaldson SP, Deacon JW (1992) Role of calcium in adhesion and germination of zoospore cysts of *Pythium*: a model to explain infection of host plants. J Gen Microbiol 138:2051–2059

Doss RP, Potter SW, Chastagner GA, Christian JK (1993) Adhesion of nongerminated *Botrytis cinerea* conidia to several substrata. Appl Environ Microbiol 59: 1786–1791

Elliot CG (1972) Calcium chloride and growth and reproduction of *Phytophthora cactorum*. Trans Br Mycol Soc 58:169–172

Elliott CG (1986) Inhibition of reproduction of *Phytophthora* by the calmodulin-interacting compounds trifluoperazine and ophiobolin A. J Gen Microbiol 132: 2781–2786

Elliott CG (1988) Stages in oosporogenesis of *Phytophthora* sensitive to inhibitors of calmodulin and phosphodiesterase. Trans Br Mycol Soc 90:187–192

Emmett RW, Parbey DG (1975) Appressoria. Annu Rev Phytopathol 13:147–167

Epstein L, Nicholson RL (1997) Adhesion of spores and hyphae to plant surfaces. In: Carroll GC, Tudzynski P (eds) The Mycota, vol 5, part A. Springer, Berlin Heidelberg New York, pp 11–25

Fletcher J (1979) Effect of calcium chloride concentration on growth and sporulation of *Saprolegnia terrestris*. Ann Bot 44:589–594

Fletcher J (1983) An analysis of ultrastructural changes during oosphere initial development in oogonia from calcium sufficient and calcium deficient cultures of *Saprolegnia terrestris*. Ann Bot 52:31–38

Fletcher J (1988) Effects of external calcium concentration and of the ionophore A23187 on development of oogonia, oospores and gemmae in *Saprolegnia diclina*. Ann Bot 62:445–448

Fletcher M (1980) Adherence of marine microorganisms to smooth surfaces. In: Beachey E (ed) Bacterial adherence. Chapman and Hall, London, pp 143–374

Gilbert RD, Johnson AM, Dean RA (1996) Chemical signals responsible for appressorium formation in the rice blast fungus *Magnaporthe grisea*. Physiol Mol Plant Pathol 48:335–346

Grant BR, Griffith JM, Irving HR (1986) A model to explain ion-induced differentiation in zoospores of *Phytophthora palmivora*. Exp Mycol 10:89–98

Griffin DH (1966) Effect of electrolytes on differentiation in *Achlya* sp. Plant Physiol 41:1254–1256

Griffith JM, Iser JR, Grant BR (1988) Calcium control of differentiation in *Phytophthora palmivora*. Arch Microbiol 149:565–571

Grinberg A, Heath IB (1997) Direct evidence for Ca^{2+} regulation of hyphal branch induction. Fungal Genet Biol 22:127–139

Gristina AJ (1987) Biomaterial-centered infection: microbial adhesion versus tissue integration. Science 237: 1588–1595

Gubler F, Hardham AR, Duniec J (1989) Characterizing adhesiveness of *Phytophthora cinnamomi* zoospores during encystment. Protoplasma 149:24–30

Hadley G, Harrold CE (1958a) The sporulation of *Penicillium notatum* Westling in submerged liquid culture I. The effects of calcium and nutrients on sporulation. J Exp Bot 9:408–417

Hadley G, Harrold CE (1958b) The sporulation of *Penicillium notatum* Westling in submerged liquid culture II. The initial sporulation phase. J Exp Bot 9: 418–425

Hamer JE, Howard RJ, Chumley FG, Valent B (1988) A mechanism for surface attachment in spores of a plant pathogenic fungus. Science 239:288–290

Hardham AR (1992) Cell biology of pathogenesis. Annu Rev Plant Physiol Plant Mol Biol 43:491–526

Hardham AR, Gubler F, Duniec J (1991) Ultrastructural and immunological studies of zoospores of *Phytophthora*. In: Lucas JA, Shattock RC, Shaw SS, Cooke LR (eds) *Phytophthora*. Cambridge Univ Press, Cambridge, pp 50–69

Harding D (1971) Stability of silica dispersions. J Coll Interface Sci 35:172–174

Harold FM (1994) Ionic and electrical dimensions of hyphal growth. In: Wessels JGH, Meinhardt F (eds) The Mycota, vol I. Springer, Berlin Heidelberg New York, pp 89–109

Heath IB (1995) Integration and regulation of hyphal tip growth. Can J Bot 73:S131–S139

Hill AE, Grayson DE, Deacon JW (1998) Suppressed germination and early death of *Phytophthora infestans* sporangia caused by pectin, inorganic phosphate, ion chelators and calcium-modulating treatments. Eur J Plant Pathol 104:367–376

Hoch HC, Staples RC (1987) Structural and chemical changes among the rust fungi during appressorium development. Annu Rev Phytopathol 25:231–247

Hoch HC, Staples RC, Bourett T (1987a) Chemically induced appressoria in *Uromyces appendiculatus* are formed aerially apart from the substrate. Mycologia 79:418–424

Hoch HC, Staples RC, Whitehead B, Comeau J, Wolf ED (1987b) Signaling for growth orientation and cell differentiation by surface topography in *Uromyces*. Science 235:1659–1662

Hoch HC, Jelinski LW, Craighead H (eds) (1996) Nanofabrication and biosystems: integrating materials science, engineering and biology. Cambridge Univ Press, Cambridge

Hyde G (1998) Calcium imaging: a primer for mycologists. Fungal Genet Biol 24:14–23

Irving HR, Grant BR (1984) The effect of calcium on zoospore differentiation in *Phytophthora cinnamomi*. J Gen Microbiol 130:1569–1576

Irving HR, Griffith JM, Grant BR (1984) Calcium efflux associated with encystment of *Phytophthora palmivora* zoospores. Cell Calcium 5:487–500

Iser JR, Griffith JM, Balson A, Grant BR (1989) Accelerated ion fluxes during differentiation in zoospores of *Phytophthora palmivora*. Cell Differ Dev 26:29–38

Israelachvili JN, McGuiggan PM (1988) Forces between surfaces in liquids. Science 241:795–800

Jackson SL, Heath IB (1989) Effects of exogenous calcium ions on tip growth intracellular calcium concentration and actin arrays in hyphae of the fungus *Saprolegnia ferax*. Exp Mycol 13:1–12

Jackson SL, Heath IB (1993) Roles of calcium ions in hyphal tip growth. Microbiol Rev 57:367–382

Jen CJ, Haug A (1981) Potassium-induced depolarization of the transmembrane potential in *Blastocladiella emersonii* zoospores precedes encystment. Exp Cell Res 131:79–87

Jones EBG (1994) Fungal adhesion. Mycol Res 98:961–981

Jones L, O'Shea P (1994) The electrostatic nature of the cell surface of *Candida albicans*: a role in adhesion. Exp Mycol 18:111–120

Kaminskyj SG, Day AW (1984) Chemical induction of infection structures in rust fungi II. Inorganic ions. Exp Mycol 8:193–201

Kendall JM, Badminton MN (1998) *Aequorea victoria* bioluminescence moves into an exciting new era. Trends Biotechnol 16:216–224

Kennedy MJ (1991) *Candida* blastospore adhesion, association, and invasion of the gastrointestinal tract of vertebrates. In: Cole GT, Hoch HC (eds) The fungal spore and disease initiation in plants and animals. Plenum Press, New York, pp 157–180

Kim Y-K, Li D, Kolattukudy PE (1998) Induction of Ca^{2+}-calmodulin signaling by hard-surface contact primes *Colletotrichum gloeosporioides* conidia to germinate and form appressoria. J Bacteriol 180:5144–5150

Krystofova S, Varecka L, Betina V (1995) The $^{45}Ca^{2+}$ uptake by *Trichoderma viride* mycelium. Correlation with growth and conidiation. Gen Physiol Biophys 14:323–337

Krystofova S, Varecka L, Betina V (1996) Effects of agents affecting Ca^{2+} homeostasis on *Trichoderma viride* growth and conidiation. Folia Microbiol 41:249–253

Kuo KC, Hoch HC (1995) Visualization of the extracellular matrix surrounding pycnidiospores, germlings, and appressoria of *Phyllosticta ampelicida*. Mycologia 87:759–771

Kuo KC, Hoch HC (1996) Germination of *Phyllosticta ampelicida* pycnidiospores: prerequisite of adhesion to the substratum and the relationship of substratum wettability. Fungal Genet Biol 20:18–29

Kwon YH, Epstein L (1993) A 90 kDa glycoprotein associated with adhesion of *Nectria haematococca* macroconidia to substrata. Mol Plant Microbe Interact 6:481–487

Lee YH, Dean RA (1993) cAMP regulates infection structure formation in the plant pathogenic fungus *Magnaporthe grisea*. Plant Cell 5:693–700

Lee YH, Dean RA (1994) Hydrophobicity of contact surface induces appressorium formation in *Magnaporthe grisea*. FEMS Microbiol Lett 115:71–75

Lenny JF, Klemmer HW (1966) Factors controlling sexual reproduction and growth in *Pythium graminicola*. Nature 209:1365–1366

Magalhaes BP, Wayne R, Humber RA, Shields EJ, Roberts DW (1991) Calcium-regulated appressorium formation of the entomopathogenic fungus *Zoophthora radicans*. Protoplasma 160:77–88

Mercure EW, Leite B, Nicholson RL (1994) Adhesion of ungerminated conidia of *Colletotrichum graminicola* to artificial hydrophobic surfaces. Physiol Mol Plant Pathol 45:421–440

Moretto KCK, Barreto M (1995) Influence of some culture media on the growth and sporulation of *Alternaria solani* and of some factors on the infection frequency on tomato. Summa Phytopathol 21:188–191

Nicholson RL, Epstein L (1991) Adhesion of fungi to the plant surface: prerequisite for pathogenesis. In: Cole GT, Hoch HC (eds) The fungal spore and disease initiation in plants and animals. Plenum Press, New York, pp 3–23

Nicholson RL, Kunoh H (1995) Early interactions, adhesion, and establishment of the infection court by *Erysiphe graminis*. Can J Bot 73:S609–S615

Ochoa JL (1978) Hydrophobic (interaction) chromatography. Biochimie 60:1–15

O'Shea P (1991) The role of electrostatic and electrodynamic forces in fungal morphogenesis and host infection. In: Latage JP, Boucias D (eds) Fungal cell wall and immune response. Springer, Berlin Heidelberg New York, pp 285–305

Parton RM, Fischer S, Malho R, Papasouliotis O, Jelitto TC, Leonard T, Read ND (1997) Pronounced cytoplasmic pH gradients are not required for tip growth in plant and fungal cells. J Cell Sci 110:1187–1198

Pascual S, Melgarejo P, Magan N (1997) Induction of submerged conidiation of the biocontrol agent *Penicillium oxalicum*. App Microbiol Biotechnol 48:389–392

Pendland JC, Boucias DG (1991) Physiochemical properties of cell surfaces from the different developmental stages of the entomopathogenic hyphomycete *Nomuraea rileyi*. Mycologia 83:264–272

Perfect SE, Hughes HB, O'Connell RJ, Green JR (1999) *Colletotrichum*: a model genus for studies on pathology and fungal-plant interactions. Fungal Genet Biol 27:186–198

Pitt D, Barnes JC (1993) Calcium homeostasis, signaling and protein phosphorylation during calcium-induced conidiation in *Penicillium notatum*. J Gen Microbiol 139:3053–3063

Pitt D, Kaile A (1990) Transduction of the calcium signal with special reference to Ca^{2+}-induced conidiation in *Penicillium notatum*. In: Kuhn PJ, Trinci APJ, Jung MJ, Goosey MW, Copping LG (eds) Biochemistry of cell walls and membranes in fungi. Springer, Berlin Heidelberg New York, pp 283–298

Pitt D, Poole PC (1981) Calcium-induced conidiation in *Penicillium notatum* in submerged culture. Trans Br Mycol Soc 76:219–230

Pitt D, Ugalde UO (1984) Calcium in fungi. Plant Cell Environ 7:467–475

Queiroz MVD, Azevedo JLD (1998) Characterization of an *Aspergillus nidulans* mutant with abnormal distribution of nuclei in hyphae, metulae, phialides and conidia. FEMS Microbiol Lett 166:49–55

Read ND, Shacklock PS, Knight MR, Trewavas AJ (1993) Imaging calcium dynamics in living plant cells and tissues. Cell Biol Int 17:111–125

Rivera RN, Rodriguez DVN (1992) Effects of calcium ions on the germination of *Sporothrix schenckii* conidia. J Med Vet Mycol 30:185–195

Robson GD, Wiebe MG, Trinci APJ (1991a) Involvement of calcium in the regulation of hyphal extension and branching in *Fusarium graminearum* A 3/5. Exp Mycol 15:263–272

Robson GD, Wiebe MG, Trinci APJ (1991b) Low calcium concentrations induce increased branching in *Fusarium graminearum*. Mycol Res 95:561–565

Roncal T, Ugalde Unai O, Irastorza A (1993) Calcium-induced conidiation in *Penicillium cyclopium*: calcium triggers cytosolic alkalinization at the hyphal tip. J Bacteriol 175:879–886

Rosenberg M, Kjelleberg S (1986) Hydrophobic interactions: role in bacterial adhesion. Adv Microb Ecol 9:353–393

Rudd JJ, Franklin-Tong VE (1999) Calcium signaling in plants. Cell Mol Life Sci 55:214–232

Rutter PR, Vincent B (1984) Physiochemical interactions of the substratum, microorganisms, and the fluid phase. In: Marshall KC (ed) Microbial adhesion and aggregation. Springer, Berlin Heidelberg New York, pp 21–38

Sanders D, Brownlee C, Harper J (1999) Communicating with calcium. Plant Cell 11:681–706

Sato N (1994) Effect of some inorganic salts and hydrogen ion concentration on indirect germination of the sporangia of *Phytophthora infestans*. Ann Phytopathol Soc Jpn 60:441–447

Schmid J, Harold FM (1988) Dual roles for calcium ions in apical growth of *Neurospora crassa*. J Gen Microbiol 134:2623–2632

Sela Burrlage MB, Epstein L, Rodriguez RJ (1991) Adhesion of ungerminated *Colletotrichum musae* conidia. Physiol Mol Plant Pathol 39:345–352

Shaw BD, Hoch HC (1999) The pycnidiospore of *Phyllosticta ampelicida*: surface properties involved in substratum attachment and germination. Mycol Res 103:915–924

Shaw BD, Hoch HC (2000) Ca²⁺ regulation of *Phyllosticta ampelicida* pycnidiospore germination and appressorium formation. Fungal Genet Biol 31:43–53

Shaw BD, Kuo KC, Hoch HC (1998) Germination and appressorium development of *Phyllosticta ampelicida* pycnidiospores. Mycologia 90:252–268

Shimomura O, Musicki B, Kishi Y (1989) Semi-synthetic aequorins with improved sensitivity to calcium ions. Biochem J 261:913–920

Smith PJS (1995) The non-invasive probes: tools for measuring transmembrane ion flux. Nature 378:645–646

Smith PJS, Sanger RH, Jaffe LF (1994) The vibrating Ca²⁺ electrode: a new technique for detecting plasma membrane regions of Ca²⁺ influx and efflux. Methods Cell Biol 40:115–134

Soll DR, Sonneborn DR (1972) Zoospore germination in *Blastocladiella emersonii* IV. Ion control over cell differentiation. J Cell Sci 10:315–333

St Leger RJ, Roberts DW, Staples RC (1989) Calcium- and calmodulin-mediated protein synthesis and protein phosphorylation during germination growth and protease production by *Metarhizium anisopliae*. J Gen Microbiol 135:2141–2154

St Leger RJ, Butt TM, Staples RC, Roberts DW (1990) Second messenger involvement in differentiation of the entomopathogenic fungus *Metarhizium anisopliae*. J Gen Microbiol 136:1779–1790

St Leger RJ, Roberts DW, Staples RC (1991) A model to explain differentiation of appressoria by germlings of *Metarhizium anisopliae*. J Invertebr Pathol 57:299–310

Staples RC, Grambow HJ, Hoch HC (1983) Potassium ion induces rust fungi to develop infection structures. Exp Mycol 7:40–46

Stumpf MA, Leinhos GME, Staples RC, Hoch HC (1991) The effect of pH and potassium on appressorium formation by *Uromyces appendiculatus* urediospore germlings. Exp Mycol 15:356–360

Suizu T, Tsutsumi H, Kawado A, Murata K, Imayasu S (1994) On the importance of calcium and magnesium ions in yeast sporulation. J Ferment Bioeng 77:274–276

Suizu T, Tsutsumi H, Kawado A, Suginami K, Imayasu S, Murata K (1995) Calcium ion influx during sporulation in the yeast *Saccharomyces cerevisiae*. Can J Microbiol 41:1035–1037

Svensson E, Unestam T (1975) Differential induction of zoospore encystment and germination in *Aphanomyces astaci*, Oomycetes. Physiol Plant 35:210–216

Terhune BT, Hoch HC (1993) Substrate hydrophobicity and adhesion of *Uromyces* urediospores and germlings. Exp Mycol 17:241–252

Thiel R, Schreurs WJA, Harold FM (1988) Transcellular ion currents during sporangium development in the water mould *Achlya bisexualis*. J Gen Microbiol 134:1089–1098

Toti L, Viret O, Chapela IH, Petrini O (1992) Differential attachment by conidia of the endophyte *Discula umbrinella* Berk. and Br. Morelet to host and non-host surfaces. New Phytol 121:469–475

Tunlid A, Johansson T, Nordbring Hertz B (1991) Surface polymers of the nematode trapping fungus *Arthrobotrys oligospora*. J Gen Microbiol 137:1231–1240

Tunlid A, Jansson HB, Nordbring Hertz B (1992) Fungal attachment to nematodes. Mycol Res 96:401–412

Ugalde UO, Pitt D (1986) Calcium uptake kinetics in relation to conidiation in submerged cultures of *Penicillium cyclopium*. Trans Br Mycol Soc 87:199–204

Ugalde UO, Virto MD, Pitt D (1990) Calcium binding and induction of conidiation in protoplasts of *Penicillium cyclopium*. J Microbiol Serol 57:43–50

Van Brunt J, Harold FM (1980) Ionic control of germination of *Blastocladiella emersonii* zoospores. J Bacteriol 141:735–744

Verwey EJW, Overbeek JTG (1948) Theory of the stability of lyphobic colloids – the interaction of sol particles having an electric double layer. Elsevier, New York

von Broembsen SL, Deacon JW (1996) Effects of calcium on germination and further zoospore release from zoospore cysts of *Phytophthora parasitica*. Mycol Res 100:1498–1504

von Broembsen SL, Deacon JW (1997) Calcium interference with zoospore biology and infectivity of *Phytophthora parasitica* in nutrient irrigation solutions. Phytopathology 87:522–528

Warburton AJ, Deacon JW (1998) Transmembrane Ca²⁺ fluxes associated with zoospore encystment and cyst germination by the phytopathogen *Phytophthora parasitica*. Fungal Genet Biol 25:54–62

Warwar V, Dickman MB (1996) Effects of calcium and calmodulin on spore germination and appressorium development in *Colletotrichum trifolii*. Appl Environ Microbiol 62:74–79

Wayne R (1985) The contribution of calcium ions and hydrogen ions to the signal transduction chain in phytochrome-mediated fern spore germination. PhD Thesis, University of Massachusetts, Amherst

Wessels JGH (1996) Fungal hydrophobins: proteins that function at an interface. Trends Plant Sci 1:9–15

Xiao JZ, Watanabe T, Kamakura T, Ohshima A, Yamaguchi I (1994) Studies on cellular differentiation of *Magnaporthe grisea*. Physicochemical aspects of substratum surfaces in relation to appressorium formation. Physiol Mol Plant Pathol 44:227–236

Xu C, Morris PF (1998) External calcium controls the developmental strategy of *Phytophthora sojae* cysts. Mycologia 90:269–275

Yang Z, Dickman MB (1997) Regulation of cAMP and cAMP dependent protein kinase during conidial germination and appressorium formation in *Col-*

letotrichum trifolii. Physiol Mol Plant Pathol 50: 117–127

Yokosawa R, Kuninaga S, Sakushima A, Sekizaki H (1995) Induction of oospore formation of *Aphanomyces euteiches* Drechsler by calcium ion. Ann Phytopathol Soc Jpn 61:434–438

Youatt J (1993) Calcium and microorganisms. Crit Rev Microbiol 19:83–97

Youatt J (1994) The toxicity of metal chelate complexes of EGTA precludes the use of EGTA buffered media for the fungi *Allomyces* and *Achlya*. Microbios 79:171–185

Youatt J, McKinnon I (1993) Manganese reverses the inhibition of fungal growth by EGTA. Microbios 74:181–196

Young DH, Kauss H (1984) Adhesion of *Colletotrichum lindemuthianum* to *Phaseolus vulgaris* hypocotyls and to polystyrene. Appl Environ Microbiol 47: 616–619

Zhou XL, Stumpf MA, Hoch HC, Kung C (1991) A mechanosensitive channel in whole cells and in membrane patches of the fungus *Uromyces*. Science 253: 1415–1417

5 Cell Biology of Fungal Infection of Plants

Adrienne R. Hardham

CONTENTS

I. Introduction

Plants are, in general, resistant to the attempts of potential fungal pathogens to infect them. Those fungi that do succeed in establishing infection and disease, however, cause widespread environmental damage and economic losses. In order to establish infection, fungal pathogens must overcome highly effective, constitutive physical and chemical barriers to pathogen ingress. They must avoid inducing additional host defenses, and they must be able to deploy mechanisms for obtaining from the plant the nutrients they need for growth and reproduction. In order to meet all these requirements, successful fungal pathogens employ a range of different infection strategies. The details of these strategies may be specific to a particular

Plant Cell Biology Group, Research School of Biological Sciences, The Australian National University, Canberra, ACT 2601, Australia

fungal species, they may differ according to the nature of the plant surface, and they may depend within a single species on the type of spore initiating the infection process. A number of important steps in the infection process are, however, common to all strategies. These include adhesion to the surface of the plant, penetration of the plant surface, and acquisition of nutrients from the plant cells. In many cases, general aspects of the infection strategies of fungal pathogens have been known for a long time. More recently, however, modern methods of cell and molecular biology have contributed greatly to our understanding of the cellular and molecular mechanisms that underlie the infection process. We still have much to learn, and it is an exciting time in a rapidly expanding area of research.

The focus of this chapter will be on recent advances in our understanding of the cell biology of the infection process, from the arrival of a fungal spore at the plant surface, through to the establishment of feeding structures within the plant tissues. A number of reviews of this topic as a whole (Howard et al. 1991a; Hardham 1992; Mendgen and Deising 1993; Deising et al. 1996; Mendgen et al. 1996; Perfect et al. 1999) and of adhesion (Kunoh et al. 1991; Nicholson and Epstein 1991; Braun and Howard 1994a; Nicholson 1996; Epstein and Nicholson 1997) and host penetration (Howard and Valent 1996; Howard 1997) have been written over the last 10 years. The reader is referred to these articles for further details and discussion.

Fungal pathogens of plants can be divided into two main categories: the **biotrophs** which manipulate the metabolism of living host cells in order to obtain the nutrients they need, and the **necrotrophs** which acquire their nutrients from dead plant cells. These two different life-styles have considerable bearing on the infection strategies employed by the fungal species involved. Important biotrophic fungi include the rust Basidiomycetes (e.g. species of *Uromyces* and *Puccinia*), the powdery mildew

The Mycota VIII
Biology of the Fungal Cell
Howard/Gow (Eds.)
© Springer-Verlag Berlin Heidelberg 2001

Ascomycetes (e.g. species of *Erysiphe*) and the downy mildews of the Oomycota (e.g. species of *Bremia*, *Peronospora* and *Plasmopara*). Important necrotrophs include the rice blast pathogen *Magnaporthe grisea*, *Cochliobolus* species, *Nectria haematococca*, and species of *Phytophthora* and *Pythium*. Some fungi, such as *Colletotrichum* species, which cause anthracnose diseases, are facultative biotrophs or hemibiotrophs. They initially establish a biotrophic interaction with living host cells before killing the plant cells and turning to a necrotrophic life-style. In any plant-pathogen interaction, the extent of progress of infection will be determined by genetic and epigenetic factors, factors that influence the degree of resistance of the host plant and the degree of virulence of the invading pathogen. A spectrum of interactions from fully resistant to fully susceptible can often be observed.

II. Adhesion

Plants may become infected by mycelia growing from a nearby diseased plant but, in the majority of cases, infection is initiated by the arrival of a fungal spore on the plant surface. In order to have a chance to establish infection, the spore must remain on the plant surface until the fungus has been able to penetrate the underlying plant cells. Thus, the first critical aspect of the infection process after spore arrival is adhesion.

Adhesion of pathogen cells to the surface of potential hosts plays an important role in a number of aspects of fungal pathogenesis. Firm adhesion of spores that have reached the plant surface will stop them from being dislodged or spatially disoriented by wind and rain before they have a chance to penetrate the plant. Close adhesion to the plant is also needed for the reception of signals – which guide pathogen growth on the plant surface, and that initiate differentiation of infection structures during pathogenesis. Tight adhesion of appressoria (and haustorial mother cells) is likely to be an absolute requirement for penetration of plant cells.

A. Fungal Spore Adhesiveness on the Plant Surface

Fungal adhesives are generally non-specific, in that they will attach the fungal spore to a range of substrata, although they may often perform better on hydrophobic surfaces, as demonstrated for *M. grisea* (Hamer et al. 1988), *Colletotrichum graminicola* (Mercure et al. 1994a), *Uromyces appendiculatus* (Terhune and Hoch 1993) and *Phyllosticta ampelicida* (Shaw and Hoch 1999). There is evidence that fungal spores can modify the properties of the surface of a plant to enhance its ability to adhere (see below).

The timing of the appearance of adhesiveness of fungal spores varies widely. In some cases, the spore is already adhesive when it reaches the plant. Conidia of *Blumeria graminis* f. sp. *hordei*, for example, are coated with a wet and sticky extracellular matrix material, which glues the conidia to the substratum as soon as they arrive (Scherer et al. 1992). In other cases, the adhesive material is rapidly released after the spore reaches the plant surface. The motile zoospores of *Phytophthora* and *Pythium* species become attached within 2 min of contacting the plant surface (Fig. 1A–C; Bartnicki-Garcia and Sing 1987; Gubler et al. 1989). The adhesive material is stored in the zoospores in small vesicles that lie beneath the grooved ventral surface from which the flagella arise (Hardham and Gubler 1990; Cope et al. 1996). These adhesive-containing vesicles are synthesized during sporangium formation (Dearnaley et al. 1996). Initially, they are randomly distributed in the cytoplasm of the multinucleated sporangia but are transported to the ventral surface as zoospores form during sporangial cytokinesis (Hyde and Hardham 1993).

Conidia of a number of plant pathogenic fungi typically become attached about 20–30 min after inoculation and hydration on the leaf surface. This is the case for the conidia of *M. grisea* that adhere 15–20 min after inoculation and which store their adhesive material in the periplasmic space between the plasma membrane and cell wall at the tip of the conidium (Fig. 1D–F; Hamer et al. 1988). During hydration, the cell wall at the conidial apex ruptures, releasing the adhesive, called spore tip mucilage, onto the leaf surface. Adhesion is not inhibited by treatment of the conidia with azide or cycloheximide, indicating that neither respiration nor protein synthesis is required for spore adhesion (Hamer et al. 1988). The periplasmic storage of pre-formed adhesive material is consistent with this observation.

Adhesive material is released from the tips of the crescent-shaped macroconidia of *N. haematococca* 10–20 min after treatment with host plant

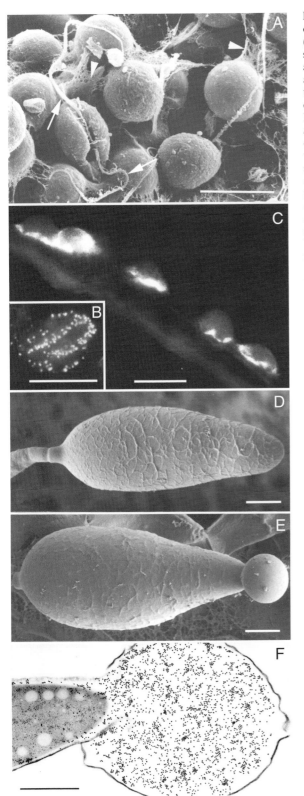

Fig. 1. A Zoospore and cysts of *Phytophthora cinnamomi* on the surface of a root of *Allium cepa*. Two flagella (*arrows*) emerge from the center of a groove on the ventral surface of the zoospores. During encystment, the flagella are detached, the cells round up and extracellular matrix and adhesive material (*arrowheads*) are secreted around the spores. Scanning electron micrograph (SEM) of critical point dried material. *Bar* 10 μm (Courtesy CSIRO Entomology). **B** Immunofluorescence micrograph of *Phytophthora cinnamomi* zoospore labelled with Vsv-1, a monoclonal antibody that reacts with the contents of ventral vesicles that occur in the cell cortex underneath the ventral surface of the spores. The contents of the ventral vesicles are secreted onto the adjacent substratum during the first 2 min of encystment, and appear to function as an adhesive. *Bar* 10 μm. **C** Immunofluorescence micrograph of a cryosection of *Phytophthora cinnamomi* cysts, 5 min after settling onto a root of *Eucalyptus seiberi*. The section has been labelled with Vsv-1 to reveal the material secreted from the ventral vesicles. The Vsv-1 antigen, the putative adhesive material, forms a pad between the spores and the root surface. *Bar* 10 μm (Hardham et al. 1991). **D** Cryo-SEM of a developing conidium of *Magnaporthe grisea*. *Bar* 2.5 μm (Howard 1994). **E** CryoSEM of a conidium of *Magnaporthe grisea* after release of adhesive mucilage material from the apex of the spore. *Bar* 2.5 μm (Braun and Howard 1994a). **F** Transmission electron micrograph (TEM) of a conidium of *Magnaporthe grisea* after release of adhesive material from the periplasmic space between the spore wall and the plasma membrane. The section has been labelled with Concanavalin A-gold that reacts with the spore tip mucilage. *Bar* 1.0 μm (Howard et al. 1991a)

extracts (Fig. 2A; Jones and Epstein 1989) and from the tips of conidia of *Cochliobolus heterostrophus* 20–40 min after inoculation and hydration (Braun and Howard 1994b). In both cases, acquisition of adhesiveness requires respiration and protein synthesis and thus the adhesive molecules associated with the spores of these species are likely to be synthesized and secreted after spore hydration.

That the strength and nature of adhesion of fungal pathogens change during the production of different infection structures is not surprising, but even spores may show different phases of adhesion to the plant surface. For example, adhesion of macroconidia *of N. haematococca* is initially temperature-dependent but becomes temperature-independent (Jones and Epstein 1989). Studies of the rust fungus, *Uromyces viciae-fabae*, have also demonstrated that dry urediniospores will initially adhere to a hydrophobic surface, probably by hydrophobic interactions (Clement et al. 1994). After hydration, the strength of attachment increases over the next 30 min, due either to the release of extracellular matrix material from the spore (Deising et al. 1992) or to capillary forces arising from water accumulation between the spore and the substratum (Clement et al. 1997). Even autoclaved spores display the initial increase in attachment, indicating that it is a passive process (Deising et al. 1992). Thereafter, the strength of attachment of living (but not dead) spores continues to increase, during which time extracellular matrix material is released from the spore wall and germ pore to form an adhesive pad between the pathogen cells and the underlying substratum (Fig. 2B; Deising et al. 1992; Clement et al. 1997). The extracellular matrix secreted by fungal spores can alter the properties of the leaf surface (Fig. 2C; Read 1991; Mercure et al. 1994b; Nicholson 1996). In two cases, urediniospores of *U. viciae-fabae* (Deising et al. 1992) and conidia of *Erysiphe graminis* (Pascholati et al. 1992; Nicholson et al. 1993), the extracellular matrix has been shown to contain esterase and cutinase activity which could change surface hydrophobicity and enhance spore adhesion.

The timing of the production of adhesive molecules and the speed with which adhesive material is released are likely to reflect the requirements of the pathogen involved. In the case of pathogens in aquatic environments, the release of adhesive may need to be rapid to prevent the spore from being swept away from the plant surface. In this

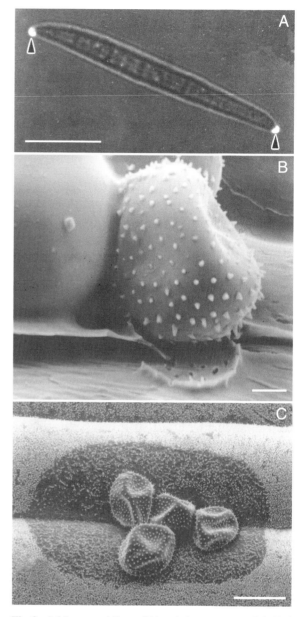

Fig. 2. A Macroconidium of *Nectria haematococca* labelled with Concanavalin A-FITC. Concanavalin A reacts with mucilage material (*arrowheads*) released from the tips of the conidium. *Bar* 20 μm (Kwon and Epstein 1993). **B** Urediniospore of *Uromyces viciae-fabae* (*right*) and glass bead (*left*) on leaf of *Vicia faba*. The pattern of spines of the urediniospore has been left in the adhesive pad of mucilage material secreted by the spore. Cryo-SEM. *Bar* 20 μm (Clement et al. 1997). **C** Urediniospores of *Puccinia hordei* on the surface of a barley leaf. Wax on the leaf surface surrounding the spores appears eroded, possibly through the action of enzymes secreted by the ungerminated spores. Cryo-SEM. *Bar* 20 μm (Read 1991)

case, the adhesive would need to be pre-formed and stored in the spores. For leaf pathogens, initial weak adhesion may be sufficient to retain the spore in place at least in the short term, while the spores have time to synthesize adhesive molecules, which are then secreted. In the case of hyphal adhesion, synthesis and release of adhesive material will be a progressive event that ensures close contact of the hyphal apex with the underlying plant surface.

B. Biochemical Characterization of Fungal Adhesive Molecules

Despite increasing recognition of the importance of adhesion in fungal pathogenesis, molecules responsible for adhesion have yet to be fully characterized. Differences in the biochemical composition of extracellular matrices associated with the surfaces of spores, germ tubes and appressoria are beginning to be determined: for example, the immunocytochemical studies of the surfaces of these cells in *Colletotrichum lindemuthianum* (O'Connell et al. 1996a,b; Pain et al. 1996). However, the extracellular matrix associated with these cells is likely to contain a mixture of components that play a variety of roles in addition to adhesion. Extracellular matrices have, for example, been suggested to function in cell protection against desiccation or toxins, in communication and signaling, as well as in modification of the properties of the underlying surface (Nicholson and Epstein 1991; Deising et al. 1992; Moloshok et al. 1993; Viret et al. 1994). While correlations have been made between the secretion of extracellular matrix material and the appearance of adhesiveness of the fungal cells, in only one case has an individual component been unambiguously demonstrated to play an adhesive role.

In a number of instances, incubation with the lectin Concanavalin A has been shown to inhibit adhesion of spores, indicating that the adhesive molecules contain mannosyl or glucosyl residues. In *M. grisea* (Hamer et al. 1988), *P. ampelicida* (Shaw and Hoch 1999) and *Discula umbrinella* (Viret et al. 1994), no further information on the nature of the molecules involved is available. In *N. haematococca*, a 90-kDa glycoprotein which reacts with Concanavalin A and snowdrop lectin is known to be a major component of the extracellular matrix material secreted from the spore

tips (Fig. 2A; Kwon and Epstein 1993, 1997a). Biochemical analysis of the 90-kDa glycoprotein has determined its amino acid composition, shown it to have hydrophobic properties and to be N-glycosylated (Kwon and Epstein 1997b). A polyclonal antiserum raised against the 90-kDa glycoprotein also labels the extracellular matrix material at the spore tips and inhibits adhesion of the macroconidia (Kwon and Epstein 1997a). Deglycosylation of the 90-kDa glycoprotein reduces antibody binding and the effectiveness of the antiserum to inhibit adhesion is reduced by incubation in mannan. These results suggest that the 90-kDa glycoprotein is the adhesive molecule and that mannosyl residues in the glycoprotein play an important role in spore adhesion. Mannose has also been found to be a predominant sugar associated with an approximately 200-kDa glycoprotein in the extracellular matrix of *C. graminicola* conidia (Sugui et al. 1998), although a role for this glycoprotein in adhesion has not been demonstrated. A monoclonal antibody, which reacts with a number of polypeptides on the surface of conidia of *C. lindemuthianum*, including a predominant 110-kDa glycoprotein, can also inhibit spore adhesion to polystyrene (Hughes et al. 1999).

In *N. haematococca*, two transglutaminase inhibitors, iodoacetamide and cystamine, reduce adhesion of the macroconidia, the amount of extracellular matrix detectable at the spore tips and the amount of 90-kDa glycoprotein extractable (Kwon and Epstein 1997a). It has also been shown that spore adhesiveness and the presence of detectable 90-kDa glycoprotein in the extracellular matrix at the spore tips are only transient: 1–2h after the onset of incubation, spores become less adhesive and the amount of 90-kDa glycoprotein bound by Concanavalin A declines. The authors suggest that these results are consistent with a model in which the spores secrete low molecular weight, water-soluble precursors of the 90-kDa glycoprotein. Rapidly, the precursor molecules are partially polymerized by a transglutaminase into the adhesive, 90-kDa glycoprotein. After 1–2h, the 90-kDa glycoprotein is further modified into a less soluble, less adhesive material (Kwon and Epstein 1997a). There is evidence for a similar phenomenon in *Phytophthora*, although adhesiveness is lost much more rapidly. *Phytophthora* cysts are adhesive for only about 5 min after secretion of material containing a 200-kDa protein from the ventrally located vesicles during encystment (Fig. 1B,C; Gubler et al. 1989; Hardham and Gubler 1990). Ca^{2+} is required for adhesion of the *Phytophthora* cysts (Gubler et al. 1989) and also enhances adhesion of pycnidiospores of *Phyllosticta ampelicida* to substrata to which the spores do not normally adhere (Shaw and Hoch 1999).

III. Growth on the Plant Surface

A. Spore Germination

Spores are designed to disperse fungal pathogens and enable them to survive adverse conditions. Fungi employ a number of mechanisms to enhance the chances of a spore being able to establish successful infection once it reaches the plant. Spores typically undergo a period of dormancy, during which time their metabolic activity is low, allowing them to survive possibly long periods of time between their production and their arrival at a suitable site for germination. Constitutive dormancy may be achieved by compounds that are secreted and which inhibit the activation of spores that are crowded together, as they are when they are formed (Macko 1981; Staples and Hoch 1997). Activation of spores may also be inhibited by compounds produced by other microbes in the environment (Lucas and Knights 1987) or by the absence of specific cues for germination. Some spores must adhere to a suitable substratum before they will germinate (e.g. Shaw and Hoch 1999). Spores need water for hydration and, in some cases, they require a nutrient source before germination is induced (see Staples and Hoch 1997).

For most fungal spores, there are three main aspects of spore activation. (1) The spores swell isotropically, at first passively during hydration and then as a result of metabolic activity. (2) Spore metabolism increases, with enhanced rates of respiration, and protein and nucleic acid synthesis. (3) The spores germinate by localized outgrowth of the cell. In general, spore germination occurs 3–8 h after hydration, although *Phytophthora* and *Pythium* cysts, which arise from active, motile zoospores, may germinate within 20 min after they are formed (Grove and Bracker 1978; Penington et al. 1989; Hardham and Gubler 1990).

Many spores are able to germinate from multiple sites on the spore surface (Figs. 3A,B). In these cases, the site at which germination actually occurs may be influenced by environmental factors, such as light, oxygen, nutrients or adhesion to the underlying substratum (Robinson 1973; Gold and Mendgen 1991; Lee and Dean 1993; Kuo and Hoch 1996; Staples and Hoch 1997; Shaw et al. 1998). Contact with another spore or proximity to a plant cell can also have a marked effect on the sites of germination. Conidia of *Geotrichum candida*, for example, form a germ tube at the site furthest away from the site of contact with another spore (Robinson 1973). Spores of *Idriella bolleyi* germinate on the side of the spore facing away from living root hairs of cereals, but towards dead root hairs (Allan et al. 1992).

For some spores, the site of germination occurs at a predictable location. In *Cochliobolus heterostrophus*, germination occurs at the ends of the crescent-shaped conidia (Braun and Howard 1994b) and in *M. grisea* it occurs at the apical and/or basal cells of the pear-shaped conidia (Fig. 3A; Hamer et al. 1988). Urediniospores, but not basidiospores (Gold and Mendgen 1991), of rusts germinate from a germination pore, a specialized area in which the spore wall is thinner than in other regions (Deacon 1997). In the Oomycota, while no structural organization indicative of the site of germination in the cyst has been recognized, there is evidence that this location is predetermined because it has been shown that cysts of *Phytophthora* and *Pythium* germinate from a region in the center of what was the ventral groove of the zoospores (Mitchell and Deacon 1986; Paktitis et al. 1986; Hardham and Gubler 1990). In this case, the region of germ-tube emergence is pre-aligned towards the plant as a result of the orientation of the motile zoospores before they encyst and adhere to the plant surface (Fig. 1C; Mitchell and Deacon 1986; Hardham and Gubler 1990).

The first morphological sign of germination is the accumulation of apical vesicles of a range of sizes (typically 40–400 nm in diameter) and morphologies, beneath the site of germ-tube emergence (Fig. 3C–E). The cluster of apical vesicles is similar to that seen in the apex of growing hyphae. The vesicles are believed to contain wall material and degradative enzymes as well as contributing membrane to the expanding plasma membrane (see Bartnicki-Garcia 1973). The apical vesicles fuse with the overlying plasma membrane in the apical dome, thus generating localized surface growth.

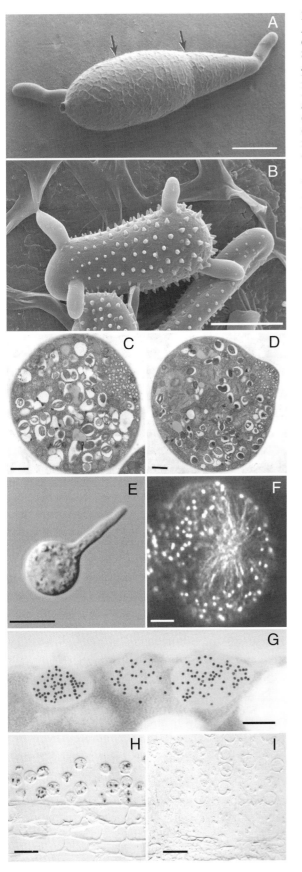

Fig. 3. A Cryo-SEM of a germinated conidium of *M. grisea*. *Arrows* indicate cross-walls between the three cells of the conidium. *Bar* 5 μm (Money and Howard 1996). **B** Cryo-SEM of a urediniospore of poplar rust pathogen (*Melampsora larici-populina*) that has germinated from four locations. *Bar* 10 μm (Unpublished micrograph courtesy of Roger Heady, ANU). **C, D** Cysts of *Phytophthora cinnamomi* prepared for TEM 5 min after encystment. In the cyst in **C**, a cluster of apical vesicles has formed in the cell cortex at a location presumed to be the future site of germination. In the cyst in **D**, the cluster of apical vesicles obviously occurs at the site of spore germination. *Bar* 1 μm. **E** Germinated cyst of *Phytophthora cinnamomi*, 60 min after encystment. Differential interference contrast micrograph. *Bar* 10 μm. **F** Hypha of *Saprolegnia ferax* labelled with rhodamine-phalloidin showing a radial array of actin microfilaments at a point of branch formation in a cell recovering from treatment with latrunculin B. *Bar* 5 μm (Bachewich and Heath 1998, reproduced with permission of Company of Biologists, Ltd.). **G** TEM post-embedment immunogold labelling of *Phytophthora cinnamomi* zoospore with monoclonal antibody Lpv-1-Au$_{18}$ that reacts with high molecular weight glycoproteins in large peripheral vesicles in the zoospore cortex. These vesicles become randomly distributed throughout the spores after encystment. *Bar* 0.25 μm (Hardham 1995). **H** Immunogold-silver enhanced light microscope labelling with Lpv-1-Au$_{18}$ of large peripheral vesicles in cysts of *Phytophthora cinnamomi* fixed 1 h after infection of a root of *Eucalyptus seiberi*. *Bar* 20 μm (Gubler and Hardham 1991). **I** Immunogold-silver enhanced labelling with Lpv-1-Au$_{18}$ of cysts of *Phytophthora cinnamomi* fixed 4 h after infection of a root of *Eucalyptus seiberi*. The large peripheral vesicles containing the Lpv-1 antigen have been degraded and are not labelled. *Bar* 20 μm (Gubler and Hardham 1991)

Localized outgrowth of the spore surface establishes the polarity of growth of the germling, and much effort has been directed towards uncovering the mechanisms operating in the establishment of the germination site. Microtubules may be involved in the distribution of organelles in the germ tube or hypha and for maintenance of the monopolar axis, but they do not seem to play a role in polarized germ-tube outgrowth or hyphal branching (Ton-That et al. 1988; Barja et al. 1993). There is, however, increasing evidence that actin microfilaments are involved in this process. Filamentous actin typically forms longitudinal bundles in the hyphal cytoplasm and extends into the apical dome at the tip of the germ tubes and hyphae (see Gow 1995). A recent study in *Saprolegnia* has revealed that radial arrays of actin microfilaments form in germinating cysts and hyphal branches before germ-tube or branch formation (Fig. 3F; Bachewich and Heath 1998). In addition, experimental removal of actin microfilaments through treatment with cytochalasins or latrunculin B inhibits germ-tube formation in germinating spores (Grove and Sweigard 1980; Barja et al. 1993; Bachewich and Heath 1998). Consistent with the involvement of actin in this process is the observation that myosin is required for the establishment of polarized growth and secretion in *Aspergillus* germlings (McGoldrick et al. 1995). These observations indicate that actin plays a critical role in the establishment of polar outgrowth during spore germination and hyphal branching. Actin could play such a role by functioning in the transport of apical vesicles to the germination or branch site, or by directing vesicle fusion at that site through regulation of the distribution or activity of ion channels or docking receptors in the plasma membrane (Levina et al. 1994; Ayscough et al. 1997). The fact that interference with Ca^{2+} homeostasis inhibits spore germination in *C. trifolii* (Warwar and Dickman 1996), and variations in Ca^{2+} concentration precede and influence the site of spore germination and branch formation in *Saprolegnia* (Hyde and Heath 1995; Grinberg and Heath 1997), supports this hypothesis. It has been suggested that Ca^{2+} ion fluxes could be responsible for organizing filamentous actin into the radial arrays (Bachewich and Heath 1998).

Spore activation is accompanied by increased rates of nucleic acid and protein synthesis, and it has been consistently shown that *de novo* protein synthesis is a prerequisite for spore germination throughout the Mycota (Lovett 1976; Van Etten et al. 1976; Penington et al. 1989). A requirement for RNA synthesis has been more controversial. On the one hand, germination of spores of *Blastocladiella emersonii* and *Allomyces arbuscula*, members of the Chytridiomycota, occurs in the absence of RNA synthesis: treatment of spores with actinomycin D to block RNA synthesis, does not inhibit germination of these spores (Lovett 1975). Among the Dikaryomycota, on the other hand, the situation has been less clear. Some species show rapid onset of RNA synthesis upon spore activation and inhibition of germination by RNA synthesis inhibitors (see Van Etten et al. 1976). Studies of other species have suggested that spore germination may not require *de novo* RNA synthesis; however, doubts concerning inhibitor penetration and effectiveness have been raised, and independence of RNA synthesis in species of the Dikaryomycota has yet to be unambiguously demonstrated (see Van Etten et al. 1976; Brambl et al. 1978). In the Oomycota, experiments using actinomycin D to inhibit RNA synthesis have demonstrated the requirement for RNA synthesis for cyst germination in *Phytophthora*, *Achlya* and *Aphanomyces* (Clark et al. 1978; MacLeod and Horgen 1979; Söderhall and Cerenius 1983; Penington et al. 1989).

Taken together, the evidence suggests that *de novo* synthesis of RNA is generally required for fungal spore germination and that the situation in the Chytridiomycota may be a special case. Those spores contain a characteristic cap of ribosomes and RNA surrounding the nucleus, and this cap may contain pre-synthesized mRNA to be used for protein synthesis during germination (Lovett 1975). Currently of more interest are the identification of individual genes that are specifically expressed during spore germination and the characterization of the proteins they encode. Potentially important candidates include proteins involved in the establishment of polarity and the initiation of wall outgrowth. Some progress has been made in this area (e.g. Sachs and Yanofsky 1991; Liu et al. 1993), and recent developments in molecular genetic techniques have paved the way for major advances in the near future.

Many spores can germinate and sustain growth of the germ tube for at least a short time in the absence of exogenous nutrients (e.g. Hemmes and Hohl 1971; Lovett 1975; Lucas and Knights 1987), indicating that they use endogenous supplies of carbon and nitrogen to support this early growth and development. Carbohydrates and lipids are the reserves most commonly recognized in spores (Gottleib 1976; Reisener 1976) but storage proteins have been documented in pycnidia of *Botryodiplodia theobromae* (Van Etten et al. 1979), sclerotia of *Sclerotinia sclerotiorum* (Russo et al. 1982) and *S. minor* (Bullock et al. 1980), and in zoospores and cysts of *Phytophthora cinnamomi* (Fig. 3G; Gubler and Hardham 1990). Storage carbohydrates may take the form of glycogen aggregates in the cytoplasm, soluble cytosolic sugars such as trehalose or mannitol, or β1,3-glucans which may be stored intracellularly in vesicles or extracellularly in the cell wall. In *Phytophthora*, storage β1,3-glucans are mycolaminarins and occur in vesicles with characteristic lamellations (Powell and Bracker 1977; Wang and Bartnicki-Garcia 1980). Lipid is found in the form of lipid globules surrounded by a single membrane. During spore germination, these carbohydrate and lipid reserves disappear (Bimpong and Hickman 1975; Wang and Bartnicki-Garcia 1980; Bullock et al. 1983; Thevelein 1996). In *B. theobromae* pycnidia and *S. sclerotiorum* sclerotia, 20- and 36-kDa polypeptides, respectively, form the major storage proteins. In *Phytophthora cinnamomi*, the storage proteins are large glycoproteins of over 500-kDa relative molecular mass (Gubler and Hardham 1988; J.S. Marshall, unpubl. observ.). In each case, the proteins are deposited in cytoplasmic vesicles (Fig. 3G,H). During germination and germling development, these protein-containing vesicles become dilated and their contents become increasingly electron-lucent as they are degraded (Bullock et al. 1983; Gubler and Hardham 1990). In *Phytophthora cinnamomi* germlings, the storage protein is no longer detectable immunocytochemically by 3–4 h after cyst germination (Fig. 3H,I).

B. Hyphal Growth on the Plant Surface

The mode of growth established by germinating spores continues as the fungus grows across the plant surface. Hyphae extend by tip growth, a process in which surface expansion is restricted to the apical dome of the hypha. This is achieved by localized fusion of cytoplasmic vesicles with the plasma membrane in the dome. Surface expansion is highest at the very apex of the dome and decreases rapidly subapically. The cylindrical walls of the hypha subtending the apex are rigid and do not expand.

The exact shape of the hyphal apex and the organization of organelles within the apical regions of the hypha vary in different species across the fungal kingdom. However, it seems likely that the basic mechanism responsible for tip growth is similar in all cases. Details of the interplay between wall extensibility, the cytoskeleton and cell turgor pressure which achieve this precisely regulated apical expansion are controversial and have given rise to a number of models of hyphal tip growth (see Bartnicki-Garcia 1996; Heath and Steinberg 1999). Some envisage that the newly formed wall at the hyphal apex is easily expandable and, in a controlled fashion, yields to the forces of turgor pressure (see Sietsma and Wessels 1994). Others suggest that turgor pressure may not participate in the extension of narrow tubes such as hyphae (see Chap. 1, this Vol.; Pickett-Heaps and Klein 1998; Heath and Steinberg 1999). Some view the newly formed wall to be so pliable that actin microfilaments are invoked to form a stabilizing and regulatory network on the cytoplasmic face of the plasma membrane in the apex (see Chap. 10, this Vol.; Heath 1994; Money 1994; Kaminskyj and Heath 1996). Others see the need for degradative enzymes to loosen the structure of the newly formed wall at the apex to allow expansion to occur (Bartnicki-Garcia 1973; Bartnicki-Garcia et al. 1989). Regardless of the details of the relationship between these factors, one thing is clear – the numerous small vesicles in the apical cytoplasm play a central and indispensable role in the delivery of wall components and membrane to the expanding apical surface.

In many cases, apical vesicles are aggregated into a cluster known as the Spitzenkörper, a structure which is visible under a light microscope (Fig. 4A–C; Girbardt 1969; Grove and Bracker 1970; Grove et al. 1970). The important role of the Spitzenkörper in tip growth was recognized soon after its discovery and, recently, elegant studies using advanced video-enhanced microscopy techniques have supported this contention (López-Franco and Bracker 1996; Riquelme et al. 1998). The Spitzenkörper appears to act as a vesicle supply center from which vesicles move out and fuse with the plasma membrane within the apical dome (Bartnicki-Garcia et al. 1989, 1990). Bartnicki-Garcia and co-workers have developed a mathematical model that accurately describes hyphal growth and that highlights two important parameters, namely, the rate of movement of the

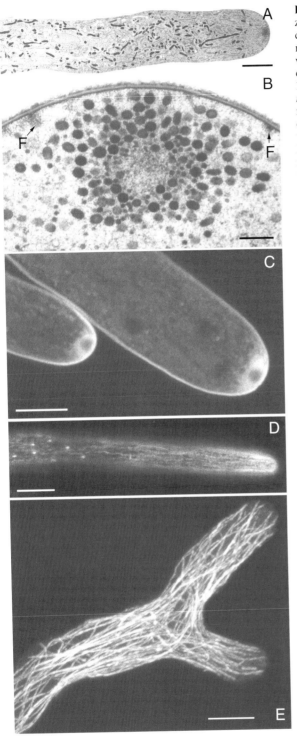

Fig. 4. A Hyphal apex of *Sclerotium rolfsii* showing zonation of organelles along the hypha. The extreme apex contains a Spitzenkörper and is free of large organelles; mitochondria and ER occur in the subapical cytoplasm; vacuoles and nuclei occur further back from the apex. TEM of freeze-substituted cell. *Bar* 4 μm (Roberson and Fuller 1988). **B** Spitzenkörper in the apex of a hypha of *Sclerotium rolfsii*. A cluster of apical vesicles surrounds a vesicle-free region in the center of the Spitzenkörper. Filasomes (*F*) occur near the plasma membrane. TEM of freeze-substituted cell. *Bar* 0.25 μm (Roberson and Fuller 1990). **C** Hyphae of *Botrytis cinerea* labelled with Concanavalin A in whole cells prepared by embedding and de-embedding in butylmethyl methacrylate resin. The plasma membrane and components of the endomembrane system, including apical vesicles, have been labelled. The core of the Spitzenkörper is unlabelled. Confocal laser scanning microscopy (CLSM) stack of optical sections. *Bar* 5 μm (Bourett et al. 1998). **D** Hyphae of *Saprolegnia ferax* labelled with rhodamine-phalloidin revealing longitudinally oriented actin microfilaments and peripheral plaques of filamentous actin. Conventional fluorescence microscopy. *Bar* 10 μm (Jackson and Heath 1990). **E** Hypha of *Rhizoctonia oryzae* labelled with antitubulin antibodies to reveal the distribution of microtubules in whole hyphae prepared by embedding and de-embedding in butylmethyl methacrylate resin. Image obtained by CLSM. *Bar* 5 μm (Bourett et al. 1998)

vesicle supply center and the rate of vesicle release from the center. The values of these two parameters determine hyphal extension rate and hyphal diameter. Cessation of movement of the vesicle supply center results in spherical tip expansion, such as occurs during appressorium, spore formation or treatment with cytochalasin (Grove and Sweigard 1980).

The size and morphology of the vesicles in the hyphal apex vary widely, a feature considered likely to reflect differences in vesicle contents and function. To date, we have little understanding of the basis for these differences, of the range of constituents in any one vesicle type, or of the distinct functions different vesicle types might serve. It is envisaged that the vesicles contain wall precursors and enzymes involved in wall construction and modification or in nutrient acquisition. Some enzymes are secreted into the wall where they generate or cleave glycosidic bonds within or between wall components. Others, such as chitin and cellulose synthetases, are membrane-bound and are likely to be located within the membrane of the vesicles so that, on fusion at the cell surface, the enzymes are automatically incorporated into the plasma membrane. Vesicles with a diameter of about 150 nm have been shown to contain polysaccharides, acid phosphatase (Hill and Mullins 1980) and cellulase (Ayscough et al. 1997); vesicles of 80 and 150 nm diameter contain inosine diphosphatase (Hill and Mullins 1980). Vesicles with a diameter of about 40–70 nm have been shown to contain chitin synthase (Bracker et al. 1976). Enzymes and other proteins, such as ion channels (e.g. Garrill et al. 1992) which function in the plasma membrane within the apical dome, may not be needed in the cylindrical part of the hyphae and may thus be retrieved and possible recycled through endocytosis. Coated vesicles, indicative of the operation of endocytosis, have been observed in the hyphal apex (Hill and Mullins 1980).

The continuous supply of vesicles to the apical cluster is critical for tip growth, and is dependent on the efficient transport of vesicles from their site of production in the Golgi apparatus (or Golgi equivalent in fungi lacking recognizable Golgi stacks) (see Welter et al. 1988; Mendgen et al. 1995; Satiat-Jeunemaitre et al. 1996) in the subapical cytoplasm. Much evidence indicates that vesicle transport is dependent on the operation of cytoskeletal elements, actin microfilaments and microtubules, but details of the mechanisms involved remain to be fully elucidated. Actin microfilaments form longitudinally oriented cables and fine filaments in the cytoplasm extending, in some cases, into the apical dome. They also form peripheral plaques (filasomes) and are associated closely with the Spitzenkörper (Fig. 4B,D; e.g. Jackson and Heath 1990; Bourett and Howard 1991; Kwon et al. 1991a; Heath 1994, 1995a; Czymmek et al. 1996; Srinivasan et al. 1996; Heath and Steinberg 1999). Microtubules are longitudinally oriented and usually do not extend into the apical dome (Fig. 4E; e.g. Heath 1994, 1995a; Roberson and Vargas 1994; Czymmek et al. 1996; Bourett et al. 1998). Disruption of the function of either actin microfilaments or microtubules (through drug treatments or disruption of genes encoding cytoskeleton-associated proteins) may or may not interfere with vesicle transport and tip growth (see Chap. 11, this Vol.; Heath 1995b; Langford 1995; Lehmler et al. 1997; Inoue et al. 1998; Riquelme et al. 1998; Yamashita and May 1998). The picture that is emerging suggests that cooperation between cytoskeletal elements and/or a range of backup systems operate to ensure that the critical process of vesicle delivery to the apex continues despite breakdown of any single mechanism (Heath 1995a; Yamashita and May 1998).

Another aspect of hyphal structure with which the cytoskeleton is involved is the stratification of organelles along the hypha (Fig. 4A). Apart from the small vesicles and cytoskeletal elements, the apical dome is largely free of other organelles (Fig. 4A,B). Beneath the dome are zones rich in mitochondria or nuclei and thereafter the hypha becomes increasingly vacuolate. Endoplasmic reticulum and Golgi bodies occur throughout the cytoplasm except within the apical dome (see, for example, Roberson and Fuller 1988; Heath and Kaminskyj 1989). Hyphae also contain a system of tubular and spherical vacuoles believed to function in the transport of material (e.g. phosphate) from the apex to more distal regions of the mycelium (Chap. 12, this Vol.; Shepherd et al. 1993; Allaway et al. 1998; Cole et al. 1998). The morphology of elements of this system varies along the length of the hypha. As in studies of the transport of apical vesicles, investigations of the role of cytoskeletal elements in organelle distribution have not resulted in a single, unified model (Heath 1995a). In some cases, actin seems to be responsible, in other cases microtubules. For example, in *Pisolithus*, microtubules, but not actin microfilaments, are responsible for establishing the morphology and movement of the tubular vacuoles (Hyde et al. 1999). In *Saprolegnia*, both microtubules and actin microfilaments are involved (Bachewich and Heath 1999).

One might ask why organelles show stratification along the hypha, or why Golgi bodies are not located, for example, in the hyphal apex close to the site of vesicle fusion with the plasma membrane. As yet, the reasons for these features are still unclear. One possible explanation for the collection of vesicles from subapically located

Golgi bodies, and their accumulation in the Spitzenkörper/vesicle supply center in the apex, is that this may be a superior method of ensuring that sufficient numbers of vesicles are available to support the rates of hyphal extension that are observed (Bartnicki-Garcia 1996).

C. Influence of the Plant Surface on Hyphal Growth

Many fungi grow across the surface of the plant in an apparently unguided fashion. However, others are capable of detecting and reacting to chemical and/or topographical signals from the plant (see reviews by Wynn 1981; Hoch and Staples 1991). Hyphae of *Rhizoctonia* and other fungal species grow towards and along the grooves overlying the anticlinal walls of epidermal cells (Fig. 5A; Armentrout et al. 1987). It is thought that this polarity is due to the leakiness of the area to exudates from within the leaf (Hoch and Staples 1991). Many pathogens that invade the leaf surface through stomatal pores show preferential growth towards stomata. This guidance may be derived from chemical signals but is also achieved by the recognition of and response to the surface topography of the leaf. The growing hypha is able to detect the ridges and grooves on the substratum and orients its growth accordingly, a process called thigmotropism (Figs. 5B,C and 6A). This phenomenon has been studied in detail in species of rusts. Many *Puccinia* species, for example, grow perpendicularly to the rows of epidermal cells (Fig. 5B,C). It has been demonstrated clearly that hyphae respond to a physical signal and not a chemical one – inert replicas of the leaf surface or even artificial gratings induce the same thigmotropic response (Wynn 1976; Dickinson 1979; Hoch et al. 1987a). This growth pattern is thought to enhance the chances of hyphae encountering stomata which, in graminaceous plants, occur in longitudinal rows (Fig. 5C). The dimensions of the topographical feature that hyphae respond to have been determined with great precision for germlings of *U. appendiculatus*. Maximum response is shown to ridges or grooves spaced 0.5–15 μm apart (Hoch et al. 1987a).

The mechanisms by which hyphae achieve directed growth are still not fully understood. The position of the Spitzenkörper or vesicle supply center in the hyphal apex is clearly involved in bringing about directed growth, but how are the topographical signals perceived and the signals transduced to bring about appropriate changes in the position of the Spitzenkörper? We do know that firm adhesion of the hypha to the substratum is mandatory (Epstein et al. 1985). However, it has been established that the topographical signals are perceived in an area of the hypha in contact with the substratum, 0–10 μm from the hyphal apex (Corrêa and Hoch 1995). Studies of *U. appendiculatus* have led to the suggestions that the cytoskeleton (Hoch et al. 1986, 1987b; Bourett et al. 1987; Hoch and Staples 1991) and/or the operation of an integrin-like protein might be involved in either signal reception or transmission (see Chap. 10, this Vol.; Corrêa et al. 1996).

These chemical and physical signals not only influence the polarity of hyphal growth, but may also induce the hyphal apex to differentiate into specialized infection structures, which penetrate the host surface.

IV. Penetration of the Plant

A. Induction of Appressorium Differentiation

Signals from the plant surface not only guide the growing germ tube or hypha, but also trigger the development of infection structures capable of penetrating the host surface. In some pathogens, the infection structures are highly specialized appressoria, which invade the underlying plant cell wall via the production of a fine penetration peg. Both physical and chemical signals play a role in the induction of appressorium differentiation (Emmett and Parbery 1975; Hoch and Staples 1991). A single cue, on its own, may be sufficient for the induction of appressorial differentiation, but it is becoming increasingly clear that maximal response is achieved when several signals are perceived by the germling (see, for example, Collins and Read 1997).

Hydrophobicity of the substratum influences appressorium induction but its exact role remains controversial. Some studies of *M. grisea* have found a correlation between surface hydrophobicity and appressorial differentiation (Jelitto et al. 1994; Lee and Dean 1994) and even that a hydrophobic surface is sufficient for appressorial induction in the absence of other cues (Gilbert et al. 1996). Other studies, however, have found that the requirement is for a hard surface, whether

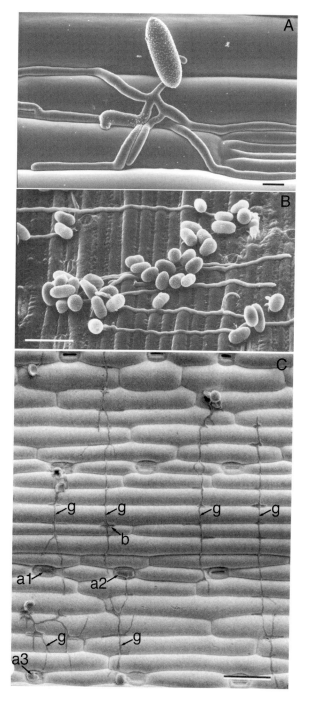

Fig. 5. A Conidium, hyphae and appressorium of *Erysiphe graminis* f. sp. *avenae* on surface of oat leaf. Hyphae grow along the grooves formed by the anticlinal walls of the epidermal cells. Cryo-SEM. *Bar* 10 μm (Carver et al. 1995). **B** Germinated urediniospores of *Puccinia graminis* f. sp. *tritici.* Hyphae grow perpendicular to the ridges and grooves on the leaf surface. Cryo-SEM. *Bar* 50 μm (Staples and Hoch 1997; reproduced with permission of Springer). **C** Urediniospores of *Puccinia hordei* on the surface of a barley leaf. Primary germ tubes (*g*) grow at right angles to the grooves formed by the anticlinal walls of the epidermal cells. On encountering a stoma the hyphal apex usually differentiates into an appressorium over the stomatal pore (*a1–3*), but occasionally does not (*b*). *Bar* 100 μm (Read et al. 1992, reproduced with permission of Cambridge University Press)

hydrophilic or hydrophobic (Xiao et al. 1994), and that the response is modulated by conidial density (Hegde and Kolattukudy 1997). These observations introduce at least two other factors into the equation, namely adhesion and chemical signals. Surface hardness and degree of hydrophobicity will influence the ability of germlings to adhere to the substratum, and the evidence consistently indicates that firm adhesion of the germling is an important requirement for appressorial differentiation (Emmett and Parbery 1975; Hoch and Staples 1991; Read et al. 1992; Jelitto et al. 1994; Liu and Kolattukudy 1999; Yamaoka and Takeuchi 1999). With regard to chemical signals, it has been postulated that at a high density of *M. grisea* conidia, appressorial differentiation is suppressed

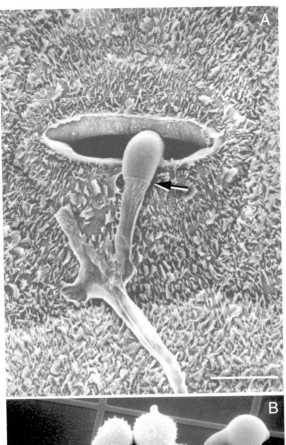

Fig. 6. A Hypha of flax rust pathogen *Melampsora lini* growing on the surface of a flax leaf. The hypha has differentiated into an appressorium over a stoma. The hypha has collapsed behind the septum (*arrow*) delineating the appressorium. Cryo-SEM. *Bar* 10 µm (Unpublished micrograph courtesy of Issei Kobayashi and CSIRO Plant Industry). **B** Urediniospores of *Uromyces appendiculatus* growing on a polystyrene replica containing ridges 0.5 µm in height. Appressoria have differentiated when the hyphal apex encountered a ridge. *Bar* 10 µm (Hoch et al. 1995, reproduced with permission of Cambridge University Press)

by self-inhibitors, which diffuse from the conidia (Liu and Kolattukudy 1999). The effect of the conidial self-inhibitors is relieved by plant surface waxes (Hegde and Kolattukudy 1997; Liu and Kolattukudy 1999).

Some pathogens, such as rust fungi in the genera *Puccinia, Uromyces* and *Melampsora*, invade leaves by forming an appressorium over a stomatal pore (Figs. 5C, 6A; Read et al. 1997). Germlings of these species can distinguish the ridges associated with the lip of the guard cells. They will also respond to stomatal impressions on plastic replicas of leaf surfaces and to ridges and grooves of similar dimensions on inert surfaces, indicating once again that surface topography can induce the response (Fig. 6B; Hoch et al. 1987a; Collins and Read 1997; Read et al. 1997). The dimensions of the topographical features that induce appressorium development have been defined precisely for the bean rust fungus, *U. appendiculatus*. A change in elevation on the substratum of 0.5 µm is optimal to trigger appressorium development, a value of similar dimensions to the height of ridges associated with the lip of the stoma (Hoch et al. 1987a; Kwon and Hoch 1991). The site of signal reception lies on the surface of the hyphae in contact with the substratum, within 20 µm from the hyphal apex (Kwon and Hoch 1991; Corrêa and Hoch 1995). In experiments where a micropipette was used to mimic the change in elevation, a period of 20 min was required for maximal response. It has also been found that the response can be inhibited by RGD-containing peptides (Corrêa et al. 1996). This sequence of three amino acids is the motif contained in extracellular matrix proteins which interact with β-integrins in animal cells (Ruoslahti 1996), as part of the reception of signals from the cell's environment (Guan and Chen 1996). In *U. appendiculatus*, appressorial development is inhibited by addition of RGD-containing peptides to a solution bathing the developing germlings or if a normally inductive glass rod is coated with RGD-peptides (Corrêa et al. 1996). One might wonder how an immobilized peptide could interact with an integrin-like receptor on the germling plasma membrane, but the cell wall is unusually thin in this receptive region and it is estimated that the β-integrin-like molecule could span the width of the wall (Corrêa et al. 1996). Antibodies to β-integrins reacted with a 95-kDa protein which had been affinity purified by reaction with RGD-containing peptides, and it will be of interest to learn more details of this molecular interaction and its role in the induction of appressorium formation (see Chap. 10, this Vol.).

Molecular genetic studies of *M. grisea* have identified another cell surface component required for appressorium induction. The gene *MPG1* encodes a small cysteine-rich protein of the hydrophobin family (Talbot et al. 1993). Hydrophobin molecules self-assemble on the surface of fungal cells and modify their hydrophobicity (Wessels 1996, 1999; Kershaw and Talbot 1998), and by doing so are likely to have a marked influence on recognition (Beckerman and Ebbole 1996) or adhesion phenomena (Talbot et al. 1996). *MPG1* null mutants undergo the initial step in appressorium induction, swelling and hook formation, but do not form appressoria on surfaces that are normally inductive (Talbot et al. 1993).

MPG1 in *M. grisea* and the putative β-integrin-like protein in *U. appendiculatus* may be involved in the first steps in fungal recognition of physical and chemical features of the plant surface. The nature of proteins that participate in this and subsequent signal transduction pathways is being unraveled through the identification of genes expressed during appressorium induction and development, and studies of the effects of their targeted deletion.

Two such proteins from *M. grisea* are thought to reside in the plasma membrane. One is a protein called Pth11p that mediates appressorium differentiation in response to cues from inductive substrata (DeZwaan et al. 1999). The sequence of Pth11p indicates that it is a transmembrane protein, and localization studies show it occurs in the plasma membrane and vacuole membranes. Mutations in *PTH11* lead to a defect in appressorium development. Conidia germinate and undergo early differentiation but, in most cases, appressoria do not mature. In *PTH11* mutants, appressoria do form at about 10–15% the frequency seen in wild-type strains indicating that the protein is not required for appressorium morphogenesis per se, but could be involved in host surface recognition (DeZwaan et al. 1999).

The other plasma membrane protein is encoded by *MagB* and is an α-subunit of a trimeric G protein (Liu and Dean 1997). G proteins function in relaying signals from activated cell surface receptors to intracellular signal transduction cascades involving adenylate cyclase, phospholipases or protein kinases (Strader et al. 1994; Bourne 1997). Deletion of the *MagB* gene significantly reduces vegetative growth, conidiation and appressorium formation, with a consequent reduction in the ability of the mutant to infect and colonize rice leaves (Liu and Dean 1997). There is evidence to support the idea that this G protein

might be signaling the induction of appressorium development by activating adenylate cyclase and/or protein kinases. *MAC1*, a gene which encodes adenylate cyclase in *M. grisea*, is also required for appressorium development (Choi and Dean 1997). Deletion of *MAC1* leads to reduction of vegetative growth; conidiation and conidium germination and no appressoria are formed on inductive surfaces. In these *MAC1* deletion mutants, appressorium formation can be restored by addition of exogenous cAMP (Choi and Dean 1997). Other studies have shown that cAMP is a second messenger that plays an integral role in appressorium formation because addition of cAMP can induce appressorium formation on non-inductive surfaces (Lee and Dean 1993; Thines et al. 1997; Yang and Dickman 1997). One of the effects of raising cAMP levels is the activation of cAMP-dependent protein kinases and induction of *M. grisea* appressorium formation by host waxes and ethylene has been shown to involve protein phosphorylation (Flaishman et al. 1995). Calyculin A, an inhibitor of protein phosphatases, induces appressorium development and protein phosphorylation, although protein kinase inhibitors (H-7 and genestein) inhibit appressorium formation and protein phosphorylation only when it is induced by ethylene and not by plant surface waxes (Flaishman et al. 1995). These and other studies suggest that there are at least two pathways involved in signal reception and transduction leading to the induction of appressorium development (Choi and Dean 1997; Dean 1997; Thines et al. 1997; Adachi and Hamer 1998).

B. Appressorium Development and Function

The first morphological sign of the initiation of appressorium development is the swelling and bending of the conidial germ tube into a hook-like structure, which proceeds to enlarge to form a domed appressorium. Ultrastructural studies show that the onset of appressorium formation is associated with the dispersal of the apical cluster of vesicles (Hoch and Staples 1987) or with the displacement of the cluster to a position close to the substratum (Mims and Richardson 1989; Bourett and Howard 1990; Howard et al. 1991a; Kwon et al. 1991b). Microtubules also become reorganized in the germ-tube tip (Kwon et al. 1991b). Continued wall deposition generates more or less isotropic expansion, although the strong adhesion to the substratum causes the base of the appressorium to be flattened (Fig. 7). Some appressoria become multi-lobed. Appressorium development and maturation is supported by nutrients stored within the conidium and transported from the conidium along the germ tube. At the end of appressorium development, a septum separates the appressorium from the germ tube and conidium; the conidium and germ tube are depleted of reserves and cytoplasm, and collapse (Figs. 6A, 7A,B).

The properties of the appressorium surface can be critical to appressorium function. In *M. grisea*, a wall approximately 150–200nm in thickness is present over the domed portion of the cell, but is absent from the base of the appressorium (Fig. 7C,D; Howard and Ferrari 1989; Bourett and Howard 1990; Howard et al. 1991a). Part of the *M. grisea* appressorium wall is a layer of melanin (Fig. 7C) that reduces the porosity of the wall to less than 1nm and renders the wall impermeable to most compounds except water (Howard et al. 1991a; Howard 1997). The appressorial pore at the base of the appressorium lacks both chitin and melanin (Fig. 7C,D), and is surrounded by a ring of wall material that adheres tightly to the underlying substratum and is presumed to function as a seal between the two surfaces (Fig. 8). Differentiation of the appressorial wall relative to this pore has also been demonstrated immunocytochemically in *C. lindemuthianum* (Pain et al. 1995). Most important, the melanized wall is impermeable to glycerol, an osmoticum that builds up to high concentrations within the appressorium. A concentration of 3.2M has been measured in appressoria of *M. grisea* (deJong et al. 1997). The high glycerol concentration leads to the generation of turgor pressures as high as 8MPa (Howard et al. 1991b; deJong et al. 1997), leading to the suggestion that physical force is used by the appressorium to breach the underlying substratum (Howard et al. 1991b; Howard 1997). Lower turgor pressures have been measured in the appressoria of other fungi, such as a value of 0.35MPa in *U. appendiculatus*, but are still sufficient to deform the lips of the underlying guard cells (Terhune et al. 1993).

The importance of melanin in host penetration by certain pathogens has been demonstrated in studies of the effects of inhibitors of melanin biosynthesis and of mutants deficient in melanin biosynthetic enzymes (Woloshuk et al. 1983; Chumley and Valent 1990; Perpetua et al. 1996; Kawamura et al. 1997). Such mutants of *M. grisea*

build up much lower concentrations of glycerol (deJong et al. 1997), develop a lower turgor pressure (Money and Howard 1996) and in both *M. grisea* and *C. lagenarium* exhibit markedly reduced pathogenicity (Chumley and Valent 1990; Perpetua et al. 1996).

In *M. grisea*, after melanization, an overlay of wall material is deposited over the appressorial pore (Fig. 8A; Bourett and Howard 1990). A fine penetration peg, about 0.7μm in diameter, forms on the pore surface and grows through the epidermal cell wall of the host (Fig. 8; Howard and Valent 1996). The cytoplasm of the penetration peg is generally devoid of organelles, including ribosomes, but contains actin microfilaments (Bourett and Howard 1991). It seems likely that the actin filaments are in some way involved in extension and function of the penetration peg, but the details of this are yet to be determined (see Howard and Valent 1996). With a turgor pressure of 8MPa, it has been calculated that the penetration peg of *M. grisea* appressoria could generate a force of 8–17μN (Money and Howard 1996), a value in good agreement with that recently measured using elastic optical wave guides for the force produced by appressorial penetration pegs of *C. graminicola* (Bechinger et al. 1999).

The fact that these pressures are sufficient for the penetration peg to force its way through a surface as hard as a plant epidermal cell wall has been elegantly demonstrated in *M. grisea* through the use of Mylar sheets which are impervious to degradation by fungal enzymes (Howard and Ferrari 1989; Bourett and Howard 1990; Howard et al. 1991a,b). These studies indicate that the fungus is capable of penetrating Mylar sheets as hard as or harder than a rice leaf surface by using force alone. However, it was also found that the fungus penetrates the rice leaf more quickly than it does Mylar sheets of similar hardness to that of the rice leaf, suggesting that, on the plant surface, the fungus uses both physical force and enzymatic digestion to penetrate the leaf surface. A number of studies have investigated the role of fungal cell wall degrading enzymes in penetration of the plant surface.

A cuticle, a major component of which is cutin, covers the aerial surface of plants. The cuticle is thus the first layer that many pathogens must breach, and one of the enzymes studied in investigations of host penetration is cutinase, the enzyme that hydrolyses ester bonds linking the fatty acids which make up the cutin polymer

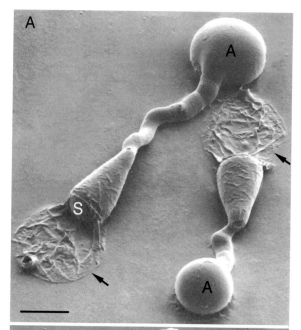

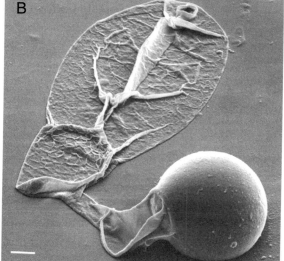

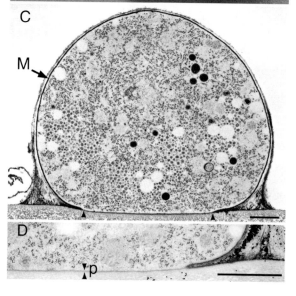

Fig. 7. A Cryo-SEM of germinated conidia of *Magnaporthe grisea*. The germ tube of each conidium has differentiated into an appressorium (*A*). The cytoplasm in the basal and middle cells of the three-celled conidia has been degraded to support growth of the appressorium; these two cells of each conidium have subsequently collapsed (*arrows*). *S* Septum. *Bar* 5.0 μm (Money and Howard 1996). **B** Mature appressorium of *Magnaporthe grisea* that has been sealed off from the subtending conidium and germ tube by a septum. The contents of the conidium and germ tube have been degraded and these cells have collapsed. Cryo-SEM. *Bar* 2.0 μm (Braun and Howard 1994a). **C** TEM of a mature *Magnaporthe grisea* appressorium. The electron dense layer of melanin (*M*) covers most of the appressorial surface but not the surface in contact with the substratum. The latter area is the appressorium pore (between *arrowheads*) and, at this stage of development, consists of a plasma membrane and appears to lack any cell wall. TEM of freeze-substituted cell. *Bar* 1.0 μm (Howard et al. 1991a). **D** High magnification image of the appressorium pore of *Magnaporthe grisea*. The section has been labelled with wheat germ agglutinin-gold to localize chitin in the fungal cell wall. No labelling is evident across the pore, reflecting the lack of cell wall material in this region. TEM of freeze-substituted cell. *P* Plasma membrane. *Bar* 1.0 μm (Howard et al. 1991a)

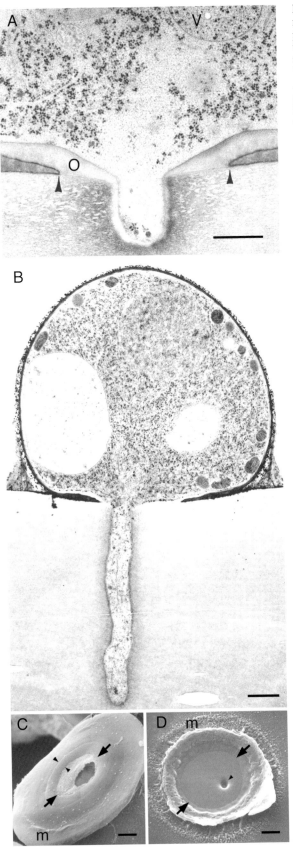

Fig. 8. A Young penetration peg formed at the appressorium pore in an appressorium of *Magnaporthe grisea*. The pore region is defined by the perimeter of the appressorium wall (*arrowheads*) and has been covered by an overlay of wall material (*O*). TEM of a freeze-substituted cell. *V* Vacuole. *Bar* 0.5 μm (Bourett and Howard 1992). **B** Appressorium of *Magnaporthe grisea* that has produced a penetration peg that has grown over 6 μm into the underlying substratum. TEM of a freeze-substituted cell labelled with wheat germ agglutinin after embedment. *Bar* 1.0 μm (Howard et al. 1991a). **C** Lower surface of an appressorium of *Magnaporthe grisea* that has been detached from the substratum. The appressorium pore (between *arrows*) is shown in the center of the appressorium base; the hole that was torn during detachment marks the site of the penetration peg. A ring (between *arrowheads*) encircles the appressorium pore. Cryo-SEM. *Bar* 1.0 μm (Howard 1997). **D** The base of an appressorium of *Magnaporthe grisea* adhering to a substratum as viewed from the inside of the appressorium after rupture of the appressorium by sonication. The hole (*arrowhead*) is an impression left in the Mylar substratum at the site of the penetration peg. Appressorium pore (between *arrows*). Extracellular matrix material (*m*). *Bar* 1.0 μm (Howard et al. 1991b)

(Martin and Juniper 1970). The results of these studies and the conclusions drawn from them regarding the importance of cutinase in host surface penetration have been controversial. On the one hand, gene disruption experiments have failed to show an essential role for cutinase (Stahl and Schäfer 1992; Sweigard et al. 1992; Yao and Koller 1995), but, on the other, the importance of the cuticle as a barrier to infection has been demonstrated in experiments using cutinase trans-genes and antibodies to cutinase. Insertion of a cutinase gene of *Fusarium solani* f. sp. *pisi* (*N. haematococca*) into *Mycosphaerella*, a pathogen that normally requires a wound in the plant surface to infect, allows the transgenic *Myco-sphaerella* to penetrate an intact surface (Dickman et al. 1989). Antibodies to cutinase inhibit infec-tion of papaya by *Mycosphaerella* (Dickman et al. 1989) and pea by *Fusarium solani* f. sp. *pisi* (Maiti and Kolattukudy 1979).

Immunolocalizations with cutinase antibodies show cutinase secretion at the infection site and association with the germ tubes produced by conidia in *Fusarium solani* f. sp. *pisi* and *Botrytis cinerea* (Shaykh et al. 1977; Coleman et al. 1993; Podila et al. 1995). The enzyme in this location could be functioning to enhance adhesion of the germlings by altering host surface properties as discussed earlier, although degradation of the cuticle at the infection site has been reported (Shaykh et al. 1977). It should also be noted that there is evidence for a role of cutinase release of cutin monomers in the induction of infection structure differentiation. Chemical inhibition of cutinase activity associated with germinating conidia of *E. graminis* f. sp. *hordei* reduced the for-mation of appressorial germ tubes and appresso-ria (Francis et al. 1996). Secretion of cutinase by the appressorial penetration peg of *C. gloeospori-oides* has also been demonstrated (Podila et al. 1995).

After breaching the cuticle, the penetration peg grows through the wall of the infected plant cell. A lack of compression, indicative of mechan-ical force, and a localized degradation of wall structure surrounding the penetration peg pro-duced by appressoria and haustorial mother cells (Fig. 9A,B) have been reported on many occasions (e.g. Chong et al. 1981; Mims et al. 1989; Clay et al. 1997) and has been taken as evidence of the action of fungal enzymes. Molecular cytochemical studies are beginning to gather evidence of both localized reduction in individual components in the plant

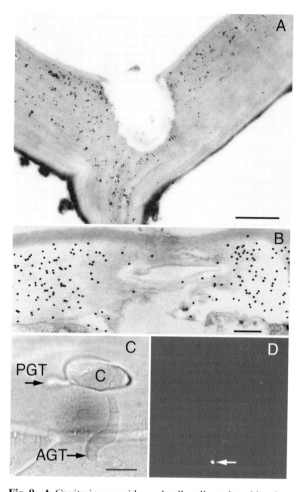

Fig. 9. A Cavity in an epidermal cell wall produced by the penetration peg of an appressorium of *Cochliobolus sativus* on a leaf of barley. The section has been immunogold-labelled with JIM5, a monoclonal antibody specific for polygalacturonic acid. The erosion of the wall suggests that the penetration peg has secreted wall-degrading enzymes. The plant cell wall contains less polygalacturonic acid near the edges of the cavity than in regions further away. *Bar* 1.0 μm (Clay et al. 1997). **B** Outer epidermal wall of *Vicia faba* that has been breached by a penetration peg formed by an appressorium of *Uromyces vignae*. The section has been labelled with JIM 7, a monoclonal antibody specific for methylesterified pectin. The wall close to the penetra-tion site has lower pectin content than in other regions. Material prepared by high pressure freezing and freeze-substitution. *Bar* 0.2 μm (Unpublished micrograph cour-tesy of H. Xu and K. Mendgen). **C** and **D** Germinated conidium of *Erysiphe graminis* f. sp. *hordei* on the surface of a barley leaf labelled with a monoclonal antibody specific for cellobiohydrolase shown in **C** bright field and **D** fluorescence illumination. The antibody (*arrow*) has labelled a highly localized area at the tip of the appressor-ial germ tube. *PGT* Primary germ tube; *AGT* appressorial germ tube. *Bar* 10 μm (Pryce-Jones et al. 1999)

cell wall at the penetration site, and of temporally and spatially restricted production and secretion of wall degrading enzymes. Reductions in pectin, polygalacturonic acid, cellulose and xyloglucan have been reported adjacent to appressorial penetration pegs produced by *Cochliobolus sativus* and *U. vignae* (Fig. 9A,B; Clay et al. 1997; Xu and Mendgen 1997). Extensive studies of the activity of cell wall degrading enzymes during infection structure formation in *U. viciae-fabae* have revealed a cascade of enzyme activities, with pectin methylesterase and cellulase activities coinciding with the differentiation of infection hyphae and haustorial mother cells (Deising and Mendgen 1992; Frittrang et al. 1992; Heiler et al. 1993; Deising et al. 1995a,b; Mendgen et al. 1996). Localization studies, one using GFP-polygalacturonase constructs, have shown polygalacturonase in germ tubes and appressoria penetration pegs of *C. lindemuthianum* (Dumas et al. 1999) and restriction of cellobiohydrolase at the appressorial penetration peg of *E. graminis* (Fig. 9C,D; Pryce-Jones et al. 1999). Antibodies directed towards subtilisin-like proteases and pectate lyase have also been shown to reduce pathogenicity of *M. poae* and *C. gloeosporioides*, respectively (Wattad et al. 1997; Sreedhar et al. 1999). Taken together, these recent studies of the production of cell wall degrading enzymes by biotrophic fungi are building evidence of the localized degradation of the plant cell wall at the penetration site.

V. Acquisition of Nutrients from the Plant

A. Haustoria and Intracellular Hyphae of Biotrophic Pathogens

The main aim of plant invasion by phytopathogenic fungi is to obtain the nutrients they need for their growth, development and reproduction. Fungi employ a variety of infection strategies to achieve this aim. Nectrotrophic fungi infect mature healthy plants, are able to overcome host defenses, and in the process obtain nutrients by killing host cells and forming expanding necrotic lesions. Biotrophic fungi, on the other hand, establish a close and stable relationship with living host cells in order to obtain the nutrients they need for their further growth and development. They do this by growing through the plant cell wall or

cuticle, and forming subcuticular hyphae, haustoria or intracellular structures, which are specialized for nutrient uptake.

If the fungus has developed an appressorium on the leaf surface, the penetration peg it produces may grow directly through the epidermal cell wall and form a haustorium or it may grow through a stomatal aperture into the substomatal cavity. In this latter case, the fungus develops a substomatal vesicle and infection hyphae that grow through the intercellular spaces in the leaf until contact is made with a parenchyma cell (Fig. 10A). The apex of the infection hypha then differentiates to form a haustorial mother cell, and it is this cell which produces a penetration peg which grows through the wall of the plant cell and develops an haustorium (Fig. 10B; see Mendgen and Deising 1993; Mendgen et al. 1996). Little is known of the regulation of the differentiation of, and penetration by, haustorial mother cells, although it is generally believed that the haustorial mother cell relies on enzymatic degradation of the plant cell wall rather than mechanical force for wall penetration. Degradation of the plant cell wall around the haustorial mother cell penetration peg has been observed (Chong et al. 1981; Taylor and Mims 1991). Penetration pegs of both appressoria and haustoria mother cells puncture the cell wall but do not breach the plasma membrane of the plant cell that is invaginated by the expanding haustorium or intracellular hypha (Fig. 10B,C).

The degree of structural and functional specialization of the nutrient-absorbing haustoria and intracellular hyphae is variable (Manners and Gay 1983). The body of the haustorium, often subtended from a narrow neck region, is approximately spherical or lobed in the downy mildews and dikaryotic rust spore infections (Fig. 10B) but is filamentous and unbranched in the monokaryotic rust spore infections. Haustoria of the powdery mildews have a high surface-to-volume ratio because of multiple projections from the haustorial body (Manners and Gay 1983). The extensions may project outwards (as in *E. graminis*) or they may wrap around the haustorial body. Hemibiotrophic fungi, such as species of *Colletotrichum* that grow biotrophically for a period of time before becoming necrotrophic, produce an infection vesicle and a primary hyphae that together constitute the intracellular hyphae (Fig. 10C). The wall and plasma membrane of the haustorium or intracellular hyphae are continuous with the wall and plasma membrane of the

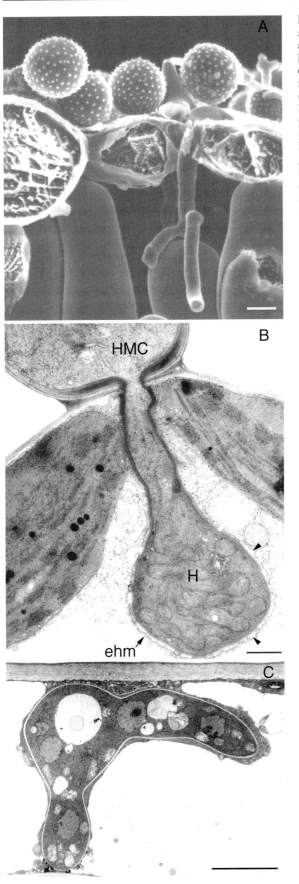

Fig. 10. A Cryo-SEM image of urediniospores of *Melampsora lini* that have germinated on the surface of a flax leaf and penetrated the leaf through a stoma. An infection hypha within the leaf has made contact with a mesophyll cell. *Bar* 1.0 μm (Hardham and Mitchell 1998). **B** TEM image of a haustorial mother cell (*HMC*) and haustorium (*H*) of *Melampsora lini* in a flax leaf mesophyll cell. The haustorium is surrounded by the invaginated host cell plasma membrane (*ehm*). The extrahaustorial matrix (*arrowheads*) lies between the haustorium wall and the host plasma membrane. *Bar* 0.5 μm (Unpublished micrograph courtesy of L. Murdoch). **C** TEM image of an infection vesicle and primary hyphae of *Colletotrichum lindemuthianum* inside a hypocotyl epidermal cell of *Phaseolus vulgaris*. *Bar* 5.0 μm (O'Connell and Bailey 1991)

appressorium or haustorial mother cell. In the powdery mildews, the haustorium becomes delineated from the appressorium by a septum in the neck, but in the downy mildews and rust fungi a septum is not formed (Manners and Gay 1983).

The invaginated host plasma membrane surrounding the haustorium is termed the extrahaustorial membrane (Fig. 10B). Over most of the haustorium surface, the wall of the haustorium (or intracellular hypha) is separated from the plant plasma membrane by an extrahaustorial or interfacial matrix layer (Figs. 10B, 11B). In the powdery mildews and dikaryotic rust infections, both the fungal and plant plasma membranes are tightly appressed against the haustorial wall in the neck of the haustorium, an arrangement that results in sealing of the extrahaustorial matrix and its separation from the apoplastic space of the plant leaf (Heath 1976). Structures in the neck region of the downy mildews are not of similar morphology to

the neckbands in the powdery mildews and rusts, but are thought to be functionally equivalent (Woods and Gay 1983).

It is widely accepted that haustoria and intracellular hyphae function in nutrient uptake from infected plants but direct evidence of this is limited (see Manners and Gay 1983). The clearest data come from experiments that monitored the uptake of ^{14}C from $^{14}CO_2$ supplied to leaves of pea plants, and that showed accumulation of labelled material from plant photoassimilates by the pea powdery mildew, $E.\ pisi$ (Manners and Gay 1978, 1980, 1982). Uptake of radioactively labelled amino acids by rust haustoria has also been demonstrated (Mendgen 1981). In recent investigations, components at the plant-fungal interface have been studied with the aim of identifying molecular specializations indicative of the role of haustoria in nutrient uptake. New information on the composition of the haustorial wall, interfacial

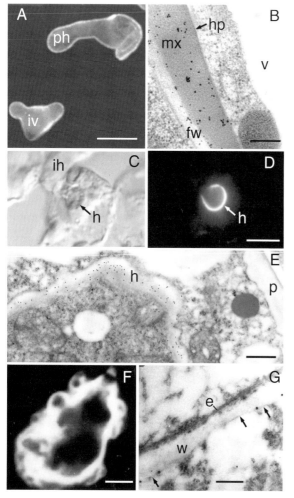

Fig. 11. A Immunofluorescence imaging of primary hypha (*PH*) and infection vesicle (*IV*) produced by *Colletotrichum lindemuthianum* within an epidermal cell of *Phaseolus vulgaris* labelled with monoclonal antibody UB25 that reacts with a 40.5-kDa glycoprotein in the intracellular hyphal walls and interfacial matrix layer. *Bar* 10 μm (Perfect et al. 1998, reproduced with permission of Blackwell Science Ltd.). **B** TEM immunogold labelling with monoclonal antibody UB25 of the wall (*FW*) of a primary hypha of *Colletotrichum lindemuthianum* and of the interfacial matrix (*Mx*) between the intracellular hypha and the invaginated host plasma membrane (*HP*) of the infected *Phaseolus vulgaris* leaf cell. *V* Host cell vacuole. *Bar* 0.25 μm (Pain et al. 1994). **C** Bright field and **D** immunofluorescence light micrographs of an infection hypha (*ih*) and a haustorium (*h*) of *Melampsora lini* in a flax leaf. The wall of the haustorium, but not of the hypha or plant cell, is labelled with monoclonal antibody, ML-1. *Bar* 10 μm (Unpublished micrographs courtesy of L. Murdoch). **E** TEM image of immunogold labelling of a section of an infected flax mesophyll cell infected by *Melampsora lini*. The haustorial wall (*h*), but not the mesophyll cell wall (*p*), is labelled with ML-1. *Bar* 0.5 μm (Murdoch et al. 1998, reproduced with kind permission of Kluwer Academic Publishers). **F** Immunofluorescence micrograph of isolated haustorium of *Erysiphe pisi* labelled with monoclonal antibody UB8 that reacts with a 62-kDa glycoprotein in the haustorium plasma membrane. *Bar* 5.0 μm (Mackie et al. 1993). **G** TEM image of immunogold labelling of a haustorium of *Erysiphe pisi* in a pea leaf cell. Monoclonal antibody UB8 labels the plasma membrane of the haustorium (*arrows*). *w* Haustorial wall. *e* extrahaustorial membrane. *Bar* 2.0 μm (Mackie et al. 1991)

matrix and plasma membranes of fungus and plant is increasing our understanding of the structure and function of haustoria. This recent research is summarized in the following discussion, starting with the plant plasma membrane surrounding the haustorium in the infected plant cell and progressing to the extrahaustorial/interfacial matrix, haustorial wall and, finally, the haustorial plasma membrane.

Early ultrastructural studies showed that the extrahaustorial membrane formed in rust, powdery mildew and downy mildew infections was morphologically distinct from the plasma membrane in other parts of the cell (Littlefield and Bracker 1972; see also Aist and Bushnell 1991; Callow and Green 1996). The extrahaustorial membrane is 1.5–2.0 times thicker than is usual for the plant plasma membrane and is heavily glycosylated. Cytochemical studies also revealed that the extrahaustorial membrane surrounding powdery mildew and dikaryotic rust haustoria lacks the ATPase activity which is generally seen in the plant plasma membrane and which is still present in the plasma membrane in other parts of the infected cell (Spencer-Phillips and Gay 1981; see also Gay and Woods 1987). In contrast, the extrahaustorial membrane surrounding filamentous haustoria of monokaryotic rust infections contains ATPase activity, although in some cases there is a gradient of reduced ATPase activity towards the tip of the haustorium, i.e. the end opposite the point of entry into the cell (Woods and Gay 1987; Baka et al. 1995). The plant membrane surrounding intracellular hyphae of C. lindemuthianum also contains ATPase activity and thus lacks this form of specialization shown by the powdery mildew and dikaryotic rust infections (O'Connell 1987). These results have been interpreted as indicating that the haustoria of monokaryotic rusts and the hemibiotrophs are less specialized than the dikaryotic haustoria (Baka et al. 1995; O'Connell 1987).

Immunocytochemical studies have also shown that a large glycoprotein, of relative molecular weight greater than 250 kDa and normally present in the plasma membrane of pea cells, was absent from the extrahaustorial membrane surrounding E. pisi haustoria during early stages of their development (Roberts et al. 1993). The antigen, labelled by monoclonal antibody UB9, appeared in the extrahaustorial membrane 5–7 days after infection – but, even at this stage, only about 20% of the haustoria was labelled. Conversely, another monoclonal antibody, UB11, revealed the presence of a 250-kDa glycoprotein in the extrahaustorial membrane surrounding E. pisi haustoria that is absent from other parts of the host plasma membrane or from any fungal membranes (Roberts et al. 1993). This is the first evidence of a novel protein in the extrahaustorial membrane.

We still have little understanding of the composition and function of the interfacial or extrahaustorial matrix between the haustorial wall and plant plasma membrane. The matrix contains carbohydrates, including β-polysaccharides (Chard and Gay 1984; O'Connell 1987), and is thought to be of gel-like consistency. The origin of material within the matrix has been the subject of controversy, but it now seems likely that components in the matrix are synthesized and secreted by both the fungus and the plant. Lectin and antibody labelling of the matrix in haustorial complexes of E. pisi and Spaerotheca pannosa var. rosae showed the absence of molecules typically present in plant or fungal walls, including esterified and unesterified pectins, cellulose, extensins and chitin (O'Connell and Ride 1990; Mackie et al. 1991). A monoclonal antibody, UB7, that reacts with a 59-kDa glycoprotein in the E. pisi mycelial and haustorial walls, also failed to label the extrahaustorial matrix (Mackie et al. 1991; Green et al. 1995). However, labelling of dikaryotic rust and powdery mildew haustorial complexes in wheat and barley with a polyclonal antibody directed towards maize threonine-hydroxyproline-rich glycoproteins has shown the presence of these molecules within the extrahaustorial matrix (Hippe-Sanwald et al. 1994). Immunocytochemical analysis of the extrahaustorial matrix of monokaryotic rust haustoria of U. vignae has also given evidence of hydroxyproline-rich glycoproteins and arabinogalactan proteins (Stark-Urnau and Mendgen 1995). In addition, monoclonal antibodies UB20, UB22 and UB25, raised against surface components on Colletotrichum cells and reacting with sets of N-linked glycoproteins, label the interfacial matrix in infected bean cells (Pain et al. 1994; O'Connell et al. 1996a). The antigens recognized by UB20 and UB22 also occur on the surface of conidia, germ tubes and appressoria (Pain et al. 1992). The antigen recognized by UB25, on the other hand, is specific for the surface of intracellular hyphae growing inside living plant cells during the biotrophic phase of growth, and is absent from the surface of primary hyphae growing intercellularly and from hyphae growing on the plant surface (Fig. 11A,B; Pain et al. 1994; O'Connell et al. 1996b). This antigen is a fungal proline-rich glycoprotein (Perfect et al. 1998).

The wall of haustoria or intracellular hyphae contains molecules common to the walls of other fungal cells, as well as molecules that are specific to the wall of these infection structures. In *E. pisi*, monoclonal antibody UB7 reacts with a 62-kDa glycoprotein present in cell walls and plasma membranes in haustoria and mycelia (Mackie et al. 1991; Callow et al. 1992). Immunocytochemical studies of haustoria in *Melampsora lini*, the flax rust pathogen, have also shown the presence of components that are common to haustorial and mycelial walls (Murdoch et al. 1998). The antibody ML4, which reacts with the *Melampsora lini* wall antigen, does not bind to the walls of *Puccinia* species, suggesting that it is unlikely to react with a common fungal wall component. It is possible, however, that ML4 targets a specific epitope on a more widespread molecule. Antibody binding was not inhibited by preincubation of the antibody with chitin or by treatment of the haustoria with chitinase, suggesting that the antigen was unlikely to be chitin (Murdoch et al. 1998). In addition, this latter study generated three monoclonal antibodies, ML1, ML2, and ML3, that react with components that are specific for the *Melampsora lini* haustorial wall (Fig. 11C,D,E; Murdoch et al. 1998). This is the first evidence of molecular specialization of the haustorial wall to substantiate the interpretations of structural and compositional wall differentiation made in early ultrastructural studies of flax rust haustoria by Littlefield and Bracker (1972). Antibodies raised against calmodulin also react with the *Melampsora lini* haustorial walls but not with mycelial or plant cell walls or with haustorial walls of maize and wheat rust fungi (Murdoch and Hardham 1998). The nature of this latter antigen remains enigmatic, but the immunolabelling gives further evidence of molecular specialization of the haustorial wall.

Finally, molecular and cell biology studies are uncovering molecular details of the specialization of the haustorial plasma membrane. Immunocytochemical investigations of *E. pisi* have identified 45-kDa and 62-kDa proteins, localized by monoclonal antibodies UB8 (Fig. 11F,G) and UB10, respectively, and demonstrated that these are restricted to the plasma membrane of haustoria (Callow et al. 1992; Mackie et al. 1993). Immunocytochemical labelling has also localized a 60-kDa putative amino acid transporter, named PIG-2, to the plasma membrane of haustorial bodies of the bean rust fungus, *U. fabae* (Hahn et al. 1997). This protein is absent from the neck region of the haustorium, from the haustorial mother cell and from intercellular hyphae. The gene encoding this putative transporter has been cloned and fully sequenced, and expression of the gene is restricted to haustoria (Hahn et al. 1997).

The PIG-2 transporter is similar to other symport amino acid permeases, which transport amino acids together with protons (Horak 1986). These carriers use the transmembrane pH gradient generated by a plasma membrane ATPase to power amino acid uptake. Until recently, the occurrence of H^+-ATPase in haustorial plasma membrane has been unclear (see Baka et al. 1995), but it has now been shown biochemically that the level of ATPase activity in plasma membranes from haustoria of *U. fabae* is eightfold higher than that in the plasma membrane from urediniospores, and fourfold higher than that in germ tubes (Struck et al. 1996). Western blots using a monoclonal antibody raised against corn plasma membrane H^+-ATPase show reaction with a 104-kDa protein in the *U. fabae* membranes. The gene encoding this enzyme has been cloned (Struck et al. 1998). There is thus firm evidence for both an amino acid transporter and H^+-ATPase activity in the plasma membrane of rust haustoria, consistent with the specialized role of haustoria in nutrient uptake from infected plant cells.

B. Colonization of Plant Tissues: the Role of Cell Wall Degrading Enzymes

Previous sections of this chapter have reviewed the role of cutinases in modifying the plant surface, in adhesion and in the induction of infection structure differentiation, and of other enzymes in penetration of plant cell walls by appressorial and haustorial mother cell penetration pegs. The highly localized nature of wall degradation associated with these enzymes in biotrophic infections contrasts with the widespread destruction of cell walls and protoplasts that occurs during necrotrophic growth. While the major role of cell wall degrading enzymes in host colonization is accepted, recent studies have addressed the question of the absolute requirement for individual categories of enzymes in the infection process.

Targeted deletion of one or more of the genes encoding cell wall degrading enzymes has generally failed to result in any detectable reduction in pathogen virulence. These studies include disruption of cutinase genes from *N. haematococca* (Stahl and Schäfer 1992), *M. grisea* (Sweigard et al. 1992) and *Botrytis cinerea* (van Kan et al. 1997); disruption of extracellular proteases, exo-β1,3-glucanase, cellulase, xylanases and polygalacturonases from *Cochliobolus carbonum* (Scott-Craig et al. 1990; Apel et al. 1993; Schaeffer et al. 1994; Sposato et al. 1995; Apel-Birkhold and Walton 1996; Murphy and Walton 1996; Scott-Craig et al. 1998); and disruption of polygalacturonase from *Cryphonectria parasitica* (Gao et al. 1996) and pectin lyase from *Glomerella cingulata* (Bowen et al. 1995). However, in almost all cases, residual enzyme activity has been present in the mutant pathogens, suggesting that other isozymes or related genes are still functioning.

In contrast to the studies listed above, the introduction of a cutinase gene into *Mycosphaerella* (Dickman et al. 1989) and of a polygalacturonase gene into *Aspergillus flavus* (Shieh et al. 1997) increased the pathogenicity of these organisms on pea and cotton, respectively. In addition, the requirement for a specific endopolygalacturonase gene, *Bcpg1*, from *Botrytis cinerea* for full virulence has recently been reported (Ten Have et al. 1998). The *Bcpg1* mutants were still pathogenic and could initiate primary infection; however, growth of the lesions beyond the point of inoculation was severely reduced. *Bcpg1* is one of a number of genes encoding endopolygalacturonases in *Botrytis cinerea*, and is constitutively expressed in planta and in vitro. The authors suggest that a product of Bcpg1 enzymatic activity could induce expression of other endopolygalacturonase genes whose encoded enzymes may be responsible for the secondary tissue maceration.

has seen the emergence of the first biochemical characterization of a fungal adhesive molecule, data that are needed to complement the cellular details of the synthesis and secretion of adhesives onto the plant surface. It has also seen a rapid advance in the identification of molecules that participate in the signal transduction pathways that regulate the polarity of hyphal growth and the induction of infection structure differentiation on the plant surface. Many details of the molecular and cellular basis of appressorium function in host penetration have been elucidated, with exciting information on the mechanisms responsible for the generation of sometimes amazingly large turgor pressures in these cells. Molecular cytology has also revealed evidence of specializations associated with host-pathogen interaction at the haustorium-plant interface, including identification of amino acid transporters in the haustorium membrane.

Many of these discoveries are the result of the development of innovative technical approaches. Digitally enhanced video microscopy has enabled the Spitzenkörper in the hyphal apex to be observed over extended periods of time in living cells; the use of freeze-substitution and low temperature scanning electron microscopy for the preparation of material for light microscopy and ultrastructural studies has provided samples of cell ultrastructure preserved in a state much closer to that of the living cells than achieved by chemical fixation; fluorescent and gold-conjugated lectins and antibodies have allowed highly specific and spatially precise localization of selected cell molecules and organelles, enabling their distribution and fate to be followed throughout development. It is clear that we have entered a new era of research into the cell biology of fungal infection of plants. The next 5 years promise to be even more productive and should be very exciting indeed!

VI. Conclusions

This chapter has reviewed our current understanding of the cellular, and to a lesser extent, molecular basis of the infection of plants by pathogenic fungi. There has been a deliberate focus on papers published over the last 5 years, and this has highlighted the wealth of new information that has become available during that time. This period

References

Adachi K, Hamer JE (1998) Divergent cAMP signaling pathways regulate growth and pathogenesis in the rice blast fungus *Magnaporthe grisea*. Plant Cell 10: 1361–1373

Aist JR, Bushnell WR (1991) Invasion of plants by powdery mildew fungi, and cellular mechanisms of resistance. In: Cole GT, Hoch HC (eds) The fungal spore and disease initiation in plants and animals. Plenum Press, New York, pp 321–345

Allan RH, Thorpe CJ, Deacon JW (1992) Differential tropism to living and dead cereal root hairs by the biocontrol fungus *Idriella bolleyi*. Physiol Mol Plant Pathol 41:217–226

Allaway WG, Ashford AE, Heath IB, Hardham AR (1998) Vacuolar reticulum in oomycete hyphal tips: an additional component of the Ca^{2+} regulatory system? Fungal Genet Biol 22:209–220

Apel-Birkhold PC, Walton JD (1996) Cloning, disruption, and expression of two endo-β1,4-xylanase genes, *XYL2* and *XYL3*, from *Cochliobolus carbonum*. Appl Environ Microbiol 62:4129–4135

Apel PC, Panaccione DG, Holden FR, Walton JD (1993) Cloning and targeted gene disruption of *XYL1*, a β1,4-xylanase gene from the maize pathogen *Cochliobolus carbonum*. Mol Plant Microbe Interact 6:467–473

Armentrout VN, Downer AJ, Grasmick DL, Weinhold AR (1987) Factors affecting infection cushion development by *Rhizoctonia solani* on cotton. Phytopathology 77:623–630

Ayscough KR, Stryker J, Pokala N, Sanders M, Crews P, Drubin DG (1997) High rates of actin filament turnover in budding yeast and roles for actin in establishment and maintenance of cell polarity revealed using the actin inhibitor latrunculin-A. J Cell Biol 137:399–416

Bachewich C, Heath IB (1998) Radial F-actin arrays precede new hypha formation in *Saprolegnia*: implications for establishing polar growth and regulating tip morphogenesis. J Cell Sci 111:2005–2016

Bachewich C, Heath IB (1999) Cytoplasmic migrations and vacuolation are associated with growth recovery in hyphae of *Saprolegnia*, and are dependent on the cytoskeleton. Mycol Res 103:849–858

Baka ZA, Larous L, Lösel DM (1995) Distribution of ATPase activity at the host-pathogen interfaces of rust infections. Physiol Mol Plant Pathol 47:67–82

Barja F, Chappuis M-L, Turian G (1993) Differential effects of anticytoskeletal compounds on the localization and chemical patterns of actin in germinating conidia of *Neurospora crassa*. FEMS Microbiol Lett 107:261–266

Bartnicki-Garcia S (1973) Fundamental aspects of hyphal morphogenesis. In: Ashworth JM, Smith JE (eds) Microbial differentiation. Cambridge Univ Press, Cambridge, pp 245–267

Bartnicki-Garcia S (1996) The hypha: unifying thread of the fungal kingdom. In: Sutton BC (ed) A century of mycology. Cambridge Univ Press, Cambridge, pp 105–133

Bartnicki-Garcia S, Sing VO (1987) Adhesion of zoospores of *Phytophthora* to solid surfaces. In: Fuller MS, Jaworski A (eds) Zoosporic fungi in teaching & research. Southeastern Publishing Corporation, Athens, GA, pp 279–283

Bartnicki-Garcia S, Hergert F, Gierz G (1989) Computer simulation of fungal morphogenesis and the mathematical basis for hyphal (tip) growth. Protoplasma 153:46–57

Bartnicki-Garcia S, Hergert F, Gierz G (1990) A novel computer model for generating cell shape: application to fungal morphogenesis. In: Kuhn PJ, Trinci APJ, Jung MJ, Goosey MW, Copping LG (eds) Biochemistry of cell walls and membranes in fungi. Springer, Berlin Heidelberg New York, pp 43–60

Bechinger C, Giebel K-F, Schnell M, Leiderer P, Deising HB, Bastmeyer M (1999) Optical measurements of invasive forces exerted by appressoria of a plant pathogenic fungus. Science 285:1896–1899

Beckerman JL, Ebbole DJ (1996) MPG1, a gene encoding a fungal hydrophobin of *Magnaporthe grisea*, is involved in surface recognition. Mol Plant Microbe Interact 9:450–456

Bimpong CE, Hickman CJ (1975) Ultrastructural and cytochemical studies of zoospores, cysts, and germinating cysts of *Phytophthora palmivora*. Can J Bot 53:1310–1327

Bourett TM, Howard RJ (1990) *In vitro* development of penetration structures in the rice blast fungus *Magnaporthe grisea*. Can J Bot 68:329–342

Bourett TM, Howard RJ (1991) Ultrastructural immunolocalization of actin in a fungus. Protoplasma 163:199–202

Bourett TM, Howard RJ (1992) Actin in penetration pegs of the fungal rice blast pathogen, *Magnaporthe grisea*. Protoplasma 168:20–26

Bourett T, Hoch HC, Staples RC (1987) Association of the microtubule cytoskeleton with the thigmotropic signal for appressorium formation in *Uromyces*. Mycologia 79:540–545

Bourett TM, Czymmek KJ, Howard RJ (1998) An improved method for affinity probe localization in whole cells of filamentous fungi. Fungal Genet Biol 24:3–13

Bourne HR (1997) How receptors talk to trimeric G proteins. Curr Opin Cell Biol 9:134–42

Bowen JK, Templeton MD, Sharrock KR, Crowhurst RN, Rikkerink EHA (1995) Gene inactivation in the plant pathogen *Glomerella cingulata*: three strategies for the disruption of the pectin lyase gene *pnlA*. Mol Gen Genet 246:196–205

Bracker CE, Ruiz-Herrera J, Bartnicki-Garcia S (1976) Structure and transformation of chitin synthetase particles (chitosomes) during microfibril synthesis in vitro. Proc Natl Acad Sci USA 73:4570–4574

Brambl R, Dunkle LD, Van Etten JL (1978) Nucleic acid and protein synthesis during fungal spore germination. In: Smith JE, Berry DR (eds) The filamentous fungi. Arnold, London, pp 94–118

Braun EJ, Howard RJ (1994a) Adhesion of fungal spores and germlings to host plant surfaces. Protoplasma 181:202–212

Braun EJ, Howard RJ (1994b) Adhesion of *Cochliobolus heterostrophus* conidia and germlings to leaves and artificial surfaces. Exp Mycol 18:211–220

Bullock S, Ashford AE, Willets HJ (1980) The structure and histochemistry of sclerotia of *Sclerotinia minor* Jagger II. Histochemistry of extracellular substances and cytoplasmic reserves. Protoplasma 104:333–351

Bullock S, Willets HJ, Ashford AE (1983) The structure and histochemistry of sclerotia of *Sclerotinia minor* Jagger III. Changes in ultrastructure and loss of reserve materials during carpogenic germination. Protoplasma 117:214–225

Callow JA, Green JR (1996) The plant plasma membrane in fungal disease. In: Smallwood M, Knox JP, Bowles DJ (eds) Membranes: specialized functions in plants. Bios Scientific Publishers, Oxford, pp 543–562

Callow JA, Mackie A, Roberts AM, Green JR (1992) Evidence for molecular differentiation in powdery mildew haustoria through the use of monoclonal antibodies. Symbiosis 14:237–246

Carver TLW, Thomas BJ, Ingerson-Morris SM (1995) The surface of *Erysiphe graminis* and the production of extracellular material at the fungus-host interface during germling and colony development. Can J Bot 73:272–287

Chard JM, Gay JL (1984) Characterization of the parasitic interface between *Erysiphe pisi* and *Pisum sativum* using fluorescent probes. Physiol Plant Pathol 25: 259–276

Choi W, Dean RA (1997) The adenylate cyclase gene *MAC1* of *Magnaporthe grisea* controls appressorium formation and other aspects of growth and development. Plant Cell 9:1973–1983

Chong J, Harder DE, Rohringer R (1981) Ontogeny of mono- and dikaryotic rust haustoria: cytochemical and ultrastructural studies. Phytopathology 71:975–983

Chumley FG, Valent B (1990) Genetic analysis of melanin-deficient, nonpathogenic mutants of *Magnaporthe grisea*. Mol Plant Microbe Interact 3:135–143

Clark MC, Melansson DL, Page OT (1978) Purine metabolism and differentiatial inhibition of spore germination in *Phytophthora infestans*. Can J Microbiol 24: 1032–1038

Clay RP, Bergmann CW, Fuller MS (1997) Isolation and characterization of an endopolygalacturonase from *Cochliobolus sativus* and a cytological study of fungal penetration of barley. Phytopathology 87:1148–1159

Clement JA, Porter R, Butt TM, Beckett A (1994) The role of hydrophobicity in attachment of urediniospores and sporelings of *Uromyces viciae-fabae*. Mycol Res 98:1217–1228

Clement JA, Porter R, Butt TM, Beckett A (1997) Characteristics of adhesion pads formed during imbibition and germination of urediniospores of *Uromyces viciae-fabae* on host and synthetic surfaces. Mycol Res 101:1445–1458

Cole L, Orlovich DA, Ashford AE (1998) Structure, function, and motility of vacuoles in filamentous fungi. Fungal Genet Biol 24:86–100

Coleman JOD, Hiscock SJ, Dewey FM (1993) Monoclonal antibodies to purified cutinase from *Fusarium solani* f. sp. *pisi*. Physiol Mol Plant Pathol 43:391–401

Collins TJ, Read ND (1997) Appressorium induction by topographical signals in six cereal rusts. Physiol Mol Plant Pathol 51:169–179

Cope M, Webb MC, O'Gara ET, Philip BA, Hardham AR (1996) Immunocytochemical comparison of peripheral vesicles in zoospores of *Phytophthora* and *Pythium* species. Mycologia 88:523–532

Corrêa A Jr, Hoch HC (1995) Identification of thigmoresponsive loci for cell differentiation in *Uromyces* germlings. Protoplasma 186:34–40

Corrêa A Jr, Staples RC, Hoch HC (1996) Inhibition of thigmostimulated cell differentiation with RGD-peptides in *Uromyces* germlings. Protoplasma 194:91–102

Czymmek KJ, Bourett TM, Howard RJ (1996) Immunolocalization of tubulin and actin in thick-sectioned fungal hyphae after freeze-substitution fixation and methacrylate de-embedment. J Microsc 181:153–161

Deacon JW (1997) Modern mycology. Blackwell Science, Oxford

Dean RA (1997) Signal pathways and appressorium morphogenesis. Annu Rev Phytopathol 35:211–234

Dearnaley JDW, Maleszka J, Hardham AR (1996) Synthesis of zoospore peripheral vesicles during sporulation of *Phytophthora cinnamomi*. Mycol Res 100:39–48

Deising H, Mendgen K (1992) Developmental control of enzyme production and cell wall modification in rust fungi, and defence reactions of the host plant. In: Stahl U, Tudzynski P (eds) Molecular biology of filamentous fungi. VCH, Weinheim, pp 27–44

Deising H, Nicholson RL, Haug M, Howard RJ, Mendgen K (1992) Adhesion pad formation and the involvement of cutinase and esterases in the attachment of urediospores to the host cuticle. Plant Cell 4:1101–1111

Deising H, Rauscher M, Haug M, Heiler S (1995a) Differentiation and cell wall degrading enzymes in the obligately biotrophic rust fungus *Uromyces viciae-fabae*. Can J Bot 73 [Suppl 1]:S624–S631

Deising H, Frittrang AK, Kunz S, Mendgen K (1995b) Regulation of pectin methylesterase and polygalacturonate lyase activity during differentiation of infection structures in *Uromyces viciae-fabae*. Microbiology 141:561–571

Deising H, Heiler S, Rauscher M, Xu H, Mendgen K (1996) Cellular aspects of rust infection structure differentiation. In: Nicole M, Gianinazzi-Pearson V (eds) Histology, ultrastructure and molecular cytology of plant-microorganism interactions. Kluwer, Dordrecht, pp 135–156

deJong JC, McCormack BJ, Smirnoff N, Talbot NJ (1997) Glycerol generates turgor in rice blast. Nature 389: 244–245

DeZwaan TM, Carroll AM, Valent B, Sweigard JA (1999) *Magnaporthe grisea* Pth11p is a novel plasma membrane protein that mediates appressorium differentiation in response to inductive substrate cues. Plant Cell 11:2013–2030

Dickinson S (1979) Growth of *Erysiphe graminis* on artificial membranes. Physiol Plant Pathol 15:219–221

Dickman MB, Podila GK, Kolattukudy PE (1989) Insertion of cutinase gene into a wound pathogen enables it to infect intact host. Nature 342:446–448

Dumas B, Centis S, Sarrazin N, Esquerré-Tugayé M-T (1999) Use of green fluorescent protein to detect expression of an endopolygalacturonase gene of *Colletotrichum lindemuthianum* during bean infection. Appl Environ Microbiol 65:1769–1771

Emmett RW, Parbery DG (1975) Appressoria. Annu Rev Phytopathol 13:147–163

Epstein L, Laccetti L, Staples RC, Hoch HC, Hoose WA (1985) Extracellular proteins associated with induction of differentiation in bean rust urediospore germlings. Phytopathology 75:1073–1076

Epstein L, Nicholson RL (1997) Adhesion of spores and hyphae to plant surfaces. In: Carroll G, Tudzynski P (eds) The Mycota, vol V, part A. Plant relationships. Springer, Berlin Heidelberg New York, pp 11–25

Flaishman MA, Hwang C-S, Kolattukudy PE (1995) Involvement of protein phosphorylation in the induction of appressorium formation in *Colletotrichum gloeosporioides* by its host surface wax and ethylene. Physiol Mol Plant Pathol 47:103–117

Francis SA, Dewey FM, Gurr SJ (1996) The role of cutinase in germling development and infection by *Erysiphe graminis* f. sp. *hordei*. Physiol Mol Plant Pathol 49:201–211

Frittrang AK, Deising H, Mendgen K (1992) Characterization and partial purification of pectinesterase, a differentiation-specific enzyme of *Uromyces viciae-fabae*. J Gen Microbiol 138:2213–2218

Gao S, Choi GH, Shain L, Nuss DL (1996) Cloning and targeted disruption of *enpg-1*, encoding the major in vitro
extracellular endopolygalacturonase of the chestnut
blight fungus, *Cryphonectria parasitica*. Appl Environ
Microbiol 62:1984–1990

Garrill A, Lew RR, Heath IB (1992) Stretch-activated Ca^{2+}
and Ca^{2+}-activated K^+ channels in the hyphal tip
plasma membrane of the oomycete *Saprolegnia ferax*.
J Cell Sci 101:721–730

Gay JL, Woods AM (1987) Induced modifications in the
plasma membranes of infected cells. In: Pegg GF,
Ayers PG (eds) Fungal infection of plants. Cambridge
Univ Press, Cambridge, pp 79–91

Gilbert RD, Johnson AM, Dean RA (1996) Chemical
signals responsible for appressorium formation in the
rice blast fungus *Magnaporthe grisea*. Physiol Mol
Plant Pathol 48:335–346

Girbardt M (1969) Die Ultrastruktur der Apikalregion von
Pilzhyphen. Protoplasma 67:413–441

Gold RE, Mendgen K (1991) Rust basidiospore germlings and disease initiation. In: Cole GT, Hoch HC
(eds) The fungal spore and disease initiation in
plants and animals. Plenum Press, New York, pp 67–
99

Gottleib D (1976) Carbohydrate metabolism and spore
germination. In: Weber DJ, Hess WM (eds) The fungal
spore. Wiley, New York, pp 141–163

Gow NAR (1995) Tip growth and polarity. In: Gow NAR,
Gadd GM (eds) The growing fungus. Chapman and
Hall, London, pp 277–299

Green JR, Pain NA, Cannell ME, Jones GL, Leckie CP,
McCready S, Mendgen K, Mitchell AJ, Callow JA,
O'Connell RJ (1995) Analysis of differentiation and
development of the specialized infection structures
formed by biotrophic fungal plant pathogens using
monoclonal antibodies. Can J Bot 73 [Suppl 1]:S408–
S417

Grinberg A, Heath IB (1997) Direct evidence for Ca^{2+}
regulation of hyphal branch induction. Fungal Genet
Biol 22:127–139

Grove SN, Bracker CE (1970) Protoplasmic organization
of hyphal tips among fungi: vesicles and Spitzenkörper. J Bacteriol 104:989–1009

Grove SN, Bracker CE (1978) Protoplasmic changes during
zoospore encystment and cyst germination in *Pythium
aphanidermatum*. Exp Mycol 2:51–98

Grove SN, Sweigard JA (1980) Cytochalasin A inhibits
spore germination and hyphal tip growth in *Gilbertella
persicaria*. Exp Mycol 4:239–250

Grove SN, Bracker CE, Morré DJ (1970) An ultrastructural
basis for hyphal tip growth in *Pythium ultimum*. Am
J Bot 57:245–266

Guan J-L, Chen H-C (1996) Signal transduction in cell-
matrix interactions. Int Rev Cytol 168:81–121

Gubler F, Hardham AR (1988) Secretion of adhesive material during encystment of *Phytophthora cinnamomi*
zoospores, characterized by immunogold labelling
with monoclonal antibodies to components of peripheral vesicles. J Cell Sci 90:225–235

Gubler F, Hardham AR (1990) Protein storage in large
peripheral vesicles in *Phytophthora* zoospores and its
breakdown after cyst germination. Exp Mycol 14:
393–404

Gubler F, Hardham AR (1991) The fate of peripheral
vesicles in zoospores of *Phytophthora cinnamomi*
during infection of plants. In: Mendgen K, Lesemann

D-E (eds) Electron microscopy of plant pathogens.
Springer, Berlin Heidelberg New York, pp 197–210

Gubler F, Hardham AR, Duniec J (1989) Characterising
adhesiveness of *Phytophthora cinnamomi* zoospores
during encystment. Protoplasma 149:24–30

Hahn M, Neef U, Struck C, Göttfert M, Mendgen K (1997)
A putative amino acid transporter is specifically
expressed in haustoria of the rust fungus *Uromyces
fabae*. Mol Plant Microbe Interact 10:438–445

Hamer JE, Howard RJ, Chumley FG, Valent B (1988) A
mechanism for surface attachment in spores of a plant
pathogenic fungus. Science 239:288–290

Hardham AR (1992) Cell biology of pathogenesis. Annu
Rev Plant Physiol Plant Mol Biol 43:491–526

Hardham AR (1995) Polarity of vesicle distribution in
oomycete zoospores: development of polarity and
importance for infection. Can J Bot 73 [Suppl 1]:
S400–S407

Hardham AR, Gubler F (1990) Polarity of attachment of
zoospores of a root pathogen and pre-alignment of
the emerging germ tube. Cell Biol Int Rep 14:947–
956

Hardham AR, Mitchell HJ (1998) Use of molecular
cytology to study the structure and biology of phytopathogenic and mycorrhizal fungi. Fungal Genet Biol
24:252–284

Hardham AR, Gubler F, Duniec J, Elliott J (1991) A review
of methods for the production and use of monoclonal
antibodies to study zoosporic plant pathogens. J
Microsc 162:305–318

Hardham AR, Cahill DM, Cope M, Gabor BK, Gubler F,
Hyde GJ (1994) Cell surface antigens of *Phytophthora*
spores: biological and taxonomic characterization.
Protoplasma 181:213–232

Heath IB (1994) The cytoskeleton in hyphal growth,
organelle movements, and mitosis. In: Wessels JGH,
Meinhardt F (eds) The Mycota, vol I. Growth, differentiation and sexuality. Springer, Berlin Heidelberg
New York, pp 43–65

Heath IB (1995a) The cytoskeleton. In: Gow NAR, Gadd
GM (eds) The growing fungus. Chapman and Hall,
London, pp 99–134

Heath IB (1995b) Integration and regulation of hyphal tip
growth. Can J Bot 73 [Suppl 1]:S131–S139

Heath IB, Kaminskyj SGW (1989) The organization
of tip-growth-related organelles and microtubules
revealed by quantitative analysis of freeze-substituted
oomycete hyphae. J Cell Sci 93:41–52

Heath IB, Steinberg G (1999) Mechanisms of hyphal tip
growth: tube dwelling amoebae revisited. Fungal
Genet Biol 28:79–93

Heath MC (1976) Ultrastructural and functional similarity
of the haustorial neckband of rust fungi and the cas-
parian strip of vascular plants. Can J Bot 54:2484–
2489

Hegde Y, Kolattukudy PE (1997) Cuticular waxes relieve
self-inhibition of germination and appressorium formation by the conidia of *Magnaporthe grisea*. Physiol
Mol Plant Pathol 51:75–84

Heiler S, Mendgen K, Deising H (1993) Cellolytic enzymes
of the obligately biotrophic rust fungus *Uromyces
viciae-fabae* are regulated differentiation-specifically.
Mycol Res 97:77–85

Hemmes DE, Hohl HR (1971) Ultrastructural aspects
of encystation and cyst-germination in *Phytophthora
parasitica*. J Cell Sci 9:175–191

Hill TW, Mullins JT (1980) Hyphal tip growth in *Achlya*. I. Cytoplasmic organization. Can J Microbiol 26:1132–1140

Hippe-Sanwald S, Marticke KH, Kieliszewski MJ, Somerville SC (1994) Immunogold localization of THRGP-like epitopes in the haustorial interface of obligate, biotrophic fungi on monocots. Protoplasma 178:138–155

Hoch HC, Staples RC (1987) Structural and chemical changes among the rust fungi during appressorium development. Annu Rev Phytopathol 25:231–247

Hoch HC, Staples RC (1991) Signaling for infection structure formation in fungi. In: Cole GT, Hoch HC (eds) The fungal spore and disease initiation in plants and animals. Plenum Press, New York, pp 25–46

Hoch HC, Bourett TM, Staples RC (1986) Inhibition of cell differentiation in *Uromyces* with D2O and taxol. Eur J Cell Biol 41:290–297

Hoch HC, Staples RC, Whitehead B, Comeau J, Wolf ED (1987a) Signaling for growth orientation and cell differentiation by surface topography in *Uromyces*. Science 235:1659–1662

Hoch HC, Tucker BE, Staples RC (1987b) An intact microtubule cytoskeleton is necessary for mediation of the signal for cell differentiation in *Uromyces*. Eur J Cell Biol 45:209–218

Hoch HC, Bojko RJ, Comeau GL, Lilienfeld DA (1995) Microfabricated surfaces in signaling for cell growth and differentiation in fungi. In: Hoch HC, Jelinski LW, Craighead H (eds) Nanofabrication and biosystems: integrating materials science, engineering, and biology. Cambridge Univ Press, Cambridge, pp 315–334

Horak J (1986) Amino acid transport in eucaryotic microorganisms. Biochem Biophys Acta 864:223–256

Howard RJ (1994) Cell biology of pathogenesis. In: Zeigler RS, Leong SA, Teng PS (eds) Rice blast disease. CAB International, Wallingford, pp 3–22

Howard RJ (1997) Breaching the outer barriers – cuticle and cell wall penetration. In: Carroll G, Tudzynski P (eds) The Mycota, vol V, part A. Plant relationships. Springer, Berlin Heidelberg New York, pp 43–60

Howard RJ, Ferrari MA (1989) Role of melanin in appressorium function. Exp Mycol 13:403–418

Howard RJ, Valent B (1996) Breaking and entering: host penetration by the fungal rice blast pathogen *Magnaporthe grisea*. Annu Rev Microbiol 50:491–512

Howard RJ, Bourett TM, Ferrari MA (1991a) Infection by *Magnaporthe*: an *in vitro* analysis. In: Mendgen K, Lesemann D-E (eds) Electron microscopy of plant pathogens. Springer, Berlin Heidelberg New York, pp 251–264

Howard RJ, Ferrari MA, Roach DH, Money NP (1991b) Penetration of hard substrates by a fungus employing enormous turgor pressures. Proc Natl Acad Sci USA 88:11281–11284

Hughes HB, Carzaniga R, Rawlings SL, Green JR, O'Connell RJ (1999) Spore surface glycoproteins of *Colletotrichum lindemuthianum* are recognized by a monoclonal antibody which inhibits adhesion to polystyrene. Microbiology 145:1927–1936

Hyde GJ, Hardham AR (1993) Microtubules regulate the generation of polarity in zoospores of *Phytophthora cinnamomi*. Eur J Cell Biol 62:75–85

Hyde GJ, Heath IB (1995) Ca^{2+}-dependent polarization of axis establishment in the tip-growing organism, *Saprolegnia ferax*, by gradients of the ionophore A23187. Eur J Cell Biol 67:356–362

Hyde GJ, Davies D, Perasso L, Cole L, Ashford AE (1999) Microtubules, but not actin microfilaments, regulate vacuole motility and morphology in hyphae of *Pisolithus tinctorius*. Cell Motil Cytoskeleton 42:114–124

Inoue S, Turgeon BG, Yoder OC, Aist JR (1998) Role of fungal dynein in hyphal growth, microtubule organization, spindle pole body motility and nuclear migration. J Cell Sci 111:1555–1566

Jackson SL, Heath IB (1990) Evidence that actin reinforces the extensible hyphal apex of the oomycete *Saprolegnia ferax*. Protoplasma 157:144–53

Jelitto TC, Page HA, Read ND (1994) Role of external signals in regulating the pre-penetration phase of infection by the rice blast fungus, *Magnaporthe grisea*. Planta 194:471–477

Jones MJ, Epstein L (1989) Adhesion of *Nectria haematococca* macroconidia. Physiol Mol Plant Pathol 35:453–461

Kaminskyj SGW, Heath IB (1996) Studies on *Saprolegnia ferax* suggest the general importance of the cytoplasm in determining hyphal morphology. Mycologia 88:20–37

Kawamura C, Moriwaki J, Kimura N, Fujita Y, Fuji S, Hirano T Koizumi S, Tsuge T (1997) The melanin biosynthesis genes of *Alternaria alternata* can restore pathogenicity of the melanin-deficient mutants of *Magnaporthe grisea*. Mol Plant Microbe Interact 10:446–453

Kershaw MJ, Talbot NJ (1998) Hydrophobins and repellents: proteins with fundamental roles in fungal morphogenesis. Fungal Genet Biol 23:18–33

Kunoh H, Nicholson RL, Kobayashi I (1991) Extracellular materials of fungal structures: their significance at prepenetration stages of infection. In: Mendgen K, Lesemann D-E (eds) Electron microscopy of plant pathogens. Springer, Berlin Heidelberg New York, pp 223–234

Kuo K, Hoch HC (1996) Germination of *Phyllosticta ampelicida* pycnidiospores: prerequisite of adhesion to the substratum and the relationship of substratum wettability. Fungal Genet Biol 20:18–29

Kwon YH, Epstein L (1993) A 90-kDa glycoprotein associated with adhesion of *Nectria haematococca* macroconidia to substrata. Mol Plant Microbe Interact 6:481–487

Kwon YH, Epstein L (1997a) Involvement of the 90-kDa glycoprotein in adhesion of *Nectria haematococca* macroconidia. Physiol Mol Plant Pathol 51:287–303

Kwon YH, Epstein L (1997b) Isolation and composition of the 90 kDa glycoprotein associated with adhesion of *Nectria haematococca* macroconidia. Physiol Mol Plant Pathol 51:63–74

Kwon YH, Hoch HC (1991) Temporal and spatial dynamics of appressorium formation in *Uromyces appendiculatus*. Exp Mycol 15:116–131

Kwon YH, Hoch HC, Staples RC (1991a) Cytoskeletal organization in *Uromyces* urediospore germling apices during appressorium formation. Protoplasma 165:37–50

Kwon YH, Hoch HC, Aist JR (1991b) Initiation of appressorium formation in *Uromyces appendiculatus*: organization of the apex, and the responses involving microtubules and apical vesicles. Can J Bot 69:2560–2573

Langford GM (1995) Actin- and microtubule-dependent organelle motors: interrelationships between the two motility systems. Curr Opin Cell Biol 7:82–88

Lee Y-H, Dean RA (1993) cAMP regulates infection structure formation in the plant pathogenic fungus *Magnaporthe grisea*. Plant Cell 5:693–700

Lee Y-H, Dean RA (1994) Hydrophobicity of contact surface induces appressorium formation in *Magnaporthe grisea*. FEMS Microbiol Lett 115:71–76

Lehmler C, Steinberg G, Snetselaar KM, Schliwa M, Kahmann R, Bölker M (1997) Identification of a motor protein required for filamentous growth in *Ustilago maydis*. EMBO J 16:3464–3473

Levina NN, Lew RR, Heath IB (1994) Cytoskeletal regulation of ion channel distribution in the tip-growth organism *Saprolegnia ferax*. J Cell Sci 107:127–134

Littlefield LJ, Bracker CE (1972) Ultrastructural specialization at the host-pathogen interface in rust-infected flax. Protoplasma 74:271–305

Liu S, Dean RA (1997) G protein α subunit genes control growth, development, and pathogenicity of *Magnaporthe grisea*. Mol Plant Microbe Interact 10:1075–1086

Liu ZL, Szabo LJ, Bushnell WR (1993) Molecular cloning and analysis of abundant and stage-specific mRNAs from *Puccinia graminis*. Mol Plant Microbe Interact 6:84–91

Liu ZM, Kolattukudy PE (1999) Early expression of the calmodulin gene, which precedes appressorium formation in *Magnaporthe grisea*, is inhibited by self-inhibitors and requires surface attachment. J Bacteriol 181:3571–3577

López-Franco R, Bracker CE (1996) Diversity and dynamics of the Spitzenkörper in growing hyphal tips of higher fungi. Protoplasma 195:90–111

Lovett JS (1975) Growth and differentiation of the water mold *Blastocladiella emersonii*: cytodifferentiation and the role of ribonucleic acid and protein synthesis. Bacteriol Rev 39:345–404

Lovett JS (1976) Regulation of protein metabolism during spore germination. In: Weber DJ, Hess WM (eds) The fungal spore. Wiley, New York, pp 190–240

Lucas J, Knights I (1987) Spores on leaves: endogenous and exogenous control of development. In: Pegg GF, Ayres PG (eds) Fungal infection of plants. Cambridge Univ Press, Cambridge, pp 45–59

Mackie AJ, Roberts AM, Callow JA, Green JR (1991) Molecular differentiation in pea powdery-mildew haustoria. Planta 183:399–408

Mackie AJ, Roberts AM, Green JR, Callow JA (1993) Glycoproteins recognised by monoclonal antibodies UB7, UB8 and UB10 are expressed early in the development of pea powdery mildew haustoria. Physiol Mol Plant Pathol 43:135–146

Macko V (1981) Inhibitors and stimulants of spore germination and infection structure formation in fungi. In: Turian G, Hohl HR (eds) The fungal spore: morphogenetic controls. Academic Press, London, pp 565–584

MacLeod H, Horgen PA (1979) Germination of the asexual spores of the aquatic fungus, *Achlya bisexualis*. Exp Mycol 3:70–82

Maiti IB, Kolattukudy PE (1979) Prevention of fungal infection of plants by specific inhibition of cutinase. Science 205:507–508

Manners JM, Gay JL (1978) Uptake of ^{14}C photosynthates from *Pisum sativum* by haustoria of *Erysiphe pisi*. Physiol Plant Pathol 12:199–209

Manners JM, Gay JL (1980) Autoradiography of haustoria of *Erysiphe pisi*. J Gen Microbiol 116:529–533

Manners JM, Gay JL (1982) Transport, translocation and metabolism of ^{14}C-photosynthates at the host-parasite interface of *Pisum sativum* and *Erysiphe pisi*. New Phytol 91:221–244

Manners JM, Gay JL (1983) The host-parasite interface and nutrient transfer in biotrophic parasitism. In: Callow JA (ed) Biochemical plant pathology. Wiley-Interscience, Chichester, pp 163–195

Martin JT, Juniper BE (1970) The cuticles of plants. Arnold, Edinburgh

McGoldrick CA, Gruver C, May GS (1995) *myoA* of *Aspergillus nidulans* encodes an essential myosin I required for secretion and polarized growth. J Cell Biol 128:577–587

Mendgen K (1981) Nutrient uptake in rust fungi. Phytopathology 71:983–989

Mendgen K, Deising H (1993) Infection structures of fungal plant pathogens – a cytological and physiological evaluation. New Phytol 124:193–213

Mendgen K, Bachem U, Stark-Urnau M, Xu H (1995) Secretion and endocytosis at the interface of plants and fungi. Can J Bot 73 [Suppl 1]:S640–S648

Mendgen K, Hahn M, Deising H (1996) Morphogenesis and mechanisms of penetration by plant pathogenic fungi. Annu Rev Phytopathol 34:367–386

Mercure EW, Leite B, Nicholson RL (1994a) Adhesion of ungerminated conidia of *Colletotrichum graminicola* to artificial hydrophobic surfaces. Physiol Mol Plant Pathol 45:421–440

Mercure EW, Kunoh H, Nicholson RL (1994b) Adhesion of *Colletotrichum graminicola* conidia to corn leaves: a requirement for disease development. Physiol Mol Plant Pathol 45:407–420

Mims CW, Richardson EA (1989) Ultrastructure of appressorium development by basidiospore germlings of the rust fungus *Gymnosporangium juniperi-virginianae*. Protoplasma 148:111–119

Mims CW, Taylor J, Richardson EA (1989) Ultrastructure of the early stages of infection of peanut leaves by the rust fungus *Puccinia arachidis*. Can J Bot 67:3570–3579

Mitchell AJ, Hutchinson KA, Pain NA, Callow JA, Green JR (1997) A monoclonal antibody that recognizes a carbohydrate epitope on N-linked glycoproteins restricted to a subset of chitin-rich fungi. Mycol Res 101:73–79

Mitchell RT, Deacon JW (1986) Chemotropism of germtubes from zoospore cysts of *Pythium* spp. Trans Br Mycol Soc 86:233–237

Moloshok TD, Leinhos GME, Staples RC, Hoch HC (1993) The autogenic extracellular environment of *Uromyces appendiculatus* urediospore germlings. Mycologia 85:392–400

Money NP (1994) Osmotic adjustment and the role of turgor in mycelial fungi. In: Wessels JGH, Meinhardt F (eds) The Mycota, vol I. Growth, differentiation and sexuality. Springer, Berlin Heidelberg New York, pp 67–88

Money NP, Howard RJ (1996) Confirmation of a link between fungal pigmentation, turgor pressure, and pathogenicity using a new method of turgor measurement. Exp Mycol 20:217–227

Murdoch LJ, Hardham AR (1998) Components in the haustorial wall of the flax rust fungus, *Melampsora lini*, are labelled by three anti-calmodulin monoclonal antibodies. Protoplasma 201:180–193

Murdoch LJ, Kobayashi I, Hardham AR (1998) Production and characterisation of monoclonal antibodies to cell wall components of the flax rust fungus. Eur J Plant Pathol 104:331–346

Murphy JM, Walton JD (1996) Three extracellular proteases from *Cochliobolus carbonum*: cloning and targeted disruption of *ALP1*. Mol Plant Microbe Interact 9:290–297

Nicholson RL (1996) Adhesion of fungal propagules. In: Nicole M, Gianinazzi-Pearson V (eds) Histology, ultrastructure and molecular cytology of plant-microorganism interactions. Kluwer, Dordrecht, pp 117–134

Nicholson RL, Epstein L (1991) Adhesion of fungi to the plant surface. In: Cole GT, Hoch HC (eds) The fungal spore and disease initiation in plants and animals. Plenum Press, New York, pp 3–23

Nicholson RL, Kunoh H, Shiraishi T, Yamada T (1993) Initiation of the infection process by *Erysiphe graminis*: conversion of the conidial surface from hydrophobicity to hydrophilicity and influence of the conidial exudate on the hydrophobicity of the barley leaf surface. Physiol Mol Plant Pathol 43:307–318

Nolan RA, Bal AK (1974) Cellulase localization in hyphae of *Achlya ambisexualis*. J Bacteriol 117:840–843

O'Connell RJ (1987) Absence of a specialized interface between intracellular hyphae of *Colletotrichum lindemuthianum* and cells of *Phaseolus vulgaris*. New Phytol 107:725–734

O'Connell RJ, Bailey JA (1991) Hemibiotrophy in *Colletotrichum lindemuthianum*. In: Mendgen K, Lesemann DE (eds) Electron microscopy of plant pathogens. Springer, Berlin Heidelberg New York, pp 211–22

O'Connell RJ, Ride JP (1990) Chemical detection and ultrastructural localization of chitin in cell walls of *Colletotrichum lindemuthianum*. Physiol Mol Plant Pathol 37:39–53

O'Connell RJ, Pain NA, Bailey JA, Mendgen K, Green JR (1996a) Use of monoclonal antibodies to study differentiation of *Colletotrichum* infection structures. In: Nicole M, Gianinazzi-Pearson V (eds) Histology, ultrastructure and molecular cytology of plant-microorganism interactions. Kluwer, Dordrecht, pp 79–97

O'Connell RJ, Pain NA, Hutchison KA, Jones GL, Green JR (1996b) Ultrastructure and composition of the cell surfaces of infection structures formed by the fungal plant pathogen *Colletotrichum lindemuthianum*. J Microsc 181:204–212

Pain NA, O'Connell RJ, Bailey JA, Green JR (1992) Monoclonal antibodies which show restricted binding to four *Colletotrichum* species: *C. lindemuthianum*, *C. malvarum*, *C. orbiculare* and *C. trifolii*. Physiol Mol Plant Pathol 40:111–126

Pain NA, O'Connell RJ, Mendgen K, Green JR (1994) Identification of glycoproteins specific to biotrophic intracellular hyphae formed in the *Colletotrichum lindemuthianum*-bean interaction. New Phytol 127:233–242

Pain NA, O'Connell RJ, Green JR (1995) A plasma membrane-associated protein is a marker for differentiation and polarisation of *Colletotrichum lindemuthianum* appressoria. Protoplasma 188:1–11

Pain NA, Green JR, Jones GL, O'Connell RJ (1996) Composition and organisation of extracellular matrices around germ tubes and appressoria of *Colletotrichum lindemuthianum*. Protoplasma 190:119–130

Paktitis S, Grant B, Lawrie A (1986) Surface changes in *Phytophthora palmivora* zoospores following induced differentiation. Protoplasma 135:119–129

Pascholati SF, Yoshioka H, Kunoh H, Nicholson RL (1992) Preparation of the infection court by *Erysiphe graminis* f. sp. *hordei*: cutinase is a component of the conidial exudate. Physiol Mol Plant Pathol 41:53–59

Penington CJ, Iser JR, Grant BR, Gayler KR (1989) Role of RNA and protein synthesis in stimulated germination of zoospores of the pathogenic fungus *Phytophthora palmivora*. Exp Mycol 13:158–168

Perfect SE, O'Connell RJ, Green EF, Doering-Saad C, Green JR (1998) Expression cloning of a fungal proline-rich glycoprotein specific to the biotrophic interface formed in the *Colletotrichum*-bean interaction. Plant J 15:273–279

Perfect SE, Hughes HB, O'Connell RJ, Green JR (1999) *Colletotrichum* – a model genus for studies on pathology and fungal-plant interactions. Fungal Genet Biol 27:186–198

Perpetua NS, Kubo Y, Yasuda N, Takano Y, Furusawa I (1996) Cloning and characterization of a melanin biosynthetic *THR1* reductase gene essential for appressorial penetration of *Colletotrichum lagenarium*. Mol Plant Microbe Interact 9:323–329

Pickett-Heaps JD, Klein AG (1998) Tip growth in plant cells may be amoeboid and not generated by turgor pressure. Proc R Soc Lond B Biol Sci 265:1453–1459

Podila GK, Rosen E, San Francisco MJD, Kolattukudy PE (1995) Targeted secretion of cutinase in *Fusarium solani* f. sp. *pisi* and *Colletotrichum gloeosporioides*. Phytopathology 85:238–242

Powell MJ, Bracker CE (1977) Isolation of zoospore organelles from *Phytophthora palmivora*. Proc 2nd Int Mycol Congr, Tampa, FL, p 533

Pryce-Jones E, Carver T, Gurr SJ (1999) The roles of cellulase enzymes and mechanical force in host penetration by *Erysiphe graminis* f. sp. *hordei*. Physiol Mol Plant Pathol 55:175–182

Read ND (1991) Low-temperature scanning electron microscopy of fungi and fungus-plant interactions. In: Mendgen K, Lesemann D-E (eds) Electron microscopy of plant pathogens. Springer, Berlin Heidelberg New York, pp 17–29

Read ND, Kellock LJ, Knight H, Trewavas AJ (1992) Contact sensing during infection by fungal pathogens. In: Callow JA, Green JR (eds) Perspectives in plant cell recognition. Cambridge Univ Press, Cambridge, pp 137–172

Read ND, Kellock LJ, Collins TJ, Gundlach AM (1997) Role of topography sensing for infection-structure differentiation in cereal rust fungi. Planta 202:163–170

Reisener HJ (1976) Lipid metabolism of fungal spores during sporogenesis and germination. In: Weber DJ, Hess WM (eds) The fungal spore. Wiley, New York, pp 166–185

Riquelme M, Reynaga-Peña CG, Gierz G, Bartnicki-García S (1998) What determines growth direction in fungal hyphae? Fungal Genet Biol 24:101–109

Roberson RW, Fuller MS (1988) Ultrastructural aspects of the hyphal tip of *Sclerotium rolfsii* preserved by freeze substitution. Protoplasma 146:143–149

Roberson RW, Fuller MS (1990) Effects of the demethylase inhibitor, cyproconazole, on hyphal tip cells of *Sclerotium rolfsii*. II. An electron microscope study. Exp Mycol 14:124–135

Roberson RW, Vargas MM (1994) The tubulin cytoskeleton and its sites of nucleation in hyphal tips of *Allomyces macrogynus*. Protoplasma 182:19–31

Roberts AM, Mackie AJ, Hathaway V, Callow JA, Green JR (1993) Molecular differentiation in the extrahaustorial membrane of pea powdery mildew haustoria at early and late stages of development. Physiol Mol Plant Pathol 43:147–160

Robinson PM (1973) Oxygen – positive chemotropic factor for fungi? New Phytol 72:1349–1356

Ruoslahti E (1996) RGD and other recognition sequences for integrins. Annu Rev Cell Biol 12:697–715

Russo GM, Dahlberg KR, Van Etten JL (1982) Identification of a development-specific protein in sclerotia of *Sclerotinia sclerotiorum*. Exp Mycol 6:259–267

Sachs MS, Yanofsky C (1991) Developmental expression of genes involved in conidiation and amino acid biosynthesis in *Neurospora crassa*. Dev Biol 148:117–128

Satiat-Jeunemaitre B, Cole L, Bourett T, Howard R, Hawes C (1996) Brefeldin A effects in plant and fungal cells: something new about vesicle trafficking? J Microsc 181:162–177

Schaeffer HJ, Leykam J, Walton JD (1994) Cloning and targeted gene disruption of *EXG1*, encoding exo-β1,3-glucanase, in the phytopathogenic fungus *Cochliobolus carbonum*. Appl Environ Microbiol 60:594–598

Scherer GFE, vom Dorp B, Schöllman C, Volkmann D (1992) Proton-transport activity, sidedness, and morphometry of tonoplast and plasma-membrane vesicles purified by free-flow electrophoresis from roots of *Lepidium sativum* L. and hypocotyls of *Cucurbita pepo* L. Planta 186:483–494

Scott-Craig JS, Panaccione DG, Cervone F, Walton JD (1990) Endopolygalacturonase is not required for pathogenicity of *Cochliobolus carbonum* on maize. Plant Cell 2:1191–1200

Scott-Craig JS, Cheng Y-Q, Cervone F, De Lorenzo G, Pitkin JW, Walton JD (1998) Targeted mutants of *Cochliobolus carbonum* lacking the two major extracellular polygalacturonases. Appl Environ Microbiol 64:1497–1503

Shaw BD, Hoch HC (1999) The pycnidiospore of *Phyllosticta ampelicida*: surface properties involved in substratum attachment and germination. Mycol Res 103:915–924

Shaw BD, Kuo K-C, Hoch HC (1998) Germination and appressorium development of *Phyllosticta ampelicida* pycnidiospores. Mycologia 90:258–268

Shaykh M, Soliday C, Kolattukudy PE (1977) Proof for the production of cutinase by *Fusarium solani* f. *pisi* during penetration into its host, *Pisum sativum*. Plant Physiol 60:170–172

Shepherd VA, Orlovich DA, Ashford AE (1993) A dynamic continuum of pleomorphic tubules and vacuoles in growing hyphae of a fungus. J Cell Sci 104:495–507

Shieh M-T, Brown RL, Whitehead MP, Cary JW, Cotty PJ, Cleveland TE, Dean RA (1997) Molecular genetic evidence for the involvement of a specific polygalacturonase, P2c, in the invasion and spread of *Aspergillus flavus* in cotton bolls. Appl Environ Microbiol 63:3548–3552

Sietsma JH, Wessels JGH (1994) Apical wall biogenesis. In: Wessels JGH Meinhardt F (eds) The Mycota, vol I. Growth, differentiation and sexuality. Springer, Berlin Heidelberg New York, pp 125–141

Söderhall K, Cerenius L (1983) Protein and nucleic acid synthesis during germination of the asexual spores of the aquatic fungus, *Aphanomyces astaci*. Physiol Plant 58:13–17

Spencer-Phillips PTN, Gay JL (1981) Domains of ATPase in plasma membranes and transport through infected plant cells. New Phytol 89:393–400

Sposato P, Ahn J-H, Walton JD (1995) Characterization and disruption of a gene in the maize pathogen *Cochliobolus carbonum* encoding a cellulase lacking a cellulose binding domain and hinge region. Mol Plant Microbe Interact 8:602–609

Sreedhar L, Kobayashi DY, Bunting TE, Hillman BI, Belanger FC (1999) Fungal proteinase expression in the interaction of the plant pathogen *Magnaporthe poae* with its host. Gene 235:121–129

Srinivasan S, Vargas MM, Roberson RW (1996) Functional, organizational, and biochemical analysis of actin in hyphal tip cells of *Allomyces macrogynus*. Mycologia 88:57–70

Stahl DJ, Schäfer W (1992) Cutinase is not required for fungal pathogenicity on pea. Plant Cell 4:621–629

Staples RC, Hoch HC (1997) Physical and chemical cues for spore germination and appressorium formation by fungal pathogens. In: Carroll G, Tudzynski P (eds) The Mycota, vol V, part A. Plant relationships. Springer, Berlin Heidelberg New York, pp 27–40

Stark-Urnau M, Mendgen K (1995) Sequential deposition of plant glycoproteins and polysaccharides at the host-parasite interface of *Uromyces vignae* and *Vigna sinensis*. Evidence for endocytosis and secretion. Protoplasma 186:1–11

Strader CD, Fong TM, Tota MR, Underwood D, Dixon RAF (1994) Structure and function of G protein-coupled receptors. Annu Rev Biochem 63:101–132

Struck C, Hahn M, Mendgen K (1996) Plasma membrane H⁺-ATPase activity in spores, germ tubes, and haustoria of the rust fungus *Uromyces viciae-fabae*. Fungal Genet Biol 20:30–35

Struck C, Siebels C, Rommel O, Wernitz M, Hahn M (1998) The plasma membrane H⁺-ATPase from the biotrophic rust fungus *Uromyces fabae*: molecular characterization of the gene (*PMA1*) and functional expression of the enzyme in yeast. Mol Plant Microbe Interact 11:458–465

Sugui JA, Leite B, Nicholson RL (1998) Partial characterization of the extracellular matrix released onto hydrophobic surfaces by conidia and conidial germlings of *Colletotrichum graminicola*. Physiol Mol Plant Pathol 52:411–425

Sweigard JA, Chumley FG, Valent B (1992) Disruption of a *Magnaporthe grisea* cutinase gene. Mol Gen Genet 232:183–190

Talbot NJ, Ebbole DJ, Hamer JE (1993) Identification and characterization of *MPG1*, a gene involved in pathogenicity from the rice blast fungus *Magnaporthe grisea*. Plant Cell 5:1575–1590

Talbot NJ, Kershaw MJ, Wakley GE, de Vries OMH, Wessels JGH, Hamer JE (1996) *MPG1* encodes a fungal hydrophobin involved in surface interactions during infection-related development of *Magnaporthe grisea*. Plant Cell 8:985–999

Taylor J, Mims CW (1991) Fungal development and host cell responses to the rust fungus *Puccinia substriata* var. *indica* in seedling and mature leaves of susceptible and resistant pearl millet. Can J Bot 69:1207–1219

Ten Have A, Mulder W, Visser J, van Kan JAL (1998) The endopolygalacturonase gene *Bcpg1* is required for full virulence of *Botrytis cinerea*. Mol Plant Microbe Interact 11:1009–1016

Terhune BT, Hoch HC (1993) Substrate hydrophobicity and adhesion of *Uromyces* urediospores and germlings. Exp Mycol 17:241–252

Terhune BT, Bojko RJ, Hoch HC (1993) Deformation of stomatal guard cell lips and microfabricated artificial topographies during appressorium formation by *Uromyces*. Exp Mycol 17:70–78

Thevelein JM (1996) Regulation of trehalose metabolism and its relevance to cell growth and function. In: Brambl R, Marzluf GA (eds) The mycota, vol III. Biochemistry and molecular biology. Springer, Berlin Heidelberg New York, pp 395–420

Thines E, Eilbert F, Sterner O, Anke H (1997) Glisoprenin A, an inhibitor of the signal transduction pathway leading to appressorium formation in germinating conidia of *Magnaporthe grisea* on hydrophobic surfaces. FEMS Microbiol Lett 151:219–224

Ton-That TC, Rossier C, Barja F, Turian G, Roos U-P (1988) Induction of multiple germ tubes in *Neurospora crassa* by antitubulin agents. Eur J Cell Biol 46:68–79

Van Etten JL, Dunkle LD, Knight RH (1976) Nucleic acids and fungal spore germination. In: Weber DJ, Hess WM (eds) The fungal spore. Wiley, New York, pp 244–299

Van Etten JL, Freer SN, McCune BK (1979) Presence of a major (storage?) protein in dormant spores of the fungus *Botryodiplodia theobromae*. J Bacteriol 138:650–652

van Kan JAL, Van't Klooster JW, Wagemakers CAM, Dees DCT, van der Vlugt-Bergmans CJB (1997) Cutinase A of *Botrytis cinerea* is expressed, but not essential, during penetration of gerbera and tomato. Mol Plant Microbe Interact 10:30–38

Viret O, Tott L, Chapela IH, Petrini O (1994) The role of the extracellular sheath in recognition and attachment of conidia of *Discula umbrinella* (Berk. & Br.) Morelet to the host surface. New Phytol 127:123–131

Wang M-C, Bartnicki-Garcia S (1980) Distribution of mycolaminarans and cell wall β-glucans in the life cycle of *Phytophthora*. Exp Mycol 4:269–280

Warwar V, Dickman MB (1996) Effects of calcium and calmodulin on spore germination and appressorium development in *Colletotrichum trifolii*. Appl Environ Microbiol 62:74–79

Wattad C, Kobiler D, Dinoor A, Prusky D (1997) Pectate lyase of *Colletotrichum gloeosporioides* attacking avocado fruits: cDNA cloning and involvement in pathogenicity. Physiol Mol Plant Pathol 50:197–212

Welter K, Müller M, Mendgen K (1988) The hyphae of *Uromyces appendiculatus* within the leaf tissue after high pressure freezing and freeze substitution. Protoplasma 147:91–99

Wessels JGH (1996) Fungal hydrophobins: proteins that function at an interface. Trends Plant Sci 1:9–14

Wessels JGH (1999) Fungi in their own right. Fungal Genet Biol 27:134–145

Woloshuk CP, Sisler HD, Vigil EL (1983) Action of the antipenetrant, tricyclazole, on appressoria of *Pyricularia oryzae*. Physiol Plant Pathol 22:245–259

Woods AM, Gay JL (1983) Evidence for a neckband delimiting structural and physiological regions of the host plasma membrane associated with haustoria of *Albugo candida*. Physiol Plant Pathol 23:73–88

Woods AM, Gay JL (1987) The interface between haustoria of *Puccinia poarum* (monokaryon) and *Tussilago farfara*. Physiol Mol Plant Pathol 30:167–185

Wynn WK (1976) Appressorium formation over stomates by the bean rust fungus: response to a surface contact stimulus. Phytopathology 66:136–146

Wynn WK (1981) Tropic and taxic responses of pathogens to plants. Annu Rev Phytopathol 19:237–255

Xiao J-Z, Watanabe T, Kamakura T, Ohshima A, Yamaguchi I (1994) Studies on cellular differentiation of *Magnaporthe grisea*. Physicochemical aspects of substratum surfaces in relation to appressorium formation. Physiol Mol Plant Pathol 44:227–236

Xu H, Mendgen K (1997) Targeted cell wall degradation at the penetration site of cowpea rust basidiosporelings. Mol Plant Microbe Interact 10:87–94

Yamaoka N, Takeuchi Y (1999) Morphogenesis of the powdery mildew fungus in water. 4. The significance of conidium adhesion to the substratum for normal appressorium development in water. Physiol Mol Plant Pathol 54:145–154

Yamashita RA, May GS (1998) Motoring along the hyphae: molecular motors and the fungal cytoskeleton. Curr Opin Cell Biol 10:74–79

Yang Z, Dickman MB (1997) Regulation of cAMP and cAMP dependent protein kinase during conidial germination and appressorium formation in *Colletotrichum trifolii*. Physiol Mol Plant Pathol 50:117–127

Yao C, Köller W (1995) Diversity of cutinases from plant pathogenic fungi: different cutinases are expressed during saprophytic and pathogenic stages of *Alternaria brassicicola*. Mol Plant Microbe Interact 8:122–130

6 Colonial Growth of Fungi

Stefan Olsson

CONTENTS

Department of Ecology, Royal Veterinary and Agricultural University, Thorvaldsensvej 40, 1871 Frederiksberg C, Copenhagen, Denmark

I. Introduction

In a previous review in Volume I of this Series, entitled "The Mycelium as an Integrated Entity", Trinci et al. (1994) treated hyphae and the mycelium from the viewpoint of growing and branching hyphae and linked these and other cellular processes to growth kinetics of the mycelium. The current chapter will focus on this integrated entity – the manifestation of fungal colony growth as the growth of a multicellular integrated organism – and consider how this organism has been studied in different growth systems and how it coordinates activities such as nutrient uptake and reallocation. The discussion revolves around plate cultures of fungi but soil plates more reminiscent of fungi in nature are also examined. Although a common means of culturing fungi for biotechnological purposes, fungal growth in submerged cultures is not dealt with here as it is far from the natural environment that most cultivated fungi have evolved to deal with and has recently been reviewed (Trinci et al. 1999).

II. What is Colonial Growth?

A. Colony in Relation to Mycelium

Fungal growth on a surface of a natural substratum results in a collection of hyphae or cells termed a **fungal colony** (Fig. 1). The word colony implies that it is made up of an aggregate of individuals (Shorter Oxford English Dictionary, 3rd edn). Among fungi this might be the case for yeast cell colonies (see also Sect. IX) but, for filamentous fungi, the whole colony might be one organism. In the latter case it is not really appropriate to think of such an entity as a colony since it is an individual, an integrated network of hyphae that resembles a colony of separate "hyphal individuals". *Mycelium* could be used to describe filamentous

The Mycota VIII
Biology of the Fungal Cell
Howard/Gow (Eds.)
© Springer-Verlag Berlin Heidelberg 2001

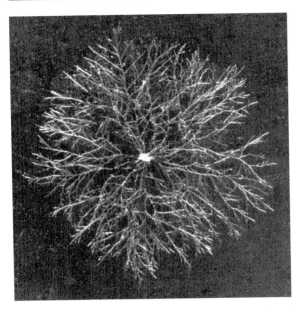

Fig. 1. The fungal colony as one organism, or a functional mycelium unit. The image shows a colony of *Rhizoctonia solani* growing on 1/20 potato dextrose agar after 54 h at 20 °C. The sclerotium used as the inoculum can be seen at the *center*

fungi rather than colony but even this term has been used to indicate a fungal colony of independent growing tips or an individualistic coordinated organism. Both these views are valid for different fungal entities and under different circumstances. In a recent review the terms **Functional Mycelium Unit** (FMU) and **Genetic Mycelium Unit** (GMU) were introduced (Olsson 1999) to describe the individuality and the genetic makeup of a mycelium. The GMU was defined as "genetically identical hyphae occupying a continuous space". The FMU was defined as "the network of functionally integrated hyphae that forms an individualistic organism". These terms are used throughout in an attempt to apply the most appropriate definitions. "Fungal colony" will only be used in its most limited sense as "a group of hyphae or yeast cells" according to Hawksworth et al. (1995).

III. Yeast Colony Growth

A. Switches in Colony Morphology

The dimorphic yeast *Candida albicans* has been found to switch spontaneously between different colony morphologies at frequencies of 1.4×10^{-4}. This is too high a frequency to be attributed to point mutations. Many different variant colony morphologies have been identified (Slutsky et al. 1985). The colony phenotypes persist in successive clonal platings. The colony phenotypes can switch back to the original or to most of the other switch phenotypes (Slutsky et al. 1985). The frequency of switching can be increased 200-fold by doses of ultraviolet light that cause limited cell death. The variability in colony phenotypes apparently does not reflect a variation in strains, but rather an inherent capacity for reversible phenotypic variability within each cell. This switch in colony morphology seems to be related to the yeast-filament transition in *C. albicans* (Radford et al. 1994). Smooth colonies are composed almost entirely of yeast cells, wrinkled colonies of branched hyphal cells, and semi-rough colonies consist of both yeast cells and elongated pseudo-hyphae but in different proportions compared to what is found in wrinkled colonies. Fuzzy colonies consist of yeast cells, pseudohyphae and true hyphae, with aerial hyphae in discrete areas. Finally, scallop colonies are composed entirely of pseudohyphal cells (Radford et al. 1994). A recent study (Pérez-Martín et al. 1999) suggests that morphological switching is controlled by an *SIR* (**Silent Information Regulator**) gene. A gene with homologies to *SIR2* in *Saccharomyces cerevisae* was found in *C. albicans*. A *sir2/sir2* mutant of the diploid fungus showed drastically increased colony switching activity. Expression of *SIR2* in the mutant made switching less rare, and an extra *SIR2* copy in a *SIR2/SIR2* strain decreased switching below the spontaneous levels. The authors also suggest the possibility that external stress could inactivate the *SIR2* gene in wild-type *C. albicans* and thereby activate phenotypic switching – thus enhancing the pathogens growth repertoire and its adaptability in vivo. In *S. cerevisiae*, *SIR2* silencing is inactivated as cells become older and it has been suggested that a different type of stress would perhaps produce a similar lifting of silencing in *C. albicans* (Pérez-Martín et al. 1999).

IV. The Study of Mycelial Colony Growth in Systems with Different Geometries

A. Growth System Geometry and Fungal Growth

It is obvious that a fungus grows in three-dimensional space; however, growth in one or two of these can be restricted so that the GMU will expand mostly by growth within the remaining space dimensions. Therefore, growth in a one-dimensional growth system will be restricted in two dimensions, and in a two-dimensional system the growth will be restricted in one dimension. The main reasons for these restrictions are practical. It might be easier to interpret results for quantitative aspects of fungal growth using one-dimensional systems or two-dimensional agar plates used routinely by most mycologists. It might appear that the three-dimensional system would be the most natural. However, this is not entirely correct since most fungal growth in nature takes place on surfaces. Soil, for example, should not be considered a purely three-dimensional system due to the particles it contains. Rather, it is a system somewhere in between two-dimensional and three-dimensional depending on the scales considered and the surfaces of the soil particles. It is, however, important to choose growth systems appropriate to the questions asked instead of relying on "standard" systems that might be far from natural environments for fungal growth and may not be appropriate for the questions being addressed.

B. One-Dimensional Growth Systems

1. Agar Systems

Fungal GMUs can be grown easily in one-dimensional systems. Pioneering work, for example that of Beadle and Tatum (1941), used growth systems such as race tubes, growth tubes or Ryan tubes (Fig. 2). The work of Ryan and colleagues (Ryan et al. 1943) represents a thorough examination of the growth of GMUs. These authors established firmly that the fungus of their study, *Neurospora crassa,* grows with a constant radial expansion of the GMU. This and more recent investigations of "experimental" fungi of the genera *Penicillium* and *Aspergillus* lead to the notion that fungal colony growth is (always) linear after an initial exponential expansion of the germ tube (Trinci et al. 1999) on an initially homogeneous medium such as agar media. It was also shown that growth at the hyphal tip is dependent on internal translocation of substances from distal areas in the mycelium since effects on growth rate were observed when hyphae at the leading edges were separated from the mycelium by incisions closer than 1.2 cm to the apex. Race tubes are still

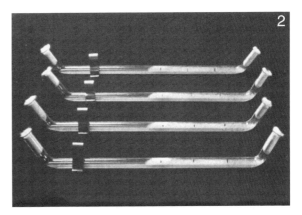

Fig. 2. Growth tubes showing the colonial growth of *Neurospora crassa.* (Ryan et al. 1943, with permission)

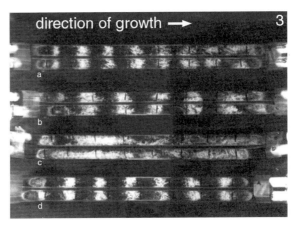

Fig. 3. Wild-type and clock mutants of *Neurospora crassa* in a *csp-1*, bd background grown in race tubes. Margins of colony growth at daily intervals are indicated on the tube to give reference points. *a* Wild-type strain in constant darkness (period = 21 h); *b frq7* mutant with a long period in constant darkness; *c frq10* null mutant exhibiting arrhythmicity; *d* wild-type strain entrained to a 12:12 LD cycle. Strains grown at 22 °C on maltose-arginine-Vogel's salts medium. (Ramsdale 1999, with permission)

in wide use even today mostly for studies of *Neurospora* or other fast-growing fungi. Rhythmic conidiation and effects of different "clock"-genes on such rhythms have been studied extensively (Fig. 3) and recently reviewed (Ramsdale 1999). Other one-dimensional growth systems introduced more recently are the long plates (Fig. 4) used by Olsson (1995) or glass fiber filter strip cultures (Olsson and Jennings 1991a,b). In both these latter systems it was possible to introduce opposing nutrient gradients. Long plates were thus used to investigate the effect of spatially heterogeneous nutrient distribution on fungal biomass densities, and allowed the conclusion that some fungi redistribute nutrients within the colony. Consequently,

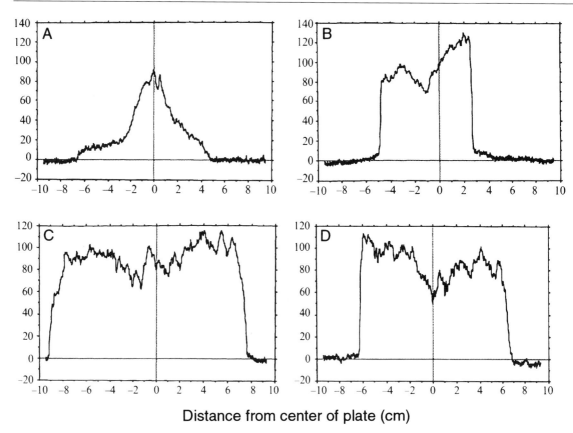

Distance from center of plate (cm)

Fig. 4. Biomass density profiles (pixel value increases) of two radically different fungi, *Geotrichum candidum* (**A, B**) and *Serpula lacrymans* (**C, D**), grown on *gradient* nutrient plates (**A, C**) and *mixed* nutrient plates (**B, D**) containing glucose-asparagine mineral (for *Geotrichum*) or glucose-NO₃ mineral medium (for *Serpula*). In gradient plates, the left half contained glucose and the right half all other nutri- ents. *G. candidum* is dependent on the introduced gradient for its growth and grows well only where all nutrients are present. *S. lacrymans* is independent of the external nutri- ent concentration and grows equally well on the gradient medium as on the mixed medium, thus implying effective and active reallocation of nutrients within the FMU (Rearranged from Olsson 1995)

the GMU was also an FMU. Some fungi, 14 out of 62 tested by Olsson (1995), showed accelerating colony radial growth rate rather than a constant rate that appears to contradict the linear growth "paradigm" for agar-grown cultures. Thus, colony radial growth rate increased with the size of the colony. Butler (1984) previously noted the same phenomenon and attributed it to physiological dif- ferentiation of hyphae within the mycelium (Butler 1984). The rapidly extending, exploratory, leading hyphae extended faster as the supply of nutrients from the nutrient-absorbing hyphae behind the colony margin increased with the increasing size of the whole colony (Butler 1984).

2. Long Soil Plates

One-dimensional plate systems (20 × 2 cm) have been used to constrain fungal growth in soil (Persson et al. 2000). This was done, for example, to enable an estimation of the spread of a nematophagous fungus into soil. This fungus does not form cords and grows as thin hyphae through the soil. Thus, it would be very difficult to investi- gate hyphal spread microscopically. The fungus received radiolabeled nutrients in trace amounts at the wood disks that served as inoculum. The label distribution in soil was recorded at different times (Fig. 5). In heat-treated soil the fungus grew a maximum distance of 7–8 cm from the food base (wood), but in non-treated soil it only grew slightly more than 1 cm. If, however, the food base was five times larger the fungus grew much further, to more than 3 cm (Fig. 6). This may reflect the con- sequences of competition for nutrients between the fungus and bacteria in non-sterile soil. Similar soil systems can be used to study the outcome of fungal interactions in the form of nutrient reallo-

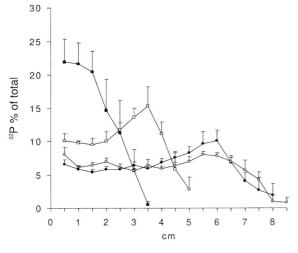

Fig. 5. Distribution of ^{32}P in dishes with heat-treated soil inoculated with one birch wood disk of *A. superba*, 0.5–10 cm from the point of inoculation on day 10 (■), 15 (□), 20 (●) and 25 (○). *Bars* indicate the standard error of the mean of five replicates. (Persson et al. 2000, with permission)

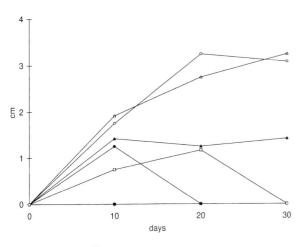

Fig. 6. Spread of ^{14}C into untreated soil measured from 1 (●, ▲) and 5 (○, △) birch wood disks colonized by *A. superba* from experiments 1 and 2. Controls with 1 (■) and 5 (□) disks show the spread of [^{14}C]-3-*O*-methylglucose from non-inoculated disks. *Bars* indicate the standard error of the mean of three replicates. (Persson et al. 2000, with permission)

cations between interacting mycelia (Lindahl et al. 1999). Here phosphorus transfer occurred from a wood-decomposing fungus to a mycorrhizal fungus. The system consisted of a long (20 × 2 cm) plate containing forest soil where a wood-decomposing fungus growing from a piece of wood met a mycorrhizal fungus in association with a plant.

C. Two-Dimensional Growth Systems

1. Still-Liquid Cultures

Still-liquid cultures in Erlenmeyer flasks or in Petri dishes are very commonly used to produce fungal biomass. This often gives a more synchronized development than in agar culture since the whole GMU floats on top of the medium and will be subjected to roughly the same nutrient concentrations during the incubation period. As a consequence, the whole GMU will enter the stationary phase at roughly the same time, resulting in a more synchronized formation of spores or sclerotia (Olsson and Nordbring-Hertz 1986) compared to GMUs grown on agar plates.

2. Agar Plates

By far the most common way of growing GMUs is on agar surfaces. Usually, researchers only note radial growth rates and the appearance of the colonies, but the agar plate culture can be used to measure many other aspects of fungal growth such as biomass density and the fractal dimensions of the hyphal network (see Sect. V).

3. Soil Plates and Saprotrophs

Extensive studies of the growth of cord-forming wood-decomposing basidiomycetes have been performed in plates or trays filled with compressed non-sterile soil. Most of this work has been summarized in a recent review (Boddy 1999). An example of how the FMU develops in such cases can be seen in Fig. 7. Early studies only gave qualitative descriptions of the development for these FMUs. With the aid of image-processing equipment these studies can also include quantitative measurements (Donnelly et al. 1999). The soil systems have been used to analyze the foraging strategies of FMUs in search of woody resources and how the growth pattern of cords is affected by environmental variables like temperature, water availability and pH, and competition with other fungi (Dawson et al. 1988; Bolton and Boddy 1993; Donnelly et al. 1995; Donnelly and Boddy 1997, 1998). The growth pattern depends on the fungal species, the nutrient status of the food resource the fungus is using, soil nutrient status and local microclimate. Species with FMUs more inclined to form discrete cords are in general less responsive than species whose cords tend to

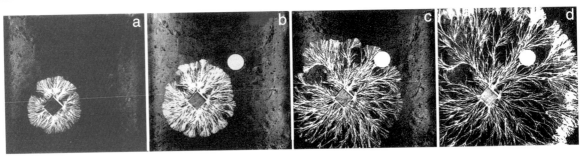

Fig. 7. Digital images showing development of *Hypholoma fasciculare* mycelial system over time in 24 × 24 cm trays of non-sterile soil: **a** 11 days, **b** 16 days, **c** 26 days, **d** 44 days. (Boddy 1999, with permission)

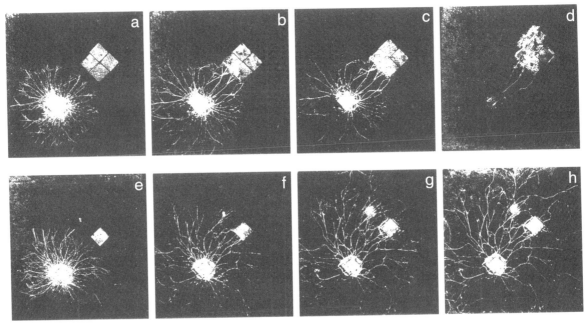

Fig. 8. Development of mycelial systems of *Phanae-rochaete velutina* from 2-cm³ beech wood inocula to 16-cm³ (**a–d**) or 2-cm³ (**e–h**) beech wood resources in 24 × 24 cm trays of non-sterile soil, after 9 (**a,-e**), 18 (**b,-f**), 31 (**c,-g**) and 39 (**d,-h**) days. (Boddy 1999, with permission)

be less discrete (Boddy 1999). Another important aspect has been the sequence of events when a FMU encounters a new nutrient-resource or several nutrient-resources in succession. If this nutrient-resource is larger than the original resource used as the inoculum, the resource-connecting cords thicken, radial extension stops and an autolysis of non-resource-connected cords occurs. Finally, an outgrowth is initiated from the newly colonized resource in the direction of travel (Fig. 8a–e).

Less drastic changes take place if the newly encountered nutrient-resource is similar or smaller in size (Fig. 8e–h). The growth system also allows the measurement of nutrient reallocation

with the aid of radioactive tracers and how this reallocation is affected by the quality of the available nutrient-resources. It has, for example, been shown that newly colonized nutrient-resources can act as temporary sinks for mineral nutrients such as phosphorus (Hughes and Boddy 1994; Wells et al. 1998a,b).

4. Soil Plates and Mycorrhizae

Soil plates (Fig. 9) are also used routinely to study ectomycorrhizal associations and the growth of the fungal partner. In these systems the outgrowth rate, influence of environmental factors, interactions between mycelia, as well as nutrient translocation, have been studied (Finlay and Read 1986; Bending and Read 1995; Timonen et al. 1996).

5. Lichens on Rock

Perhaps the most obvious two-dimensional natural substratum for fungal growth is a piece of rock covered by lichens (Fig. 10). Here, manifestations of GMU growth and interactions can be seen easily. Some GMUs extend while others defend their territories with chemical antagonists. Black phenolics are often formed in the "war zones" between adjacent colonies. If it were possible to visualize natural populations, of for example soil fungi in situ, a similar picture would likely be observed although it would be in three dimensions and more dynamic.

6. Fractals to Describe Growth in Two-Dimensional Systems

Several recent studies have described mycelial systems as fractal (Obert et al. 1990; Ritz and Crawford 1990; Mihail et al. 1995). Objects that are fractal are non-Euclidian in that they cannot be described using 1, 2, 3, etc. dimensional geometrical dimensions but are better described in fractal geometry with dimensions lying between the whole-number dimensions. The essential feature of fractals is the way that the material composing them is distributed in space. For an FMU, a high fractal dimension would represent high space-filling but less area than could be covered by the same FMU-biomass in lower fractal dimensions. Thus, fractal growth and changes in the fractal dimension in different parts of the mycelium might therefore optimize FMU nutrient capture by optimizing the balance between exploratory and exploitative modes of growth. Exploratory growth is taken to indicate growth in which colonial expansion occurs over barren zones with sparse mycelial branching with or without little assimilation of nutrients, while, in exploitative growth, assimilation and branching frequency are high. For a roughly circular colony this can be described as:

$$M(r) = kr^D$$

where $M(r)$ is the mass contained within the radius r, and k and D are both constants (Ritz and Crawford 1990). The fractal dimension is represented by D and can be estimated from the slope of $\ln(r)$ plotted against $\ln M(r)$.

D. Three-Dimensional Growth Systems

Although three-dimensional growth systems are common in nature they have been studied most in stirred liquid cultures (Trinci et al. 1999). The three-dimensional distribution of FMUs in soil or compost and the dynamics of this distribution have not been studied to the same extent. Although these are of great interest and have ecological significance, such studies are difficult due to the opacity of the environment and the fragility and dimensions of the hyphal structures. New methods are desperately needed since the soil mycologist/ecologist in most cases does not know the size of the organisms being studied. Is it a louse or an elephant? Most ecologists concerned with the macroflora or macrofauna at least know the size of their organisms of study!

V. Changes in Colony Morphology and the Medium Due to Fungal Activities

A. Nutrient Depletion Under the Colony

When GMUs grow on agar media they deplete the nutrients under the GMU. This was predicted theoretically (Pirt 1967) and then shown for *Rhizoctonia cerealis* GMUs growing on a glucose-containing medium (Robson et al. 1987). A further refined method able to demonstrate depletion both for glucose and phosphorus was later developed and used for FMUs of *Fusarium oxysporum* growing on defined media (Olsson 1994). The depletion of glucose and phosphorus under colonies growing on the same glucose concentration but with varying mineral medium content or under colonies with the same mineral medium concentration but varying glucose concentration curves were compared (Fig. 11). The depletion curves were identical and indicate that the FMU depletes the respective nutrients in the immediate surroundings over a large range of nutrient ratios despite sometimes an expected surplus of the nutrient in question.

B. Triggering of Conidiation

Nutrient limitation is a common trigger of asexual and sexual sporulation. When FMUs grow on agar and deplete nutrients under the colony, conidia-

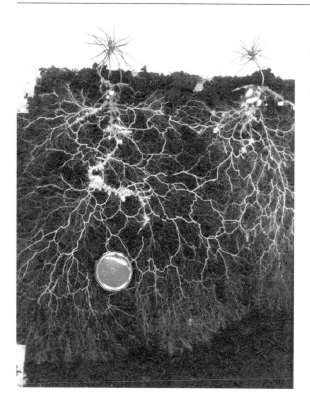

Fig. 9. Soil tray with developing mycelium of the mycor-rhizal fungus *Paxillus involutus* growing together with seedlings of *Pinus silvestris*. The round cap was used to administer radiolabeled phosphorus to the system. (Photograph supplied by Professor Roger Finlay, SLU, Sweden)

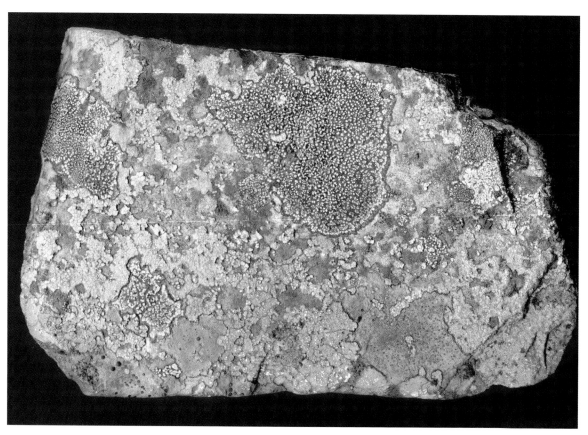

Fig. 10. Lichen community on a piece of exposed rock from the very north coast of Norway (Hamningsberg), not too distant from the North Pole

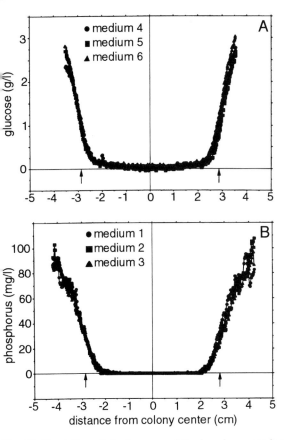

Fig. 11. Depletion of **A** glucose and **B** phosphorus under colonies of *Fusarium oxysporum* grown on the same glucose concentration but different mineral medium concentrations (medium 4–6) or on the same mineral medium concentration but different glucose concentrations (medium 1–3). (Olsson 1994, with permission)

tion is usually triggered at some distance behind the FMU edge. In this way the FMUs' own activities trigger sporulation. In *Neurospora crassa* the reason for these environmental triggers has been attributed to growth checks or other environmental changes that result in an intracellularly hyperoxidant state (Hansberg et al. 1993; Toledo and Hansberg 1994, 1995). This hyperoxidant state leads to the oxidation of cell components that in turn trigger conidiation. Although submerged cultures normally initiate conidial development after exhaustion of either carbon or nitrogen sources, colonies growing on agar plates often sporulate without an apparent complete nutrient exhaustion (Navarro-Bordonaba and Adams 1994). On agar media, with high and low concentrations of nitrogen and carbon in different proportions, it took roughly the same time to initiate conidiophore formation in the resulting colonies (Pastushok and

Axelrod 1976). The problem here is that the concentration gradients developing under the colony may not have been fully considered. Most likely steep concentration gradients developed under all colonies (see Sect. V.A). In these cases there would be little difference in the distance from the hyphal tips to the zone of the agar where nutrients are exhausted irrespective of the nutrient concentration in the medium. In another series of experiments presented in the same study, colonies grown on membranes were transferred to fresh medium every second hour. This had little effect on the time before conidiation initiated. If anything, the time to initiation was shortened compared to cultures that were not transferred. Again, this suggests that medium exhaustion was not directly responsible for triggering conidiation and that the conidiation is under the control of a fixed developmental program. There is a clear requirement for hyphae to become competent for differentiation (Axelrod 1972; Axelrod et al. 1973; Gealt and Axelrod 1974) but the signal to conidiate is most likely an environmental one. The signal could represent a hyperoxidant state caused by many different forms of stress, light, starvation, drying, air exposure, etc., which are all known to initiate conidiation/sporulation in many fungi. The most common way to initiate conidiation in *Aspergillus nidulans* is to expose a submerged culture to air. This results in a sudden local drop in external nutrient concentration, due to uptake by FMUs that, in turn, results in a drastically slower delivery of nutrients to the surface of the most exposed hyphae. This sudden drop in externally available nutrients will result in a change in metabolism from one derived from external nutrients to one dependent on internal stores and a drop in the rate of oxygen usage through oxidative phosphorylation. Partial pressure of oxygen and the mass transfer of oxygen to a submerged FMU are much lower than for a surface-grown FMU. An oxygen-protective enzyme system like superoxide dismutase does not seem to be produced in excess but is just sufficient to provide protection from oxidation under the present growth conditions (Gort and Imlay 1998). Sudden increase in oxygen exposure (or light exposure) combined with a decreased oxygen removal rate though respiration will lead to an increased level of reactive oxygen species and thus an increased oxidation of cell components. This in itself could trigger the morphological changes associated with sporulation (Toledo and Hansberg 1995).

C. Cord Formation

When growing over nutrient-poor areas many basidiomycetes form linear structures made up of many parallel hyphae. These structures can be more or less differentiated, and can represent a loose collection of hyphae to specialized structures like rhizomorphs of *Armillaria* sp. Fungi that form less specialized cords shift between cord growth and diffuse growth depending on local environmental conditions. In areas of high nutrient availability the growth is diffuse and in areas of low availability the growth becomes corded. This has been demonstrated using checkerboard partitioned culture dishes in which adjacent compartments contain higher or lower concentrations of assimilable nutrients (Fig. 12).

D. Autolysis

Many fungal FMUs exhibit autolysis of older parts of the FMU especially when growing on nutrient-poor media. Other fungi, like heavily sporulating *Aspergillus* species, autolyse the vegetative mycelium when conidiophores and conidia are being formed. In both these cases it is believed that the material in the autolysed parts are reused

Fig. 12. *Coprinus picaceus*, grown from an inoculum in a high nutrient compartment, on a matrix of compartments alternating between high (2% malt agar) and low (tap water) nutrient concentrations. (Rayner 1996, with permission)

in the living parts. In a series of experiments with *Schizophyllum commune*, it was shown that transfer of FMUs to nutrient-deficient medium resulted in autolysis, increased proteolytic activity and a transfer of label from labeled proteins in older parts to protein in newer growing parts of the mycelium (Sessom and Lilly 1986; Lilly et al. 1991). The increased proteolytic activity could in principle lead to a non-specific release of labeled amino acids to the medium and a subsequent uptake by the growing parts of the FMU. In other studies by the same authors it was shown that labeled non-incorporable 2-aminoisobutyric acid was not released to the medium but was translocated to hyphal apices. This supports the existence of intracellular pathways for reallocation of and reuse of products of autolysis within the FMU. Other obvious examples of seemingly FMU-controlled dying-off of parts of the mycelium are evident in experiments in which certain basidiomycetes colonize a series of isolated wood blocks in a progressive fashion. After assimilating nutrients locally in the soil the mycelium is resorbed as the advancing cells encounter new large sources of nutrients in the form of wood blocks (Fig. 8). These FMU-controlled modes of dying might be similar to apoptosis in other eukaryotic organisms. Conclusive evidence for this is still lacking although *Mucor racemosus* has recently been shown to respond with signs of apoptosis to lovastatin, a compound that induces apoptosis in other cell types (Roze and Linz 1998).

E. Paramorphogens

In a liquid culture all hyphae are in contact with approximately the same concentration of medium. This is especially true for stirred media. If the rate-limiting step for growth of the FMU is nutrient uptake this will be dependent on the surface area of the colony. Thus, in liquid culture, it is essential for the FMU to have a large surface-to-volume ratio to optimize exposure to nutrients. Conversely, on a surface, uptake of nutrients by the FMU will deplete the nutrients under the FMU and the FMU growth rate will be dependent on how much new medium the colony comes in contact with – and this is related to the extension rate of hyphae at the margin of the colonies. Liquid culture would thus favor (or not disfavor) small highly branched FMUs with high surface area and low vacuolization. In other words an

FMU with a high fractal dimension (see Sect. IV.C.6) can be expected to have an advantage under these conditions (Wiebe et al. 1991, 1992). The most efficient cellular morphology for the FMU in a nutrient-rich liquid would thus be to grow as a single small cell that swells and divides or buds (i.e. yeast cells). In contrast, growth on a low nutrient surface should favor economy in materials used, little branching and a fractal geometry substantially less than two, and high vacuolization. The higher the fractal dimension value of the branching, the more exploitive the growth, and the lower the fractal dimension, the more exploratory the growth form. The growth rate of the FMU thus depends on balancing the exploratory and the exploitive growth modes to suit the availability of nutrients (Toledo and Hansberg 1995; Ritz and Crawford 1999). Thus, the switch observed in *C. albicans* and many other dimorphic fungi between yeast growth to filamentous growth (see Chap. 3, this Vol.) could be seen as a switch between exploitative growth in a rich liquid medium to a more exploratory (invasive) growth where the nutrient acquisition depends on the hyphae contacting new nutrient resources (Brown and Gow 1999). Therefore, some paramorphogens that affect branching frequency, but have no nutrient value, inhibit growth by inducing inappropriate colony morphology. The paramorphogen validamycin A, which is used to control the plant pathogen *Rhizoctonia cerealis*, probably works in this way (Robson et al. 1991). It has also recently been discovered that viscosinamide, a peptide produced by a bacterial biological control agent against *Rhizoctonia solani*, acts like a paramorphogen of colonial morphology (Thrane et al. 1999). Such paramorphogens may therefore provide alternatives for controlling fungal growth of pathogens (Robson et al. 1991).

F. Differential Insulation Theory

Fungi growing in heterogeneous environments branch vigorously and achieve high mycelial densities in areas of high nutrient availability. Branching frequency decreases in areas of lower nutrients and fungi may be induced to form cords. For example, *Coprinus picaceus* can be grown over a culture plate with a checkerboard arrangement of high and low levels of nutrients (Fig. 12). The junctions between the squares are narrow slits in the plastic walls of the dish sections. As a result,

highly branched mycelia develop in the sections with high nutrient levels and cords with poorly branched mycelia are formed in the sections with low nutrient levels. This type of behavior is difficult to explain by linear growth models (Rayner et al. 1995) but can be explained if one assumes that the fungal hyphae can be differently insulated towards the environment through modification of the cell wall, as shown in Fig. 13. Insulation should be minimized when external nutrient availability is maximal and vice versa. In this way input can be maximized in high nutrient areas and minimized in low nutrient areas (Rayner et al. 1995). The frequency and pattern of branching (fractal dimension) will depend on the degree of insulation of the hyphal boundaries, whether the hyphae are in exploitative (i.e., optimized for nutrient uptake) or exploratory mode (designed to translocate nutrients through the mycelium) (Rayner et al. 1995). There are, however, as yet no data relating changes in cell wall permeability in regions of a FMU to the fractal dimension in the same region. A possible mechanism for these permeability changes may be mediated by changes in phenoloxidase and peroxidase activity. This is often seen in areas of stress and in a variety of morphological changes that would be characterized by an increased hydrophobicity. The activity of these enzymes has the potential to cause polymerization reactions forming hydrophobic substances (Rayner 1996).

VI. Nutrient Translocation

A. Translocation Mechanisms

Four different mechanisms for nutrient translocation can be identified (Olsson 1999).

1. Passive

Nutrients are taken up by hyphae according to the needs of hyphae in the region of uptake. Translocation inside hyphae occurs by simple diffusion as in any water phase. This type of translocation would not require any extra energy.

2. Passive–Active Uptake

This mechanism is similar to the passive mechanism but nutrients are taken up in excess of local

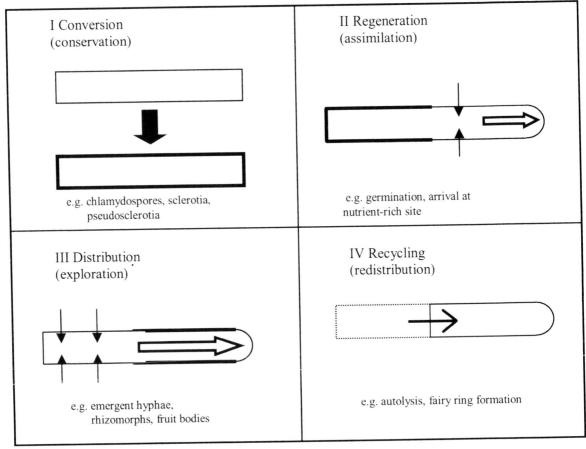

Fig. 13. Four fundamental processes in elongated hydrodynamic systems, as determined by boundary deformability, permeability and internal partitioning. Rigid boundaries are shown as *straight lines*, deformable boundaries as *curves*, impermeable boundaries as *thicker lines*, degenerating boundaries as *broken lines* and protoplasmic disjunction by an *internal dividing line*. *Simple arrows* indicate input across permeable boundaries into metabolically active protoplasm; *double arrows* represent throughput due to displacement. (Redrawn from Rayner et al. 1995)

needs (Olsson and Jennings 1991a) creating a steep nutrient gradient inside the hyphae. This results in higher nutrient flux as mediated by diffusion and thus in a faster delivery of nutrients to remote parts of the FMU. This type of translocation requires extra energy for uptake, but not for the actual translocation process that is driven by diffusion.

3. Active, Cytoplasmic

Nutrient translocation has often been associated with active movements in the cytoplasm, either through movements of organelles (Schütte 1956; Cooper and Tinker 1981; see also Chap. 11, this Vol.) or through peristaltic vacuole systems (Shepherd et al. 1993a,b; see also Chap. 12, this Vol.).

4. Active, Pressure-Driven Bulk Flow

If the active uptake mechanism outlined above (mechanism 2) results in a high osmotic concentration inside the hyphae, water will flow in through the semipermeable cell membrane. This will create a higher turgor inside the hyphae and water and nutrients will flow inside the FMU in the direction of least resistance (Jennings 1994). Resistance comes from either the cell wall resistance to deformation or from resistance against water flow within the hyphae – friction (Cairney 1992). This type of translocation requires energy for both the active uptake and to overcome the resistance to water flow within the hyphae.

Reallocation of nutrients within an FMU might involve components of all the above simultaneously.

It has been demonstrated that phosphorus and a non-metabolizable [14C]-labeled amino acid analogue ([14C]-2-aminoisobutyric acid, [14C]-AIB) used to trace the amino acid pools in fungi (Lilly et al. 1990) moved actively through FMUs of several fungi (Olsson and Gray 1998). The translocation showed characteristic patterns of movements in the FMUs. For example, in *Pleurotus ostreatus*, there was a quick radial movement of label to and from the center of the colony along specific pathways in the FMU. These pathways were invisible macroscopically and seemed to represent a functional specialization of the mycelium not followed by readily recognizable morphological changes. There was also a tangential movement of label along the hyphal front indicating that there must be a great deal of integration of the FMU through anastomoses relatively close to the hyphal front. The label also moved with the growing hyphal front. What was most surprising in this study was that the same label was seen moving in opposite directions in many parts of the mycelium. This could indicate that nutrients are "circulated" in the FMU, in a similar manner to blood circulation, with a sink region in the FMU removing nutrients from the stream passing by (Olsson and Gray 1998).

VII. Effect of Nutritional Heterogeneities on the Growth of Colonies on Agar

In most natural environments nutrients are distributed heterogeneously in space. From recent studies it has become apparent that the nature and extent of such heterogeneities can play an important governing role in fungal growth and development (Olsson 1995; Cairney and Burke 1996; Rayner 1996; Ritz et al. 1996). Gradients can be generated in agar as shown above (Sect. IV.B.1) but the agar can also be divided into discrete resources. This can be done by using "checkerboard" agar plates (Fig. 12) or by tessellations of agar plates with differing composition (Ritz 1995). In this latter technique diffusion is limited between the blocks by having an air-gap between the tiles. A tessellation of 3×3 tiles was used and the central tile was inoculated with a single FMU; the FMU development was then mapped by regular observations. In general, the hyphal

density was high on high nutrient tiles and low on low nutrient tiles. There was, however, a tendency for higher hyphal density values on the low nutrient tiles when there were high nutrient tiles in the tessellation, thus indicating reallocation of resources within the FMU-network. The technique has been refined even more to give better visualization of colony responses (Ritz et al. 1996).

VIII. Selected Models for Describing Mycelial Growth and Their Principal Advantages and Disadvantages

We have seen previously that a fungal colony can be modeled using fractal geometry. As pointed out by Ritz and Crawford (1999), this implies that the distribution of a hypha at one point is correlated with the distribution at another. The colony thus grows coherently suggesting that some form of communication exists between distal parts of the colony – that the colony is an integrated functional mycelium unit (FMU). The dominating mathematical models for fungal growth assume that mycelium growth is a result of local tip growth processes and branch initiation close to tips (Prosser and Trinci 1979; Bartnicki-Garcia 1990). These processes at the tips are assumed to be independent of each other and therefore these models cannot accommodate the observed coherence (fractal nature) of the colony. The growth responses seen in heterogeneous systems and the knowledge of the extensive translocation in some FMUs cannot be explained either. Stochastic models have also been put forward (Kotov and Reshetnikov 1990; Obert 1994). These models are empirically based, describing the growth of the colony by using distribution functions determined from laboratory experiments. Thus, the models are species-specific and do not address the underlying mechanism of colony form and cannot at present deal with heterogeneously distributed nutrients.

The only models available that can produce spatially coherent structures and deal with heterogeneities in the medium are reaction-diffusion models. Growth patterns are, in these models, thought to originate by autocatalysis where the presence of one element like biomass promotes the production of more biomass by uptake and synthesis. In the model of Regalado et al. (1996), an autocatalytically produced activator and an inhibitor govern tip positioning. Tip extension arises from interactions with external nutrients of a defined distribution. The uptake of nutrients causes changes in this distribution. The model output is in agreement with the observed fractal nature of mycelia in that a higher fractal dimension was produced on high nutrient medium and a lower fractal dimension on a lower nutrient medium.

The model of Davidson et al. (1996) takes a different approach and considers biomass and nutrient distribution without reference to hyphal growth details. Instead it concerns the distribution of biomass in space and considers distribution of activator over length-scales larger than what corresponds to individual hyphae. Although the model does not invoke individual hyphae, different network topologies are implicitly accounted for. If substrate is not renewed, or renewed at a very low rate, the model predicts a constant radial expansion of the colony with active growth only at the margin. The model can generate and predict phenomena like concentric ring formation and autodegeneration of the colony. A development of the model also accounts for nutrient translocation by diffusion inside the mycelium since it treats intra-hypha diffusion separately from inter-hypha diffusion and has not been carried out elsewhere. This can predict the biomass distributions observed in colonies and in agar tile tessellation experiments (Davidson 1998). The latest addition to the model is the incorporation of active translocation of nutrients (Davidson and Olsson 2000) to be able to model the distribution of added label to *Arthrobotrus superba* (Fig. 5) growing out into soil from a food base (Persson et al. 2000).

So far these models describe many of the features seen in colonies that cannot be described by other models. The model output of the colony-scale model (Davidson et al. 1996) is in biomass distribution and position of the colony edge, substrate depletion outside the colony, and intracellular substrate concentration. Most of these parameters can be measured relatively easily experimentally. This makes it possible to use experimental results from one set of conditions, relating to heterogeneous nutrient distribution for a particular fungus, to the set of fungal specific parameters of the model (the capabilities of specific fungi). In the next step, with these parameters fixed, one can change the environmental parameters by considering radical differences in the

heterogeneity of nutrient availability and hence predict the behavior of the fungus in this system. The third step is then of course to check the predictions against new experimental data. If discrepancies arise these can be used to refine the model. This is needless to say a standard procedure in modeling approaches but, compared with others, reaction-diffusion models are more "experimentalist friendly". The model can, for example, be "cornered" so that it produces outputs with large differences for the conditions to be tested so that the experimental testing does not have to be performed with unrealistically high precision. It is important to note that these reaction-diffusion models also give good descriptions for growth under specific conditions and for fungal species used previously in model exercises. Reaction-diffusion models offer more general descriptions of fungal growth and also account for the coherent properties of colonies and their individuality as FMUs.

IX. The Individualistic Mycelium Concept

Several individual mycelia with similar genetic composition (a GMU) can join through anastomoses to form a single large FMU. Buller (1931) highlighted the significance of this in the formation of fruit bodies in his account of "The social organisation in *Coprinus sterquilinus* and other fungi". Rayner (1991) further stressed the concept of the mycelium as an individual organism based on his studies of interactions between mycelia of similar and different strains and species. The whole mycelium acts as a coordinated organism and is equivalent to a FMU – a physiologically integrated organism. Rock-inhabiting fungi growing in very hostile environments form microcolonies that grow even more like a multicellular organism. Here the leading hyphae grow very slowly but more growth occurs within the colony. The outermost hyphae (the leading hyphae) die and form a protective layer of dead cells (in three dimensions) behind which growth can occur (Gorbushina et al. 1999). The whole process is not dissimilar to the growth of bark on trees or skin on animals. Although fungal hyphae are linked physiologically, through anastomoses into a FMU, it has become evident that a cytoplasmic linkage or a common "nervous system" is not necessary for a colony to behave in concert as a single organism. Bacterial colonies formerly considered "just a heap of cells" are now more and more considered to be multicellular individuals. Titles of recent reviews such as "Thinking about bacterial populations as multicellular organisms" (Shapiro 1998), "Bacteria as modular organisms" (Andrews 1998) and "Cooperative organization of bacterial colonies: from genotype to morphotype" (Ben-Jacob et al. 1998) reinforce this viewpoint. Some bacterial colonies also show features reminiscent of fungal colonies on plates (Ben-Jacob et al. 1998). Similar considerations and processes implicated in the cooperative nature of bacterial colonies might also be valid for yeast colonies and we should perhaps think of these colonies as "multicellular" individuals. Thus it appears that all types of microbial colonies, both bacterial and fungal, are individual functional units – albeit with different degrees of cooperation and integration of the constituent parts – and not colonies built up of separate entities competing with each other. From these considerations the word "colony" may create an inappropriate representation of the social interactions between component members. For mycelial fungi I have suggested the concept of the functional mycelium unit (FMU) as I think it fits the observations well. For a more general term that includes the FMUs it may be appropriate to speak of functional cell agglomerations or functional units of cells. Thus this chapter is not about fungal colonies; it is about functional fungal cell agglomerations.

X. Conclusions

The main conclusion of this chapter is not new and is perfectly summarized by Buller (1931). "A large vegetative mycelium without any hyphal fusions and entirely un-netted would be about as unfitted for conducting food materials to a large fruit-body as a single un-netted railway system would be for carrying goods and passengers in a country such as England. A finely netted mycelium provides many different channels through which food materials can be conducted to a developing fruit-body or sclerotium and therein lies a very important part of its biological significance". For a netted mycelium (FMU) this concept of physiological individuality is especially obvious but physiological coordination also seems to apply for

simpler cell aggregations, such as those found in colonies of uni-cells, and might be a general phenomenon for all cell aggregations.

FMUs are not all the same and there are large variations in FMU behavior and organization between species, sizes of the FMU and the environments in which they live. Therefore one should be careful in extrapolating fungal mycelium functions from one species to another, from one size mycelium to another, and from one growth condition to another.

New mathematical models are being developed that are more easily applied and tested in experimental investigations and these will be of great use in understanding FMU behavior.

Finally, there is a scarcity of knowledge of FMU sizes and behavior outside the laboratory and especially in soil systems. More microcosm and field experiments are required, especially in the context of the eco-physiology of most fungi, even for those that are considered to be "well studied".

References

Andrews JH (1998) Bacteria as modular organisms. Annu Rev Microbiol 52:105–126

Axelrod DE (1972) Kinetics of differentiation of conidiophores and conidia by colonies of *Aspergillus nidulans*. J Gen Microbiol 73:181–184

Axelrod DE, Gealt M, Pastushok M (1973) Gene control of developmental competence in *Aspergillus nidulans*. Dev Biol 34:9–15

Bartnicki-Garcia S (1990) Role of vesicles in apical growth and a new mathematical model of hyphal morphogenesis. In: Heath IB (ed) Tip growth in plant and fungal cells. Academic Press, London, pp 211–232

Beadle GW, Tatum EL (1941) Genetic control of biochemical reactions in *Neurospora*. Proc Natl Acad Sci USA 27:499–506

Ben-Jacob E, Cohen I, Gutnick GL (1998) Cooperative organization of bacterial colonies: from genotype to morphotype. Annu Rev Microbiol 52:779–806

Bending GD, Read DR (1995) The structure and function of the vegetative mycelium of ectomycorrhizal plants. V. The foraging behavior of ectomycorrhizal mycelium and the translocation of nutrients from exploited organic matter. New Phytol 130:401–409

Boddy L (1999) Saprotrophic cord-forming fungi: meeting the challenge of heterogeneous environments. Mycologia 91:13–32

Bolton RG, Boddy L (1993) Characterization of the spatial aspects of foraging mycelial cord systems using fractal geometry. Mycol Res 97:762–768

Brown AJP, Gow NAR (1999) Regulatory networks controlling *Candida albicans* morphogenesis. Trends Microbiol 7:333–338

Buller AHR (1931) Social organisation in *Coprinus sterquilinus* and other fungi. In: Researches on Fungi IV. Hafner, New York, pp 139–186

Butler GM (1984) Colony ontogeny in basidiomycetes. In: Jennings DH, Rayner ADM (eds) The Ecology and the Physiology of the Fungal Mycelium. Cambridge Univ Press, Cambridge, pp 53–71

Cairney JWG (1992) Translocation of solutes in ectomycorrhizal and saprotrophic rhizomorphs. Mycol Res 96:135–141

Cairney JWG, Burke RM (1996) Physiological heterogeneity within fungal mycelia: an important concept for a functional understanding of the ectomycorrhizal symbiosis. New Phytol 134:685–695

Cooper KM, Tinker PB (1981) Translocation and transfer of nutrients in vesicular-arbuscular mycorrhizas. IV. Effect of environmental variables on movement of phosphorus. New Phytol 88:327–339

Davidson FA (1998) Modelling the qualitative response of fungal mycelia to heterogeneous environments. J Theor Biol 195:281–292

Davidson FA, Olsson S (2000) Translocation induced outgrowth of fungi in nutrient-free environments. J Theor Biol 205:73–84

Davidson FA, Sleeman BD, Rayner ADM, Crawford JW, Ritz K (1996) Context-dependent macroscopic patterns in growing and interacting mycelial networks. Proc R Soc Lond B Biol Sci 263:873–880

Dawson CG, Rayner ADM, Boddy L (1988) The form and outcome of mycelial interactions involving cord-forming decomposer basidiomycete in homogeneous and heterogeneous environments. New Phytol 109:423–432

Donnelly DP, Boddy L (1997) Development of mycelial systems of *Stropharia caerulea* and *Phanaerochaete velutina* on soil: effect of temperature and water potential. Mycol Res 101:705–713

Donnelly DP, Boddy L (1998) Repeated damage results in polarised development of foraging mycelial systems of *Phanerochaete velutina*. FEMS Microbiol Ecol 26:101–108

Donnelly DP, Wilkins MF, Boddy L (1995) An integrated image analysis approach for determining biomass, radial extent and box-count fractal dimension in macroscopic mycelial systems. Binary 7:19–28

Donnelly DP, Boddy L, Wilkins MF (1999) Image analysis – a valuable tool for recording and analysing development of mycelial systems. Mycologist 13:120–125

Finlay RD, Read DJ (1986) The structure and function of the vegetative mycelium of ectomycorrhizal plants. I. Translocation of 14C-labelled carbon between plants interconnected by a common mycelium. New Phytol 103:143–156

Gealt M, Axelrod AE (1974) Coordinate regulation of enzyme inducibility and developmental competence in *Aspergillus nidulans*. Dev Biol 41:224–232

Gorbushina AA, Krumbein WE, von Ossietzky C (1999) The poikilotrophic micro-organism and its environment. In: Seckbach J (ed) Enigmatic microorganisms and life in extreme environments. Kluwer, Dordrecht, pp 175–185

Gort AS, Imlay JA (1998) Balance between endogenous superoxide stress and antioxidant defences. J Bacteriol 180:1402–1410

Hansberg W, De-Groot H, Helmut S (1993) Reactive oxygen species associated with cell differentiation in *Neurospora crassa*. Free Radic Biol Med 14:287–293

Hawksworth DL, Kirk PM, Sutton BC, Pegler DN (1995) Ainsworth and Bisby's dictionary of fungi, 8th edn. CAB International, Wallingford

Hughes CL, Boddy L (1994) Translocation of ^{32}P between wood resources recently colonised by mycelial cord systems of *Phanaerochaete velutina*. FEMS Microbiol Ecol 14:201–212

Jennings DH (1994) Translocation in mycelia. In: Wessels JGH, Meinhardt F (eds) The Mycota, vol I. Growth, differentiation and sexuality. Springer, Berlin Heidelberg New York, pp 163–173

Kotov V, Reshetnikov S (1990) A stochastic model for early mycelial growth. Mycol Res 94:577–586

Lilly WW, Higgins SM, Wallweber GJ (1990) Uptake and translocation of 2-aminoisobutyric acid by *Schizophyllum-commune*. Exp Mycol 14:169–177

Lilly WW, Wallweber GJ, Higgins SM (1991) Proteolysis and amino acid recycling during nitrogen deprivation in *Schizophyllum-commune*. Curr Microbiol 23:27–32

Lindahl BJ, Stenlid J, Olsson S, Finlay R (1999) Translocation of ^{32}P between interacting mycelia of *Hypholoma fasciculare* and ectomycorrhizal fungi in microcosm systems. New Phytol 144:183–193

Mihail JD, Obert M, Bruhn JN, Taylor SJ (1995) Fractal geometry of diffuse mycelia and rhizomorphs of *Armillaria* species. Mycol Res 99:81–88

Navarro-Bordonaba J, Adams TH (1994) Development of conidia and fruiting bodies in ascomycetes. In: Wessels JGH, Meinhardt F (eds) The Mycota, vol I. Growth, differentiation and sexuality. Springer, Berlin Heidelberg New York, pp 333–350

Obert M (1994) Microbial growth patterns: fractal and kinetic characteristics of patterns generated by a computer model to simulate fungal growth. Fractals 1:354–374

Obert M, Pfeifer P, Sernetz M (1990) Microbial growth patterns described by fractal geometry. J Bacteriol 172:1180–1185

Olsson S (1994) Uptake of glucose and phosphorus by growing colonies of *Fusarium oxysporum* as quantified by image analysis. Exp Mycol 18:33–47

Olsson S (1995) Mycelial density profiles of fungi on heterogeneous media and their interpretation in terms of nutrient reallocation patterns. Mycol Res 99:143–153

Olsson S (1999) Nutrient translocation and electric signalling in mycelia. In: Gow NAR, Robson GD, Gadd GM (eds) The Fungal Colony. Cambridge Univ Press, Cambridge, pp 25–48

Olsson S, Gray SN (1998) Patterns and dynamics of ^{32}P-phosphate and labelled 2-aminoisobutyric acid (^{14}C-AIB) translocation in intact basidiomycete mycelia. FEMS Microbiol Ecol 26:109–120

Olsson S, Jennings DH (1991a) Evidence for diffusion being the mechanism of translocation in the hyphae of three molds. Exp Mycol 15:303–309

Olsson S, Jennings DH (1991b) A glass fiber filter technique for studying nutrient uptake by fungi: the technique used on colonies grown on nutrient gradients of carbon and phosphorus. Exp Mycol 15:292–301

Olsson S, Nordbring-Hertz B (1986) Microsclerotial germination of *Verticillium dahliae* as affected by rape rhizosphere. FEMS Microbiol Ecol 31:293–300

Pastushok M, Axelrod DE (1976) Effect of glucose, ammonium and media maintenance on the time of conidiophore initiation by surface colonies of *Aspergillus nidulans*. J Gen Microbiol 94:221–224

Persson C, Olsson S, Jansson H-B (2000) Growth of *Arthrobotrys superba* from a birch wood resource base into soil determined by radioactive tracing. FEMS Microbiol Ecol 31:47–51

Pérez-Martín J, Uría JA, Johnson AD (1999) Phenotypic switching in *Candida albicans* is controlled by a *SIR2* gene. EMBO J 18:2580–2592

Pirt SJ (1967) A kinetic study of the mode of growth of surface colonies of bacteria and fungi. J Gen Microbiol 66:137–143

Prosser JI, Trinci APJ (1979) A model for hyphal growth and branching. J Gen Microbiol 111:153–164

Radford DR, Challacombe SJ, Walter JD (1994) A scanning electron microscopy investigation of the structure of colonies of different morphologies produced by phenotypic switching of *Candida albicans*. J Med Microbiol 40:416–423

Ramsdale M (1999) Circadian rhythms in filamentous fungi. In: Gow NAR, Robson GD, Gadd GM (eds) The fungal colony. Cambridge Univ Press, Cambridge, pp 75–107

Rayner ADM (1991) The challenge of the individualistic mycelium. Mycologia 83:48–71

Rayner ADM (1996) Interconnectedness and individualism in fungal mycelia. In: Sutton BC (ed) A century of mycology. Cambridge Univ Press, Cambridge, pp 193–232

Rayner ADM, Griffith GS, Ainsworth AM (1995) Mycelial interconnectedness. In: Gow NAR, Gadd GM (eds) The growing fungus. Chapman and Hall, London, pp 21–40

Regalado CM, Crawford JW, Ritz K, Sleeman BD (1996) The origins of spatial heterogeneity in vegetative mycelia: a reaction-diffusion model. Mycol Res 100:1138–1142

Ritz K (1995) Growth responses of some soil fungi to spatially heterogeneous nutrients. FEMS Microbiol Ecol 16:269–280

Ritz K, Crawford JW (1990) Quantification of the fractal nature of colonies of *Trichoderma viride*. Mycol Res 94:1138–1141

Ritz K, Crawford JW (1999) Colony development in nutritionally heterogeneous environments. In: Gow NAR, Robson GD, Gadd GM (eds) The Fungal Colony. Cambridge Univ Press, Cambridge, pp 49–74

Ritz K, Millar SM, Crawford JW (1996) Detailed visualisation of hyphal distribution in fungal mycelia growing in heterogeneous nutritional environments. J Microbiol Methods 25:23–28

Robson GD, Bell SD, Kuhn PJ, Trinci APJ (1987) Glucose and penicillin concentrations in the medium below fungal colonies. J Gen Microbiol 133:361–367

Robson GD, Kuhn PJ, Trinci APJ (1991) Antagonism by sugars of the effects of validamycin a on growth and morphology of *Rhizoctonia cerealis*. Mycol Res 95:129–134

Roze LV, Linz JE (1998) Lovastatin triggers an apoptosis-like cell death process in the fungus *Mucor racemosus*. Fungal Genet Biol 25:119–133

Ryan FJ, Beadle GW, Tatum EL (1943) The tube method of measuring the growth rate of *Neurospora*. Am J Bot 30:784–799

Schütte KH (1956) Translocation in the fungi. New Phytol 55:164–182

Sessom DB, Lilly WW (1986) Derepressible proteolytic activity in homokaryotic hyphae of *Schizophyllum commune*. Exp Mycol 10:294–300

Shapiro JA (1998) Thinking about bacterial populations as multicellular organisms. Annu Rev Microbiol 52:81–104

Shepherd VA, Orlovich DA, Ashford AE (1993a) A dynamic continuum of pleiomorphic tubules and vacuoles in growing hyphae of a fungus. J Cell Sci 104:495–507

Shepherd VA, Orlovich DA, Ashford AE (1993b) Cell-to-cell transport via motile tubules in growing hyphae of a fungus. J Cell Sci 105:1173–1178

Slutsky B, Buffo J, Soll DR (1985) High-frequency switching of colony morphology in *Candida albicans*. Science 230:666–669

Thrane C, Olsson S, Nielsen TH, SØrensen J (1999) Vital fluorescent stains for detection of fungal stress in *Pythium ultimum* and *Rhizoctonia solani* challenged with viscosinamide from *Pseudomonas fluorescence* DR54. FEMS Microbiol Ecol 30:11–23

Timonen S, Finlay RD, Olsson S, Söderström B (1996) Dynamics of phosphorus translocation in intact ectomycorrhizal systems: non-destructive monitoring using beta-scanner. FEMS Microbiol Ecol 19:171–180

Toledo VJA, Hansberg W (1994) Enzyme inactivation related to a hyperoxidant state during conidiation of *Neurospora crassa*. Microbiology 140:2391–2397

Toledo VPR, Hansberg W (1995) Redox inbalance at the start of each morphogenetic step of *Neurospora crassa* conidiation. Arch Biochem Biophys 319:519–524

Trinci APJ, Wiebe MG, Robson GD (1994) The mycelium as an integrated entity. In: Wessels JGH, Meinhardt F (eds) The Mycota, vol I. Growth, differentiation and sexuality. Springer, Berlin Heidelberg New York, pp 175–194

Trinci APJ, Bocking S, Swift RJ, Withers JM, Robson GD, Weibe MG (1999) Growth, branching and enzyme production by filamentous fungi in submerged culture. In: Gow NAR, Robson GD, Gadd GM (eds) The Fungal Colony. Cambridge Univ Press, Cambridge, pp 108–125

Wells JM, Boddy L, Donnelly DP (1998a) Temporary phosphorus partitioning in mycelial systems of the cord-forming basidiomycete *Phanerochaete velutina*. New Phytol 140:283–293

Wells JM, Boddy L, Donnelly DP (1998b) Wood decay and phosphorus translocation by the cord-forming basidiomycete *Phanaerochaete velutina*: the significance of local nutrient supply. New Phytol 138:607–617

Wiebe MG, Trinci APJ, Cunliffe B, Robson GD, Oliver SG (1991) Appearance of morphological colonial mutants of *Fusarium graminearum* a3-5 in glucose-limited continuous flow cultures. Mycol Res 95:1284–1288

Wiebe MG, Robson GD, Trinci APJ, Oliver SG (1992) Characterization of morphological mutants generated spontaneously in glucose-limited continuous flow cultures of *Fusarium graminearum* a3-5. Mycol Res 96:555–562

Structural Continuum

7 Fungal Hydrophobins

Nicholas J. Talbot

CONTENTS

I. Introduction

Fungal hydrophobins are small, secreted proteins produced by fungi during a variety of developmental processes. The production of hydrophobins may be fundamental to the fungal lifestyle and it seems likely that hydrophobins are present in most filamentous fungal species. Hydrophobins have so far been found only in fungi and represent one of the few examples of a protein class confined to a single kingdom of eukaryotes. These unusual proteins play important roles in the ability of fungi to adhere to surfaces and to develop aerial hyphae. In basidiomycete species, hydrophobins play a part in the formation of complex fruit bodies, while in pathogenic fungal species hydrophobins may be important in bringing about plant infection. Hydrophobins are therefore critical determinants of fungal development. This chapter will discuss the unusual characteristics of hydrophobins and the roles these proteins play in diverse morphogenetic processes in fungi.

II. Identification of Fungal Hydrophobins

Hydrophobins have been characterized mainly as a result of discovering the genes that encode them. The reasons for this are twofold: first, fungal hydrophobin genes are highly expressed, often accounting for a significant proportion of the mRNA population during a given developmental process (Wessels 1996, 1997), and, secondly, because hydrophobin proteins cannot be detected using conventional protein purification techniques (Wessels et al. 1991). Hydrophobin genes were first recognized in the gill mushroom fungus *Schizophyllum commune* during a differential cDNA screen to identify genes that were expressed preferentially during the monokaryotic and dikaryotic phases of growth (Dons et al. 1984; Mulder and Wessels 1986). A number of highly expressed genes were found, and three of these genes showed homology to one another, but were not similar to any sequences in the GenBank database at that time (Schuren and Wessels 1990). Subsequently, similar genes were discovered during differential screens to identify genes expressed during conidiogenesis in *Aspergillus nidulans* (Stringer et al. 1991) and *Neurospora crassa* (Bell-Pederson et al. 1992; Lauter et al. 1992), and during plant infection by the rice blast fungus *Magnaporthe grisea* (Talbot et al. 1993). The subsequent development of procedures to purify these water-repellent proteins led to the first biochemical characterization of fungal hydrophobins (Wessels et al. 1991).

School of Biological Sciences, University of Exeter, Exeter EX4 4QJ, UK

The Mycota VIII
Biology of the Fungal Cell
Howard/Gow (Eds.)
© Springer-Verlag Berlin Heidelberg 2001

III. Physical Characteristics of Fungal Hydrophobins

Fungal hydrophobins are secreted, hydrophobic proteins and can be defined by the presence of eight cysteine residues which are spaced in a particular manner within the amino acid sequence (see Table 1). The spacing of cysteine residues, and the distribution of hydrophobic and hydrophilic amino acids within the polypeptide sequences of hydrophobins, led to the classification of hydrophobins into two different classes, I and II (Wessels 1994, 1996, 1997). All eight of the cysteine residues in hydrophobins appear to take part in intramolecular disulfide bridge formation, based on analysis of the class II hydrophobin cerato-ulmin and the class I hydrophobin SC3 (Yaguchi et al. 1993; de Vocht et al. 1998). The cysteine linkages identified in cerato-ulmin are between Cys_1 and Cys_2, Cys_3 and Cys_4, Cys_5 and Cys_6, and Cys_7 and Cys_8. This predicts a protein with four loops, two of which are principally composed of hydrophobic residues (Wessels 1997; Kershaw and Talbot 1998). The characteristic spacing of the eight cysteine residues in hydrophobins distinguishes them from other cysteine-rich fungal and plant proteins such as the peptide elicitors (Van den Ackervecken et al. 1992; Templeton et al. 1994; Rohe et al. 1995), chitin-binding proteins, lipid-binding proteins (Sterk et al. 1991; Désormeaux et al. 1992; José-Estanyol et al. 1992; Castonguay et al. 1994) and extracellular toxins.

Hydrophobin genes all contain putative signal peptide sequences at the N-terminus and it is likely that they are all secreted from hyphal tips during distinct developmental processes. Hydrophobin purification has confirmed that hydrophobins are secreted by a number of fungi (Wessels et al. 1991; de Vries et al. 1993; Lora et al. 1995; Talbot et al. 1996) and the cleavage sites for signal peptides have been confirmed for the *S. commune* hydrophobins (Wessels et al. 1991; Wessels 1997) and a number of others (Bidochka et al. 1995; Templeton et al. 1994; de Vries et al. 1999). Immunolocalization studies, using antibodies raised to purified hydrophobins or epitope tagging, have also shown the spatial distribution of hydrophobins in fungal structures (Wösten et al. 1994b; Lugones et al. 1996, 1998, 1999).

A. Class I Hydrophobins

A selection of the hydrophobins described to date is given in Table 2. Hydrophobins can be clearly divided into two groups based on their amino acid sequences, and biochemical analysis of members of each group has confirmed the different natures of these proteins. Class I hydrophobins are highly insoluble and can only be purified from fungal culture filtrates or cell wall extracts by treating these with agents to dissociate aggregated hydrophobin protein. Full aggregation of hydrophobins is normally carried out by vigorously aerating extracts and recovering insoluble material by centrifugation. Aggregated hydrophobin is insoluble in hot detergent solutions (such as sodium dodecyl sulfate) such that most other proteins can be removed from samples by hot detergent extraction and recovery of the

Table 1. Cysteine spacing within the fungal hydrophobins

Name of hydrophobin	Cysteine spacing[a]
Consensus (class I)	$C\text{-}X_{5-7}\text{-}C\text{-}C\text{-}X_{19-39}\text{-}C\text{-}X_{8-23}\text{-}C\text{-}X_5\text{-}C\text{-}C\text{-}X_{6-18}\text{-}C\text{-}X_{2-13}$
MPG1	$C\text{-}X_7\text{-}C\text{-}C\text{-}X_{22}\text{-}C\text{-}X_{19}\text{-}C\text{-}X_5\text{-}C\text{-}C\text{-}X_{11}\text{-}C\text{-}X_5$
rodA	$C\text{-}X_7\text{-}C\text{-}C\text{-}X_{39}\text{-}C\text{-}X_{18}\text{-}C\text{-}X_5\text{-}C\text{-}C\text{-}X_{17}\text{-}C\text{-}X_7$
Sc3	$C\text{-}X_6\text{-}C\text{-}C\text{-}X_{33}\text{-}C\text{-}X_{12}\text{-}C\text{-}X_5\text{-}C\text{-}C\text{-}X_{12}\text{-}C\text{-}X_6$
ssgA	$C\text{-}X_5\text{-}C\text{-}C\text{-}X_{19}\text{-}C\text{-}X_{15}\text{-}C\text{-}X_5\text{-}C\text{-}C\text{-}X_{12}\text{-}C\text{-}X_5$
Sc1	$C\text{-}X_6\text{-}C\text{-}C\text{-}X_{33}\text{-}C\text{-}X_{12}\text{-}C\text{-}X_5\text{-}C\text{-}C\text{-}X_{12}\text{-}C\text{-}X_7$
Sc4	$C\text{-}X_6\text{-}C\text{-}C\text{-}X_{33}\text{-}C\text{-}X_{12}\text{-}C\text{-}X_5\text{-}C\text{-}C\text{-}X_{12}\text{-}C\text{-}X_5$
Eas	$C\text{-}X_8\text{-}C\text{-}C\text{-}X_{25}\text{-}C\text{-}X_8\text{-}C\text{-}X_5\text{-}C\text{-}C\text{-}X_{18}\text{-}C\text{-}X_2$
Consensus (class II)	$C\text{-}X_{9-10}\text{-}C\text{-}C\text{-}X_{11}\text{-}C\text{-}X_{16}\text{-}C\text{-}X_{8-9}\text{-}C\text{-}C\text{-}X_{10}\text{-}C\text{-}X_{6-7}$
Cerato-ulmin	$C\text{-}X_9\text{-}C\text{-}C\text{-}X_{11}\text{-}C\text{-}X_{16}\text{-}C\text{-}X_9\text{-}C\text{-}C\text{-}X_{10}\text{-}C\text{-}X_6$
Cryparin	$C\text{-}X_9\text{-}C\text{-}C\text{-}X_{11}\text{-}C\text{-}X_{16}\text{-}C\text{-}X_8\text{-}C\text{-}C\text{-}X_{10}\text{-}C\text{-}X_6$

[a] C, Cysteine; X, any amino acid. The number of amino acid residues preceding the first cysteine cannot be accurately stated because signal peptide cleavage sites have not been determined in all cases.

Table 2. A selection of fungal hydrophobins

Name	Hydrophobin class	Taxonomic class	Organism	Mutant phenotype determined[a]	Biological function	Reference(s)
SC1	I	Basidiomycotina	*Schizophyllum commune*	–	Unknown. Expressed in dikaryotic phase	Schuren and Wessels (1990); Wessels et al. (1991)
SC3	I			+	Involved in aerial hyphae formation and ability to attach to hydrophobic surfaces	Wösten et al. (1994b); Van Wetter et al. (1996)
SC4	I			+	Required for lining gas channels in plectenchyma of fruit body	Schuren and Wessels (1990); Wessels et al. (1991)
SC6	I			–	Unknown. Expressed in dikaryotic phase	Wessels et al. (1991)
CoH1	I		*Coprinus cinereus*	–	Monokaryon-specific. Potential SC3 ortholog	Ásgeirsdóttir et al. (1997)
HydPt-1	I		*Pisolithus tinctorius*	–	Unknown. Expressed in mycorrhiza	Tagu et al. (1996)
HydPt-2	I			–	Unknown. Expressed in mycorrhiza	Tagu et al. (1996)
ABH1	I		*Agaricus bisporus*	–	Found lining gas channels in plectenchyma of fruit bodies	Lugones et al. (1996, 1999); de Groot et al. (1996)
ABH2	I			–	Unknown. Expressed in dikaryotic phase	de Groot et al. (1996); Lugones et al. (1996)
ABH3	I			–	Potential role in aerial hyphae formation	Lugones et al. (1998, 1999)
Fbh1	I		*Pleurotus ostreatus*	–	Localizes to fruit body. Potential SC4 ortholog	Penas et al. (1998)
POH1	I				Fruit-body expressed	Ásgeirsdóttir et al. (1998)
POH2	I				Expressed in vegetative hyphae. Potential SC3 ortholog	Ásgeirsdóttir et al. (1998)
POH3	I				Expressed in vegetative hyphae. Potential SC3 ortholog	Ásgeirsdóttir et al. (1998)
HCf-1	I	Ascomycotina	*Cladosporium fulvum*	+	Surface hydrophobicity of conidiating cultures	Spanu (1997, 1998)
SSGA	I		*Metarhizium anisopliae*	–	Unknown. Expressed during appressorium development	St. Leger et al. (1992)

Table 2. *Continued*

Name	Hydrophobin class	Taxonomic class	Organism	Mutant phenotype determined[a]	Biological function	Reference(s)
MPG1	I		*Magnaporthe grisea*	+	Conidial rodlet protein. Involved in conidium and appressorium formation. Required for full pathogenicity	Talbot et al. (1993, 1996); Kershaw et al. (1998)
RodA	I		*Aspergillus nidulans*	+	Conidial rodlet protein	Stringer et al. (1991)
DewA	I		*Aspergillus nidulans*	+	Conidial spore wall protein	Stringer and Timberlake (1995)
HYP1	I		*Aspergillus fumigatus*	+	Conidial rodlet protein	Parta et al. (1994); Thau et al. (1994)
Eas	I		*Neurospora crassa*	+	Conidial rodlet protein	Bell-Pederson et al. (1992); Lauter et al. (1992)
CU	II		*Ophiostoma ulmi*	+	Cerato-ulmin is involved in conditioning hydrophobicity of yeast-like propagules and attachment to insect disease vector	Temple et al. (1997)
CRYP	II		*Cryphonectria parasitica*	+	Unknown. Abundant expression in host plant. A cell wall protein	Zhang et al. (1994); McCabe and Van Alfen (1999)
QID3	II		*Trichoderma harzianum*	–	Unknown. Highly expressed in chitin-containing medium.	Lora et al. (1995)
HFB1	II		*Trichoderma reesei*	–	Unknown. Highly expressed in glucose-containing medium	Nakari-Setälä et al. (1996)
HFB2	II		*Trichoderma reesei*	–	Unknown	Nakari-Setälä and Pentilla (unpubl.)
cpa3	II		*Claviceps purpurea*	–	Unknown. Unusual "tri-hydrophobin" structure. Expressed in alkaloid-producing cultures	DeVries et al. (1999)

[a] + denotes that the phenotype has been determined by analysis of a naturally occurring mutant or by targeted gene deletion; – denotes that it has not.

insoluble material. Hydrophobin aggregates are dissociated with trifluoroacetic acid, or by formic and performic acid extraction (Wessels et al. 1991; Wösten et al. 1993). The dissociated hydrophobin monomers can then be dissolved in detergent and fractionated by polyacrylamide gel electrophoresis. When extracted with trifluoroacetic acid, the hydrophobins remarkably maintain their biological activity and monomeric hydrophobin can still aggregate to form polymerized structures (Wösten et al. 1993). Using the procedures just described, a survey of fungal species was carried out and revealed that the presence of class I hydrophobins was widespread among fungi (de Vries et al. 1993).

Class I hydrophobins can be modified prior to secretion and evidence exits to suggest certain hydrophobins can be glycosylated or associated with lipid. The SC3 hydrophobin, for example, is a glycosylated protein and contains 17–20 mannose residues (de Voght et al. 1998). The EAS hydrophobin, produced as a spore coat protein by *Neurospora crassa*, contains three potential myristoylation motifs indicating that lipid interactions may be involved in its linkage to the underlying spore cell wall (Bell-Pederson et al. 1992; Lora et al. 1995).

B. Class II Hydrophobins

The most extensively characterized class II hydrophobin, cerato-ulmin, was first studied as a putative phytotoxin long before it was recognized as a hydrophobin. Cerato-ulmin is produced in large quantities by *Ophiostoma ulmi*, the causal agent of Dutch elm disease (for review see Temple and Horgen 2000). Throughout the 1970s and early 1980s, Takai and colleagues carried out a detailed analysis of the protein in order to determine its function in Dutch elm disease (Takai 1974, 1980; Richards and Takai 1993). They found that cerato-ulmin could be purified readily from culture filtrates of *O. ulmi* by vigorous aeration that led to aggregation of the protein. Cerato-ulmin aggregates, however, are relatively unstable in comparison to those formed by class I hydrophobins and can be disrupted by application of pressure, or by extraction with 60% ethanol (Richards 1993; Richards and Takai 1993). Cerato-ulmin was only recognized as a hydrophobin after analysis of its amino acid sequence by keen-eyed hydrophobin researchers (Stringer and

Timberlake 1993), although the similarities in the proteins were immediately apparent; cerato-ulmin aggregates into rods and fibrils in the same way as the *S. commune* hydrophobins analyzed by Wessels and co-workers (1991). Class II hydrophobins were later classified as a separate group by comparison of amino acid sequences, and in particular the distribution of cysteine residues and hydrophobic amino acids in the polypeptide sequence (Wessels 1994).

The precise role of cerato-ulmin in the biology of *Ophiostoma ulmi* is still not completely clear, and is the subject of considerable speculation (for an excellent discussion of this, see Temple and Horgen 2000). A large amount of evidence was accumulated throughout the 1970s and 1980s to suggest that cerato-ulmin acts as a wilt toxin. The protein is capable of causing wilt symptoms in elm trees, for example, at concentrations as low as $2\,\mathrm{ng\,ml^{-1}}$ (Stevenson et al. 1979), and numerous studies have shown that application of cerato-ulmin to plants is sufficient to cause plant disease (Takai et al. 1983; Takai and Hiratsuka 1984; Scala et al. 1997). Significantly, the expression of the cerato-ulmin gene (*CU*) is also higher in more aggressive isolates of *O. novo-ulmi*, which has been responsible for the most recent epidemics of Dutch elm disease (Jeng et al. 1996). In spite of the weight of evidence implicating cerato-ulmin in the virulence of the fungus genetic studies, when the *CU* gene was mutated by targeted deletion, showed cerato-ulmin not to be a pathogenicity determinant (Bowden et al. 1996; Temple et al. 1997). This is consistent with the observation that natural isolates of *O. novo-ulmi*, which produce only small amounts of cerato-ulmin, can often be as pathogenic as wild-type isolates (Brasier et al. 1995). Furthermore, over-expression of the *CU* gene in a non-aggressive isolate of *O. novo-ulmi* does not affect the virulence of transformants (Temple et al. 1997).

Recent studies have, however, provided evidence that cerato-ulmin does contribute to the pathogenic fitness of *Ophiostoma*. Using a Δ*cu* mutant, a cerato-ulmin over-expressing strain, and wild-type strains of *O novo-ulmi* that vary in their aggressiveness, Temple et al. (1997) carried out a detailed study of the role of cerato-ulmin in pathogenicity. They found that cerato-ulmin contributes to the surface hydrophobicity of the yeast-like cells produced by *O. novo-ulmi*. These propagules are important to the fungus because they adhere to the bark beetles that transmit

Dutch elm disease. The disease is normally transmitted when beetles carry infectious yeast-like cells to feeding sites in the twig crotches of healthy elm trees (Temple and Horgen 2000). If yeast-like propagules of *O. novo-ulmi* are unable to adhere efficiently to the beetles then the disease is unable to become established in new elm trees. Cerato-ulmin is known to coat the surface of yeast cells produced by *O. novo-ulmi* (Takai and Hiatsuka 1980) and, therefore, may be required for dispersal, and perhaps propagule survival, during dry periods. Taken together these features probably account for the large amounts of cerato-ulmin produced during active growth of *Ophiostoma* species and suggest that it contributes to the spread of the disease (Temple and Horgen 2000). The importance of cerato-ulmin is further highlighted by a recent report that introduction of the *CU* gene is sufficient to allow *Ophiostoma quercus* (a non-pathogen of elm) to cause Dutch elm disease (DelSorbo et al. 2000).

Abundantly produced class II hydrophobins with similar features to cerato-ulmin have now been recognized in a number of species. Cryparin, for example, is produced in abundance by the horse chestnut blight fungus *Cryphonectria parasitica*. Cryparin shows a high degree of similarity to cerato-ulmin, except for the presence of a long N-terminus composed almost exclusively of serine and glycine residues (Zhang et al. 1994). Long N-terminal sequences have also been found in other class II hydrophobins, including most strikingly in the cpa3 hydrophobin from *Claviceps fusiformis* (Arntz and Tudzynski 1997; de Vries et al. 1999). The cpa3 hydrophobin is a large protein containing three individual hydrophobin-type domains (each with eight cysteine residues) separated by glycine and asparagine-rich tracts of amino acids. Class II hydrophobins have been found, based either on protein purification or differential gene expression studies, in a number of fungi including the tomato leaf mould fungus *Cladosporium fulvum* and the *Trichoderma* species *T. reesei* and *T. harzianum* (Lora et al. 1995; Nakari-Setälä et al. 1996; Segers et al. 1999).

C. The SC3 Hydrophobin from *Schizophyllum* and Interfacial Self-Assembly

The most intensively studied hydrophobin is the class I hydrophobin SC3 from *Schizophyllum commune* which is required for aerial hyphae for-

mation by the fungus. Structural studies have shown that the fungus secretes hydrophobin as a monomeric protein that contains four intramolecular disulfide bridges (Wessels et al. 1991; de Vries et al. 1993; de Voght et al. 1998). Monomers of SC3 self-assemble when they become exposed to an interface between water and air, or between water and a hydrophobic substrate like mineral oil (Wösten et al. 1993, 1994a). This remarkable process leads to spontaneous aggregation of SC3 whenever *S. commune* encounters the aerial surface of its liquid environment, or a hydrophobic solid substrate. When self-assembled, the SC3 hydrophobin forms an amphipathic membrane-like structure. Amphipathic aggregates of SC3 are extremely stable and can only be disrupted by trifluoroacetic acid extraction, as described previously. The amphipathic nature of SC3 aggregates is only achieved through self-assembly. This induces a conformational change in SC3 resulting in an increase in β-sheet structure (de Vocht et al. 1998). Assembled SC3 hydrophobin is extremely hydrophobic on one side and produces contact angles with water of approximately 110° (Wösten et al. 1994a). This is similar in nature to the non-stick surface polytetrafluoroethylene (PTFE, Teflon; DuPont). In contrast, the hydrophilic side of assembled SC3 aggregates produces contact angles with water of 36° (Wösten et al. 1994a), which is at least in part due to SC3 glycosylation and the exposure of mannose residues on the hydrophilic side of aggregated SC3.

IV. Biological Roles of Fungal Hydrophobins

A. Aerial Morphogenesis

Hydrophobins serve many roles in fungi due to their ability to self-assemble in response to interfaces. In the case of SC3 a clear role in the formation of aerial hyphae, and in the ability of hyphae to adhere to hydrophobic surfaces, has been found (Wösten et al. 1994b, 1999; van Wetter et al. 1996). A targeted deletion of the SC3 gene resulted in $\Delta sc3$ mutants that were deficient in their ability to produce aerial hyphae. This was particularly apparent when plate cultures were sealed to create a humid environment; under such conditions, $\Delta sc3$ mycelial cultures were entirely devoid of aerial growth (van Wetter et al. 1996). The way in which

SC3 brings about aerial growth of hyphae has now been examined in detail (see Fig. 1). SC3 can act as a surfactant, causing a large fall in the surface tension of water when it undergoes self-assembly at the surface. In a simple experiment, Wösten and colleagues (1999) showed that the normal surface tension of water ($72\,\text{mJ}\,\text{m}^{-2}$) could be reduced to as low as $24\,\text{mJ}\,\text{m}^{-2}$ by adding purified SC3. This makes SC3 the most powerful surface-active protein known and suggests that secretion of hydrophobins at high concentrations can lower surface tension dramatically, allowing fungal hyphae to escape the liquid and grow into the air (Talbot 1999). Continued secretion of SC3 is, however, also important for sustained aerial growth (Fig. 1). Adding SC3 to the growth medium of a $\Delta sc3$ mutant, for example, produces hydrophilic aerial hyphae that lack the hydrophobic coating produced by the wild type (Wösten et al. 1999). This means that SC3 production by hyphae occurs throughout aerial growth, and is consistent with gene expression studies that showed that the *SC3* transcript was abundant during aerial morphogenesis (Mulder and Wessels 1986). The surfactant activity of SC3 is among the highest reported (Van der Vegt et al. 1996). This is an unusual characteristic for a protein, because most natural surfactants are lipids, or lipid-associated proteins. Structural studies have shown, however, that SC3 lacks a close association with lipids either in its monomeric or assembled form (de Vocht et al. 1998).

SC3 also plays an important role in attachment of *S. commune* to hydrophobic surfaces (see Fig. 2). This is because of the ability of the protein to self-assemble in response to hydrophobic surfaces (Wösten et al. 1994b). Attachment to hydrophobic surfaces is important for most fungal species, which have to form intimate associations with substrates in order to secrete extracellular enzymes and take up the resulting simple sugars and amino acids. Hydrophobic surfaces, such as wood, leaf litter, and plant surfaces, must be adhered to firmly before (and during) penetration and colonization by hyphae. The ability of $\Delta sc3$ mutants to attach to Teflon membranes was found to be severely impaired in comparison with an isogenic wild-type strain but could be complemented by addition of purified SC3 protein (Wösten et al. 1994b). Moreover, the distribution of SC3 hydrophobin was shown by immunolocalization to be associated with the interface between the hyphae and the hydrophobic surface during hyphal attachment (Wösten et al. 1994a). The ability of hydrophobins to self-assemble at hydrophobic surfaces may also be of significance for their role in pathogenic fungi, as described below.

B. Fruit-Body Morphology in Basidiomycetes

Hydrophobins have been found in a number of mushroom-forming fungal species including both the edible mushroom *Agaricus bisporus* and the gill mushroom *S. commune*. There are known to be at least three hydrophobins produced by *A. bisporus* encoded by the *ABH1*, *ABH2* and *ABH3* genes (Table 2). The *ABH1* gene has been studied

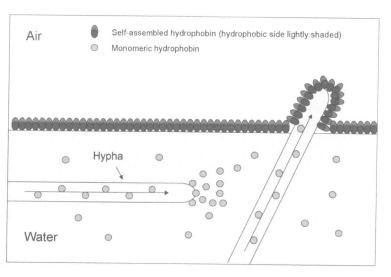

Fig. 1. The role of hydrophobins in erection of aerial hyphae. Hydrophobins are secreted as monomeric proteins. These self-assemble when they encounter the air–water interface. Hydrophobin self-assembly decreases the surface tension allowing the hypha to emerge into the air. Continued secretion of hydrophobin produces a hydrophobic coating for the aerial structure. Based on a model formulated by Wessels (1999), devised from the results of experiments by Wösten et al. (1999)

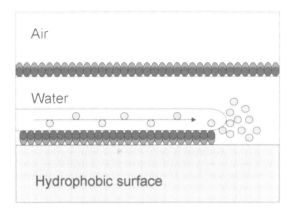

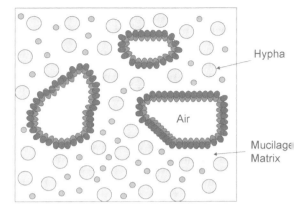

Fig. 2. Hydrophobin self-assembly mediates attachment of hyphae to hydrophobic surfaces. Hydrophobins are secreted as monomeric proteins and self-assemble upon encountering a hydrophobic surface. This provides a hydrophilic surface for the hyphal wall to adhere to, normally using accessory adhesive molecules such as mucilages. Based on a model formulated by Wösten et al. (1994b)

Fig. 3. Hydrophobins line gas channels in the plectenchyma of mushrooms. Basidiomycetes utilize hydrophobins to line the gas channels in their fruiting bodies. Monomeric hydrophobins are secreted and fill the gaps between hyphae. Hydrophobins self-assemble along the interior of gas channels due to the air-water interface encountered there. Hydrophobins may be required to prevent water-soaking of fruiting bodies, which is vital for basidiospore dispersal. They will also act in this context to mediate efficient gas exchange. Model first formulated by Lugones et al. (1999)

in great detail and appears to play a similar role to *SC4* in *S. commune*. In both cases the hydrophobins appear to line the gas channels that occur within the mushroom fruit body between the aggregated hyphae. These hydrophobins probably play a role in cementing hyphae together during production of these large structures, but their main role may be in ensuring that air can pass freely throughout the mushroom fruit body allowing adequate gas exchange for all hyphae (Lugones et al. 1999; van Wetter 2000). This function probably occurs due to the hydrophobin preventing water from entering, or remaining within, the channels during periods of rain or flooding (Fig. 3). This role is thus likely to be of key importance for the survival of fruit bodies. To investigate this process a targeted mutation was made at the *SC4* locus and a homozygous Δ*sc4* dikaryon was then generated. In an elegant experiment, the homozygous Δ*sc4* dikaryon was shown to produce mushrooms that could become easily water-soaked and would sink when placed in water. An isogenic wild-type strain meanwhile produced mushrooms that floated in water (van Wetter 2000).

In addition to their role in allowing free passage of air throughout the fruit body, hydrophobins probably serve to prevent water soaking of mushrooms by forming a hydrophobic coating on the surface of the pileus (Wessels 1999). Electron microscopy shows the presence of interwoven

rodlets on the pileus surface of many mushrooms (Fig. 4) and immunolocalization has confirmed the presence of hydrophobin fulfilling this function in *Agaricus bisporus* (deGroot et al. 1996; Lugones et al. 1996; Kershaw and Talbot 1998).

C. Conidiogenesis

The presence of interwoven rodlets on the surface of conidia has been known for some time and was used in the 1960s as a taxonomic feature for classification of *Penicillium* species, for example (Hess et al. 1968). Hydrophobins were found to form conidial rodlet proteins, first in *A. nidulans* where the *rodA* gene was identified as a gene expressed highly during conidiogenesis. Targeted deletion of *rodA* resulted in *rodletless* mutants that were hydrophilic and lacked their characteristic spore coating (Stringer et al. 1991). Subsequent analysis of the *EAS* gene, which was identified both as a conidiation-specific gene and a gene under circadian control (named *ccg-2*), showed that the spore coat of *N. crassa* was also hydrophobin-encoded (Bell-Pederson et al. 1992; Lauter et al. 1992). This provided a molecular explanation for the *easily wettable* phenotype observed in *N. crassa* in the late 1970s (Beever and Dempsey 1978). The purpose of the rodlet layer on spores has not been

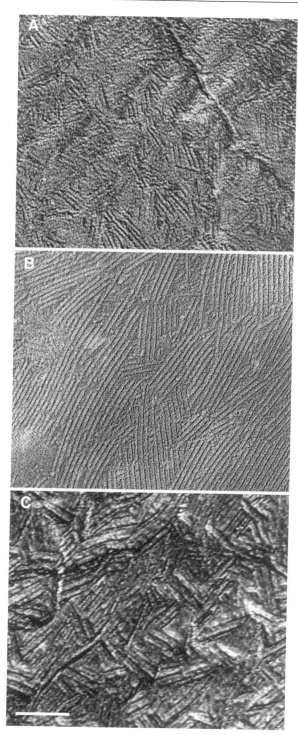

Fig. 4. Ultrastructural analysis of hydrophobin-encoded rodlet layers. Transmission electron micrographs of carbon-platinum replicas of the surfaces of fungal structures. **A** Surface of the pileus of an *Agaricus excellens* mushroom showing rodlet layer. **B** Surface of conidium of *Aspergillus nidulans* showing the *rodA*-encoded rodlet layer, freeze-fractured. **C** Surface of conidium of *Neurospora crassa* showing the *EAS*-encoded rodlet layer, freeze-fractured. *Bar* 100 nm

investigated in detail, but it seems likely that spore dispersal, which often involves splash dispersion or wind scattering, requires a hydrophobic spore coat, although definitive experimental proof has been lacking (Fig. 4). Another potential role for hydrophobic rodlet layers is to prevent water soaking, and perhaps water loss from the spore. Spore rodlet layers may involve the contribution of more than one hydrophobin protein. The *dewA* gene, for example, contributes to spore hydrophobicity in *A. nidulans* – Δ*dewA* mutants are detergent-wettable in contrast to wild-type strains – but does not appear to encode rodlets on *A. nidulans* spores in the absence of *rodA*. The dewA protein therefore contributes to the spore wall, perhaps forming some sort of hybrid layer in concert with the more major rodA protein (Stringer and Timberlake 1995).

D. Infection Structure Formation

Hydrophobins may contribute to the formation of the specialized infection structures that many pathogenic fungi use to infect plants and animals. Two lines of evidence suggest such a role. In the rice blast fungus *Magnaporthe grisea*, it was found that the MPG1 hydrophobin is required for efficient production of the infection cells produced by the fungus to breach the rice leaf cuticle (Talbot et al. 1993). These cells, called appressoria, form in response to a hard hydrophobic surface and can be induced to form in the absence of a plant on inert, hydrophobic plastic surfaces (for review see Talbot 1995). *MPG1* was first isolated as a gene that is highly expressed during the infection process, and was found to be expressed during appressorium formation itself. A targeted deletion of *MPG1* resulted in mutants that were unable to form appressoria in large numbers; Δ*mpg1* mutants typically produce 18–20% of the numbers of appressoria generated by an isogenic wild-type *M. grisea* strain (Talbot et al. 1993). *MPG1* was found to encode a class I hydrophobin that is also responsible for production of the conidial rodlet protein in *M. grisea* and is highly expressed during conidiogenesis and also during nutrient starvation stress. The protein appears to undergo self-assembly at hydrophobic surface interfaces, based on the observation that an SDS-insoluble protein contributes to the attachment of appressoria to Teflon surfaces in *M. grisea* wild-type strains but not in Δ*mpg1* mutants (Talbot et al. 1996). The role of

MPG1 in appressorium development may, therefore, simply be due to its secretion during germ-tube formation and self-assembly in response to the hydrophobic surface interface. The increased wettability resulting from this process may act as an efficient primer for the action of hydrophilic mucilages, thereby allowing firm attachment of the germ-tube tip, prior to appressorium differentiation. The attachment process may therefore act as a developmental signal for cellular differentiation. Appressorium differentiation in *M. grisea* is known to be regulated, at least in part, by a cyclic AMP-dependent signaling pathway and the fact that Δ*mpg1* mutants can be induced to undergo appressorium formation in the presence of exogenous cAMP is consistent with the action of MPG1 prior to cellular differentiation (Beckerman and Ebbole 1996; Talbot et al. 1996). Interestingly, it appears that many class I hydrophobin genes are able to substitute for *MPG1* during appressorium formation, and to form rodlet proteins on the surface of conidia. In a comprehensive set of cross-species complementation experiments, Kershaw et al. (1998) showed that *SC1*, *SC4*, *ssgA*, *EAS*, *dewA* and *rodA* could all substitute for *MPG1* at least in part, restoring the pathogenicity and ability to make appressoria to a Δ*mpg1* mutant, when introduced under the control of the *MPG1* promoter. What this indicates is that the action of *MPG1* is probably fairly non-specific and is likely to be a consequence of its ability to self-assemble at hydrophobic surface interfaces (Kershaw et al. 1998). The second line of evidence indicating that hydrophobins play a role in infection structure formation comes from the identification of *ssgA*, a gene from the entomopathogenic fungus *Metarhizium anisopliae*. *SsgA* is highly expressed during appressorium development, even though appressoria of this species are protease-secreting structures that are markedly different from the appressoria produced by *M. grisea*.

How widespread the role of hydrophobins is as virulence determinants in pathogenic fungi is unclear although they have now been found in many pathogenic species. Multiple hydrophobins have been found, for example, in the tomato leaf mould fungus *Cladosporium fulvum*, including both class I and class II hydrophobins (Segers et al. 1999). The presence of such a diversity of hydrophobins indicates that many functions might be fulfilled by these proteins (Spanu 1998), a fact further supported by the different expression profiles found for each of the genes. Conversely,

hydrophobins are secreted in abundance by fungi and thus have been implicated as molecules that might become recognized by host plants in order to stage a defense against pathogenic fungi (Templeton et al. 1994). The role of hydrophobins as elicitors of plant defense responses has not been investigated in detail although it is interesting that the avirulence gene product NIP1 produced by *Rhynchosporium secalis* shows homology (at least in cysteine residue spacing) with hydrophobins (Rohe et al. 1995).

V. Functional Relatedness of Fungal Hydrophobins

Hydrophobins appear to fulfill many functions in fungal development, but how closely related they are, as a class of proteins, remains an open question. At the amino acid level there is barely any significant homology between hydrophobins in different species (Kershaw and Talbot 1998; Wessels 1999), but the spacing of cysteine residues and the conservation in distribution of hydrophobic and hydrophilic residues makes them instantly recognizable based on hydropathy plots. The only detailed genetic analysis of hydrophobin function to date has been the study of Kershaw et al. (1998) that showed close functional relationships between class I hydrophobins produced during quite distinct developmental processes. Most of the hydrophobins were able to form a rodlet layer of some kind on the surface of *M. grisea* conidia when expressed under the correct developmental control. The rodlet layer was not, however, complete in any case suggesting that the linkage of hydrophobins to the underlying cell wall may be more specific to individual hydrophobins. This is consistent with studies in *S. commune* that have shown that SC3 has lectin-binding activity that is required for its linkage to the underlying cell wall during aerial hyphae formation (de Vocht et al. 1998).

The study by Kershaw et al. (1998) did, however, define two functional classes of hydrophobins because SC3 was unable to complement any of the Δ*mpg1* mutant phenotypes and could not elaborate a rodlet layer on spore surfaces. Since this protein forms normally on the surface of undifferentiated hyphae, it is possible that class I hydrophobins may broadly group into those that function in differentiated fungal structures, such as appressoria, conidia, and mushrooms, and those that play a part in aerial morphogenesis and adhesion of undifferentiated hyphae. This hypothesis could be tested by complementation experiments with the *COH1* hydrophobin gene from *C. cinereus*, the *POH2* and *POH3* hydrophobins from *Pleurotus ostreatus*, and the *ABH3* hydrophobin gene from *A. bisporus* that all appear to play similar roles to *SC3* (Ásgeirsdóttir et al. 1997, 1998; Lugones et al. 1998; Penas et al. 1999). The functional relationship of *SC3* and *SC4* has been recently investigated in *S. commune* and here it is also clear that the two hydrophobins differ functionally (van Wetter 2000). Expressing the class I hydrophobin under the *SC3* promoter in monokaryotic *S. commune* did not allow complementation of all of the mutant phenotypes of $\Delta sc3$ mutants. Aerial hyphae were formed by *SC3p:SC4* transformants but they were not able to adhere tightly to hydrophobic surfaces because the *SC4*-encoded rodlet layer was not attached sufficiently strongly to the hyphal cell wall (van Wetter 2000).

The functional relationship between class I and II hydrophobins is also an interesting and largely unexplored area for future investigation. Clearly, class II hydrophobins are different in their biochemical characteristics and yet they are produced at similar levels of abundance to class I hydrophobins during fungal development (Temple and Horgen 2000). The fact that many species appear to have both classes of proteins indicates that distinct functions may be present and this is further highlighted by recent biochemical characterization of cryparin (McCabe and Van Alfen 1999). It has been demonstrated that the cryparin protein, for example, becomes attached to the cell walls of submerged hyphae *after* secretion, but prior to emergence into the air. A different process is therefore likely to mediate the attachment of class II hydrophobins to fungal hyphae in comparison with self-assembling class I hydrophobins. The aggregation of class II hydrophobins may thus be a fundamentally different process compared with self-assembly of class I hydrophobins.

VI. Evolution of Fungal Hydrophobins

One of the most fascinating areas for future research is likely to be the determination of the origin of fungal hydrophobins and their significance in fungal evolution. Hydrophobins have been found in Ascomycetes and Basidiomycetes, but, although proteins with hydrophobin-like properties are present in zygomycetes (de Vries et al. 1993), they have yet to be purified and characterized. Hydrophobin genes have also, as yet, not been identified from zygomycete or chytridiomycete fungi. The identification of hydrophobins among these taxa will be important in determining whether evolution of hydrophobins played a part in terrestrial colonization by fungi, particularly if they turn out to be absent in the Chytridiomycota. It seems likely that the spread of fungi to terrestrial ecosystems would have required the action of hydrophobin-like proteins in order for fungi to adhere to substrates and disperse their spores on aerial structures. In this context, it will be important to determine whether a single primary function can be defined for hydrophobins (for example, substrate attachment) that pre-dates other adaptations. Did hydrophobin gene families, for example, evolve by gene duplications and subsequent evolution of paralogs? Conversely, was the evolution of hydrophobins polyphyletic, resulting in the two functional classes of proteins observed today? Addressing these questions by using the contemporary tools of comparative genomics is likely to prove exciting in understanding the place of these fascinating proteins in the evolution of the Mycota.

VII. Industrial Applications of Fungal Hydrophobins

The remarkable properties of fungal hydrophobins have suggested a number of likely applications in industrial processes (Wessels 1997; Kersaw and Talbot 1998). The ability of class I hydrophobins to self-assemble offers the potential to coat hydrophobic plastic surfaces with a protein monolayer, for example. This can offer the chance to change the surface properties of a substrate and provide molecular anchor points for other biochemical functions, such as enzymatic activities. This type of property is sought after for a variety of products, including biosensors and physiological testing agents such as pregnancy diagnostic kits. Medical applications of class I hydrophobins might also take advantage of their ability to attach to surfaces in order to coat catheters, prostheses and artificial organs with a natural protein layer during transplantation surgery.

Self-assembly of class I hydrophobins also produces very stable foams (Wösten et al. 1993; Talbot, unpubl. observ.). This is a useful attribute for many food manufacturing processes, including production of confectionery and carbonated drinks. The surface activity of hydrophobins may therefore be of significant utility for a large number of processes, and the widespread distribution of hydrophobins in edible fungi such as *A. bisporus* suggests they are natural surfactants that are safe and non-toxic.

VIII. Conclusions

The production of hydrophobins is one of the only characteristics that so far can be said to be truly specific to fungi. Although fungi have coopted and adapted numerous biochemical and genetic processes for specific purposes, there are no other examples of an entire class of protein that is completely restricted to the kingdom (insects and crustaceans, for example, also carry out chitin synthesis). Research to date has clearly established that hydrophobins are involved in aerial morphogenesis, conidiogenesis and surface attachment, but there is still very much that is unknown about these extraordinary proteins. Do all filamentous basidiomycete and ascomycete fungi, for example, require hydrophobins in order to produce aerial hyphae? The role of SC3 has been established pretty conclusively, and it is even sufficient to bring about aerial hyphae formation in filamentous bacteria (Tillotson et al. 1998). But do all higher fungi possess an *SC3* ortholog, or are there non-hydrophobin-mediated routes for aerial hyphae production? And, if there is a degree of functional redundancy among hydrophobins, as suggested by studies so far (Kershaw et al. 1998), then why do some species contain many hydrophobin genes (Segers et al. 1999)? These are but a selection of the questions that will need to be addressed in the next few years to try and fully appreciate the functional significance of hydrophobins in fungal development. Detailed structure and function studies of the best-known hydrophobins, coupled with the rapidly emerging analytical processes of computational sequence examination and comparative genomics, will hopefully provide the answers.

References

Arntz C, Tudzynski P (1997) Identification of genes induced in alkaloid-producing cultures of *Claviceps* sp. Curr Genet 31:357–360

Ásgeirsdóttir SA, Halsall JR, Casselton LA (1997) Expression of two closely linked hydrophobin genes of *Coprinus cinereus* is monokaryon-specific and downregulated by the *oid-1* mutation. Fungal Genet Biol 22:54–63

Ásgeirsdóttir SA, de Vries OMH, Wessels JGH (1998) Identification of three differentially expressed hydrophobins in *Pleurotus ostreatus* (oyster mushroom). Microbiology 144:2961–2969

Beckerman JL, Ebbole DJ (1996) *MPG1*, a gene encoding a fungal hydrophobin of *Magnaporthe grisea*, is involved in surface recognition. Mol Plant Microbe Interact 9:450–456

Beever RE, Dempsey GP (1978) Function of rodlets on the surface of fungal spores. Nature 272:608–610

Bell-Pederson D, Dunlap JC, Loros JJ (1992) The *Neurospora* circadian clock-controlled gene, *ccg-2*, is allelic to *eas* and encodes a fungal hydrophobin required for formation of the conidial rodlet layer. Genes Dev 6:2382–2394

Bidochka MJ, St Leger RJ, Joshi L, Roberts DW (1995) The rodlet layer from aerial and submerged conidia of the entomopathogenic fungus *Beauvaria bassiana* contains hydrophobin. Mycol Res 99:403–406

Bowden CG, Smalley E, Guries RP, Hubbes M, Horgen PA (1996) Lack of association between cerato-ulmin production and virulence in *Ophiostoma novo-ulmi*. Mol Plant Microbe Interact 9:556–564

Brasier CM, Kirk SA, Tegli S (1995) Naturally occurring non cerato-ulmin producing mutants of *Ophiostoma novo-ulmi* are pathogenic but lack aerial mycelium. Mycol Res 99:436–440

Carpenter CE, Mueller RJ, Kazmierczak P, Zhang L, Villalon DK, Van Alfen NK (1992) Effect of a virus on the accumulation of a tissue-specific cell-surface protein of the fungus *Cryphonectria parasitica*. Mol Plant Microbe Interact 4:55–61

Castonguay Y, Laberge S, Nadeau P, Vézina L-P (1994) A cold-induced gene from *Medicago sativa* encodes a bimodular protein similar to developmentally regulated proteins. Plant Mol Biol 24:799–804

de Groot PWJ, Schaap PJ, Sonnenberg ASM, Visser J, Van Griensven LJLD (1996) The *Agaricus bisporus HypA* gene encodes a hydrophobin and specifically accumulates in peel tissue of mushroom caps during fruit body development. J Mol Biol 257:1008–1018

DelSorbo G, Scala F, Parrella G, Lorito M, Coparini C, Ruocco M, Scala A (2000) Functional expression of the cu gene, encoding the phytotoxic hydrophobin cerato-ulmin, enables *Ophiostoma quercus*, a non-pathogen of elm, to cause symptoms of Dutch elm disease. Mol Plant Microbe Interact 13:43–53

de Vocht ML, Scholtmeijer K, van der Vegte EW, de Vries OMH, Sonveaux N, Wösten HAB, Ruysschaert J-M, Hadziioannou G, Wessels JGH, Robillard GT (1998) Structural characterization of the hydrophobin SC3, as a monomer and after self-assembly at hydrophobic/hydrophilic interfaces. Biophys J 74:2059–2068

de Vries OMH, Fekkes MP, Wosten HAB, Wessels JGH (1993) Insoluble hydrophobin complexes in the walls of *Schizophyllum commune* and other filamentous fungi. Arch Microbiol 159:330–335

De Vries OMH, Moore S, Arntz C, Wessels JGH, Tudzynski P (1999) Identification and characterization of a tri-partite hydrophobin from *Claviceps fusiformis* – a novel type of class II hydrophobin. Eur J Biochem 262:377–385

Désormeaux A, Blochet J-E, Pézolet M, Marion D (1992) Amino acid sequence of a non-specific wheat phospholipid transfer protein and its conformation as revealed by infrared and Raman spectroscopy. Role of disulfide bridges in the stabilization of the α-helix structure. Biochim Biophys Acta 1121:137–152

Dons JJM, Springer J, de Vries OMH, Wessels JGH (1984) Molecular cloning of a gene abundantly expressed during fruiting body initiation in *Schizophyllum commune*. J Bacteriol 157:802–808

Hess WM, Sassen MMA, Remson CC (1968) Surface characteristics of *Penicillium* conidia. Mycologia 60:290–303

Jeng RS, Hintz WE, Bowden CG, Horgen PA, Hubbes M (1996) A comparison of the nucleotide sequence of the cerato-ulmin gene and the rDNA ITS between aggressive and non-aggressive isolates of *Ophiostoma ulmi* sensu lato, the causal agent of Dutch elm disease. Curr Genet 29:168–173

José-Estanyol M, Ruiz-Avila L, Puigdomenech P (1992) A maize embryo-specific gene encodes a proline-rich and hydrophobic protein. Plant Cell 4:413–423

Kershaw MJ, Talbot NJ (1998) Hydrophobins and repellents: proteins with fundamental roles in fungal morphogenesis. Fungal Genet Biol 23:18–33

Kershaw MJ, Wakley GE, Talbot NJ (1998) Complementation of the MPG1 mutant phenotype in *Magnaporthe grisea* reveals functional relatedness between fungal hydrophobins. EMBO J 17:3838–3849

Lauter FR, Russo VEA, Yanofsky C (1992) Developmental and light regulation of *Eas*, the structural gene for the rodlet protein of *Neurospora crassa*. Genes Dev 6:2373–2381

Lora JM, Pinto-Toro JA, Bentítez T, Romero LC (1995) Qid3 protein links plant bimodular proteins with fungal hydrophobins. Mol Microbiol 18:380–382

Lugones LG, Bosscher JS, Scholtmeyer K, de Vries OMH, Wessels JGH (1996) An abundant hydrophobin (*ABH1*) forms hydrophobic rodlet layers in *Agaricus bisporus* fruiting bodies. Microbiology 142:1321–1329

Lugones LG, Wösten HAB, Wessels JGH (1998) A hydrophobin (ABH3) specifically secreted by vegetatively growing hyphae of *Agaricus bisporus* (common white button mushroom). Microbiology 144:2345–2353

Lugones LC, Wösten HAB, Birkenkamp KU, Sjollema KA, Zagers J, Wessels JGH (1999) Hydrophobins line air channels in fruiting bodies of *Schizophyllum commune* and *Agricus bisporus*. Mycol Res 103:635–640

McCabe PM, Van Alfen NK (1999) Secretion of cryparin, a fungal hydrophobin. Appl Environ Microbiol 65:431–435

Mendgen K, Deising H (1993) Infection structures of fungal plant pathogens – a cytological and physiological evaluation. New Phytol 124:193–213

Mulder GH, Wessels JGH (1986) Molecular cloning of RNAs differentially expressed in monokaryons and dikaryons of *Schizophyllum commune* in relation to fruiting. Exp Mycol 10:214–227

Nakari-Setälä T, Aro N, Kalkkinen N, Alatalo E, Penttila M (1996) Genetic and biochemical characterization of the *Trichoderma reesei* hydrophobin *HFB1*. Eur J Biochem 235:248–255

Parta M, Chang Y, Rulong S, Pinto da Silva P, Kwon Chung KJ (1994) *Hyp1*, a hydrophobin gene from *Aspergillus fumigatus* complements the rodletless phenotype in *Aspergillus nidulans*. Infect Immun 62:4389–4395

Penas MM, Asgeirsdottir SA, Lasa I, CulianezMacia FA, Pisabarro AG, Wessels JGH, Ramirez L (1998) Identification, characterization, and in situ detection of a fruit-body-specific hydrophobin of *Pleurotus ostreatus*. Appl Environ Microbiol 64:4028–4034

Richards WC (1993) Cerato-ulmin: a unique wilt toxin of instrumental significance in the development of Dutch elm disease. In: Sticklen MB, Sherald JL (eds) Dutch elm disease research, cellular and molecular approaches. Springer, Berlin Heidelberg New York, pp 89–151

Richards WC, Takai S (1993) Amino acid sequence and spectroscopic studies of Dutch elm disease toxin, cerato-ulmin. In: Sticklen MB, Sherald JL (eds) Dutch elm disease research, cellular and molecular approaches. Springer, Berlin Heidelberg New York, pp 152–170

Rohe M, Gierlich A, Hermann H, Hahn M, Schmidt B, Rosahl S, Knogge W (1995) The race-specific elicitor, *NIP 1*, from the barley pathogen, *Rhynchosporium secalis*, determines avirulence on host plants of the Rrsl resistance genotype. EMBO J 14:4168–4177

Russo PS, Blum FD, Ipsen JD, Abul-Hajj YJ, Miller WG (1982) The surface activity of the phytotoxin cerato-ulmin. Can J Bot 60:1414–1422

Schuren FHJ, Wessels JGH (1990) Two genes specifically expressed in fruiting dikaryons of *Schizophyllum commune*: homologies with a gene not regulated by mating type genes. Gene 90:199–205

Segers GC, Hamada W, Oliver RP, Spanu PD (1999) Isolation and characterisation of five different hydrophobin-encoding cDNAs from the fungal tomato pathogen *Cladosporium fulvum*. Mol Gen Genet 261:644–652

Selitrennikof CP (1976) *Easily wettable*, a new mutant. Neurospora Newslett 23:23

Spanu P (1997) HCF-1, a hydrophobin from the tomato pathogen *Cladosporium fulvum*. Gene 193:89–96

Spanu P (1998) Deletion of HCf-1, a hydrophobin gene of *Cladosporium fulvum*, does not affect pathogenicity in tomato. Physiol Mol Plant Pathol 52:323–334

St Leger RJ, Staples RC, Roberts DW (1992) Cloning and regulatory analysis of starvation-stress gene, *ssgA*, encoding a hydrophobin-like protein from the entomopathogenic fungus *Metarhizium anisopliae*. Gene 120:119–124

Sterk P, Booij H, Schellekens GA, Van Kammen A, de Vries SC (1991) Cell-specific expression of the carrot lipid transfer protein gene. Plant Cell 3:907–921

Stevenson KJ, Slater RC, Takai S (1979) Cerato-ulmin – a wilting toxin of Dutch elm disease fungus. Phytochemistry 18:235–238

Sticklen MB, Bolyard M (1994) Refinement of physiological roles for cerato-ulmin by analogy with other hydrophobins. Trends Microbiol 2:213–215

Stringer MA, Timberlake WE (1993) Cerato-ulmin, a toxin involved in Dutch elm disease, is a fungal hydrophobin. Plant Cell 5:145–146

Stringer MA, Timberlake WE (1995) *dewA* encodes a fungal hydrophobin component of the *Aspergillus* spore wall. Mol Microbiol 16:33–44

Stringer MA, Dean RA, Sewell TC, Timberlake WE (1991) Rodletless, a new *Aspergillus* developmental mutant induced by directed gene inactivation. Genes Dev 5:1161–1171

Tagu D, Nasse B, Martin F (1996) Cloning and characterization of hydrophobin-encoding cDNAs from the ectomycorrhizal basidiomycete *Pisolithus tinctorius.* Gene 168:93–97

Takai S (1974) Pathogenicity and cerato-ulmin production in *Ceratocystis ulmi.* Nature 252:124–126

Takai S (1980) Relationship of the production of toxin, cerato-ulmin, to synnemata formation, pathogenicity, mycelial habit, and growth of *Ceratocystis ulmi* isolates. Can J Bot 58:658–662

Takai S, Hiratsuka Y (1980) Accumulation of the material containing the toxin cerato-ulmin on the hyphal surface of *Ceratocystis ulmi.* Can J Bot 58:658–662

Takai S, Hiratsuka Y (1984) Scanning electron microscope observations of internal symptoms of white elm following *Ceratocystis ulmi* infection and cerato-ulmin treatment. Can J Bot 62:1365–1371

Takai S, Richards WC (1978) Cerato-ulmin, a wilting toxin of *Ceratocystis ulmi*: isolation and some properties of cerato-ulmin from the culture of *C. ulmi.* J Phytopathol 91:129–146

Takai S, Richards WC, Stevenson KJ (1983) Evidence for the involvement of the *Ceratocystis ulmi* toxin cerato-ulmin in the development of Dutch elm disease. Physiol Plant Pathol 23:275–280

Talbot NJ (1995) Having a blast: exploring the pathogenicity of *Magnaporthe grisea.* Trends Microbiol 3:9–16

Talbot NJ (1997) Fungal biology: growing into the air. Curr Biol 7:R78–82

Talbot NJ (1999) Coming up for air and sporulation. Nature 398:95–296

Talbot NJ, Ebbole DJ, Hamer JE (1993) Identification and characterization of *MPG1*, a gene involved in pathogenicity from the rice blast fungus *Magnaporthe grisea.* Plant Cell 5:1575–1590

Talbot NJ, Kershaw MJ, Wakley GE, de Vries OMH, Wessels JGH, Hamer JE (1996) *MPG1* encodes a fungal hydrophobin involved in surface interactions during infection-related development by *Magnaporthe grisea.* Plant Cell 8:985–989

Temple B, Horgen PA (2000) Biological roles for cerato-ulmin, a hydrophobin secreted by the elm pathogens, *Ophiostoma ulmi* and *O. novo-ulmi.* Mycologia 92:1–9

Temple B, Horgen PA, Bernier L, Hintz WE (1997) Cerato-ulmin, a hydrophobin secreted by the causal agent of Dutch elm disease, is a parasitic fitness factor. Fungal Genet Biol 22:39–53

Templeton MD, Rikkerink EHA, Beever RE (1994) Small, cysteine-rich proteins and recognition in fungal-plant interactions. Mol Plant Microbe Interact 7:320–325

Thau N, Monod M, Crestani B, Rolland C, Tronchin G, Latge JP, Paris S (1994) Rodletless mutants of *Aspergillus fumigatus.* Infect Immun 62:4380–4388

Tillotson RD, Wösten HAB, Richter M, Willey JM (1998) A surface active protein involved in aerial hyphae formation in the filamentous fungus *Schizophyllum commune* restores the capacity of a bald mutant of the filamentous bacterium *Streptomyces coelicolor* to erect aerial structures. Mol Microbiol 30:595–602

Van den Ackerveken GFJM, Van Kan JAL, De Wit PJGM (1992) Molecular analysis of the avirulence gene *avr9* of the fungal tomato pathogen *Cladosporium fulvum* fully supports the gene-for-gene hypothesis. Plant J 2:359–366

Van der Vegt W, Van der Mei HC, Wösten HAB, Wessels JGH, Busscher HJ (1996) A comparison of the surface activity of the fungal hydrophobin Sc3p with those of other proteins. Biophys Chem 51:253–260

Van Wetter M-A (2000) Functions of hydrophobins in *Schizophyllum commune.* PhD Thesis, University of Groningen, The Netherlands

Van Wetter M-A, Schuren FHJ, Schuurs TA, Wessels JGH (1996) Targeted mutation of the *Sc3* hydrophobin gene of *Schizophyllum commune* affects formation of aerial hyphae. FEMS Microbiol Lett 140:265–269

Wessels JGH (1992) Gene expression during fruiting in *Schizophyllum commune.* Mycol Res 96:609–620

Wessels JGH (1994) Developmental regulation of fungal cell wall formation. Annu Rev Phytopathol 32:413–437

Wessels JGH (1996) Fungal hydrophobins: proteins that function at an interface. Trends Plant Sci 1:9–15

Wessels JGH (1997) Hydrophobins: proteins that change the nature of the fungal surface. Adv Microbial Physiol 38:1–45

Wessels JGH (1999) Fungi in their own right. Fungal Genet Biol 27:134–145

Wessels JGH, de Vries OMH, Ásgeirsdóttir SA, Schuren FHJ (1991) Hydrophobin genes involved in formation of aerial hyphae and fruit bodies in *Schizophyllum.* Plant Cell 3:793–799

Willey J, Santamaria R, Guijarro J, Geistlich M, Losick R (1991) Extracellular complementation of a developmental mutation implicates a small sporulation protein in aerial mycelium formation by *S. coelicolor.* Cell 65:641–650

Willey J, Schwedock J, Losick R (1993) Multiple extracellular signals govern production of a morphogenetic protein involved in aerial mycelium formation by *Streptomyces coelicolor.* Genes Dev 7:895–903

Williams SJ, Bradshaw RM, Costerton JW, Forge A (1972) Fine structure of the spore sheath of some *Streptomyces* species. J Gen Microbiol 72:249–258

Wösten HAB, de Vries OMH, Wessels JGH (1993) Interfacial self-assembly of a fungal hydrophobin into a hydrophobic rodlet layer. Plant Cell 5:1567–1574

Wösten HAB, ásgeirsdóttir SA, Krook JH, Drenth JHH, Wessels JGH (1994a) The fungal hydrophobin Sc3p self-assembles at the surface of aerial hyphae as a protein membrane constituting the hydrophobic rodlet layer. Eur J Cell Biol 63:122–129

Wösten HAB, Schuren FHJ, Wessels JGH (1994b) Interfacial self-assembly of a hydrophobin into an amphipathic membrane mediates fungal attachment to hydrophobic surfaces. EMBO J 13:5848–5854

Wösten HAB, Ruardy TG, Van der Mei HC, Busscher HJ, Wessels JGH (1995) Interfacial self-assembly of a *Schizophyllum commune* hydrophobin into an insoluble amphipathic membrane depends on surface hydrophobicity. Colloids Surfaces B 5:189–195

Wösten HAB, vanWetter M-A, Lugones LG, van der Mei HC, Busscher HJ, Wessels JGH (1999) How a fungus escapes the water to grow into the air. Curr Biol 9:85–88

Yaguchi M, Pusztai-Carey M, Roy C, Surewicz WK, Carey PR, Stevenson KJ, Richards WC, Takai S (1993) Amino acid sequence and spectroscopic studies of Dutch elm disease toxin, cerato-ulmin. In: Sticklen MB, Sherald JL (eds) Dutch elm disease research, cellular and molecular approaches. Springer, Berlin Heidelberg New York, pp 152–170

Zhang L, Villalon D, Sin Y, Kazmierczak P, Van Alfen NK (1994) Virus-associated down-regulation of the gene encoding cryparin, an abundant cell surface protein of the chestnut blight fungus *Cryphonectria parasitica*. Gene 139:59–64

8 Cell Wall of Human Fungal Pathogens and its Interaction with Host Extracellular Matrix

W. LaJean Chaffin

CONTENTS

I. Introduction

Animal cells exist within a mesh of intercellular secreted proteins and polysaccharides, the extracellular matrix (ECM). The interactions of cells with this protein network have profound effects on the organization, migration, differentiation, and function of cells. Proteins of the ECM include fibronectin, laminin, collagen, entactin, tenascin, and vitronectin. There may be multiple forms of these proteins which may be derived from different genes, e.g. collagen, or from alternative RNA splicing, e.g. fibronectin. Typically, ECM proteins have multiple domains with specific binding sites

for other matrix molecules and cells. These binding sites help organize the matrix and attachment of cells to it. (For more detailed discussion of ECM proteins, organization, and tissue and organ specificity, see textbooks on cell biology such as Alberts et al. 1994.) Recently, controlled proteolysis of ECM proteins by matrix metalloproteinases has come to be recognized as an important regulator of adherence (Horwitz and Werb 1998).

Animal cells interact with ECM proteins through transmembrane cell-surface molecules. The principal receptors for ECM proteins belong to a family of proteins called integrins. Some integrins recognize several ECM proteins and conversely most ECM proteins bind to several integrins. Integrins are heterodimeric proteins that signal between the ECM and the cell (for a recent review see Giancotti and Ruoslahti 1999; see also Chap. 10, this Vol.). Loss of ECM attachment may lead to apoptosis in some cell types while the attachment requirement is lost in neoplastic cells. For cells anchored to the ECM, proliferation, differentiation, or changes in cell shape may ensue. The identity of the ECM protein to which the cell is attached may also determine the type of response. The recognition between integrin and an ECM protein is frequently through an arginine-glycine-aspartic acid (RGD) motif in the ECM protein. After binding ligand, integrins cluster and through the cytoplasmic tails interact with the cytoskeleton and initiate signal cascades.

Some fungal pathogens also use ECM proteins as ligands for attachment to a human host. These include *Aspergillus fumigatus*, *Candida albicans*, *Histoplasma capsulatum*, *Paracoccidioides brasiliensis*, *Pneumocystis carinii*, *Penicillium marneffei*, and *Sporothrix schenckii* (Table 1).

Recognition that binding to ECM protein was also a mechanism by which a fungal pathogen could adhere to a human host is largely attributable to Klotz (Klotz 1987, 1990; Klotz and Maca 1988). Klotz showed that *C. albicans* binds to con-

Department of Microbiology and Immunology, Texas Tech University Health Sciences Center, Lubbock, Texas 79430, USA

The Mycota VIII
Biology of the Fungal Cell
Howard/Gow (Eds.)
© Springer-Verlag Berlin Heidelberg 2001

Table 1. Recognition of ECM components by fungal human pathogens

Pathogen	ECM component								Proteolysis
	Laminin	Fibronectin	Collagen type I	Collagen type IV	Collagen type II	Entactin	Tenascin	Vitronectin	ECM component(s)
A. fumigatus	+	+	+	+					+
C. albicans	+	+	+	+		+	+	+	+
H. capsulatum	+								+
P. brasiliensis	+								+
P. carinii		+						+	
P. marneffei	+	+							+
S. schenckii	+	+			+				

Fig. 1. Schematic representation of increased binding of *C. albicans* yeast cells to extracellular matrix (*ECM*) exposed by contraction of cultured endothelial cells (*HC*)

fluent monolayers of endothelial cells at cell junctions and that exposure of the underlying ECM enhances binding (Fig. 1). Since that time, the ability of other fungi to bind to individual components of the ECM has been demonstrated. Some of these binding reactions, like integrin recognition, may be mediated through the RGD motif. The parallel in ligand recognition and the RGD-binding motif between fungi and animal cells raised the possibility that there may also be a parallel between the fungal ECM-binding proteins and integrins. This possibility has been incorporated into some experimental approaches to characterize and identify the fungal interactions and binding proteins. However, the evidence for an in vivo role for adherence to ECM in pathogenesis remains indirect and scant. Two lines of in vivo evidence are derived from demonstrating that an ECM protein is bound to organisms in vivo and that receptors for ECM proteins are expressed in vivo. These studies are discussed in the following sections with the individual ECM proteins and various organisms. The third line of indirect evidence arises from the demonstration that the ECM component of peptides containing the RGD motif can reduce infection. In a rabbit model of candidal infection, Klotz et al. (1992) reported than an RGD-containing peptide reduced the

burden of fungal cells in several tissues 4 h after intravenous inoculation. They suggested that the effect was due to blockage of fungal cell binding to host ECM proteins. In another contemporaneous study performed with a perfused murine liver model, Sawyer et al. (1992) reported that treatment of *C. albicans* yeast cells with RGD-containing peptides increased the trapping and killing of yeast cells when the treated cells were infused into the liver. The authors suggested that bound peptide might serve as an opsonin that increased phagocytosis by Kupffer cells. From the perspective of adherence, another possibility would be that the proportion of non-adhered cells was increased and that these non-adhered yeast cells were available for phagocytosis and killing. Additional support is derived from a hamster testicular model of *P. brasiliensis* infection. Animals inoculated with laminin-coated yeast cells showed enhanced pathology compared with uncoated cells (Vicentini et al. 1994). Treatment of yeast cells with anti-receptor antibody and laminin reduced the effect (Vicentini et al. 1997).

The binding of host ECM ligands by various pathogenic fungi is presented in more detail in the following sections.

II. *Candida albicans*

Not surprisingly, the binding of ECM proteins to *C. albicans* has been the most extensively examined. These interactions have been summarized in a recent review of cell wall proteins that includes binding proteins (Chaffin et al. 1998). This organism binds collagen, entactin, fibronectin, laminin, tenascin, and vitronectin (Table 1). The binding to fibronectin, laminin, and collagen has been studied more extensively than interactions with the other

ECM proteins. In addition to binding to ECM, *C. albicans* can also degrade ECM components. The organism contains a family of genes, *SAP1–9*, that encode secreted aspartyl proteinases. The expression of the various genes is regulated by strain, growth conditions and morphology. Proteinase production is thought to enhance the ability of the organism to colonize and invade the host. Proteinase-secreting organisms can degrade subendothelial matrix (Morschhauser et al. 1997). Both soluble and immobilized laminin and fibronectin were degraded. Thus, not only can the organism be localized in the host by adherence to ECM proteins, but degradation of the ECM may also provide access to deeper penetration and spread.

A. Fibronectin

Adherence to fibronectin was the first of the ECM-binding activities described for *C. albicans* more than 15 years ago (Skerl et al. 1984). However, subsequent studies to identify the adhesin(s) and characterize the interactions between the fungus and the host ligand have yielded numerous reports without arriving at a consensus. Fibronectin is a 440-kDa dimeric glycoprotein plasma protein as well as an ECM component. It contains binding domains that are identified as fibrin, collagen, DNA, cell (containing an RGD sequence), and heparin.

The binding of soluble fibronectin to *C. albicans* is saturable with either rapid binding to 8000 sites (Klotz and Smith 1991) or slower binding to 5000 high affinity and 30,000 low-affinity binding sites per cell (Negre et al. 1994) depending on the study. It has been reported that binding is enhanced by divalent cations (Jakab et al. 1993; Klotz et al. 1993) or is cation independent (Negre et al. 1994). Calcium ion-mediated binding is a characteristic of mammalian integrins. The role of the cell-binding or collagen-binding domains in binding of fibronectin to yeast cells is contradictory. Binding of soluble fibronectin has been reported to be inhibited by a proteolytic fragment containing the cell-binding domain and an RGD peptide (Jakab et al. 1993; Penn and Klotz 1994). Other related peptides such as GRGESP were less effective while GRGDSP was ineffective (Klotz and Smith 1991). However, one study found RGD to be non-inhibitory (Negre et al. 1994). The immobilized cell-binding domain fragment also promoted adherence and binding was inhibited by

GRGDSP but not GRGESP (Santoni et al. 1994). Another study found that GRGESP was an effective inhibitor of binding to immobilized fibronectin (Klotz and Smith 1991). Negre et al. (1994) reported that a cell-binding domain with an altered RGD sequence was as effective as the native sequence in promoting adherence to the immobilized fragment, although neither was as active as other domains. GRGDSP but not GRGESP is an effective inhibitor of mammalian integrins and thus the activity of these peptides may reflect the extent to which candidal adhesins have the same interactions as integrins. Various studies also conflict with respect to the role of the collagen- and heparin-binding domains.

Both the collagen- and the heparin-binding domain failed to bind to fungal cells (Penn and Klotz 1994) and the collagen-binding domain was a less effective inhibitor of fibronectin binding than was the cell-binding domain (Jakab et al. 1993). On the other hand, in another study, the collagen-binding domain was the most active domain in promoting fungal binding followed by the cell-binding domain, fibrin and heparin domains (Negre et al. 1994). Heparin was reported to inhibit binding of fibronectin (Jakab et al. 1993) but not binding of the cell-binding fragment (Penn and Klotz 1994). Heparin and heparan sulfate inhibited binding of the fungus to fibronectin apparently by binding to the fibronectin (Klotz and Smith 1992).

The expression of fibronectin-binding activity appears to be regulated, at least in part, by environmental factors. Differences in binding have been noted with growth of the organism in different batches of medium (Negre et al. 1994) and in different media (greatest when grown on Sabouraud medium) (Yan et al. 1996). Growth in defined medium supplemented with hemoglobin increased binding 20- to 80-fold (Yan et al. 1996). The extent of enhancement varied among strains. Binding was greater for organisms grown in suspension culture than on agar (Jakab et al. 1993), at higher (37 °C) temperature (Jakab et al. 1993) and metabolically active (Klotz and Smith 1991) or with greater cell surface hydrophobicity (Silva et al. 1995). Hydrophobic yeast cells grown at 23 °C bound more abundantly to immobilized fibronectin than did hydrophilic yeast cells grown at 37 °C (Silva et al. 1995). The difference between hydrophobic and hydrophilic cells was greater for adherence to laminin and fibrinogen and similar for type IV collagen and no difference was observed for entactin and vitronectin. Pretreatment of emerging germ tubes with a monoclonal antibody (MAb) to any of three hydrophobic proteins reduced adherence to laminin but only two of these reduced adherence to fibronectin (Masuoka et al. 1999). The authors hypothesized that ECM proteins and fungal proteins interact through hydrophobic domains.

Two major methods used to identify the fungal fibronectin-binding proteins rely on recognition of the receptor either by antigenic similarity using antibodies to mammalian integrins that recognize the same ligand or by ligand affinity blotting. Cloning and affinity chromatography have also been employed. Despite these efforts the number and identity of the receptors remain elusive. Antiserum and MAbs to the $\alpha_5\beta_1$ integrin have been employed. One of two anti-α_5 MAbs and anti-receptor antiserum bound to yeast cells (Santoni et al. 1994) and were able to inhibit partially or completely binding to immobilized fibronectin. Germ tubes showed a greater reactivity (97% positive) than yeast cells as well as reactivity with anti-β_1 MAbs (about 50% positive). Western blot analysis of an undialyzed cell wall extract with the antiserum revealed three major species in the 40–50 kDa range and a polydisperse reactivity above 143 kDa (Glee et al. 1996). Ligand affinity blotting revealed additional species above and below the 35–50 kDa range. Depending on the preparation of the extract, differences were noted in the reactive species.

Fibronectin affinity chromatography was employed to isolate binding proteins from a detergent extract of yeast cells (Klotz et al. 1993). Two species of 60 and 105 kDa were detected by non-reducing sodium dodecyl sulfate polyacrylamide electrophoresis (SDS-PAGE). These species were recovered from cell wall and cell membrane but not cytosol extracts as well as from germ-tube extracts. The 60-kDa moiety in the extract reacted with antibody to vitronectin and fibronectin integrin receptors. Additional reactivity was noted with a 50-kDa band and broad reactivity in the 120–135 kDa range. The moieties obtained by fibronectin affinity chromatography were further purified by high-performance liquid chromatography (HPLC) and analysis suggested that the 105-kDa species was a dimer or aggregate of the 60-kDa species (Klotz et al. 1994). The candidate adhesin was approximately 80% carbohydrate by weight and blocked to Edman degradation.

As noted above, the presence of hemoglobin in defined medium during yeast growth increased subsequent adherence of yeast cells to fibronectin. The induced receptor was saturable with about 27,000 sites per cell (Yan et al. 1996). This appeared to correspond to the low affinity receptor previously described for organisms grown in complex medium (Yan et al. 1998b). A subsequent study demonstrated that adherence was also increased to immobilized laminin, fibrinogen and type IV collagen but not type I collagen or thrombospondin (Yan et al. 1998a). Each nonradiolabeled ligand inhibited binding to iodinated ligand. Fibrinogen also was equally active in inhibiting binding of fibronectin or laminin, suggesting a common receptor. On the other hand, laminin was a better inhibitor of laminin binding than fibronectin binding suggesting an additional receptor. Several proteins appeared more abundant in cell wall extracts of induced cells compared to the uninduced cells. Fibronectin protection assays suggested that a 55-kDa species was the induced receptor. Further, a 55-kDa moiety was the major component of material obtained from fibronectin affinity chromatography of the extract. The fragments of fibronectin that were the most effective inhibitors of intact fibronectin binding were derived from the cell-binding domain (Yan et al. 1998b). This was in contrast to the effectiveness of the collagen-domain fragment in inhibited binding to cells grown in complex medium as noted above (Negre et al. 1994). A modified fragment lacking the RGD sequence in the cell-binding domain was as effective as the native sequence. The immobilized fragments containing the cell-binding domain also promoted adherence (Yan et al. 1998b). Ligand protection assays with the cell-binding domain again implicated the 55-kDa protein as the receptor.

In other studies, a candidal antigen recognized by patient sera was demonstrated to be the 33-kDa glyceraldehyde-3-phosphate dehydrogenase (GAPDH). This proved to be present not only in the cytoplasm but also on the cell surface and was enzymatically active (Gil-Navarro et al. 1997). In *Streptococcus pyogenes*, this protein is also surface localized where it binds fibronectin, lysozyme, actin, and myosin (Pancholi and Fischetti 1992). In *C. albicans* this protein was also a fibronectin- and laminin-binding protein (Gozalbo et al. 1998). Antibody to candidal GAPDH inhibited binding of yeast cells to immobilized fibronectin and laminin by 65–72%. Addition of the *Saccha-*

romyces cerevisiae GAPDH protein also reduced binding to the two ligands by 66–85%. In addition, binding of fibronectin or laminin reduced binding of antibody to the cell surface. Purified cytoplasmic protein (Gozalbo et al. 1998) and the recombinant protein produced in *Escherichia coli* bound fibronectin and laminin in ligand affinity blots (Villamon et al. 1999). Surface-associated GAPDH was detected on clinical isolates (Gil et al. 1999). Immunohistochemical analysis of tissue sections from patients with disseminated candidiasis also revealed surface expression of GADPH.

Other *Candida* species also bind fibronectin. Binding of fibronectin to *C. tropicalis* was saturable and reversible with less than 1000 receptors per cell (DeMuri and Hostetter 1996). A 125-kDa moiety was detected by ligand affinity blotting and Western blotting with anti-$\alpha_5\beta_1$ and anti-β_1 antiserum but not an anti-α_5 MAb. A 105-kDa species was detected by immunoprecipitation. A different anti-α_5 MAb and anti-fibronectin receptor antiserum reacted with *C. tropicalis* and to a lesser extent with *C. stellatoidea* and *C. glabrata* and minimally with *C. krusei* (Santoni et al. 1995). These three organisms also bound to immobilized fibronectin and the 120-kDa cell-binding domain fragment. Binding was reduced by the presence of GRGDSP, anti-α_5 MAbs or anti-fibronectin integrin receptor antiserum. Growth in the presence of hemoglobin increased binding of soluble radiolabeled fibronectin to strains of *C. tropicalis* 55- to 78-fold and *C. krusei* and *C. glabrata* 40- to 50-fold (Rodrigues et al. 1998). Increased binding was saturable and inhibited by unlabeled fibronectin with about 8000 sites per *C. tropicalis* or *C. krusei* cell. Binding to immobilized fibronectin was also enhanced. In addition to the high affinity binding, *C. krusei* and *C. tropicalis* showed the presence of non-saturable low-affinity sites non-displaced by unlabeled fibronectin. This property was shared with *S. cerevisiae* and *Trichosporon beigelii*. Binding to immobilized fibronectin was associated with the high-affinity site as these two organisms did show increased adherence.

Another fibronectin adhesin candidate has been identified by cloning a *C. albicans* DNA sequence that conferred adherence to fibronectin, type IV collagen, and laminin on non-adherent *S. cerevisiae* (Gaur and Klotz 1997). *S. cerevisiae* cells expressing the sequence bound to magnetic beads coated with each of the ECM components. Sequence analysis indicated that the cloned gene, *ALA1/ALS5*, was a member of the *ALS* gene family with a predicted size of 150 kDa. This gene family shares structural features with yeast agglutinins. The ALS family proteins contain serine- and threonine-rich tandem repeats and consensus sequences for secretion and glycophosphatidyl inositol anchors. Thus, unlike other candidates for ECM adhesins, this protein is likely to be attached to cell wall glucan (Smits et al. 1999). However, as yet, the function of this gene in *C. albicans* as encoding a multifunctional ECM adhesin has not been reported.

Although the discussion so far has focused on adherence to soluble and immobilized fibronectin, several studies support a role for fibronectin in cellular adherence. Buccal and vaginal epithelial cells have surface fibronectin and adherence of *C. albicans* to these cells correlated with the level of surface fibronectin and was reduced by addition of fibronectin (Skerl et al. 1984; Kalo et al. 1988). Epithelial cells with higher levels of surface fibronectin are more abundant in the first and fourth week of the menstrual cycle, during pregnancy and in diabetic patients (Kalo et al. 1988). In a model of non-bacterial thrombotic endocarditis, where fibronectin was present on damaged but not normal heart valvular tissue, a correlation between fibronectin-binding organisms and infection was noted (Scheld et al. 1985). Fibronectin, laminin, and type I and IV collagen expression was detected on the surface of HEp-2 cells (derived from an epidermoid carcinoma of the larynx) (Cotter et al. 1998). Monoclonal antibodies to each of the ECM components inhibited binding of yeast cells to HEp-2 cells by 30–50%. Fibronectin and type IV collagen reduced binding of hydrophobic yeast cells to murine spleen sections in an ex vivo assay, primarily by reducing binding in the white pulp and not the marginal zone (Silva et al. 1995).

B. Laminin

Laminin is able to mediate the attachment of whole cells to solid substrates. Yeast cells grown in Sabouraud dextrose medium bind to laminin immobilized in the wells of microtiter plates and binding is inhibited by fibronectin, RGD-containing peptides, and anti-laminin antibody (Klotz 1990; Klotz and Smith 1991). When keratinocytes were used as the substrate, adhesion was inhibited by several sugars and peptides, including a synthetic peptide from the laminin B chain, although laminin itself was not very effective (Ollert et al. 1993).

The initial description of a binding protein for laminin was reported by Bouchara et al. (1990). Binding of the laminin as detected by indirect immunofluorescence was observed to germ tubes induced in medium 199 but not to yeast cells grown in Lee medium. The binding was confined to the outer fibrillar layer of the cell. The laminin binding was saturable and reversible with about 8000 binding sites per organism. Binding competition was observed with fibrinogen but not fibronectin. Ligand affinity blotting of proteins present in a dithiothreitol-iodoacetamide extract localized the major activity to a 68-kDa moiety and to a lesser extent with a doublet at 60 and 62 kDa. The finding of similar sized moieties that reacted with fibrinogen, and C3d, and bound to plastic (Tronchin et al. 1988, 1989), raised the possibility of a receptor with multiple ligand recognition. A recent study with a different strain found that radiolabeled laminin bound to germ tubes and only weakly to yeast cells (Sakata et al. 1999). Binding to germ tubes was inhibited by fibrinogen but not fibronectin, a GRGDS peptide or a laminin-derived peptide, YIGSR. Ligand affinity blotting of a cell extract identified a single reactive 21-kDa moiety. Another binding parameter is cell surface hydrophobicity (CSH), as hydrophobic cells showed increased adherence to laminin compared to hydrophilic cells (Silva et al. 1995) and adherence was reduced by MAbs to three different hydrophobic proteins (Masuoka et al. 1999).

In another study, ligand affinity blotting of a yeast surface protein extract revealed two apparently different moieties of 37 and 67 kDa, the larger of which reacted with Concanavalin A (Lopez-Ribot et al. 1994). The human 67-kDa high-affinity laminin receptor and its 37-kDa precursor are of similar size and antibody to the C-terminus of this receptor reacted with 37- and 67-kDa species from yeast cells. Antibody to an internal domain of the human 37-kDa precursor reacted with 37-kDa moieties in extracts from both yeast cells and germ tubes. However, the germ-tube species did not bind laminin and neither of the yeast species bound fibrinogen, fibronectin, or collagen type IV. This protein was localized by indirect immunofluorescence to the cell surface of about 10% of the yeast cells where its distribution was heterogeneous. The laminin receptor was not co-localized with a 58-kDa fibrinogen-binding protein that was also heterogeneously distributed on the cell surface (Lopez-Ribot et al. 1996). However, the 58-kDa fibrinogen-binding protein and the 37-kDa laminin-binding protein did share two characteristics. Both proteins contained collagen-like domains (Sepulveda et al. 1995) and were ubiquitinated (Sepulveda et al. 1996). Immunohistochemical analysis of tissue sections from patients with candidiasis using antibody to the 37-kDa moiety showed that it was expressed in vivo in a pattern similar to the in vitro pattern as reactivity was essentially restricted to yeast cells and only about half of these were reactive (Lopez-Ribot et al. 1996). In a more recent study, a cDNA encoding a 37-kDa protein was isolated using anti-cell wall antiserum (Montero et al. 1998). The deduced sequence of the protein had 60% sequence identity with the 37-kDa human high-affinity laminin-binding protein and 60% identity with Yst ribosomal proteins of *S. cerevisiae*. Antibodies to the fusion protein reacted with a 37-kDa moiety localized in membranes, ribosomes and, to a lesser extent, in the cytosol but failed to react with a cell wall extract. Laminin binding was not detected to the fusion protein. The cDNA complemented *S. cerevisiae* YST defects.

A study that used ligand affinity blotting to identify laminin-binding proteins detected an even larger number of binding proteins, perhaps more than ten (Glee et al. 1996). In this study the extract was obtained after a limited glucan hydrolysis of organisms grown in yeast nitrogen base and the extract was used without further treatment. The authors showed that dialysis altered the profile of proteins recovered in the extract. In the same study, multiple fibronectin-binding proteins were demonstrated and the profile of reactive species differed between dialyzed and undialyzed extracts.

As discussed in the previous section on fibronectin-binding proteins, *C. albicans* binding to laminin was increased when the organism was cultured in defined medium supplemented with hemoglobin (Yan et al. 1998a). The increased laminin binding was associated with an induction of a 55-kDa species that recognized multiple ECM components as well as an additional species that was not further described.

A candidate gene for a multifunctional adhesin with a predicted 150-kDa size that recognizes laminin, fibronectin, and type IV collagen was discussed in the fibronectin section.

C. Collagen

C. albicans yeast cells bound to immobilized type IV collagen found in basement membrane (Klotz 1990). The organism bound less strongly to type I collagen and was reported to have either reduced or similar binding to denatured type I collagen (gelatin) (Klotz 1990; Klotz and Smith 1995). Binding was reduced by removal of divalent cations (Klotz 1990; Klotz et al. 1993). Hydrophobic cells grown at 23 °C showed greater binding to type IV collagen than hydrophilic cells grown at 37 °C (Silva et al. 1995). Addition of fibronectin or the peptide GRGESP effectively (about 70–95%) inhibited binding to type IV and type I collagen (Klotz and Smith 1991). On the other hand, GRGDTP and RGD peptides inhibited binding by about 40% to type I collagen and type IV collagen, respectively. Heparin inhibited binding to both collagens and when tested with type I collagen, heparan sulfate and dextran sulfate were more effective inhibitors (about 70%) (Klotz and Smith 1992). The inhibition appeared to be attributable to binding of the glycosaminoglycans to the target and blocking of the fungal recognition site. An unfractionated tryptic digest of gelatin almost completely abolished yeast cell adherence to gelatin, type I and type IV collagen, fibronectin and laminin (Klotz and Smith 1995). Among several synthetic peptides derived from the gelatin sequence that were tested, one decapeptide inhibited yeast cell binding to gelatin by about 70% and was suggested as a potential biocompatible therapeutic candidate.

Two potential binding proteins of 60 and 105 kDa were obtained by gelatin affinity chromatography of a detergent extract of yeast cells (Klotz et al. 1993). Species of the same size were obtained from extracts of germ tubes or from fibronectin affinity chromatography. This candidate adhesin is discussed under fibronectin.

Supplementation of defined medium with hemoglobin during culture of *C. albicans* increased binding of ECM components, as noted in other sections (Yan et al. 1998a). Binding to type IV collagen was increased more than 10-fold, similar to fibronectin binding, while adherence to gelatin increased less. No enhancement was observed for native type I collagen. Adherence to immobilized type IV collagen but not to type I collagen or gelatin was observed. Whether the enhanced type IV collagen binding is associated with the hemoglobin-induced 55-kDa promiscu-

ous receptor for other ECM components has not been reported. A candidate gene for a multifunctional adhesin with a predicted 150-kDa size that recognizes laminin, fibronectin, and type IV collagen was discussed in the fibronectin section.

D. Vitronectin

Vitronectin is present in serum, vascular walls, and dermis. It is a multi-domain protein that contains an RGD sequence in a domain that interacts with mammalian integrins and a glycosaminoglycan-binding domain that interacts with carbohydrates. Binding capacity of yeast cells for vitronectin increased during late exponential growth of *C. albicans* (Jakab et al. 1993). The effects of growth temperature and cell surface hydrophobicity (CSH) on binding are unclear as, in one study, organisms grown at higher temperature (37 °C) and with greater CSH bound more vitronectin than cells grown at a lower temperature or with less CSH (Jakab et al. 1993). In the second study there appeared to be no effect of growth temperature or CSH (Silva et al. 1995). Binding was optimal at acidic pH and in the presence of calcium ions (Jakab et al. 1993). Analysis of binding of plasma vitronectin showed both high-affinity and low-affinity receptors although the dissociation constant for the high-affinity receptor (98,000 sites per cell) was similar to low-affinity receptors reported for some other ligands (Limper and Standing 1994). Binding of vitronectin was reduced by treatment of cells with proteases or heat (Jakab et al. 1993). Addition of fibronectin reduced binding (64%) while a minimal effect was observed with addition of fibrinogen, type I collagen, or GRGDS peptide and no effect was observed with addition of type IV collagen, gelatin, or other RGD-containing peptides. The most effective inhibitor was heparin (50–85%) (Jakab et al. 1993; Limper and Standing 1994). Binding to immobilized vitronectin was stimulated by divalent cations and inhibited by anti-vitronectin antibody (Spreghini et al. 1999). Binding was inhibited (at 50%) by GRGDSP and GRGDS peptides with the former more effective at low concentrations.

Two classes of receptors for vitronectin, one protein and one carbohydrate, have been reported. Ligand affinity blotting revealed a 30-kDa binding species in an extract obtained from cells with SDS and β-mercaptoethanol (βME)

(Limper and Standing 1994). Binding was abolished by addition of heparin. Another study assumed that receptors for vitronectin shared a common motif(s) and used antiserum to the human integrin receptor as a probe in a Western blot analysis of an octylglucoside extract of isolated cell walls (Klotz et al. 1993). Moieties of 50, 60, and 90 kDa were most reactive with lesser reactivity detected with other species in the same size range. A similar sized 60-kDa moiety was recovered by fibronectin or gelatin affinity chromatography and reacted with anti-fibronectin receptor antiserum as discussed above for fibronectin-binding protein candidates. In a very recent study, antibodies to integrin subunits, α_v, β_3, β_5, and receptors, $\alpha_v\beta_3$ and $\alpha_v\beta_5$, were used to assess the antigenic similarity of candidal binding proteins (Spreghini et al. 1999). All antibodies stained yeast cells as determined by fluorescence flow cytometry. Immunoprecipitation of reactive material from a cell lysate with anti-α_v or anti-β_3 antisera yielded three proteins of 130, 110, and 100 kDa, while anti-β_3 antiserum yielded a major 84-kDa species. Subsequent analysis suggested that the 130-kDa band, the 110- and 100-kDa doublet, and the 84-kDa species corresponded to α_v, β_3, and β_5, respectively. MAbs to the $\alpha_v\beta_3$ and $\alpha_v\beta_5$ receptors alone or in suboptimal concentration combinations were able to inhibit binding of yeast cells to immobilized vitronectin. Binding of C. albicans yeast cells to a human endothelial cell line that expresses surface vitronectin was inhibited by more than 60% by addition of soluble vitronectin.

β-Glucan, that is abundant in the fungal cell wall, also binds vitronectin (Olson et al. 1996). β-Glucan from S. cerevisiae was used to demonstrate concentration-dependent and specific binding of vitronectin to the polysaccharide. Binding of vitronectin to the candidal surface appears to increase binding to a macrophage cell line and phagocytosis of the fungus (Limper and Standing 1994). Vitronectin-coated β-glucan particles at low concentrations stimulated release of tumor necrosis factor alpha (TNF-α) from macrophages while higher concentrations suppressed release. Since β-glucan is a common component of fungi, this mechanism may also be common to other fungi. However, C. albicans appears to bind to macrophages through multiple mechanisms as mannan structures also mediate adherence (see recent review, Chaffin et al. 1998).

E. Entactin

Entactin is a glycoprotein that contains an RGD sequence and forms a tight complex with laminin and type IV collagen in basal lamina. Indirect immunofluorescence demonstrated that entactin bound to about 10% of yeast cells in a population and along most hyphal extensions of germ tubes in a pattern of heterogeneous distribution (Lopez-Ribot and Chaffin 1994). Cell wall material in a βME extract of intact cells of both morphologies bound to immobilized entactin. The binding was inhibited by antibody either to entactin or cell wall material. Extracts from yeast cells completely inhibited binding of an extract from germ tubes, and vice versa. An RGDS peptide inhibited binding approximately by 50%, suggesting both RGD-dependent and RGD-independent binding. Preincubation of the extract with immobilized laminin or fibronectin reduced the entactin-binding capacity. Three reactive proteins of 25, 44, and 65 kDa in cell wall extracts from both yeast cells and germ tubes reacted with entactin in a ligand blot. There appeared to be no effect of growth temperature or CSH as hydrophilic cells grown at 37 °C bound as well as hydrophobic cells grown at 23 °C (Silva et al. 1995).

F. Tenascin

The tenascins are large multimer proteins that contain repeated structural motifs that include fibronectin type III, epidermal growth factor (EGF)-like repeats, and a globular fibrinogen-like domain. The four family members are tenascin-C, tenascin-R, tenascin-X and tenascin-Y. Tenascin-C is found in a large number of developing tissues and is frequently overexpressed in tumor cells. Binding of soluble tenascin-C to germ tubes (but not yeast cells) was demonstrated by indirect immunofluorescence (Lopez-Ribot et al. 1999). The binding was heterogeneously distributed on the surface. However, material in cell wall extracts from both morphologies bound to immobilized tenascin-C. This binding was inhibited by addition of anti-tenascin or anti-cell-wall antiserum but not by addition of anti-laminin antibody. Binding was also inhibited by fibronectin (62%) but not by fibrinogen or RGD peptide; in this case, the RGD peptide enhanced (20%) binding. The RGD equence is present in the fibronectin repeat. Binding also appeared to be cation independent. These observations suggest that tenascin and fibronectin share a binding protein(s).

G. *C. albicans* ECM Adhesins

The organization of the *C. albicans* cell wall appears similar to that of *S. cerevisiae*, with β-1,3-glucan, β-1,6-glucan, and chitin present as structural polysaccharides with cross-links formed between the major β-1,3-glucan and the other two polymers (see Chap. 9, Fig. 1, this Vol.). Covalently attached to β-1,6-glucan are the glycosylphosphoidyl inositol dependent cell wall proteins (GPI-CWP). *S. cerevisiae* and *C. albicans* contain a second group of proteins with internal repeats (PIR-CWP) linked to β-1,3-glucan. In addition there are non-covalently attached proteins. *C. albicans* recognizes the glycoprotein components of the host's extracellular matrix as well as these components on host cell surfaces. This recognition involves multiple proteins including covalently and non-covalently attached candidates and β-glucan (Table 2). Some of the binding proteins share antigenic determinants with integrins and ligand binding may be inhibited by anti-integrin antibody. Although structural homology to mammalian integrins has not been demonstrated for the candidal receptors, the binding proteins do share common characteristics in that generally more than one binding activity is observed for each ECM component and some binding proteins bind more than one ECM component. Despite all the reports there are still significant differences among various studies that have examined the interaction of the fungus with the same ligand. Also a relationship between binding proteins that recognize the same ECM component has not been demonstrated biochemically or genetically. The differences may reflect the presence of multiple adhesins, method of size determination, use of different strains, different growth conditions, different assays, and examination of different parameter sets. Several studies noted above, and studies of other candidal adhesins as reviewed in Chaffin et al. (1998), demonstrate that some of these differences contribute to the lack of agreement between studies.

The more recent introduction of molecular approaches should facilitate analysis of specific protein receptor candidates and their role in mediating interactions of the intact cell with the ligand(s) in vitro and in vivo. However, the first study to employ such an approach to clone an adhesin gene through its function identified a new candidate that had not been suggested by prior approaches and thus initially has added to rather than resolved differences. The purification of endogenous and recombinant *C. albicans* GAPDH and a demonstration that the protein binds laminin and fibronectin (Villamon et al. 1999) are important steps towards more rigorous demonstration of the role of the putative adhesin.

III. *Aspergillus fumigatus*

The infectious airborne conidia of *A. fumigatus* are able to bind about tenfold more abundantly to the ECM laid down by a pulmonary epithelial cell line than to the cells themselves (Bromley and Donaldson 1996). Similar observations for adherence to ECM compared to host cells were noted above for *C. albicans* yeast cells. The RGD peptide did not reduce conidial binding. However, pretreatment of the ECM with hydrogen peroxide to simulate oxidant stress increased conidial binding by about one-third. This and other studies discussed below have demonstrated that conidia can bind to the individual ECM components, laminin, fibronectin, and type I and IV collagen (Table 1). In addition to the binding activity, during germination, the organism produces serine and aspartic proteases and metalloproteases that can degrade elastin, laminin, fibronectin, and type I and type III collagen ECM proteins (Tronchin et al. 1993; Jaton-Ogay et al. 1994; Lee and Kolattukudy 1995; Iadarola et al. 1998). In an animal infection model, a mutant strain deficient in the serine protease and metalloprotease and the wild type showed no difference in pathogenicity (Jaton-Ogay et al. 1994). On the other hand, in another study, mice intratracheally injected with serine protease showed significant lower respiratory tract destruction (Iadarola et al. 1998). Both the binding to ECM proteins and the proteolytic degradation of ECM proteins may contribute to the establishment of pulmonary infection.

Table 2. Human fungal pathogen adhesins recognizing human ECM ligands[a]

Adhesin	Organism(s)	Ligand(s)	Identification[b]	Reference(s)
60 kDa	C. albicans	Fibronectin, collagen, vitronectin	Affinity chromatography, Ab Fn and Vn integrins	Klotz et al. (1993, 1994)
50, 90 kDa	C. albicans	Fibronectin, vitronectin	Ab Fn and Vn integrins	Klotz et al. (1993)
120–135 kDa	C. albicans	Fibronectin	Ab Fn integrin	
55 kDa	C. albicans	Fibronectin, laminin, fibrinogen	Hemoglobin-induced protein profile, ligand protection	Yan et al. (1996, 1998a,b)
33 kDa (GAPDH)	C. albicans	Fibronectin, laminin	Ligand blotting, purified protein	Gozalbo et al. (1998); Villamon et al. (1999)
150 kDa predicted (ALA1/ALS5)	C. albicans	Fibronectin, laminin, collagen	Function in S. cerevisiae	Gaur and Klotz (1997)
3 Species 40–50 kDa, >143 kDa	C. albicans	Fibronectin	Ab Fn receptor	Glee et al. (1996)
105 kDa	C. tropicalis	Fibronectin	Western blot and IP with Ab Fn integrin	DeMuri and Hostetter (1996)
68, 62, 60 kDa	C. albicans	Laminin, fibrinogen, plastic	Ligand blot	Bouchara et al. (1990)
37 kDa	C. albicans	Laminin	Ligand blot, Ab human high-affinity receptor; ligand blot negative with recombinant protein	Lopez-Ribot et al. (1994); Montero et al. (1998)
30 kDa	C. albicans	Vitronectin	Ligand blot	Limper and Standing (1994)
130, 110–100, 84 kDa	C. albicans	Vitronectin	IP and Western blot with anti-integrin Ab, correspond to integrin α_v, β_3, β_5, respectively	Spreghini et al. (1999)
50, 60, 90 kDa	C. albicans	Vitronectin	Ab integrin Vn receptor	Klotz et al. (1993)
65, 44, 25 kDa	C. albicans	Entactin	Ligand blot	Lopez-Ribot and Chaffin (1994)
23, 30 kDa	A. fumigatus	Fibronectin	Ligand blot	Penalver et al. (1996)
37 kDa	A. fumigatus	Laminin, fibronectin?	Ligand blot extract	Gil et al. (1996)
37 kDa Asp f2 allergen			Ligand blot recombinant protein	Banerjee et al. (1998)
72 kDa		Laminin, fibrinogen	Ligand blot	Tronchin et al. (1997)
120 kDa gpA (gp120, MSG-1)	P. carinii	Fibronectin	Ligand blot, affinity chromatography	Pottratz et al. (1991); Wisniowski et al. (1994)
33 kDa	P. carinii	Fibronectin, laminin	Sequence homology human non-integrin laminin receptor, ligand blot recombinant protein	Narasimhan et al. (1994)
43 kDa (gp43 antigen)	P. brasiliensis	Laminin	Bind purified protein, Western blot Ab to bacterial and human non-integrin receptors	Vicentini et al. (1994, 1997)
20 kDa	P. marneffei	Laminin, fibronectin by sialic acid recognition	Ligand blot	Hamilton et al. (1998, 1999)
50 kDa	H. capsulatum	Laminin	Ligand blot, Western blot Ab to human 67-kDa laminin receptor	McMahon et al. (1995)

[a] Does not include multiple fibronectin- and laminin-binding proteins determined from ligand blot by Glee et al. (1996).

[b] Ab, Polyclonal antiserum; Fn, fibronectin; Vn, vitronectin.

A. Fibronectin

Conidia bound soluble fibronectin with some degree of cell-to-cell variability as determined by indirect immunofluorescence microscopy (Penalver et al. 1996). On germinated conidia the fluorescence was reduced. No fluorescence was observed on long hyphae although a weak fluorescence was noted on some young, short outgrowths. Confocal fluorescence microscopy showed that reactivity was concentrated at conidial protrusions, as was noted for laminin (Gil et al. 1996). In germinated organisms, this method also showed reduced reactivity of parent conidia and the absence of hyphal reactivity. Binding of fibronectin was dose-dependent and saturable, as revealed by fluorescence activated flow cytometry (FACS) analysis (Penalver et al. 1996; Bouchara et al. 1997). Penalver et al. (1996) observed, at saturation, two cell populations that bound different amounts of fibronectin. Treatment with trypsin reduced ligand binding. Binding of fibronectin was inhibited by laminin (74%) and the RGD peptide (25–37%) (Gil et al. 1996). Glucose, galactose, mannose, N-acetylglucosamine, N-acetylgalactosamine amine, lactose, and maltose had little effect on fibronectin binding (Bouchara et al. 1997). On the other hand, mucin was an effective inhibitor (95%) with sialyllactose (23%) and asialomucin (29%) less effective.

Conidia also bound to immobilized fibronectin and the extent of binding was dependent on the amount of ECM component used to coat the wells of a microtiter plate (Bromley and Donaldson 1996; Gil et al. 1996; Penalver et al. 1996). Even at maximum binding, adherence of conidia to fibronectin was three- to tenfold less than binding to laminin and fourfold less than to collagen (Bromley and Donaldson 1996; Coulot et al. 1994). Adherence was inhibited by soluble fibronectin (88%), anti-fibronectin antibody (92%), and laminin (74%) (Penalver et al. 1996). The RGD peptide reduced binding to immobilized fibronectin by about 40–50% (Bromley and Donaldson 1996; Gil et al. 1996).

Ligand affinity blotting was employed to identify a potential fibronectin adhesin (Penalver et al. 1996). Cell-free conidial extracts, a βME and an SDS extract of isolated conidial walls were analyzed. Two polypeptides of 23 and 30 kDa were reactive in the cell-free homogenate and the βME extract.

B. Laminin

A. fumigatus conidia bind both soluble and immobilized laminin. Immunoelectron microscopy, indirect immunofluorescence confocal microscopy, and scanning electron microscopy demonstrated that laminin bound over the entire conidial surface with a concentration in the area of protrusions (Tronchin et al. 1993, 1997; Gil et al. 1996). Immunoelectron microscopy of thin sections also revealed numerous cytoplasmic binding sites while the plasma membrane and inner cell wall were not labeled (Tronchin et al. 1993). As conidia swelled and germinated, the emerging projection did not bind laminin and the binding to the parent conidia became less and more diffuse, as demonstrated by confocal microscopy and FACS (Gil et al. 1996; Tronchin et al. 1997). The binding of soluble laminin was dose-dependent and saturable, as demonstrated by FACS analysis (Gil et al. 1996; Bouchara et al. 1997). Binding of laminin increased with age reaching a maximum with approximately 5-day-old conidia (Bouchara et al. 1997). The binding was reduced by pretreatment of conidia with trypsin (Gil et al. 1996; Tronchin et al. 1997). Pretreatment with fibronectin reduced binding by 74% (Gil et al. 1996) and treatment with fibrinogen also markedly reduced laminin binding (Bouchara et al. 1997). Binding was not inhibited by addition of glycosaminoglycans, hyaluronic acid, heparan sulfate, keratan sulfate, or chondroitan sulfate A or B (Bouchara et al. 1997). The RGD peptide (Gil et al. 1996; Bouchara et al. 1997) and other synthetic peptides based on adhesive motifs of laminin and fibrinogen also did not reduce binding. However, several proteolytic fragments derived from laminin and fibrinogen as well as the laminin P1 fragment supported binding to immobilized fragments (Tronchin et al. 1993; Bouchara et al. 1997). Bouchara et al. (1997) considered the possibility that the adhesin had lectin-like activity and recognized a carbohydrate ligand shared by oligosaccharides at glycosylation sites on the peptides. The authors proposed a sialic acid dependent recognition as laminin binding (monitored by FACS) was reduced by N-acetylneuraminic acid and to a lesser extent by sialyl lactose and by mucin but not asialomucin.

Conidia also bound to laminin immobilized in wells of a microtiter plate (Tronchin et al. 1993, 1997; Coulot et al. 1994; Bromley and Donaldson 1996; Gil et al. 1996). This binding was inhibited by addition of soluble laminin (77%), anti-laminin antibody (92%), fibronectin (61%), and galactose (20%) but not by glucose or mannose (Gil et al. 1996). Binding was not inhibited by the RGD peptide (Bromley and Donaldson 1996; Gil et al. 1996). The adherence of swollen conidia was about three-times less than that of resting conidia (Tronchin et al. 1997).

Two potential laminin adhesins have been identified. Gil et al. (1996) examined cell-free homogenates, a βME extract of isolated conidial walls, and an SDS extract of conidial walls by ligand affinity blotting. A single reactive 37-kDa moiety was observed in the cell-free homogenate. The authors suggested that perhaps this binding protein either was not solubilized or poorly solubilized from the conidial walls by the methods used. More recently, Banerjee et al. (1998) have cloned a sequence encoding a major *A. fumigatus* allergen, Asp f2, and expressed the sequence in a bacterial system. The sequence showed homology to an *A. nidulans* protein and a *C. albicans* fibrinogen-binding protein. The allergen was also recovered from *A. fumigatus* culture filtrates. Compared to the predicted size of 29kDa, the isolated protein migrated at 37kDa on SDS-PAGE. Laminin bound to both the recombinant and native proteins immobilized in microtiter plate wells. The authors suggest that Asp f2 may be an adhesin for ECM. Although it remains to be tested, it is reasonable to speculate that the 37-kDa species identified in the two studies is the same. A second candidate emerged from a study by Tronchin and colleagues (1997). In this study, conidial wall proteins were extracted by boiling conidia in buffer containing SDS and dithiothreitol. Ligand affinity blotting identified a single 72-kDa reactive species. This component also appeared to react with Concanavalin A, suggesting the putative adhesin was a glycoprotein.

C. Collagens

Adherence of conidia to collagens has received less attention than adherence to fibrinogen, laminin, or fibronectin. *A. fumigatus* conidia adhere to immobilized type I and type IV collagen (Bromley and Donaldson 1996; Gil et al. 1996). As demonstrated for the other ECM components, binding to either collagen was dependent on the amount of ligand used to coat the well of a microtiter plate and the binding to the two collagens was similar. Laminin inhibited binding to

both collagens (70–77%) while fibronectin was a weak inhibitor (17%) (Gil et al. 1996). Two studies have examined inhibition of binding by RGD and found that the peptide did not alter conidial binding to type IV collagen (Bromley and Donaldson 1996; Gil et al. 1996). On the other hand, the studies differ in that one study found approximately 50% inhibition of binding to type I collagen (Bromley and Donaldson 1996) while the other noted no inhibition (Gil et al. 1996).

D. Conidial Adhesins for ECM

In summary, *A. fumigatus* resting conidia can adhere to ECM components. As conidia swell and germinate the adhesive interaction does not appear on hyphae and is reduced on the parent conidium. The identification of the adhesins that mediate this binding is less clear (Table 2). The ability of one component to reduce the binding of other components points to an adhesin(s) that recognizes multiple ligands. However, the inhibition while often more than 50%, is not complete and thus suggests that there may be more than one adhesin that recognizes each ligand. Resolution of this point has not emerged from the three studies that have identified potential candidates by ligand affinity blotting. Two candidates were identified for fibronectin binding in one study and two studies on laminin binding proposed two different candidates. If there is a multifunctional adhesin, the differences in size suggest that it has not yet been identified. A lectin-like adhesin has been proposed for laminin and fibrinogen. The differences in the effect of the RGD peptide on adherence to the various ECM components also suggest that there are multiple adhesins.

IV. *Pneumocystis carinii*

Pneumocystis carinii interacts with alveolar type I but not type II epithelial cells. ECM proteins are among the mediators of this adherence (Table 1). Both the microbe and host cell respond to adherence. Adherence of the fungal organism to fibronectin has been the most extensively studied.

A. Fibronectin

Binding of fibronectin to *P. carinii* was first reported by Pottratz and Martin (1990b), who demonstrated that the ^{51}Cr-labeled organisms (amoeboid-shaped yeast cells, trophozoites) bound the ECM component in a specific and saturable manner with about 640,000 binding sites per cell. Binding of the ^{51}Cr-labeled organisms was reduced about 60% in the presence of an RGDS peptide. Binding was partially dependent on Ca^{2}+, enhanced by Mn^{2+}, inhibited by NaCl and showed decreased binding when the pH was lowered from 8 to 6.8 (Wisniowski et al. 1994). *P. carinii* also bound to immobilized fibronectin and binding was inhibited by RGDS peptide and monoclonal antibody to the fibronectin cell binding domain that contains the RGD sequence (Pottratz et al. 1994a). Binding was reduced by the removal of Ca^{2+}.

Organisms derived from both rat and mouse adhere to cell lines, as determined by binding ^{51}Cr-labeled organisms to an alveolar epithelial cell line A549 (Pottratz and Martin 1990b) or by microscopic analysis of adherence to Vero or MRC-5 cell lines (Aliouat et al. 1993). Binding was inhibited 30–50% by addition of anti-fibronectin antibody (Aliouat et al. 1993; Pottratz and Martin 1990b) or by addition of soluble fibronectin (Aliouat et al. 1993), or GRDS peptide (Aliouat et al. 1993). Trypsin but not neuraminidase treatment reduced *P. carinii* adherence to Vero and MRC-5 culture cells (Aliouat et al. 1993). Reactivity of anti-fibronectin antibody with the surface of MRC-5 cells demonstrated the presence of fibronectin on the target cell surface (Aliouat et al. 1993). In another study, addition of fibronectin or laminin had no effect on binding of *P. carinii* to A549 cells (Fishman et al. 1991). However, if the organisms were first cultured with a lung fibroblast line, preincubation with fibronectin slightly increased adherence to lung epithelial cells in buffer while laminin decreased binding. After 3 days culture with MRC-5 cells, *P. carinii* filopodia penetrated the target cell apparently anchoring the interaction without evidence of membrane fusion (Aliouat et al. 1993). Killed *P. carinii* did not adhere. Formation of cytoskeleton within these extensions has been observed (Itatani and Marshall 1988). Inhibition of *P. carinii* cytoskeleton function with colchicine, a microtubule inhibitor (Limper 1990), or with cytochalasin B, a microfilament inhibitor (Limper 1990; Aliouat et al. 1993), inhibited binding.

As noted in the introduction, integrins are mammalian cell surface receptors for ECM proteins and potential candidates for fibronectin-binding proteins on these cells. Antibody to α_5 and α_v, subunits of fibronectin-binding integrins, inhibited *P. carinii* binding to A549 cells by 36% and 23%, respectively, while antibody to α_2, a subunit of a non-fibronectin binding integrin, had no effect (Pottratz et al. 1994b). After incubation of *P. carinii* with A549 cells for 24h, the reactivity with anti-α_5 increased threefold while no change was observed with the anti-α_v antibody. Addition of a major surface glycoprotein, gpA (gp120 or MSG-1), increased the reactivity tenfold and anti-gpA antibody reduced the increase by 50%. There was no change in the level of α_5 subunit mRNA. The organism adhered to type I alveolar epithelial cells and, in culture, type II alveolar epithelial cells acquired type I phenotypes; there was also a corresponding increase in *P. carinii* adherence (Pottratz and Weir 1995). Addition of anti-fibronectin antibody decreased adherence while anti-laminin and anti-type IV collagen antibody had no effect. Western blot analysis showed increased α_5 and α_v reactivity and antibody to these two integrin subunits decreased adherence of *P. carinii* to the older cells. A549 cells exposed to γ-interferon, a major host defense factor, showed a concentration- and time-dependent decrease in attachment of *P. carinii* (Pottratz and Weir 1997). The expression of $\alpha_5\beta_1$ integrin and the corresponding mRNA was decreased while that of the α_v integrin was not changed. *P. carinii* also induces interleukin 6 (IL-6) from A549 cells (Pottratz et al. 1998). Treatment of A549 cells with IL-6 resulted in an increased production of cell fibronectin and attachment of *P. carinii*. Alveolar macrophages are involved in clearance of *P. carinii*. Attachment to these cells was reduced by about 50% by pretreatment of the macrophage with a monoclonal antibody to the cell-binding domain of fibronectin, 84% by RGDS peptide or 70% by calcium removal (Pottratz and Martin 1990a). There was no effect of antibody to the heparin-binding domain of fibronectin. Binding of *P. carinii* through fibronectin did not result in phagocytosis of the fungus. Release of the cytokine TNF-α is enhanced in macrophage recovered from *P. carinii* infected animals. Compared with untreated *P. carinii*, organisms treated with fibronectin or vitronectin stimulated TNF-α release from macrophage (Neese et al. 1994).

The surface protein gpA has been identified as a fibronectin-binding protein. Adherence of the organism to a A549 cell line was inhibited 50% by addition of purified gpA, which is highly glycosylated and contains both mannose and *N*-acetylglucosamine residues in the oligosaccharides (Pottratz et al. 1991). Antibody to gpA inhibited binding of the *P. carinii* to the cell line. Polyclonal and monoclonal anti-gpA inhibited binding of soluble fibronectin to *P. carinii* (Pottratz et al. 1991; Wisniowski et al. 1994) and binding of the fungus to immobilized fibronectin (Pottratz et al. 1994a). A ligand affinity blot of proteins obtained from the fungus detected a single 120-kDa reactive species (Pottratz et al. 1991). Fibronectin affinity column purification of a ligand-binding moiety(ies) from an extract yielded gpA (Wisniowski et al. 1994). A second receptor candidate was reported in a gene encoding a 33-kDa laminin- and fibronectin-binding moiety with similarity to the mammalian non-integrin laminin receptor (Narasimhan et al. 1994). The gene was initially detected in a screen of a cDNA library with a probe for a glucan synthase gene and confirmed to be of *P. carinii* and not rat origin. A ligand affinity blot of the recombinant protein over-expressed in bacteria showed reactivity with laminin and fibronectin. Antibodies prepared to the recombinant protein reacted with a 33-kDa protein in total lysates of *P. carinii*.

B. Vitronectin

Vitronectin binds to mammalian integrins and is also reported to bind to *P. carinii*. Vitronectin bound to the organism (trophozoites and cysts together) in a dose-dependent and saturable manner with an estimated 547,000 binding sites per organism (Limper et al. 1993). Vitronectin has an RGD sequence in its cell attachment domain and RGD, but not RGES, can inhibit vitronectin binding to mammalian cells. However, binding of vitronectin to *P. carinii* was not inhibited by addition of the RGDS peptide but rather binding was increased about 2.2-to 3.5-fold by RGDS and RGES, respectively. No effect was observed with n-MeGRGDSP (Wisniowski and Martin 1995). Inhibition of binding by heparin (52–64%) suggested that the glycosaminoglycan domain was involved in binding (Limper et al. 1993; Wisniowski and Martin 1995). Binding was reduced by 0.2 M NaCl and eliminated by 1 M NaCl (Wisniowski and Martin 1995). There was a decreasing binding with increasing pH. Binding was enhanced by addition of Mn^{2+} and Ca^{2+}. Sodium periodate or heparitinase treatment of *P. carinii* inhibited subsequent vitronectin binding by 90% (Wisniowski and Martin 1995) or 50%,

respectively (Limper et al. 1993). A ligand affinity blot of a *P. carinii* extract showed reactivity with high molecular weight material >250 kDa that disappeared following treatment with heparitinase. The parameters of vitronectin binding suggested that the interaction was a lectin-like interaction between carbohydrate on the fungal surface and the glycosaminoglycan domain of vitronectin. As discussed for *C. albicans*, fungal β-glucan binds vitronectin and may be the source of fungal carbohydrate binding to vitronectin.

Binding of the fungus to A549 cells was stimulated by addition of vitronectin and reduced by addition of anti-vitronectin antiserum (Limper et al. 1993). Vitronectin and fibronectin are abundant in the lungs of infected rats and a Western blot of a total *P. carinii* extract of isolated organisms revealed the presence of vitronectin and fibronectin supporting in vivo binding of these components (Neese et al. 1994).

C. Bridging Between Fungi and Lung Epithelia

P. carinii has been demonstrated to bind fibronectin in vivo and in vitro. The binding of fibronectin appears to be through a protein-protein interaction between the cell-binding domain of fibronectin and a surface protein of the fungus (gpA) (Table 2). Integrins on the cell surface also bind the fibronectin through the cell-binding domain, thus fibronectin bridges between the fungus and the host cell. In contrast, the vitronectin binding appears to be a lectin-like interaction between fungal surface carbohydrate and the glycosaminoglycan-binding domain of vitronectin while the cell-binding domain is the likely attachment site with the cell. Binding appears to trigger a response of the fungus in cytoskeletal rearrangements and a host cytokine response, both of which modulate the outcome of fungal infection.

V. *Histoplasma capsulatum*

Like most of the other fungi, the intracellular pathogen of the reticuloendothelial system *Histoplasma capsulatum* not only binds to a component of ECM but also possesses proteolytic activity towards ECM (Table 1). Among several protease activities found in culture supernatants, a low molcular weight species was active against collagen (Okeke and Muller 1991). To date, adherence only to laminin has been observed for this dimorphic fungus (McMahon et al. 1995). Yeast cells attached to both immobilized laminin and Matrigel (contains primarily laminin, as well as collagen IV and ntactin) although adherence to the latter was proportional to the laminin content. In solution, laminin bound to yeast cells in a time-dependent and saturable manner with 30,000 binding sites per cell estimated. Binding was not inhibited by sugars found in laminin oligosaccharide side chains. Two peptides, one from domain III of the B1 chain and one from domain I of the A chain, have been implicated in mammalian cell adherence to laminin. The A-chain peptide, IKVAV, inhibited binding approximately 70%. A ligand affinity blot with cell wall proteins solubilized from isolated cell walls identified a single reactive 50-kDa species (Table 2). In common with the 37-kDa laminin-binding protein described for C. albicans, antibody to the human 67-kDa high-affinity laminin receptor reacted with the 50-kDa moiety. Antibody prepared to this 50-kDa protein bound to the surface of H. capsulatum yeast cells demonstrating its surface localization and reduced laminin binding to the cells.

While adherence to laminin may promote pathogenesis, adherence of host macrophage to collagen may contribute to host defense (Newman et al. 1997). Macrophage adhered to type I collagen gels but not to non-gelled collagen, fibronectin, laminin, or vitronectin coated surfaces. Macrophage immediately acquired increased fungistatic activity upon adherence. The increased activity was not associated with production of reactive oxygen or nitrogen species or iron restriction. Compared to macrophage adhered to plastic, the collagen-bound macrophage had increased phagolysosome fusion.

VI. *Paracoccidioides brasiliensis*

This dimorphic fungus can both adhere to and degrade components of the basement membrane (Table 1). *Paracoccidioides brasiliensis* produces an exocellular protease activity that will hydrolyze laminin, type IV collagen, and fibronectin but not fibrinogen or type I collagen (Puccia et al. 1998). Among the basement membrane components, adherence of the fungus to laminin has been described (Vicentini et al. 1994). A major antigenic component of *P. brasiliensis*, gp43, was a laminin-binding protein. Binding to a purified gp43 was specific and saturable. Laminin bound to the yeast cell surface. Laminin-coated yeast cells showed a fourfold increased binding to Madin-Darby canine kidney cell monolayers. In a hamster model of testicular infection, histological examination showed that tissue destruction and granuloma formation were more intense in animals infected with laminin-coated yeast cells than in those with uncoated cells.

A monoclonal antibody to the laminin-binding protein of *Staphylococcus aureus* reacted with gp43, as determined by Western blot analysis (Vicentini et al. 1997; Table 2). When antibody was added to a yeast cell suspension along with laminin, the subsequent binding of the yeast cells to mammalian cell monolayers was reduced, as was the extent of tissue damage in the hamster testicular model. Along with the cross-reactivity of antibody to the human 67-kDa laminin-binding protein and fungi, cross-reactivity antibody to a bacterial receptor suggests that laminin-binding proteins may share common motifs. Monoclonal antibodies to gp43 itself when assessed in an in vivo infection model either enhanced or inhibited granuloma formation (Gesztesi et al. 1996). Those antibodies that reduced adherence to laminin also reduced the pathogenicity of laminin-coated cells while one antibody that slightly increased laminin binding enhanced tissue destruction in the infection model.

VII. *Penicillium marneffei*

Two recent studies have demonstrated interaction of *Penicillium marneffei* conidia with laminin and fibronectin (Table 1). Binding of laminin and fibronectin to the conidia of *P. marneffei* was demonstrated by indirect immunofluorescence microscopy (Hamilton et al. 1998, 1999). No binding was observed on hyphae. Conidia also bound to immobilized laminin or fibronectin. Adherence to laminin was inhibited by soluble laminin and anti-laminin antibody (Hamilton et al. 1998) and to fibronectin by soluble fibronectin and anti-fibronectin antibody (Hamilton et al. 1999). At all concentrations of bound ligand, binding to laminin was greater than to fibronectin (Hamilton et al. 1999). In addition, binding to fibronectin was inhibited about 85% by addition of laminin and likewise addition of fibronectin inhibited binding to laminin. Conidia from older cultures (4 and 8 days) were more adherent to either ligand than conidia from younger cultures (1 and 2 days). Binding to laminin or fibronectin was eliminated by *N*-acetylneuraminic acid (Hamilton et al. 1998, 1999). The effectiveness of this compound suggested that the binding protein was a sialic acid specific lectin. One of the laminin- and fibrinogen-binding proteins of *A. fumigatus* was noted above as having a similar lectin-like activity.

Prolonged treatment of conidia with chymotrypsin reduced subsequent binding to fibronectin or laminin (Hamilton et al. 1999). A ligand affinity blot of a conidial wall extract and conidial homogenate showed that both ligands reacted with a 20-kDa moiety in the homogenate (Table 2). The apparent anomaly in location may reflect the low recovery of material in the conidial wall extract. This protein appeared to be a common receptor for both laminin and fibronectin.

VIII. *Sporothrix schenckii*

Sporothrix schenckii infection is primarily a subcutaneous infection, although more generalized infection may arise from dissemination or from pulmonary infection. Both conidia and yeast cells of *S. schenckii* adhere to ECM components (Lima et al. 1999; Table 1). Binding to immobilized ligands was monitored in ELISA with rabbit anti-*S. schenckii* antiserum. Both yeast cells and conidia adhered to type II collagen, fibronectin and laminin. Yeast cells also bound to fibrinogen. No binding of either morphological form was observed to thrombospondin. Binding of *S. schenckii* cells to type I and type III collagen was slight. While the binding of yeast cells and conidia to laminin showed a similar enhancement compared with the control, conidia showed a greater enhancement than yeast cells when fibronectin or type II collagen was the target.

IX. Conclusions

The studies discussed in the previous sections make the case that human fungal pathogens recognize host ECM components and that recogni-

tion occurs with both soluble and immobilized ligand. *C. albicans* recognizes more ligands than other fungi. However, this may reflect fewer and less extensive studies with other organisms. The identification of the specific surface components that mediate this recognition is still very much a work in progress. However, several generalizations can be drawn from the present information. The dominant recognition pattern is a protein-protein recognition although there are several examples of carbohydrate-protein recognition with the carbohydrate supplied by either the fungus or the host ligand (Table 3; Fig. 2). *β*-Glucan is a potential common fungal-binding moiety for interaction with vitronectin while some other interactions implicate a fungal lectin recognizing a glycosyl motif on the ECM component. From early in the investigation of fungal-ECM interactions, the hypothesis that fungal ECM-binding proteins shared characteristics with the integrin family of mammalian binding proteins has contributed to the experimental approach to identifying fungal adhesins. The ability of some antibodies to mammalian integrins to bind to fungal surfaces and block binding to the ligand provides support for some common epitopes. However, evidence for structural parallels is tenuous. In more general terms, there are similarities between the host and fungal recognition of ECM proteins. First, as noted above, the dominant recognition is a protein-protein recognition. Secondly, multiple proteins have been identified that recognize the same host ligand. Third, more than one ECM component may be recognized by the same binding protein. Fourth, in some studies with *A. fumigatus*, *C. albicans*, and *P. carinii*, the RGD motif appears to contribute to binding. Use of the RGD motif in adherence has been implicated in other cell adhesion systems such as that of the basidiomycete *Lentinus edodes* (Yasuda and Shishido 1997). The identification of the specific surface components

Fig. 2. Schematic depiction of motifs on ECM proteins recognized by fungal adhesins: oligosaccharide motifs (*solid circles*), RGD amino acid sequence (*RGD*) and amino acid sequences other than RGD (*solid triangles*)

Table 3. Receptor-ligand interactions

Protein ligand (ECM) – Protein receptor (fungus)	Protein ligand (ECM) – Carbohydrate receptor (fungus)	Carbohydrate ligand (ECM) – Protein receptor (fungus)

Fibronectin – various proteins
  C. albicans, A. fumigatus, P. carinii
Laminin – various proteins
  C. albicans, A. fumigatus?, P. brasiliensis
Collagen – various proteins
  C. albicans
Entactin – various proteins
  C. albicans
Tenascin – protein(s)
  C. albicans
Vitronectin – various proteins
  C. albicans
	Vitronectin – β-glucan
	  C. albicans, P. carinii, others
		Sialic acid on laminin/fibronectin
		  P. marneffii, A. fumigatus

that mediate ECM recognition is still confusing, particularly for the more extensively studied pathogens. Resolving this confusion constitutes the major near-term challenge. The demonstration that purified GAPDH of *C. albicans*, Asp f2 of *A. fumigatus*, gpA of *P. carinii* and gp43 of *P. brasiliensis* can bind their proposed ligands is an important step in the move from identification of proteins in a complex mixture by ligand blotting or reactivity with anti-integrin antibodies to determination of the biochemical and genetic relationship among multiple adhesins. The longer and perhaps more difficult challenge will be the elucidation of how recognition of ECM components by fungi – both adherence and degradation – contributes to the colonization and invasion of host tissue.

Acknowledgements. This work was supported in part by a Public Health Service grant (AI23416) from the National Institutes of Health.

References

Alberts B, Bray D, Lewis J, Raff M, Roberts K, Watson JD (1994) Molecular biology of the cell, 3rd edn. Garland Publishing, New York

Aliouat EM, Dei-Cas E, Ouaissi A, Palluault F, Soulez B, Camus D (1993) In vitro attachment of *Pneumocystis carinii* from mouse and rat origin. Biol Cell 77:209–217

Banerjee B, Greenberger PA, Fink JN, Kurup VP (1998) Immunological characterization of Asp f2, a major allergen from *Aspergillus fumigatus* associated with allergic bronchopulmonary aspergillosis. Infect Immun 66:5175–5182

Bouchara JP, Tronchin G, Annaix V, Robert R, Senet JM (1990) Laminin receptors on *Candida albicans* germ tubes. Infect Immun 58:48–54

Bouchara JP, Sanchez M, Chevailler A, Marot-Leblond A, Lissitzky JC, Tronchin G, Chabasse D (1997) Sialic acid-dependent recognition of laminin and fibrinogen by *Aspergillus fumigatus* conidia. Infect Immun 65:2717–2724

Bromley IM, Donaldson K (1996) Binding of *Aspergillus fumigatus* spores to lung epithelial cells and basement membrane proteins: relevance to the asthmatic lung. Thorax 51:1203–1209

Chaffin WL, Lopez-Ribot JL, Casanova M, Gozalbo D, Martinez JP (1998) Cell wall and secreted proteins of *Candida albicans*: identification, function, and expression. Microbiol Mol Biol Rev 62:130–180

Cotter G, Weedle R, Kavanagh K (1998) Monoclonal antibodies directed against extracellular matrix proteins reduce the adherence of *Candida albicans* to HEp-2 cells. Mycopathologia 141:137–142

Coulot P, Bouchara JP, Renier G, Annaix V, Planchenault C, Tronchin G, Chabasse D (1994) Specific interaction of *Aspergillus fumigatus* with fibrinogen and its role in cell adhesion. Infect Immun 62:2169–2177

DeMuri GP, Hostetter MK (1996) Evidence for a beta 1 integrin fibronectin receptor in *Candida tropicalis*. J Infect Dis 174:127–132

Fishman JA, Samia JA, Fuglestad J, Rose RM (1991) The effects of extracellular matrix (ECM) proteins on the attachment of *Pneumocystis carinii* to lung cell lines in vitro. J Protozool 38:34S–37S

Gaur NK, Klotz SA (1997) Expression, cloning, and characterization of a *Candida albicans* gene, *ALA1*, that confers adherence properties upon *Saccharomyces cerevisiae* for extracellular matrix proteins. Infect Immun 65:5289–5294

Gesztesi JL, Puccia R, Travassos LR, Vicentini AP, de Moraes JZ, Franco MF, Lopes JD (1996) Monoclonal antibodies against the 43,000 Da glycoprotein from *Paracoccidioides brasiliensis* modulate laminin-mediated fungal adhesion to epithelial cells and pathogenesis. Hybridoma 15:415–422

Giancotti FG, Rusolatti E (1999) Integrin signaling. Science 285:1028–1032

Gil ML, Penalver MC, Lopez-Ribot JL, O'Connor JE, Martinez JP (1996) Binding of extracellular matrix proteins to *Aspergillus fumigatus* conidia. Infect Immun 64:5239–5247

Gil ML, Villamon E, Monteagudo C, Gozalbo D, Martinez JP (1999) Clinical strains of *Candida albicans* express the surface antigen glyceraldehyde-3-phosphate dehydrogenase in vitro and in infected tissues. FEMS Immunol Med Microbiol 23:229–234

Gil-Navarro I, Gil ML, Casanova M, O'Connor JE, Martinez JP, Gozalbo D (1997) The glycolytic enzyme glyceraldehyde-3-phosphate dehydrogenase of *Candida albicans* is a surface antigen. J Bacteriol 179: 4992–4999

Glee PM, Masuoka J, Ozier WT, Hazen KC (1996) Presence of multiple laminin- and fibronectin-binding proteins in cell wall extract of *Candida albicans*: influence of dialysis. J Med Vet Mycol 34:57–61

Gozalbo D, Gil-Navarro I, Azorin I, Renau-Piqueras J, Martinez JP, Gil ML (1998) The cell wall-associated glyceraldehyde-3-phosphate dehydrogenase of *Candida albicans* is also a fibronectin and laminin binding protein. Infect Immun 66:2052–2059

Hamilton AJ, Jeavons L, Youngchim S, Vanittanakom N, Hay RJ (1998) Sialic acid-dependent recognition of laminin by *Penicillium marneffei* conidia. Infect Immun 66:6024–6026

Hamilton AJ, Jeavons L, Youngchim S, Vanittanakom N (1999) Recognition of fibronectin by *Penicillium marneffei* conidia via a sialic acid-dependent process and its relationship to the interaction between conidia and laminin. Infect Immun 67:5200–5205

Horowitz AR, Werb Z (1998) Cell adhesion and the extracellular matrix: recent progress and emerging themes. Curr Opin Cell Biol 10:563–565

Iadarola P, Lungarella G, Martorana PA, Viglio S, Guglielminetti M, Korzus E, Gorrini M, Cavarra E, Rossi A, Travis J, Luisetti M (1998) Lung injury and degradation of extracellular matrix components by *Aspergillus fumigatus* serine proteinase. Exp Lung Res 24:233–251

Itatani CA, Marshall GJ (1988) Ultrastructural morphology and staining characteristics of *Pnemocystis carinii* in situ and from bronchoalveolar lavage. J Parasit 74:700–712

Jakab E, Paulsson M, Ascencio F, Ljungh A (1993) Expression of vitronectin and fibronectin binding by *Candida albicans* yeast cells. Apmis 101:187–193

Jaton-Ogay K, Paris S, Huerre M, Quadroni M, Falchetto R, Togni G, Latge JP, Monod M (1994) Cloning and disruption of the gene encoding an extracellular metalloprotease of *Aspergillus fumigatus*. Mol Microbiol 14:917–928

Kalo A, Segal E, Sahar E, Dayan D (1988) Interaction of *Candida albicans* with genital mucosal surfaces: involvement of fibronectin in adherence. J Infect Dis 157:1253–1256

Klotz SA (1987) The adherence of *Candida* yeast to human and bovine vascular endothelium and subendothelial extracellular matrix. FEMS Microbiol Lett 48:201–205

Klotz SA (1990) Adherence of *Candida albicans* to components of the subendothelial extracellular matrix. FEMS Microbiol Lett 56:249–254

Klotz SA, Maca RD (1988) Endothelial cell contraction increases *Candida* adherence to exposed extracellular matrix. Infect Immun 56:2495–2498

Klotz SA, Smith RL (1991) A fibronectin receptor on *Candida albicans* mediates adherence of the fungus to extracellular matrix. J Infect Dis 163:604–610

Klotz SA, Smith RL (1992) Glycosaminoglycans inhibit *Candida albicans* adherence to extracellular matrix proteins. FEMS Microbiol Lett 78:205–208

Klotz SA, Smith RL (1995) Gelatin fragments block adherence of *Candida albicans* to extracellular matrix proteins. Microbiology 141:2681–2684

Klotz SA, Smith RL, Stewart BW (1992) Effect of an arginine-glycine-aspartic acid-containing peptide on hematogenous candidal infections in rabbits. Antimicrob Agents Chemother 36:132–136

Klotz SA, Rutten MJ, Smith RL, Babcock SR, Cunningham MD (1993) Adherence of *Candida albicans* to immobilized extracellular matrix proteins is mediated by calcium-dependent surface glycoproteins. Microb Pathog 14:133–147

Klotz SA, Hein RC, Smith RL, Rouse JB (1994) The fibronectin adhesin of *Candida albicans*. Infect Immun 62:4679–4681

Lee JD, Kolattukudy PE (1995) Molecular cloning of the cDNA and gene for an elastinolytic aspartic proteinase from *Aspergillus fumigatus* and evidence of its secretion by the fungus during invasion of the host lung. Infect Immun 63:3796–3803

Lima OC, Figueiredo CC, Pereira BA, Coelho MG, Morandi V, Lopes-Bezerra LM (1999) Adhesion of the human pathogen *Sporothrix schenckii* to several extracellular matrix proteins. Braz J Med Biol Res 32:651–657

Limper AHM, Martin WJ II (1990) *Pneumocystis carinii*: inhibition of lung cell growth mediated by parasite attachment. J Clin Invest 85:391–397

Limper AH, Standing JE (1994) Vitronectin interacts with *Candida albicans* and augments organism attachment to the NR8383 macrophage cell line. Immunol Lett 42:139–144

Limper AH, Standing JE, Hoffman OA, Castro M, Neese LW (1993) Vitronectin binds to *Pneumocystis carinii* and mediates organism attachment to cultured lung epithelial cells. Infect Immun 61:4302–4309

Lopez-Ribot JL, Chaffin WL (1994) Binding of the extracellular matrix component entactin to *Candida albicans*. Infect Immun 62:4564–4571

Lopez-Ribot JL, Casanova M, Monteagudo C, Sepulveda P, Martinez JP (1994) Evidence for the presence of a high-affinity laminin receptor-like molecule on the surface of *Candida albicans* yeast cells. Infect Immun 62:742–746

Lopez-Ribot JL, Monteagudo C, Sepulveda P, Casanova M, Martinez JP, Chaffin WL (1996) Expression of the fibrinogen binding mannoprotein and the laminin receptor of *Candida albicans* in vitro and in infected tissues. FEMS Microbiol Lett 142:117–122

Lopez-Ribot JL, Bikandi J, San Millan R, Chaffin WL (1999) Interactions between *Candida albicans* and the human extracellular matrix component tenascin-C. Mol Cell Biol Res Commun 2:58–63

Masuoka J, Wu G, Glee PM, Hazen KC (1999) Inhibition of *Candida albicans* attachment to extracellular matrix by antibodies which recognize hydrophobic cell wall proteins. FEMS Immunol Med Microbiol 24:421–429

McMahon JP, Wheat J, Sobel ME, Pasula R, Downing JF, Martin WJ II (1995) Murine laminin binds to

Histoplasma capsulatum. A possible mechanism of dissemination. J Clin Invest 96:1010–1017

Montero M, Marcilla A, Sentandreu R, Valentin E (1998) A *Candida albicans* 37 kDa polypeptide with homology to the laminin receptor is a component of the translational machinery. Microbiology 144:839–847

Morschhauser J, Virkola R, Korhonen TK, Hacker J (1997) Degradation of human subendothelial extracellular matrix by proteinase-secreting *Candida albicans*. FEMS Microbiol Lett 153:349–355

Narasimhan S, Armstrong MY, Rhee K, Edman JC, Richards FF, Spicer E (1994) Gene for an extracellular matrix receptor protein from *Pneumocystis carinii*. Proc Natl Acad Sci USA 91:7440–744

Neese LW, Standing JE, Olson EJ, Castro M, Limper AH (1994) Vitronectin, fibronectin, and gp120 antibody enhance macrophage release of TNF-alpha in response to *Pneumocystis carinii*. J Immunol 152:4549–4556

Negre E, Vogel T, Levanon A, Guy R, Walsh TJ, Roberts DD (1994) The collagen binding domain of fibronectin contains a high affinity binding site for *Candida albicans*. J Biol Chem 269:22039–22045

Newman SL, Gootee L, Kidd C, Ciraolo GM, Morris R (1997) Activation of human macrophage fungistatic activity against *Histoplasma capsulatum* upon adherence to type 1 collagen matrices. J Immunol 158:1779–1786

Okeke CN, Muller J (1991) Production of extracellular collagenolytic proteinases by *Histoplasma capsulatum* var. *duboisii* and *Histoplasma capsulatum* var. *capsulatum* in the yeast phase. Mycoses 34:453–460

Ollert MW, Sohnchen R, Korting HC, Ollert U, Brautigam S, Brautigam W (1993) Mechanisms of adherence of *Candida albicans* to cultured human epidermal keratinocytes. Infect Immun 61:4560–4568

Olson EJ, Standing JE, Griego-Harper N, Hoffman OA, Limper AH (1996) Fungal beta-glucan interacts with vitronectin and stimulates tumor necrosis factor alpha release from macrophages. Infect Immun 64:3548–3554

Pancholi V, Fischetti VA (1992) A major surface protein on group A streptococci is a glyceraldehyde-3-phosphate-dehydrogenase with multiple binding activity. J Exp Med 176:415–426

Penalver MC, O'Connor JE, Martinez JP, Gil ML (1996) Binding of human fibronectin to *Aspergillus fumigatus* conidia. Infect Immun 64:1146–1153

Penn C, Klotz SA (1994) Binding of plasma fibronectin to *Candida albicans* occurs through the cell binding domain. Microb Pathog 17:387–393

Pottratz ST, Martin WJ II (1990a) Mechanism of *Pneumocystis carinii* attachment to cultured rat alveolar macrophages. J Clin Invest 86:1678–1683

Pottratz ST, Martin WJ II (1990b) Role of fibronectin in *Pneumocystis carinii* attachment to cultured lung cells. J Clin Invest 85:351–356

Pottratz ST, Weir AL (1995) Attachment of *Pneumocystis carinii* to primary cultures of rat alveolar epithelial cells. Exp Cell Res 221:357–362

Pottratz ST, Weir AL (1997) Gamma-inteferon decreases *Pneumocystis carinii* attachment to lung cells by decreasing expression of lung cell suface integrins. Eur J Clin Invest 27:17–22

Pottratz ST, Paulsrud J, Smith JS, Martin WJ II (1991) *Pneumocystis carinii* attachment to cultured lung cells by

pneumocystis gp 120, a fibronectin binding protein. J Clin Invest 88:403–407

Pottratz ST, Paulsrud JR, Smith JS, Martin WJ II (1994a) Evidence for *Pneumocystis carinii* binding to a cell-free substrate: role of the adhesive protein fibronectin. J Lab Clin Med 123:273–281

Pottratz ST, Weir AL, Wisniowski PE (1994b) *Pneumocystis carinii* attachment increases expression of fibronectin-binding integrins on cultured lung cells. Infect Immun 62:5464–5469

Pottratz ST, Reese S, Sheldon JL (1998) *Pneumocystis carinii* induces interleukin 6 production by an alveolar epithelial cell line. Eur J Clin Invest 28:424–429

Puccia R, Carmona AK, Gesztesi JL, Juliano L, Travassos LR (1998) Exocellular proteolytic activity of *Paracoccidioides brasiliensis*: cleavage of components associated with the basement membrane. Med Mycol 36:345–348

Rodrigues RG, Yan S, Walsh TJ, Roberts DD (1998) Hemoglobin differentially induces binding of *Candida*, *Trichosporon*, and *Saccharomyces* species to fibronectin. J Infect Dis 178:497–502

Sakata N, Yamazaki K, Kogure T (1999) Identification of a 21 kDa laminin-binding component of *Candida albicans*. Zentralbl Bakteriol 289:217–225

Santoni G, Gismondi A, Liu JH, Punturieri A, Santoni A, Frati L, Piccoli M, Djeu JY (1994) *Candida albicans* expresses a fibronectin receptor antigenically related to alpha 5 beta 1 integrin. Microbiology 140:2971–2979

Santoni G, Birarelli P, Hong LJ, Gamero A, Djeu JY, Piccoli M (1995) An alpha 5 beta 1-like integrin receptor mediates the binding of less pathogenic *Candida* species to fibronectin. J Med Microbiol 43:360–367

Sawyer RT, Garner RE, Hudson JA (1992) Arg-Gly-Asp (RGD) peptides alter hepatic killing of *Candida albicans* in the isolated perfused mouse liver model. Infect Immun 60:213–218

Scheld WM, Strunk RW, Balian G, Calderone RA (1985) Microbial adhesion to fibronectin in vitro correlates with production of endocarditis in rabbits. Proc Soc Exp Biol Med 180:474–482

Sepulveda P, Murgui A, Lopez-Ribot JL, Casanova M, Timoneda J, Martinez JP (1995) Evidence for the presence of collagenous domains in *Candida albicans* cell surface proteins. Infect Immun 63:2173–2179

Sepulveda P, Lopez-Ribot JL, Gozalbo D, Cervera A, Martinez JP, Chaffin WL (1996) Ubiquitin-like epitopes associated with *Candida albicans* cell surface receptors. Infect Immun 64:4406–4408

Silva TM, Glee PM, Hazen KC (1995) Influence of cell surface hydrophobicity on attachment of *Candida albicans* to extracellular matrix proteins. J Med Vet Mycol 33:117–122

Skerl KG, Calderone RA, Segal E, Sreevalsan T, Scheld WM (1984) In vitro binding of *Candida albicans* yeast cells to human fibronectin. Can J Microbiol 30:221–227

Smits GJ, Kapteyn JC, van den Ende H, Klis FM (1999) Cell wall dynamics in yeast. Curr Opin Microbiol 2:348–352

Spreghini E, Gismondi A, Piccoli M, Santoni G (1999) Evidence for alphavbeta3 and alphavbeta5 integrin-like vitronectin (VN) receptors in *Candida albicans* and their involvement in yeast cell adhesion to VN. J Infect Dis 180:156–166

Tronchin G, Bouchara JP, Larcher G, Lissitzky JC, Chabasse D (1993) Interaction between *Aspergillus fumigatus* and basement membrane laminin: binding and substrate degradation. Biol Cell 77:201–208

Tronchin G, Bouchara JP, Robert R, Senet JM (1988) Adherence of *Candida albicans* germ tubes to plastic: ultrastructural and molecular studies of fibrillar adhesins. Infect Immun 56:1987–1993

Tronchin G, Bouchara JP, Robert R (1989) Dynamic changes of the cell wall surface of *Candida albicans* associated with germination and adherence. Eur J Cell Biol 50:285–290

Tronchin G, Esnault K, Renier G, Filmon R, Chabasse D, Bouchara JP (1997) Expression and identification of a laminin-binding protein in *Aspergillus fumigatus* conidia. Infect Immun 65:9–15

Vicentini AP, Gesztesi JL, Franco MF, de Souza W, de Moraes JZ, Travassos LR, Lopes JD (1994) Binding of *Paracoccidioides brasiliensis* to laminin through surface glycoprotein gp43 leads to enhancement of fungal pathogenesis. Infect Immun 62:1465–1469

Vicentini AP, Moraes JZ, Gesztesi JL, Franco MF, de Souza W, Lopes JD (1997) Laminin-binding epitope on gp43 from *Paracoccidioides brasiliensis* is recognized by a monoclonal antibody raised against *Staphylococcus aureus* laminin receptor. J Med Vet Mycol 35:37–43

Villamon E, Gozalbo D, Martinez JP, Gil ML (1999) Purification of a biologically active recombinant glyceraldehyde 3-phosphate dehydrogenase from *Candida albicans*. FEMS Microbiol Lett 179:61–65

Wisniowski P, Martin WJ II (1995) Interaction of vitronectin with *Pneumocystis* carinii: evidence for binding via the heparin binding domain. J Lab Clin Med 125:38–45

Wisniowski P, Pasula R, Martin WJ II (1994) Isolation of *Pneumocystis carinii* gp120 by fibronectin affinity: evidence for manganese dependence. Am J Respir Cell Mol Biol 11:262–269

Yan S, Negre E, Cashel JA, Guo N, Lyman CA, Walsh TJ, Roberts DD (1996) Specific induction of fibronectin binding activity by hemoglobin in *Candida albicans* grown in defined media. Infect Immun 64: 2930–2935

Yan S, Rodrigues RG, Cahn-Hidalgo D, Walsh TJ, Roberts DD (1998a) Hemoglobin induces binding of several extracellular matrix proteins to *Candida albicans*. Identification of a common receptor for fibronectin, fibrinogen, and laminin. J Biol Chem 273:5638–5644

Yan S, Rodrigues RG, Roberts DD (1998b) Hemoglobin-induced binding of *Candida albicans* to the cell-binding domain of fibronectin is independent of the Arg-Gly-Asp sequence. Infect Immun 66:1904–1909

Yasuda T, Shishido K (1997) Aggregation of yeast cells induced by the Arg-Gly-Asp motif-containing fragment of high-molecular-mass cell-adhesion protein MFBA, derived from the basidiomycetous mushroom *Lentinus edodes*. FEMS Microbiol Lett 154:195–200

9 Molecular Organization and Construction of the Fungal Cell Wall

Hans De Nobel,[1] J. Hans Sietsma,[2] Herman Van Den Ende,[1] and Frans M. Klis[1]

CONTENTS

I. Introduction

The fungal wall accounts for about 20–30% of the cellular dry weight and thus represents a major investment for the cell in terms of metabolic energy. It is responsible for the shape of the cell, offers protection against mechanical damage, and functions as a molecular sieve. Furthermore, it may contain adhesive proteins involved in bundling hyphae into macroscopic structures, in recognizing mating partners or host cells, or in binding to the substratum. Cell wall proteins may also confer hydrophobic properties to the cell surface. Various reviews and books have discussed the fungal cell wall and its properties (Kuhn et al. 1990; Peberdy 1990; Fleet 1991; Ruiz-Herrera 1992; Sentandreu et al. 1994; Sietsma and Wessels 1994; Wessels 1994; Gooday 1995; for earlier work see Wessels and Sietsma 1981). In this review, we will focus on recent developments and try to present them against the background of older work. As the best-studied fungal organism to date is the budding yeast *Saccharomyces cerevisiae*, we will discuss to which extent our knowledge of its wall is relevant for other fungi, in particular the mycelial species of the Ascomycotina. We hope that this review will stimulate researchers in answering their own particular questions in this rapidly developing field.

II. A Molecular Model of the Cell Wall of the Budding Yeast *Saccharomyces cerevisiae*

The cell wall of the budding yeast *S. cerevisiae* is probably the best-studied cell wall of any fungus to date (Orlean 1997; Kapteyn et al. 1999a). It contains four classes of components, namely, chitin, β1,3-glucan, β1,6-glucan, and mannoproteins. However, in contrast to many other fungi, it does not contain α-glucan. Electron microscopy reveals two layers. The inner layer is electron-transparent and represents the chitin→β-glucan complex, which is responsible for the mechanical strength of the cell wall. The electron-dense outer layer consists of proteins that extend radially from the cell surface (Baba and Ohsumi 1987) and limit the permeability of the cell wall (Zlotnik et al.

[1] Swammerdam Institute for Life Sciences, University of Amsterdam, Nieuwe Achtergracht 166, 1018 WV Amsterdam, The Netherlands
[2] Molecular Plant Biology Laboratory, Groningen Biomolecular Sciences and Biotechnology Institute (GBB), University of Groningen, Kerklaan 30, 9751 NN Haren, The Netherlands

The Mycota VIII
Biology of the Fungal Cell
Howard/Gow (Eds.)
© Springer-Verlag Berlin Heidelberg 2001

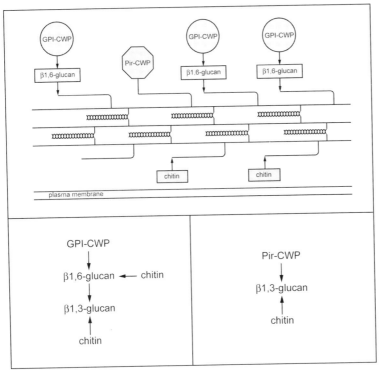

Fig. 1. Molecular model of the cell wall of *S. cerevisiae*. **A** Comprehensive model of the cell wall of *S. cerevisiae* (Smits et al. 1999). β1,3-Glucan molecules kept together by hydrogen bridges form the skeletal network. The nonreducing ends of the β1,3-glucan molecules function as attachment sites for chitin chains at the interior side of the cell wall and for β1,6-glucan molecules and cell wall proteins at the exterior side of the wall. There are two types of cell wall proteins present. The GPI cell wall proteins (*GPI-CWPs*) are tethered through a highly branched, water-soluble β1,6-glucan molecule to the β1,3-glucan network, whereas the Pir cell wall proteins (*Pir-CWPs*) are directly linked to the network. Under normal conditions the GPI-CWPs are much more abundant than the Pir-CWPs. Some chitin chains may be linked to β1,6-glucan, but this is omitted for the sake of clarity. (Reprinted from Smits et al. 1999 with permission from Elsevier Science) **B** The two main structural subunits of the cell wall of *S. cerevisiae*. The normally infrequent linkage of chitin to β1,6-glucan is now included (*left panel*). The linkage between GPI-CWPs and β1,6-glucan has been described by Kollár et al. (1997): -AA$_\omega$-Et-P-Man$_5$→Glc-, in which AA$_\omega$ represents the carboxy-terminal amino acid of the mature protein linked by an amide linkage to ethanolamine (*Et*), which in turn is linked through a phosphodiester bridge (*P*) to the Man$_5$ sequence of the GPI anchor; the Man$_5$ sequence is linked to a nonreducing end of β1,6-glucan. The presence of a phosphodiester bridge in the GPI-anchor remnant might explain the effectiveness of hot water extraction in releasing fungal cell wall proteins. The precise linkage between Pir-CWPs and β1,3-glucan is unknown

1984; De Nobel et al. 1990). Depending on the growth conditions, one may find another (electron-dense) protein layer close to the plasma membrane, representing periplasmic proteins such as invertase and acid phosphatase. These are soluble proteins that are apparently too large to pass the cell wall and get trapped in the inner regions of the cell wall (De Nobel et al. 1989; De Nobel and Barnett 1991). Recently, a molecular model of the cell wall of *S. cerevisiae* has been presented (Fig. 1; Smits et al. 1999). Its main characteristics are:

1. An internal skeletal framework corresponding to the electron-transparent layer of the wall. It consists of an elastic three-dimensional network of β1,3-glucan molecules that surrounds the entire cell. Mature β1,3-glucan molecules are branched (Manners et al. 1973a) and thus possess multiple nonreducing ends. These may function as acceptor sites for the other components of the cell wall such as β1,6-glucan, chitin and Pir-proteins (see below). Chitin in the lateral walls is mainly found close to the plasma membrane, where it is deposited after cytokinesis (Schekman and Brawley 1979; Shaw et al. 1991). In other words, walls of growing buds do not contain chitin. In wild-type cells most of the chitin is linked to β1,3-glucan and only a minor amount is connected to β1,6-glucan. The chitin content is normally

about 1–2%, but in some cell wall defective mutants chitin levels may rise to 20% and more due to increased deposition of chitin in the lateral walls (Dallies et al. 1998; Kapteyn et al. 1999a,b). This is accompanied by a strong increase in chitin bound to $\beta1,6$-glucan (Kapteyn et al. 1997). It is uncertain if all chitin chains become covalently linked to other wall polymers (Kollár et al. 1995).

2. An outer layer of glycoproteins corresponding to the electron-dense layer of the wall. Two classes of covalently linked cell wall proteins (CWPs) can be distinguished, differing from each other in the way they are linked to the $\beta1,3$-glucan network. They are designated as glycosylphosphatidylinositol dependent cell wall proteins (GPI-CWPs) and Pir-CWPs (Table 1). Whereas GPI-CWPs are indirectly linked through a highly branched and water-soluble $\beta1,6$-glucan molecule, Pir-CWPs seem to be directly linked to $\beta1,3$-glucan molecules. In contrast to the structural complex GPI-CWP$\rightarrow\beta1,6$-glucan$\rightarrow\beta1,3$-glucan, which has been extensively investigated (Kapteyn et al. 1996; Kollár et al. 1997; Fujii et al. 1999), the Pir-CWP–$\beta1,3$-glucan complex is as yet ill defined.

Many fractionation studies of the cell wall make use of extraction with alkali. The alkali-soluble fraction contains (degraded) mannoproteins, which are often designated as the mannan fraction of the yeast cell wall due to the presence of extended N-chains in yeast mannoproteins (see below). The alkali extract of the yeast cell wall also contains a $\beta1,3/\beta1,6$-glucan heteropolymer, which is similar in structure to the alkali-insoluble $\beta1,3/\beta1,6$-glucan heteropolymer. Using pulse-chase techniques, Hartland et al. (1994) have shown that the alkali-soluble $\beta1,3/\beta1,6$-glucan in time becomes partially alkali-insoluble. Since after enzymatic degradation of chitin the alkali-insoluble $\beta1,3/\beta1,6$-glucan becomes largely soluble in alkali (Mol and Wessels 1987), the

linkage to chitin apparently explains the alkali-insolubility of the $\beta1,3/\beta1,6$-glucan.

In the following sections we will discuss to which extent the organization of the cell wall of *S. cerevisiae* may help to understand the organization of the cell wall of the mycelial fungi and in particular species belonging to the Ascomycotina.

III. Organization of the Cell Wall of Mycelial Species of the Ascomycotina

As mentioned above, electron micrographs of the cell wall of the budding yeast *S. cerevisiae* generally show a layered structure. Electron micrographs of the cell walls of mycelial species of the Ascomycotina generally reveal a similar organization both in the hyphal tip and in the subapical regions, and an apparent wall thickness of about 40–70 nm (Hunsley and Burnett 1970; Barran et al. 1975; Trinci and Collinge 1975; Hunsley and Kay 1976; Heath 1994; Kurtz et al. 1994; Schoffelmeer et al. 1999). Binding of Concanavalin A (Con A), which recognizes α-linked mannose and glucose residues in sugar sequences, is widespread among fungi. Barkai-Golan et al. (1978) found that except for melanin-forming species most of the mycelial species they tested from the Ascomycotina, the Basidiomycotina, and the Zygomycota, reacted with Con A. This suggests that these fungal cell walls generally possess an external glycoprotein layer. Binding of wheat germ agglutinin (WGA), which recognizes N-acetylglucosamine sequences, is also widespread, but seems to be limited to hyphal tips at least in the Ascomycotina (Mirelman et al. 1975; Barkai-Golan et al. 1978; Von Sengbusch et al. 1983). This indicates that chitin in the hyphal tips is more accessible to WGA and further indicates that composition and organization of the apical wall differs from the mature wall.

Table 1. Covalently bound cell wall proteins in *S. cerevisiae*. The C-terminal region of Pir-CWPs has four cysteine residues at invariant positions ($CX_{66}CX_{16}CX_{12}C$) and is highly conserved. *GPI* GPI anchor addition signal; *Pir* protein with internal repeats; *SP* signal peptide

Class	Characteristics	References
GPI-CWPs[a]	SP...repeats...GPI	Caro et al. (1997); Hamada et al. (1998, 1999)
Pir-CWPs	SP...repeats...$CX_{66}CX_{16}CX_{12}C$	Kapteyn et al. (1999b); Costanzo et al. (2000)

[a] Not all GPI-CWPs contain repeats.

Howard and Aist (1979) have studied the ultrastructure of *Fusarium* hyphal tips using freeze-substitution. They found that the cell wall of *Fusarium* becomes slightly thicker in the subapical region. This has also been observed in *Aspergillus flavus* (Kurtz et al. 1994) and in *Neurospora crassa* (Hunsley and Kay 1976), and is consistent with the incorporation pattern of glucose seen in an autoradiographic study (Gooday 1971), namely strong incorporation in the hyphal tip, but also considerable incorporation in the subapical region. It implies that, in a growing hypha, cell wall construction is a carefully orchestrated process in both space and in time. This observation is also consistent with the finding of various types of secretory vesicles in the hyphal tip (Howard and Aist 1979).

IV. Fungal Cell Wall Proteins

Cell wall proteins form an external layer covering the skeletal layer and thus largely determine the surface properties of the cell. Here, we find proteins involved in hyphal aggregation or anastomosis or involved in recognition of sexual partners or host cells. The protein layer also limits the permeability of the cell wall for foreign compounds such as cell wall degrading enzymes and thus may play a protective role (De Nobel et al. 1990; De Nobel and Barnett 1991). Possibly, cell wall proteins, which in general are heavily glycosylated and thus heavily hydrated, may also function as 'lubricant' to a growing hypha when it burrows through soil or wood. Finally, they also determine the antigenic properties of the cell. The walls of aerial hyphae form a special case. They may become covered with a special class of proteins, designated as hydrophobins, to allow them to grow into the air (Wösten et al. 1999).

A. Identification and Isolation

Isolated fungal cell walls are generally heavily contaminated with proteins. These either are from adherent membrane fragments or become adsorbed to the cell wall during homogenization. For example, extraction of isolated walls of *S. cerevisiae* with hot sodium dodecyl sulfate (SDS) releases about 60 low-molecular-weight proteins (Valentin et al. 1984); similar results have been obtained with *Candida* walls (Molloy et al. 1989).

To identify genuine cell wall associated proteins, two methods have been developed based on preferential labeling of surface proteins of intact cells with either [^{125}I] (Molloy et al. 1989) or a negatively charged and thus membrane-impermeable biotinylation reagent (Casanova et al. 1992; Mrsa et al. 1997; Cappellaro et al. 1998). Using these methods, only nine protein bands were identified in SDS extracts of isolated walls of *S. cerevisiae* (Molloy et al. 1989; Cappellaro et al. 1998). Both labeling methods can of course also be used to detect cell wall proteins that are resistant to extraction with hot SDS and can only be released by other means.

Several methods are available to liberate covalently linked cell wall proteins, such as enzymatic digestion of the cell wall using pure β1,3- or β1,6-glucanase or chitinase (Klis et al. 1998), extraction with mild alkali (Mrsa et al. 1997; Kapteyn et al. 1999b), which releases Pir-like proteins, or aqueous hydrogen fluoride (HF), which specifically cleaves phosphodiester bonds thereby releasing glycosylphosphatidylinositol (GPI)-dependent cell wall proteins (Kapteyn et al. 1996; Schoffelmeer 1999). Biotinylation of cell wall proteins also makes it relatively easy to purify them as a group for further study after their release from the cell wall. Finally, hydrophobins may be released from the cell wall by using formic acid, which disrupts hydrophobic interactions (Wösten et al. 1999).

B. Glycosylation

Classical fractionation procedures of the fungal wall often start with alkali extraction followed by neutralization. The polysaccharides found in the alkali-extractable, water-soluble fraction are presumably in many cases protein-bound. Unfortunately, often no attempt is made to establish whether this is indeed the case or not. In addition, well-defined chemical and enzymatic methods such as beta-elimination for the isolation of *O*-chains and enzymatic release of *N*-chains are only rarely used.

One of the main features of the secretory pathway is the synthesis and glycosylation of cell surface proteins and their transport to the cell surface. During their passage through the secretory pathway, they usually receive *N*-chains and/or *O*-chains, whereas a subset of secretory proteins may also receive a GPI-anchor (for an up-to-date overview of all main aspects of glycosylation in *S. cerevisiae*, see Orlean and Krag 1999). *N*-Glycosylation in yeasts and presumably also in mycelial fungi follows similar lines to those in mammalian cells, particularly in the endoplasmic reticulum (Gemmill and Trimble 1999). Potential *N*-glycosylation sites (Asn-Xxx-Ser/Thr) may receive a preassembled oligosaccharide (Glc$_3$-Man$_9$GlcNAc$_2$) in the endoplasmic reticulum, resulting in the formation of Glc$_3$Man$_9$GlcNAc$_2$-Asn. A highly conserved protein complex, the oligosaccharyl transferase complex (Knauer and Lehle 1999), mediates this step. The added *N*-chains are immediately trimmed in the endoplasmic reticulum (Gemmill and Trimble 1999; Herscovics 1999), forming a core-chain. *N*-Glycosylation in the Golgi is vastly different from mammalian cells. In the Golgi of fungal cells, the core-chains are often extended with an outer chain of considerable length (Dean 1999). In the older literature, mature *N*-chains are often considered to represent an independent carbohydrate polymer designated as 'mannan' or 'galactomannan', etc. In *S. cerevisiae*, the outer chain has an α1,6-linked mannose backbone to which short mannose side chains are attached (Dean 1999). In the protein-bound galactomannan fraction of the cell walls of *N. crassa* a similar α1,6-linked mannose backbone has been found decorated with terminal galactofuranosyl units (Nakajima et al. 1984; Leal et al. 1996).

O-Glycosylation in ascomycotinous yeasts begins in the endoplasmic reticulum with the transfer of a single mannose residue from Dol-P-Man to serine and threonine residues (Strahl-Bolsinger et al. 1999). A family of protein mannosyl-transferases mediates this step. Homologous genes have been found in the higher plant *Arabidopsis thaliana* and in mammalian cells, suggesting that this step is possibly not unique to fungi as originally thought. *O*-Chains are extended in the Golgi, but compared to their counterparts, the *N*-chains, they remain short and have a relatively simple structure at least in yeasts (Gemmill and Trimble 1999). In mycelial fungi, however, both short and long *O*-linked carbohydrate chains have been observed (Nakajima et al. 1984; Jikibara et al. 1992). Interestingly, hydroxy amino acids are often concentrated in specific domains of cell wall proteins. Extensive decoration with short *O*-chains probably confers on them a highly extended, semi-rodlike character (Jentoft 1990; Williamson et al. 1992).

C. Glycosylphosphatidylinositol-Dependent Cell Wall Proteins (GPI-CWPs)

Anchoring of secreted proteins into the external face of the plasma membrane through glycosylphosphatidylinositol (GPI) anchors is common among all eukaryotes. In the yeast *S. cerevisiae*, however, it has been shown that the GPI anchor in a subset of these proteins is processed, resulting in a covalent attachment of these proteins to the cell wall β-glucan by way of a remnant of the GPI anchor (De Nobel and Lipke 1994; Lu et al. 1994; Kollár et al. 1997; Fujii et al. 1999). GPI proteins enter the endoplasmic reticulum where they receive a preassembled GPI anchor that replaces the carboxy-terminal GPI-anchor addition signal at the C-terminus and links them to the lumenal side of the membrane. One group of GPI proteins ends up in the plasma membrane whereas other GPI proteins are incorporated in the cell wall (GPI-CWPs). Genomic analysis in *S. cerevisiae* has shown that there are about 20 GPI proteins that when expressed remain associated with the plasma membrane, whereas about 40 GPI proteins are released from the plasma membrane to become incorporated into the external protein layer of the cell wall (Caro et al. 1997; Hamada et al. 1998, 1999). Likewise, genes encoding cell wall proteins containing a putative GPI-anchor addition signal have been cloned from the human pathogens *Candida albicans* (Bailey et al. 1996; Gaur and Klotz 1997; Hoyer et al. 1998a,b; Staab and Sundstrom 1998), *C. glabrata* (Cormack et al. 1999), *Coccidioides immitis* (Zhu et al. 1996; Dugger et al. 1996) and *Penicillium marneffii* (Cao et al. 1998) as well as from the plant pathogens *Fusarium oxysporum* (Schoffelmeer 1999) and *Glomerella cingulata* (Hwang and Kolattukudy 1995; Table 2). The GPI-CWPs Ala1 (Als5) and Hwp1 from *C. albicans* and Epa1 from *C. glabrata* deserve extra attention because these proteins are responsible for adherence of these fungal pathogens to epithelial cells of the host, thus identifying these GPI-CWPs as potential antifungal targets (Gaur and Klotz 1997; Cormack et al. 1999; Staab et al. 1999).

The proteins present in the external protein layer largely determine the surface properties of a fungal cell. The introduction of foreign proteins in this layer is therefore an obvious means to change the surface properties of fungal cells. In *S. cerevisiae*, fusion proteins consisting of the desired protein extended with a C-terminal part of known

Table 2. Putative GPI-CWPs in the Ascomycotina[a]

Organism	Gene	Accession number	Reference(s)
Candida albicans	*ALS1*	P46590	Hoyer et al. (1995)
	ALS2	O74657	Hoyer et al. (1998a)
	ALS3	O74623	Hoyer et al. (1998b)
	ALS4	O74660	Hoyer et al. (1998a)
	ALA1/ALS5	O13368	Gaur and Klotz (1997)
	ALS6	AAD42033.1	Hoyer and Hecht (2000)
	ALS7	AAD02580	Hoyer and Hecht (2000)
	CHT2	P40953	McCreath et al. (1995)
	HWP1	P46593	Staab et al. (1996); Staab and Sundstrom (1998)
	HYR1	P46591	Bailey et al. (1996)
Candida glabrata	*EPA1*	AAD34013	Cormack et al. (1999)
Coccidioides immitis	*Ag2*	U32518	Dugger et al. (1996); Zhu et al. (1996)
Fusarium oxysporum	*FEM1*		Schoffelmeer (1999)
Glomerella cingulata	*CAP22*	Q00371	Hwang and Kolattukudy (1995)
Penicillium marneffii	*MP1*	122964	Cao et al. (1998)
Trichoderma harzianum	QID3	P52755	Lora et al. (1994)

[a] GPI-CWPs from *S. cerevisiae* are discussed in Caro et al. (1997), Hamada et al. (1998, 1999), and Smits et al. (1999).

GPI-dependent cell wall proteins have been used to achieve this (Schreuder et al. 1996). As GPI-CWPs seem to be widespread among the Ascomycotina (Table 2), it seems likely that genetic engineering of the cell surface properties of mycelial fungi may be achieved in the same way as in yeast.

D. Pir-Like Cell Wall Proteins (Pir-CWPs)

Homologues of *S. cerevisiae* Pir-CWPs (Pir = protein with internal repeats) have been identified in *C. albicans* by a homology search, cross-hybridization and immunologically (Kapteyn et al. 2000), in *Kluyveromyces lactis* and *Zygosaccharomyces rouxii* using cross-hybridization (Toh-e et al. 1993), and in *Torulopsis delbrueckii*, *K. marxianus* and *Schizosaccharomyces pombe* using Pir2 antibodies (Russo et al. 1992). Cell wall association has only been shown for *S. cerevisiae*, *C. albicans* and *S. pombe* (Kapteyn et al. 1999b; Russo et al. 1992). As discussed below, the cell wall of *Aspergillus fumigatus* does seem to lack β1,6-glucan. This might mean that all cell wall proteins in this species and presumably other mycelial species might be linked to the β1,3-glucan network in a Pir-like manner. Alternatively, other, as yet unidentified, means of linking cell wall proteins to the β1,3-glucan network might come into play.

E. Hydrophobins

Hydrophobins are small, secretory proteins that cover aerial hyphae, conidia, and fruiting bodies with a hydrophobic layer that may mediate attachment of fungi to hydrophobic surfaces on their hosts (see Chap. 7, this Vol.). They are approximately 100 amino acids long and contain eight cysteine residues arranged in a characteristic pattern (Wessels 1996, 1997; Talbot 1997). Several hydrophobins are now recognized and they may well prove to be ubiquitous in mycelial fungi (Wessels 1997). They seem, however, to be absent in *Saccharomyces*, *Candida* and *Schizosaccharomyces* (J.H. Sietsma, unpubl.).

The most characteristic feature of hydrophobins is their capacity to aggregate into insoluble amphipathic membranes at hydrophilic-hydrophobic interfaces such as the interface between water and air (Wösten et al. 1993). Such membranes are 10 nm thick, and present a rodlet structure at the hydrophobic surface but appear smooth at the hydrophilic surface. Similar rodlets are seen at the surface of aerial hyphae, conidia and fruiting bodies (Stringer et al. 1991; Bell-Pedersen et al. 1992; Wösten et al. 1993). The best-studied hydrophobin is SC3 produced by the basidiomycete *Schizophyllum commune*. Submerged hyphae secrete SC3 at their growing tips as soluble monomers into the medium, where they aggregate at the water-air interface of the medium thus reducing surface tension and allowing hyphal tips to breach the surface. When hyphae grow into the air, newly secreted SC3 aggregates at the interface of the hydrophilic hyphal wall and air, thus covering the hypha with a hydrophobic layer. This layer is located at the outside of the wall and hydrophobic interactions are involved in binding it to the wall (Wösten et al. 1994).

Self-assembly of hydrophobins between the hyphal wall and a hydrophobic surface may also facilitate the attachment of fungi to their host, as judged by the abundant expression of the hydrophobin genes *ssgA* in the insect-pathogenic fungus *Metarhizium anisopliae* (St. Leger et al. 1992), of *mpg1* during appressoria formation by the rice blast fungus *Magnaporthe grisea* (Talbot et al. 1993), and when the mycorrhizal fungus *Pisolithus tinctorius* attaches itself to the roots of the eucalyptus tree (Martin et al. 1999).

V. Carbohydrate Polymers in the Cell Wall

As we have seen, the cell wall of the budding yeast *S. cerevisiae* contains only four types of macromolecules, namely mannoproteins, β1,6-glucan,

β1,3-glucan and chitin. This also holds for *C. albicans* (Shepherd 1987) and presumably for other yeast species as well (Villa et al. 1980; Nguyen et al. 1998). However, in fission yeast and in the mycelial fungi of the Ascomycotina, additional carbohydrate polymers have been identified (Table 3; see also Fleet and Phaff 1981).

A. Chitin

Chitin is a characteristic component of the cell wall in the fungal kingdom. In hyphal walls of the Ascomycotina, chitin and β1,3-glucan form a molecular network that is located at the inside of cell wall and is masked by amorphous (proteinaceous) material (Hunsley and Kay 1976). Autoradiographic studies have convincingly shown that the incorporation of radioactively labeled N-acetylglucosamine takes place mainly at the hyphal tip and sharply declines directly afterwards (Gooday 1971). Glucose is also largely incorporated in the hyphal tip, but considerable labeling is also observed in adjacent subapical regions. In other words, in mycelial fungi, the incorporation of chitin into the cell wall and, to a lesser extent, β1,3-glucan also (and α1,3-glucan) seems to be

Table 3. Carbohydrate polymers in cell walls of the Ascomycotina. The polymers are not necessarily found together in a single cell wall; for example, the cell wall of *S. cerevisiae* contains only chitin, β1,3-glucan, β1,6-glucan and protein-bound mannan. This table is not meant to be complete: carbohydrate polymers that seem to be restricted to a limited number of species (e.g., nigeran, or capsular polysaccharides like cinerean) are not included

Component	Comments
β1,3-Glucan	Synthesized at the plasma membrane (Orlean 1997); present in hyphal tips; forms a three-dimensional network kept together by hydrogen bridges (Smits et al. 1999); the mature form is generally branched (Manner et al. 1973a; Fontaine et al. 2000); skeletal element
Chitin	Synthesized at the plasma membrane (Orlean 1997); present in hyphal tips; may become linked to the β1,3-glucan network (Kollár et al. 1995; Fontaine et al. 2000) and to β1,6-glucan (Kollár et al. 1997); skeletal element
β1,6-Glucan	Presumably mainly synthesized at the plasma membrane (Montijn et al. 1999); highly branched molecule (Manners et al. 1973b); described in *S. cerevisiae* and *C. albicans*; presence in mycelial species uncertain (Fontaine et al. 2000); covalently linked to β1,3-glucan and chitin (Kapteyn et al. 1996; Kollár et al. 1997)
β1,3/1,4-Glucan	Site of synthesis unknown; covalently linked to β1,3-glucan in *A. fumigatus* (Fontaine et al. 2000)
α1,3-Glucan	Site of synthesis presumably at the plasma membrane (Hochstenbach et al. 1998; Katayama et al. 1999); absent from *S. cerevisiae*, but present in the fission yeast *S. pombe* and in many mycelial species (Wessels and Sietsma 1981; Hearn and Sietsma 1994; Schoffelmeer et al. 1999); presence in hyphal tips unknown; skeletal element in *S. pombe* (Hochstenbach et al. 1998), but not in *A. nidulans* (Polachek and Rosenberger 1977)
Galactosamine polymer	Site of synthesis unknown; found in *A. fumigatus* (Fontaine et al. 2000)
(Galacto)mannan	Protein-bound; synthesized in the secretory pathway (Orlean 1997)
Acidic polysaccharides	Protein-bound (Jikibara et al. 1992); *Fusarium* hyphal walls may contain 10% of uronic acids (Schoffelmeer et al. 1999)

limited to the hyphal tips (Gooday 1971). In *S. cerevisiae*, however, chitin incorporation into the lateral walls is delayed until after cytokinesis (Schekman and Brawley 1979; Shaw et al. 1991).

Chitin contributes significantly to the mechanical strength of the cell wall, particularly in mycelial fungi, which generally contain much higher chitin levels than normally found in yeasts. When chitin synthesis is affected, the hyphae form pronounced bulges and tend to lyse unless the osmolarity of the medium is increased (Benitez et al. 1976; Gooday 1990; Bago et al. 1996; Specht et al. 1996). Similarly, inhibition of postsynthetic steps such as crystallization of chitin or coupling of chitin to other cell wall polymers either by dyes (Roncero and Duran 1985) or by chitin-binding proteins may have a deleterious effect on growth (Mirelman et al. 1975; Broekaert et al. 1989). In *S. cerevisiae*, chitin also seems to be used to shore up the wall in case of damage (Popolo et al. 1997; Ram et al. 1998; Osmond et al. 1999). Chitin levels may rise to about 20% of the wall dry weight (Dallies et al. 1998; Kapteyn et al. 1999a,b). The mechanism behind cell wall stress-related chitin synthesis is still unknown, but might be generally used by fungi.

Extensive studies of chitin synthesis (*CHS*) enzymes and their genes in yeast have provided the basis for analyzing the corresponding genes in mycelial fungi. In agreement with their increased developmental complexity, mycelial fungi seem to use more chitin synthases than yeasts. Based on structural homologies, chitin synthases are currently classified into five classes (Bowen et al. 1992; Bulawa 1993; Horiuchi et al. 1999; Park et al. 1999). In *S. cerevisiae* and *C. albicans* three structural genes for chitin synthase have been detected (Bulawa 1993; Mio et al. 1996). The best-studied mycelial *CHS* genes are the five genes found in *N. crassa* and *A. nidulans* and the seven genes in the human pathogen *A. fumigatus* (Borgia et al. 1996). In the basidiomycete *Ustilago maydis*, six *CHS* genes have been identified (Xoconostle-Cazares et al. 1997).

Chitin synthases are not only found at the plasma membrane, but some of them are also found in small cytoplasmic vesicles both in *S. cerevisiae* (Martinez and Schwencke 1988; Leal-Morales et al. 1994; Chuang and Schekman 1996; Ziman et al. 1996), in the dimorphic fungus *C. albicans* (Gozalbo et al. 1987, 1992), and in mycelial fungi (Ruiz-Herrera 1992 for a review of the older literature; Sietsma et al. 1996). These vesicles seem to arise by endocytosis (Chuang and Schekman 1996; Ziman et al. 1996). As mentioned above, *S. cerevisiae* is able to incorporate massive amounts of chitin into the lateral walls in response to wall damage (Kapteyn et al. 1999a,b; Smits et al. 1999). Possibly, the presence of chitosomes in the cytosol allows the cell to rapidly respond to cell wall stress.

B. *β*1,3-Glucan

*β*1,3-Glucan is a major compound of fungal cell walls. When glucan synthesis is inhibited, growth is severely restricted (Ruiz-Herrera 1992; Kurtz et al. 1994; Douglas et al. 1997; Kurtz and Douglas 1997 for a review of the older literature; Thompson et al. 1999). *β*1,3-Glucan may crystallize in the form of triple helices displaying a characteristic X-ray diffraction pattern (Jelsma and Kreger 1975; Marchessault and Deslandes 1979). However, in its mature form, *β*1,3-glucan is branched with *β*1,6-linkages at the branching points (Manners et al. 1973a; Fontaine et al. 2000), which interfere with crystallization. Thus, in the untreated cell wall, an X-ray diffraction pattern characteristic of *β*1,3-glucan is seldom seen; only after treatment of the wall with dilute acid, which probably hydrolyses the branching points, does this diffraction pattern appear (Kreger 1954). In addition, in the mature wall, chitin chains and other macromolecules are linked to the *β*1,3-glucan which may also prevent crystallization (Sietsma and Wessels 1981; Mol and Wessels 1987; Kollár et al. 1995; Kapteyn et al. 1999b; Fontaine et al. 2000). In *S. cerevisiae*, the reducing end of the chitin chains is linked to the nonreducing ends of the *β*1,3-glucan (Fig. 1). Recently, a similar linkage type was found in *A. fumigatus* (Fontaine et al. 2000), indicating that this type of linkage probably also occurs in mycelial fungi. In *C. albicans*, however, a linkage between the reducing end of the glucan with C6 of an *N*-acetylglucosamine residue in the chitin chains has been proposed (Surarit et al. 1988). There is strong evidence in yeast that other wall components such as *β*1,6-glucan and glycoproteins are also linked to *β*1,3-glucan (Kollár et al. 1997; Smits et al. 1999).

It has been well documented that *β*1,3-glucan synthesis occurs at the cytoplasmic membrane by an integral membrane protein that accepts UDP-glucose at the cytoplasmic site and extrudes a linear β1,3-glucan at the outside (Shematek et al. 1980). After synthesis, branching enzymes in the wall process the chain and other wall components are coupled to it. In *S. cerevisiae*, two genes (*FKS1* and *FKS2*) have been identified that are believed to encode the catalytic subunit for β1,3-glucan synthase (Orlean 1997). A third *FKS* homologue in *S. cerevisiae*, which was discovered as a result of the genome sequencing effort, seems dispensable. *FKS* homologues have been identified in *C. albicans* (*FKS1/GSC1*, *GSL1* and *GSL2*, Douglas et al. 1997; Mio et al. 1997), in *S. pombe* (cps1+, Ishiguro et al. 1997), in the mycelial species *A. nidulans* (Kelly et al. 1996) and *A. fumigatus* (GenBank Accession no. U79728), and in the basidiomycete *Cryptococcus neoformans* (Thompson et al. 1999).

β1,3-Glucan synthase activity is regulated by a GTP-binding protein identified as the essential Ras-homologue Rho1p, a member of a family of GTP-ases (Drgonova et al. 1996; Mazur and Baginsky 1996; Qadota et al. 1996; Cabib et al. 1998). This mechanism for regulation of β1,3-glucan synthase is also found in *C. albicans* (Kondoh et al. 1997) and *S. pombe* (Arellano et al. 1996), and is probably general in fungi (Szaniszlo et al. 1985; Tanaka et al. 1999). Intriguingly, Rho1p has other functions as regulator of cell cycle events and morphogenesis (Cabib et al. 1998).

C. α1,3-Glucan

Although the cell walls of the budding yeast *S. cerevisiae* and the dimorphic fungus *C. albicans* do not contain α1,3-glucan, it is present as a prominent alkali-soluble component in the walls of many species from the Ascomycotina and the Basidiomycotina. It is always in a crystalline form with an easily recognizable X-ray diffraction pattern (Bacon et al. 1968). Cytochemical studies of *S. pombe* and *S. commune* have shown that α1,3-glucan is predominantly located in the outside regions of the wall (Van der Valk et al. 1977; Horisberger and Rouvet-Vauthey 1985).

It is not yet known if α1,3-glucan becomes linked to other cell wall components after being deposited in the cell wall. Recently, two groups have independently identified a gene (ags1/mok1) in the fission yeast *S. pombe* that is required for the synthesis of α1,3-glucan and probably encodes the α1,3-glucan synthase (Hochstenbach et al. 1998; Katayama et al. 1999). Interestingly, the predicted protein is probably not only responsible

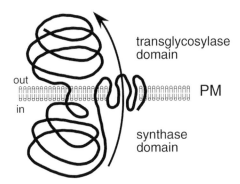

Fig. 2. Putative domain structure of the aga1+ gene in *S. pombe* (Hochstenbach et al. 1998). The N-terminal part of the protein is predicted to be located outside of the plasma membrane and to be involved in remodeling α1,3-glucan. The large intracellular loop is proposed to be responsible for the synthesis of α1,3-glucan, and the C-terminal part of the protein is believed to be involved in the export of α1,3-glucan. *PM* Plasma membrane. (Reprinted from Hochstenbach et al. 1998; copyright 1998 National Academy of Sciences, USA)

for the synthesis of α1,3-glucan, but might also contain a pore-like domain for exporting α1,3-glucan and an external domain involved in remodeling α1,3-glucan (Fig. 2). These have been designated as ags1 and mok1, respectively. Two more genes were detected but they do not seem to be expressed and may represent silent copies. Loss of ags1/mok1 function results in cell lysis, which is partially remediable by increasing the osmolarity of the medium, whereas partial loss of function results in spherical cells with reduced amounts of α1,3-glucan in their walls. These observations strongly indicate that in *S. pombe* α1,3-glucan is essential for the mechanical strength of the cell wall. This finding may not be generalized. In the mycelial species *A. nidulans*, α1,3-glucan is not essential to the mechanical strength or shape of the hyphal wall (Polachek and Rosenberger 1977). Interestingly, in *A. nidulans*, it may even be reused when sexual fruiting structures (cleistothecia) are formed (Zonneveld 1972). Possibly, in *S. pombe*, α1,3-glucan takes over the function of chitin because, in contrast to *A. nidulans*, the cell walls of *S. pombe* have very low chitin levels (Sietsma and Wessels 1990).

α1,3-Glucan is also found in several medically important fungi such as *A. fumigatus*, *Blastomyces dermatitidis*, *Coccidioides immitis*, *Histoplasma capsulatum*, and *Paracoccidioides brasiliensis* (Domer et al. 1967; Bacon et al. 1968; Kanetsuna and Carbonell 1970, 1971; Hearn and Sietsma 1994). Interestingly, for several of these species a correlation between virulence and the levels of α1,3-glucan in the cell wall has been noted (San Blas and San Blas 1977; Klimpel and Goldman 1988; Hogan and Klein 1994).

D. β1,6-Glucan

β1,6-Glucan plays an important role in the construction of the cell wall of *S. cerevisiae* and the dimorphic fungus *C. albicans*, where it interconnects GPI-CWPs to the β1,3-glucan network (Kapteyn et al. 1996, 2000; Kollár et al. 1997), and probably also in other yeast species such as *Pichia anomala*, *Nadsonia elongata*, and *Ashbya gossypii* (Tanaka and Watanabe 1995) and in the pleiomorphic pathogen *Exophiala dermatitidis* (Montijn et al. 1997). Recently, evidence has emerged indicating that in *S. cerevisiae* the bulk of β1,6-glucan synthesis takes place at the plasma membrane (Montijn et al. 1999). It is uncertain if all mycelial fungi of the Ascomycotina produce β1,6-glucan. Although in some mycelial species such as *A. niger* (Brul et al. 1997) and *F. oxysporum* (Schoffelmeer et al. 1999) cell wall proteins released by β1,3-glucanase react strongly with β1,6-glucan antibodies, the walls of *A. fumigatus* seem to lack β1,6-glucan (Fontaine et al. 2000) and their cell wall proteins react only weakly with β1,6-glucan antibodies (A.W.R.H. Teunissen and F.M. Klis, unpubl. data). This raises the interesting question whether GPI-dependent cell wall proteins in some mycelial species may be connected to the β1,3-glucan network in a different way either directly or through another carbohydrate polymer such as β1,3/1,4-glucan (Table 3). Finally, in the basidiomycete *S. commune*, there is evidence for the presence of β1,6-glucan in the cell wall (Sietsma and Wessels 1977).

VI. Remodeling and Coupling Enzymes

As shown above for *S. cerevisiae*, all the individual components of the cell wall may be covalently interconnected forming a supramolecular complex. As chitin and β1,3-glucan are synthesized by individual enzyme complexes located in the plasma membrane, and as GPI-CWPs are transported through the secretory pathway to the cell surface where they become linked to β1,6-glucan, it is clear that several extracellular cell wall

construction enzymes are needed to interconnect the various macromolecules. In addition, remodeling of cell wall β-glucans has to take place to introduce branches in otherwise linear structures that subsequently may serve as acceptor sites for the addition of other polymers (Manners et al. 1973a; Kollár et al. 1995, 1997; Fontaine et al. 2000). One may also predict a role for remodeling enzymes to allow insertion of new cell wall material in the growing bud of yeasts (see Sect. VII.A) or in the subapical region of a growing hypha. Finally, remodeling enzymes are probably needed during cell fusion, spore germination and hyphal branching. Detailed knowledge about remodeling and coupling enzymes is, however, limited. Some known transglycosylases are Bgl2p/Bgt1p and Gas1p, but this list is likely to expand in the near future.

A. β1,3-Glucan-Processing Enzymes

Bgl2p/Bgt1p is a 29-kDa β1,3-glucan-processing enzyme that is strongly associated with the cell wall of *S. cerevisiae*. In vitro, it binds tightly to β1,3-glucan and chitin (Klebl and Tanner 1989). Antibodies raised against Bgl2p cross-react with a cell wall protein of similar size from *Hansenula wingei*, *Torulopsis glabrata*, *Pichia guilliermondii* and *K. lactis*, suggesting the presence of Bgl2-like proteins in these yeasts as well (Herrero and Boyd 1986; Herrero et al. 1987; Klebl and Tanner 1989). Bgl2p homologues with similar transferase activity have also been isolated from cell walls of *C. albicans* (Hartland et al. 1991; Yu et al. 1993) and the mycelial species *A. fumigatus* (Hartland et al. 1996).

The precise function of Bgl2p is still unknown. No distinct phenotype is associated with the loss of Bgl2p in *S. cerevisiae* (Klebl and Tanner 1989) and *A. fumigatus* (Mouyna et al. 1998). It acts as a glucanase at low concentrations of acceptor molecules, but, at higher concentrations, it transfers the newly generated reducing end to the nonreducing end of another β1,3-glucan forming a new intrachain β1,6-linkage (Goldman et al. 1995). The intrachain β1,6-linkage results in a segmented chain. Bgl2p requires a β1,3-glucan-chain with a free reducing end on which to act (Hartland et al. 1991; Yu et al. 1993; Goldman et al. 1995). In this context the question becomes relevant whether a β1,3-glucan-chain emerges from the synthase complex in the plasma membrane with its reducing end first or the other way around. If the reducing end appears first, as has been shown for cellulose chains (Koyama et al. 1997), Bgl2p/Bgt1p might be involved in covalently linking a new β1,3-glucan-chain to the existing β1,3-glucan network.

Two more β1,3-glucan-processing enzymes with transglycosylase activity have been biochemically characterized, namely Gel1p and Bgt2p, both from *A. fumigatus*. Both act as endoglucanases without requiring a free reducing end for their action and transfer the newly generated reducing end to the nonreducing end of another β1,3-glucan chain through a β1,3-linkage thus resulting in the elongation of β1,3-glucan chains (Hartland et al. 1996; Mouyna et al. 1999). Gel1p belongs to the GAS1 family, originally identified in *S. cerevisiae* (Popolo and Vai 1999). Gas1p is the most abundant GPI protein in yeast and is attached through a GPI anchor to the external face of the plasma membrane. Although Gas1p is not essential for cell viability, loss of Gas1p leads to a variety of phenotypes such as hypersensitivity to the cell wall disturbing compound Calcofluor white, swollen round cells, and a decrease in β-glucan levels (Ram et al. 1998). *GAS1*-related sequences (*PHR1* and *PHR2*) have further been found in *C. albicans* (Saporito-Irwin et al. 1995; Mühlschlegel and Fonzi 1997) and in *C. maltosa* (Nakazawa et al. 1998). Recently, Fonzi (1999) has proposed that Phr1p and Phr2p create acceptor sites in the β1,3-glucan network for the attachment of β1,6-glucan molecules.

B. β1,6-Glucan-Processing Enzymes

The *S. cerevisiae* gene *KRE1* encodes a GPI-anchored plasma membrane protein (Roemer and Bussey 1995) that seems to be involved in the elongation of β1,6-glucan. Cells lacking Kre1p produce a β1,6-glucan with a smaller average polymer size resulting in cell walls that contain about 40% less β1,6-glucan than walls of wild-type cells (Boone et al. 1990). A *KRE1* homologue isolated from *C. albicans* complements the *S. cerevisiae kre1* deletant, suggesting conservation of the *KRE1* function in *C. albicans* (Boone et al. 1991). Kre9p is a secretory protein that is also required for β1,6-glucan of normal length. Loss of *KRE9* alone results in a much reduced polymer size and an 80% reduction of β1,6-glucan levels in the wall (Brown and Bussey 1993). The *kre9* deletant is

partially rescued by the activity of a Kre9 homologue, Knh1p (Dijkgraaf et al. 1996). Loss of both homologues is lethal in *S. cerevisiae* (Dijkgraaf et al. 1996). The simultaneous loss of Kre1p and Kre9p is also lethal (Boone et al. 1990; Brown and Bussey 1993), stressing the importance of β1,6-glucan synthesis to cell wall construction. Interestingly, synthesis of β1,6-glucan is also essential in two human pathogenic fungi. The *KRE9* homologue in *C. albicans* is essential for growth on glucose and for induction of hyphal formation when grown in serum (Lussier et al. 1998). In addition, *C. glabrata* contains homologues of *KRE9* and *KNH1*, and gives similar phenotypes upon loss of these β1,6-glucan-processing enzymes, as found in *S. cerevisiae* (Nagahashi et al. 1998).

VII. Dynamics of the Fungal Cell Wall

A. Yeast and Hyphal Mode of Growth

Although on first sight the yeast and hyphal modes of growth seem to differ greatly, the underlying mechanisms show clear similarities (see Chap. 10, Sect. IV.A; Chap. 2, this Vol.). Staebell and Sol (1985) and later Rico et al. (1991) studied cell wall expansion in the dimorphic fungus *C. albicans* and found that, in both modes of growth, surface expansion involves an apical growth zone. However, whereas in hyphae the contribution of the apical growth zone to surface expansion predominates as long as the hyphae continue to grow, in yeast cells an apical growth zone is observed only in small- to medium-sized buds, but not in larger buds (Staebell and Sol 1985). In larger buds, the cell wall expands by insertion of new material over the entire surface. Subsequently, a new apical site appears elsewhere in both the mother and the daughter cell. Rigidification of the yeast cell wall begins only after cytokinesis has taken place by deposition of chitin in the walls (Schekman and Brawley 1979; Shaw et al. 1991) and by coupling it to β1,3-glucan (Kollár et al. 1997).

B. Maturation of the Cell Wall

In the Basidiomycotina, maturation and rigidification of the cell wall behind the growing hyphal tip have been extensively analyzed in *S. commune* (Wessels and Sietsma 1981; Sietsma and Wessels 1994; Wessels 1994). These data, which relate only to the skeletal polysaccharides chitin and β1,3-glucan and not to α1,3-glucan or cell wall proteins, are consistent with the following sequence of events. Both β1,3-glucan and chitin are synthesized as individual linear chains at the growing tip. Subsequently, the β1,3-glucan chains become branched, as reflected in an increased number of 1,6-linkages and resulting in an increased number of potential acceptor sites. Finally, chitin and β1,3-glucan molecules are covalently connected, forming the alkali-insoluble β-glucan-chitin complex. As the skeletal network of the cell wall of both the Ascomycotina and the Basidiomycotina generally consists of a chitin-β1,3-glucan network, it is tempting to extend this concept to both taxonomic groups.

In *N. crassa*, secondary wall deposition has been observed in the older regions of the hyphae (Trinci and Collinge 1975). For example, in undifferentiated mycelia, the wall in the hyphal tip and in the adjacent subapical region is about 65 μm thick. However, from between 10 to about 200 μm behind the hyphal tip, the wall becomes progressively thicker, as if the original wall is covered by a new layer of wall material until a final thickness of about 120–140 μm is reached. The formation of an additional wall layer in the older parts of the hyphae might be related to the changing environmental conditions from the periphery of a colony to the center of the colony when growing on a solid surface (Trinci et al. 1994). It is not certain whether this is a general phenomenon, because in *Geotrichum candidum* a similar increase in thickness was not observed. Nevertheless, it illustrates the importance of choosing the right growth conditions for biochemical studies.

C. Compensation Mechanisms

1. Detection of Cell Wall Defects

Cell wall mutants and, in general, cells with a weakened cell wall can be easily identified using Calcofluor or Congo Red. These are related anionic dyes that preferentially bind to hexopyranose polymers in the β-configuration, such as cellulose and chitin and, to a lesser extent, also β1,3-glucan (Maeda and Ishida 1967; Wood 1980). Because both dyes possess two sulfonic acid groups, they do not pass the cell membrane unless the external pH drops to a level at which the sul-

fonic acid groups lose their charge. Their inhibitory effect on fungal growth (Elorza et al. 1983; Roncero and Duran 1985; Fèvre et al. 1990) seems therefore due to their perturbing action on cell wall construction, probably by preventing the crystallization of chitin and cellulose and perhaps also β1,3-glucan, thereby weakening the cell wall (Elorza et al. 1983; Vermeulen and Wessels 1986; Fèvre et al. 1990; Kopecka and Gabriel 1992; Bartnicki-Garcia et al. 1994). Consistent with this explanation, the growth-inhibitory effects of Calcofluor White and Congo Red can be counteracted by increasing the osmolarity of the medium (Roncero and Duran 1985; Frost et al. 1995). This makes these cell wall perturbing compounds ideally suited to specifically detect cells with a defective wall because such cells show altered sensitivity to Calcofluor White and Congo Red (Ram et al. 1994; Lussier et al. 1997). In addition, both compounds are effective towards a wide range of fungi, including representatives from the Ascomycotina, Basidiomycotina, Zygomycota, and the Oomycota (Elorza et al. 1983; Roncero and Duran 1985; Fèvre et al. 1990). Interestingly, cells with low chitin levels become more resistant to Calcofluor and probably to Congo Red, facilitating the identification of this specific class of cell wall mutants (Roncero et al. 1988; Valdivieso et al. 1991). Other methods to identify cell wall defects have been reviewed elsewhere (Orlean 1997).

2. Compensatory Changes in the Cell Wall

Systematic screening of deletants of *S. cerevisiae* for alterations in cell wall composition and structure has shown that about 20% of all deletants show a cell wall related phenotype (P. de Groot, pers. comm.). This impressive number of genes reflects the fact that formation of the cell wall is an intricate part of growth and must therefore be intimately coordinated with processes such as cell cycle progression, polarity establishment, and secretion, and also indicates that its composition and structure are highly regulated. There is no reason to believe that the situation in mycelial fungi is any different.

Fungi often live in a hostile environment in which they are confronted with cell wall degrading enzymes secreted by other organisms. To some extent, fungi have learned to cope with such threats. In budding yeast, a signaling pathway, designated as the cell wall integrity pathway, operates that is activated by cell wall stress (Gustin et al. 1998; Jung and Levin 1999; Smits et al. 1999). Homologues of components of this pathway have been found in *C. albicans*, *S. pombe*, and in *A. niger* and *Trichoderma reesei* (Morawetz et al. 1996; Navarro-Garcia et al. 1998; Toda et al. 1996). Activation of the cell wall integrity pathway results in up-regulation of a group of cell wall maintenance genes (Jung and Levin 1999) and compensatory changes in cell wall structure. For example, a wide variety of cell wall mutants in *S. cerevisiae* (Popolo et al. 1997; Dallies et al. 1998; Kapteyn et al. 1999a,b) show elevated chitin levels in their walls, indicating that induction of chitin synthesis is a general response to cell wall weakening and functions as a means to compensate for cell wall damage (Ketela et al. 1999). Similarly, *C. albicans* cells treated with inhibitors of the synthesis of β1,3-glucan (Hector 1993), and cell wall mutants of *C. glabrata* (Nagahashi et al. 1998) and *A. nidulans* (Polacheck and Rosenberger 1977), have enhanced chitin levels in their walls. This leads to the prediction that cell wall mutants with higher chitin content in the lateral walls are hypersensitive to inhibitors of chitin synthase activity and that combinations of cell wall active compounds might act synergistically (Hector 1993). Indeed, *S. cerevisiae gas1Δ* cells are hypersensitive to nikkomycin (Popolo et al. 1997). Interestingly, when *A. flavus* was treated with an inhibitor of β1,3-glucan synthesis, the thickness of the cell wall increased dramatically. Whereas the apical wall of untreated hyphae had an apparent thickness of about 40 nm including a very thin electron-dense outer layer, the wall of treated hyphae had a lighter-staining inner layer of about 90 nm and a darkly staining outer layer of about 40–70 nm (Kurtz et al. 1994). Finally, the cell wall perturbants Calcofluor White, Rylux BSU, and Congo Red also cause the chitin levels in the cell wall to rise (Roncero et al. 1988; Haplová et al. 1994). Cell wall weakening is expected to lead to membrane stress. Indeed, it is tempting to predict that local membrane stress may cause localized activation of chitin synthesis (Gooday and Schofield 1995). A related example of compensatory changes in the fungal cell wall is presented by the fission yeast *S. pombe*, in which both α1,3-glucan and β1,3-glucan contribute to cell wall integrity (Hochstenbach et al. 1998). When it was treated with an inhibitor of β1,3-glucan synthesis, the production of α1,3-glucan was enhanced (Miyata et al. 1985).

VIII. Conclusions

The three best-studied species from the Ascomycotina in terms of cell wall composition and molecular architecture are currently *S. cerevisiae*, *C. albicans*, and *A. fumigatus*. We have seen that the knowledge with respect to the cell wall of *S. cerevisiae* can often successfully be extrapolated to other (mycelial) species from this taxonomic group:

1. The general organization of the cell wall seems to be similar with an external layer consisting of glycoproteins that are covalently linked to the underlying skeletal layer composed of β1,3-glucan and chitin.
2. Homologues of the genes that are involved in the synthesis of chitin and β1,3-glucan have been found.
3. Activation of β1,3-glucan synthase by the GTPase Rho1p as observed in *S. cerevisiae* seems widespread.
4. In both yeasts and mycelial fungi β1,3-glucan seems to be synthesized as linear chains, which subsequently become branched, resulting in the formation of multiple nonreducing ends that may function as acceptor sites for the addition of other cell wall polymers such as chitin and cell wall proteins.
5. Homologues of β1,3-glucan- and β1,6-glucan-remodeling enzymes have been identified.
6. GPI-dependent cell wall proteins have been found in various mycelial species.
7. The cell wall is an extremely dynamic entity. In addition, compensatory changes occur in the cell wall in response to cell wall damage, and homologues of proteins functioning in the cell wall integrity pathway have been identified in various fungi.

It is equally clear that hyphal species have their own unique features in terms of cell wall composition and possibly organization. For example, the hyphal cell wall of *A. fumigatus* and possibly many other mycelial species seems to lack β1,6-glucan, but, in contrast to the wall of *S. cerevisiae*, contains α1,3-glucan, β1,3/1,4-glucan, and polygalactosamine. Many questions still remain unanswered, but with the avalanche of genomic and proteomic data now in the pipeline and to be released in the public domain in the near future, an exciting time is awaiting us.

Acknowledgements. We would like to thank all the members in the Klis laboratory for stimulating discussions. H. de Nobel acknowledges the financial support of the Dutch Ministry of Economic Affairs and of Unilever, and F. M. Klis acknowledges the support of the EU program EUROFAN II.

References

Arellano M, Duran A, Perez P (1996) Rho 1 GTPase activates the (1–3)beta-D-glucan synthase and is involved in *Schizosaccharomyces pombe* morphogenesis. EMBO J 15:4584–4591

Baba M, Osumi M (1987) Transmission and scanning electron microscopic examination of intracellular organelles in freeze-substituted *Kloeckera* and *Saccharomyces cerevisiae* yeast cells. J Electron Microsc Tech 5:249–261

Bacon JSD, Jones D, Farmer VC, Webley DM (1968) The occurrence of α-(1-3)-glucan in *Cryptococcus*, *Schizosaccharomyces* and *Polyporus* species and its hydrolysis by a streptomycete culture filtrate lysing cell walls of *Cryptococcus*. Biochim Biophys Acta 158:313–315

Bago B, Chamberland H, Goulet A, Vierheilig H, Lafontaine J-G, Piché E (1996) Effect of nikkomycin Z, a chitin synthase inhibitor, on hyphal growth and cell wall structure of two arbuscular-mycorrhizal fungi. Protoplasma 192:80–92

Bailey DA, Feldmann PJ, Bovey M, Gow NA, Brown AJ (1996) The *Candida albicans HYR1* gene, which is activated in response to hyphal development, belongs to a gene family encoding yeast cell wall proteins. J Bacteriol 178:5353–5360

Barkai-Golan R, Mirelman D, Sharon N (1978) Studies on growth inhibition by lectins of *Penicillia* and *Aspergilli*. Arch Microbiol 116:119–124

Barran LR, Schneider EF, Wood PJ, Madhosingh C, Miller RW (1975) Cell wall of *Fusarium sulphureum*. I. Chemical composition of the hyphal cell wall. Biochim Biophys Acta 392:148–158

Bartnicki-Garcia S, Persson J, Chanzy H (1994) An electron microscope and electron diffraction study of the effect of Calcofluor and Congo Red on the biosynthesis of chitin in vitro. Arch Biochem Biophys 310:6–15

Bell-Pedersen D, Dunlap JC, Loros JJ (1992) The *Neurospora* circadian clock-controlled gene, ccg-2, is allelic to eas and encodes a fungal hydrophobin required for formation of the conidial rodlet layer. Genes Dev 6:2382–2394

Benitez T, Villa TG, Acha IG (1976) Effects of polyoxin D on germination, morphological development and biosynthesis of the cell wall of *Trichoderma viride*. Arch Microbiol 108:183–188

Boone C, Sommer SS, Hensel A, Bussey H (1990) Yeast *KRE* genes provide evidence for a pathway of cell wall beta-glucan assembly. J Cell Biol 110:1833–1843

Boone C, Sdicu A, Laroche M, Bussey H (1991) Isolation from *Candida albicans* of a functional homolog of the *Saccharomyces cerevisiae KRE1* gene, which is involved in cell wall beta-glucan synthesis. J Bacteriol 173:6859–6364

Borgia PT, Iartchouk N, Riggle PJ, Winter KR, Koltin Y, Bulawa CE (1996) The chsB gene of *Aspergillus nidulans* is necessary for normal hyphal growth and development. Fungal Genet Biol 20:193–203

Bowen AR, Chen-Wu JL, Momany M, Young R, Szaniszlo PJ, Robbins PW (1992) Classification of fungal chitin synthases. Proc Natl Acad Sci USA 89:519–523

Broekaert WF, Van Parijs J, Leyns F, Joos H, Peumans WJ (1989) A chitin-binding lectin from stinging nettle rhizomes with antifungal properties. Science 245: 1100–1102

Brown JL, Bussey H (1993) The yeast *KRE9* gene encodes an *O*-glycoprotein involved in cell surface beta-glucan assembly. Mol Cell Biol 13:6346–6356

Brul S, King A, Van Der Vaart JM, Chapman J, Klis F, Verrips CT (1997) The incorporation of mannoproteins in the cell wall of *S. cerevisiae* and filamentous Ascomycetes. Antonie van Leeuwenhoek 72:229–237

Bulawa CE (1993) Genetics and molecular biology of chitin synthesis in fungi. Annu Rev Microbiol 47:505–534

Bulawa CE, Slater M, Cabib E, Au-Young J, Sburlati A, Adair WL Jr, Robbins PW (1986) The *S. cerevisiae* structural gene for chitin synthase is not required for chitin synthesis in vivo. Cell 46 213–225

Cabib E, Drgonova J, Drgon T (1998) Role of small G proteins in yeast cell polarization and wall biosynthesis. Annu Rev Biochem 67:307–333

Cao L, Chan CM, Lee C, Wong SS, Yuen KY (1998) MP1 encodes an abundant and highly antigenic cell wall mannoprotein in the pathogenic fungus *Penicillium marneffei*. Infect Immun 66:966–973

Cappellaro C, Mrsa V, Tanner W (1998) New potential cell wall glucanases of *Saccharomyces cerevisiae* and their involvement in mating. J Bacteriol 180:5030–5037

Caro LH, Tettelin H, Vossen JH, Ram AF, van den Ende H, Klis FM (1997) In silicio identification of glycosylphosphatidyl inositol-anchored plasma-membrane and cell wall proteins of *Saccharomyces cerevisiae*. Yeast 13:1477–89

Casanova M, Lopez-Ribot JL, Martinez JP, Sentandreu R (1992) Characterization of cell wall proteins from yeast and mycelial cells of *Candida albicans* by labeling with biotin: comparison with other techniques. Infect Immun 60:4898–4906

Chuang JS, Schekman RW (1996) Differential trafficking and timed localisation of two chitin synthase proteins, Chs2p and Chs3p. J Cell Biol 135:597–610

Cormack BP, Ghori N, Falkow S (1999) An adhesin of the yeast pathogen *Candida glabrata* mediating adherence to human epithelial cells. Science 285:578–582

Costanzo MC, Hogan JD, Cusick ME, Davis BP, Fancher AM, Hodges PE, Kondu P, Lengieza C, Lew-Smith JE, Linger C, Roberg-Perez KJ, Tillberg M, Brooks JE, Garrels JI (2000) The yeast proteome database (YPD) and *Caenorhabditis elegans* proteome database (WormPD): comprehensive resources for the organization and comparison of model organism protein information. Nucleic Acids Res 28:73–76

Dallies N, Francois J, Paquet V (1998) A new method for quantitative determination of polysaccharides in the yeast cell wall. Application to the cell wall defective mutants of *Saccharomyces cerevisiae*. Yeast 14:1297–1306

Dean N (1999) Asparagine-linked glycosylation in the yeast Golgi. Biochim Biophys Acta 1426:309–322

De Nobel JGH, Barnett JA (1991) Passage of molecules through yeast cell walls: a brief essay-review. Yeast 7:313–323

De Nobel JG, Lipke PN (1994) Is there a role for GPIs in yeast cell-wall assembly? Trends Cell Biol 4:42–45

De Nobel JG, Dijkers C, Hooijberg E, Klis FM (1989) Increased cell wall porosity in Saccharomyces cerevisiae after treatment with dithiothreitol and EDTA. J Gen Microbiol 135:2077–2084

De Nobel JG, Klis FM, Priem J, Munnik T, Van Den Ende H (1990) The glucanase-soluble mannoproteins limit cell wall porosity in Saccharomyces cervisiae. Yeast 6:491–499

Dijkgraaf GJ, Brown JL, Bussey H (1996) The KNH1 gene of Saccharomyces cerevisiae is a functional homolog of KRE9. Yeast 12:683–692

Domer JE, Hamilton JG, Harkin JC (1967) Comparative study of the cell walls of the yeastlike and mycelial phases of Histoplasma capsulatum. J Bacteriol 94:466–474

Douglas CM, D'Ippolito JA, Shei GJ, Meinz M, Onishi J, Marrinan JA, Li W, Abruzzo GK, Flattery A, Bartizal K, Mitchell A, Kurtz MB (1997) Identification of the FKS1 gene of Candida albicans as the essential target of 1,3-beta-D-glucan synthase inhibitors. Antimicrob Agents Chemother 41:2471–2479

Drgonova J, Drgon T, Tanaka K, Kollar R, Chen GC, Ford RA, Chan CS, Takai Y, Cabib E (1996) Rho1p, a yeast protein at the interface between cell polarization and morphogenesis. Science 272:277–279

Dugger KO, Villareal KM, Ngyuen A, Zimmermann CR, Law JH, Galgiani JN (1996) Cloning and sequence analysis of the cDNA for a protein from Coccidioides immitis with immunogenic potential. Biochem Biophys Res Commun 218:485–489

Elorza MV, Rico H, Sentandreu R (1983) Calcofluor White alters the assembly of chitin fibrils in Saccharomyces cerevisiae and Candida albicans cells. J Gen Microbiol 129:1577–1582

Fèvre M, Girard V, Nodet P (1990) Cellulose and β-glucan synthesis in Saprolegnia. In: Kuhn PJ, Trinci APJ, Jung MJ, Goosey MW, Copping LG (eds) Biochemistry of cell walls and membranes in fungi. Springer, Berlin Heidelberg New York, pp 97–107

Fleet GH (1991) Cell walls. In: Rose AH, Harrison JS (eds) The yeasts, 2nd edn, vol 4. Yeast organelles. Academic Press, London, pp 199–277

Fleet GH, Phaff HJ (1981) Fungal glucans. Structure and metabolism. In: Tanner W, Loewus FA (eds) Encyclopedia of plant physiology, vol 13B. Plant carbohydrates II, extracellular carbohydrates. Springer, Berlin Heidelberg New York, pp 416–440

Fontaine T, Simenel C, Dubreucq G, Adam O, Delepierre M, Lemoine J, Vorgias CE, Diaquin M, Latgé J-P (2000) Molecular organization of the alkali-insoluble fraction of Aspergillus fumigatus cell wall. J Biol Chem 275:27594–27607

Fonzi WA (1999) PHR1 and PHR2 of Candida albicans encode putative glycosidases required for proper cross-linking of beta-1,3- and beta-1,6-glucans. J Bacteriol 181:7070–7079

Frost DJ, Brandt KD, Cugier D, Goldman R (1995) A whole-cell Candida albicans assay for the detection of inhibitors towards fungal cell wall synthesis and assembly. J Antibiot (Tokyo) 48:306–310

Fujii T, Shimoi H, Iimura Y (1999) Structure of the glucan-binding sugar chain of Tip1p, a cell wall protein of Saccharomyces cerevisiae. Biochim Biophys Acta 1427:133–144

Gaur NK, Klotz SA (1997) Expression, cloning, and characterization of a Candida albicans gene, ALA1, that confers adherence properties upon Saccharomyces cerevisiae for extracellular matrix proteins. Infect Immun 65:5289–5294

Gemmill TR, Trimble RB (1999) Overview of N- and O-linked oligosaccharide structures found in various yeast species. Biochim Biophys Acta 1426:227–237

Goldman RC, Sullivan PA, Zakula D, Capobianco JO (1995) Kinetics of beta-1,3-glucan interaction at the donor and acceptor sites of the fungal glucosyltransferase encoded by the BGL2 gene. Eur J Biochem 227:372–378

Gooday GW (1971) An autoradiographic study of hyphal growth of some fungi. J Gen Microbiol 67:125–133

Gooday GW (1990) Inhibition of chitin metabolism. In: Kuhn PJ, Trinci APJ, Jung MJ, Goosey MW, Copping LG (eds) Biochemistry of Cell Walls and Membranes in Fungi. Springer, Berlin Heidelberg New York, pp 61–79

Gooday GW (1995) Cell walls. In: Gow NAR, Gadd GM (eds) The Growing Fungus. Chapman and Hall, London, pp 43–62

Gooday GW, Schofield DA (1995) Regulation of chitin synthesis during growth of fungal hyphae: the possible participation of membrane stress. Can J Bot 73 [Suppl 1]:S114–S121

Gozalbo D, Dubon F, Schwenke J, Sentandreu R (1987) Characterization of chitosomes in Candida albicans protoplasts. Exp Mycol 11:331–337

Gozalbo D, Dubon F, Sentandreu R (1992) Chitin synthetase activity is bound to the plasma membrane and to a cytoplasmic particulate fraction in Candida albicans germ tube cells. FEMS Microbiol Lett 97:255–259

Gustin MC, Albertyn J, Alexander M, Davenport K (1998) MAP kinase pathways in the yeast Saccharomyces cerevisiae. Microbiol Mol Biol Rev 62:1264–1300

Hamada K, Terashima H, Arisawa M, Kitada K (1998) Amino acid sequence requirement for efficient incorporation of glycosylphosphatidylinositol-associated proteins into the cell wall of Saccharomyces cerevisiae. J Biol Chem 9:26946–26953

Hamada K, Terashima H, Arisawa M, Yabuki N, Kitada K (1999) Amino acid residues in the omega-minus region participate in cellular localization of yeast glycosylphosphatidylinositol-attached proteins. J Bacteriol 181:3886–3889

Haplová J, Farkas V, Hejtmánek M, Kodousek R, Malinsky J (1994) Effect of the new fluorescent brightener Rylux BSU on morphology and biosynthesis of cell walls in Saccharomyces cerevisiae. Arch Microbiol 161:340–344

Hartland RP, Emerson GW, Sullivan PA (1991) A secreted β-glucan-branching enzyme from Candida albicans. Proc R Soc Lond B Biol Sci 246:155–160

Hartland RP, Vermeulen CA, Klis FM, Sietsma JH, Wessels JG (1994) The linkage of (1-3)-beta-glucan to chitin during cell wall assembly in Saccharomyces cerevisiae. Yeast 10:1591–1599

Hartland RP, Fontaine T, Debeaupuis J-P, Simenel C, Delepierre M, Latgé J-P (1996) A novel β-(1-3)-glucanosyltransferase from the cell wall of Aspergillus fumigatus. J Biol Chem 271:26843–26849

Hearn VM, Sietsma JH (1994) Chemical and immunological analysis of the *Aspergillus fumigatus* cell wall. Microbiology 140:789–795

Heath IB (1994) The cytoskeleton. In: Wessels JGH, Meinhardt F (eds) The Mycota, vol 1. Springer, Berlin Heidelberg New York, pp 43–65

Hector RF (1993) Compounds active against cell walls of medically important fungi. Clin Microbiol Rev 6:1–21

Herscovics A (1999) Processing glycosidases of *Saccharomyces cerevisiae*. Biochim Biophys Acta 1426: 275–286

Herrero E, Boyd A (1986) Mannoproteins from the cell wall of *Kluyveromyces lactis*. FEMS Microbiol Lett 36:207–211

Herrero E, Sanz P, Sentandreu R (1987) Cell wall proteins liberated by zymolyase from several ascomycetous and imperfect yeasts. J Gen Microbiol 133:2895–2903

Hochstenbach F, Klis FM, Van Den Ende H, Van Donselaar E, Peters PJ, Klausner RD (1998) Identification of a putative alpha-glucan synthase essential for cell wall construction and morphogenesis in fission yeast. Proc Natl Acad Sci USA 95:9161–9166

Hogan LH, Klein BS (1994) Altered expression of surface alpha-1,3-glucan in genetically related strains of *Blastomyces dermatitidis* that differ in virulence. Infect Immun 62:3543–3546

Horisberger M, Rouvet-Vauthey M (1985) Cell wall architecture of the fission yeast *Schizosaccharomyces pombe*. Experientia 41:748–750

Horiuchi H, Fujiwara M, Yamashita S, Ohta A, Takagi M (1999) Proliferation of intrahyphal hyphae caused by disruption of csmA, which encodes a class V chitin synthase with a myosin motor-like domain in *Aspergillus nidulans*. J Bacteriol 181:3721–3729

Howard RJ, Aist JR (1979) Hyphal tip cell ultrastructure of the fungus *Fusarium*: improved preservation by freeze-substitution. J Ultrastruct Res 66:24–234

Hoyer LL, Hecht JE (2000) The ALS6 and ALS7 genes of *Candida albicans*. Yeast 16:847–855

Hoyer LL, Scherer S, Shatzman AR, Livi GP (1995) *Candida albicans ALS1*: domains related to a *Saccharomyces cerevisiae* sexual agglutinin separated by a repeating motif. Mol Microbiol 15:39–54

Hoyer LL, Payne TL, Hecht JE (1998a) Identification of *Candida albicans ALS2* and *ALS4* and localization of Als proteins to the fungal cell surface. J Bacteriol 180:5334–5343

Hoyer LL, Payne TL, Bell M, Myers AM, Scherer S (1998b) *Candida albicans ALS3* and insights into the nature of the *ALS* gene family. Curr Genet 33:451–459

Hunsley D, Burnett JH (1970) Ultrastructural architecture of the walls of some hyphal fungi. J Gen Microbiol 62:203–218

Hunsley D, Kay D (1976) Wall structure of the *Neurospora* hyphal apex: immunofluorescent localization of wall surface antigens. J Gen Microbiol 95:233–248

Hwang CS, Kolattukudy PE (1995) Isolation and characterization of genes expressed uniquely during appressorium formation by *Colletotrichum glueosporioides* conidia induced by the host surface wax. Mol Gen Genet 247:282–294

Ishiguro J, Saitou A, Duran A, Ribas JC (1997) cps1+, a *Schizosaccharomyces pombe* gene homolog of *Saccharomyces cerevisiae FKS* genes whose mutation confers hypersensitivity to cyclosporin A and papulacandin B. J Bacteriol 179:7653–7662

Jelsma J, Kreger DR (1975) Ultrastructural observation on (1-3)-β-D-glucan from fungal cell walls. Carbohydr Res 43:200–203

Jentoft N (1990) Why are proteins O-glycosylated? Trends Biochem Sci 15:291–294

Jikibara T, Takegawa K, Iwahara S (1992) Studies on the uronic acid-containing glycoproteins of Fusarium sp. M7-1. III. The primary structures of the acidic polysaccharides of the glycoproteins. J Biochem 111:236–243

Jung US, Levin DE (1999) Genome-wide analysis of gene expression regulated by the yeast cell wall integrity signalling pathway. Mol Microbiol 34:1049–1057

Kanetsuna F, Carbonell LM (1970) Cell wall glucans of the yeast and mycelial forms of *Paracoccidioides brasiliensis*. J Bacteriol 101:675–681

Kanetsuna F, Carbonell LM (1971) Cell wall composition of the yeast-like and mycelial forms of *Blastomyces dermatitidis*. J Bacteriol 106:946–953

Kapteyn JC, Montijn RC, Vink E, De La Cruz J, Llobell A, Douwes JE, Shimoi H, Lipke PN, Klis FM (1996) Retention of *Saccharomyces cerevisiae* cell wall proteins through a phosphodiester-linked β-1,3/β-1,6-glucan heteropolymer. Glycobiology 6:337–345

Kapteyn JC, Ram AFJ, Groos EM, Kollar R, Montijn RC, Van Den Ende H, Llobell A, Cabib E, Klis FM (1997) Altered extent of cross-linking of beta 1,6-glucosylated mannoproteins to chitin in S *accharomyces cerevisiae* mutants with reduced cell wall beta-1,3-glucan content. J Bacteriol 179:6279–6284

Kapteyn JC, Van Den Ende H, Klis FM (1999a) The contribution of cell wall proteins to the organization of the yeast cell wall. Biochim Biophys Acta 6:373–383

Kapteyn JC, Van Egmond P, Sievi E, Van Den Ende H, Makarow M, Klis FM (1999b) The contribution of the *O*-glycosylated protein Pir2p/Hsp150 to the construction of the yeast cell wall in wild-type cells and beta-1,6-glucan-deficient mutants. Mol Microbiol 31: 1835–1844

Kapteyn JC, Hoyer LL, Hecht JE, Müller WH, Verkley AJ, Andel A, Makarow M, Van Den Ende H, Klis FM (2000) The cell wall architecture of *Candida albicans* wild-type cells and cell-wall-defective mutants. Mol Microbiol 35:601–611

Katayama S, Hirata D, Arellano M, Pérez P, Toda T (1999) Fission yeast α-glucan synthase Mok1 requires the actin cytoskeleton to localize the sites of growth and plays an essential role in cell morphogenesis downstream of protein kinase C function. J Cell Biol 144:1173–1186

Kelly R, Register E, Hsu MJ, Kurtz M, Nielsen J (1996) Isolation of a gene involved in β-1,3-glucan synthesis in *Aspergillus nidulans* and purification of the corresponding protein. J Bacteriol 178:4381–4391

Ketela T, Green R, Bussey H (1999) *Saccharomyces cerevisiae* Mid2p is a potential cell wall stress sensor and upstream activator of the *PKC1-MPK1* cell integrity pathway. J Bacteriol 181:3330–3340

Klebl F, Tanner W (1989) Molecular cloning of a cell wall exo-beta-1,3-glucanase from *Saccharomyces cerevisiae*. J Bacteriol 171:6259–6264

Klimpel KR, Goldman WE (1988) Cell walls from avirulent variants of *Histoplasma capsulatum* lack alpha-(1,3)-glucan. Infect Immun 56:2997–3000

Klis FM, Ram AFJ, Montijn RC, Kapteyn JC, Caro LHP, Vossen JH, Van Berkel MAA, Brekelmans SSC, Van

Den Ende H (1998) Posttranslational modifications of secretory proteins. Methods Microbiol 26:224–238

Knauer R, Lehle L (1999) The oligosaccharyltransferase complex from yeast. Biochim Biophys Acta 1426:259–274

Kollár R, Petrakova E, Ashwell G, Robbins PW, Cabib E (1995) Architecture of the yeast cell wall: the linkage between chitin and β-1,3-glucan. J Biol Chem 270:1170–1178

Kollár R, Reinhold BB, Petráková E, Yeh HJC, Ashwell G, Drgonová J, Kapteyn JC, Klis FM, Cabib E (1997) Architecture of the yeast cell wall: β-1,6-glucan interconnects mannoprotein, β-1,3-glucan, and chitin. J Biol Chem 272:17762–17788

Kondoh O, Tachibana Y, Ohya Y, Arisawa M, Watanabe T (1997) Cloning of the *RHO1* gene from *Candida albicans* and its regulation of beta-1,3-glucan synthesis. J Bacteriol 179:7734–7741

Kopecka M, Gabriel M (1992) The influence of Congo Red on the cell wall and (1-3)-beta-D-glucan microfibril biogenesis on *Saccharomyces cerevisiae*. Arch Microbiol 158:115–126

Koyama M, Helbert W, Tomoya I, Sugiyama J, Henrissat B (1997) Parallel-up structure evidences the molecular directionality during biosynthesis of bacterial cellulose. Proc Natl Acad Sci USA 94:9091–9095

Kreger DR (1954) Observations on cell walls of yeasts and some other fungi by X-ray diffraction and solubility tests. Biochim Biophys Acta 13:1–9

Kuhn PJ, Trinci APJ, Jung MJ, Goosey MW, Copping LG (1990) Biochemistry of cell walls and membranes in fungi. Springer, Berlin Heidelberg New York

Kurtz MB, Douglas CM (1997) Lipopeptide inhibitors of fungal glucan synthase. J Med Vet Mycol 35:79–86

Kurtz MB, Heath IB, Marrinan J, Dreikorn S, Onishi J, Douglas C (1994) Morphological effects of lipopeptides against *Aspergillus fumigatus* correlate with activities against (1,3)-β-D-glucan synthase. Antimicrob Agents Chemother 38:1480–1489

Leal J-A, Jiménez-Barbero J, Gómez-Miranda B, Prieto A, Domenech J, Bernabé M (1996) Structural investigation of a cell-wall galactomannan from *Neurospora crassa* and *N. sitophila*. Carbohydr Res 283:215–222

Leal-Morales CA, Bracker CE, Bartnicki-Garcia S (1994) Subcellular localization, abundance and stability of chitin synthetases 1 and 2 from *Saccharomyces cerevisiae*. Microbiology 140:2207–2216

Lora JM, de la Cruz J, Benitez T, Llobell A, Pintor-Toro JA (1994) A putative catabolite-repressed cell wall protein from the mycoparasitic fungus *Trichoderma harzianum*. Mol Gen Genet 242:461–466

Lu CF, Kurjan J, Lipke PN (1994) A pathway for cell wall anchorage of *Saccharomyces cerevisiae* α-agglutinin. Mol Cell Biol 9:4825–4833

Lussier M, White AM, Sheraton J, di Paolo T, Treadwell J, Southard SB, Horenstein CI, Chen-Weiner J, Ram AF, Kapteyn JC, Roemer TW, Vo DH, Bondoc DC, Hall J, Zhong WW, Sdicu AM, Davies J, Klis FM, Robbins PW, Bussey H (1997) Large scale identification of genes involved in cell surface biosynthesis and architecture in *Saccharomyces cerevisiae*. Genetics 147:435–450

Lussier M, Sdicu AM, Shahinian S, Bussey H (1998) The *Candida albicans KRE9* gene is required for cell wall beta-1,6-glucan synthesis and is essential for growth on glucose. Proc Natl Acad Sci USA 95:9825–9830

Maeda H, Ishida N (1967) Specificity of binding of hexopyranosyl polysaccharides with fluorescent brightener. J Biochem (Tokyo) 62:276–278

Manners DJ, Masson AJ, Patterson JC (1973a) The structure of a β-(1-3)-D-glucan from yeast cell walls. Biochem J 135:19–30

Manners DJ, Masson AJ, Patterson JC, Bjorndal H, Lindberg B (1973b) The structure of a β-(1-6)-D-glucan from yeast cell walls. Biochem J 135:31–36

Marchessault RH, Deslandes Y (1979) Fine structure of (1-3)-β-glucans: curdlan and paramylon. Carbohydr Res 75:231–242

Martin F, Laurent P, de Carvalho D, Voiblet C, Balestrini R, Bonfante P, Tagu D (1999) Cell wall proteins of the ectomycorrhizal basidiomycete *Pisolithus tinctorius*: identification, function and expression in symbiosis. Fungal Genet Biol 27:161–174

Martinez FA, Schwencke J (1988) Chitin synthetase activity is bound to chitosomes and to the plasma membrane in protoplasts of *Saccharomyces cerevisiae*. Biochim Biophys Acta 946:328–336

Mazur P, Baginsky W (1996) In vitro activity of 1,3-beta-D-glucan synthase requires the GTP-binding protein Rho1. J Biol Chem 271:14604–14609

McCreath KJ, Specht CA, Robbins PW (1995) Molecular cloning and characterization of chitinase genes from *Candida albicans*. Proc Natl Acad Sci USA 92:2544–2548

Mio T, Yabe T, Sudoh M, Satoh Y, Nakajima T, Arisawa M, Yamada-Okabe H (1996) Role of three chitin synthase genes in the growth of *Candida albicans*. J Bacteriol 178:2416–2419

Mio T, Adachi-Shimizu M, Tachibana Y, Tabuchi H, Inoue S, Yabe T, Yamada-Okabe H, Arisawa M, Watanabe T, Yamada-Okabe H (1997) Cloning of the *Candida albicans* homolog of *Saccharomyces cerevisiae GSC1/ FKS1* and its involvement in β-1,3-glucan synthesis. J Bacteriol 179:4096–4105

Mirelman D, Galun E, Sharon N, Loan R (1975) Inhibition of fungal growth by wheat germ agglutinin. Nature 256:414–416

Miyata M, Kanbe T, Tanaka K (1985) Morphological alterations of the fission yeast *Schizosaccharomyces pombe* in the presence of aculeacin A: spherical wall formation. J Gen Microbiol 131:611–621

Mol PC, Wessels JGH (1987) Linkages between glucosaminoglycan and glucan determine alkali-insolubility of the glucan in walls of *S cerevisiae*. FEMS Microbiol Lett 41:95–99

Molloy C, Shepherd MG, Sullivan PA (1989) Identification of envelope proteins of *Candida albicans* by vectorial iodination. Microbios 57:73–84

Montijn RC, Van Wolven P, De Hoog S, Klis FM (1997) Beta-glucosylated proteins in the cell wall of the black yeast *Exophiala* (*Wangiella*) *dermatitidis*. Microbiology 143:1673–1680

Montijn RC, Vink E, Müller WH, Verkley AJ, Van Den Ende H, Henrissat B, Klis FM (1999) Localization of the synthesis of β1,6-glucan in *Saccharomyces cerevisiae*. J Bacteriol 181:7414–7420

Morawetz R, Lendenfeld T, Mischak H, Muhlbauer M, Gruber F, Goodnight J, de Graaff LH, Visser J, Mushinski JF, Kubicek CP (1996) Cloning and characterisation of genes (pkc1 and pkcA) encoding protein kinase C homologues from *Trichoderma reesei* and *Aspergillus niger*. Mol Gen Genet 250:17–28

Mouyna I, Hartland RP, Fontaine T, Diaquin M, Simenel C, Delepierre M, Henrissat B, Latge JP (1998) A 1,3-beta-glucanosyltransferase isolated from the cell wall of *Aspergillus fumigatus* is a homologue of the yeast Bgl2p. Microbiology 144:3171–3180

Mouyna I, Vai T, Fontaine M, Monod M, Fonzi WA, Diaquin M, Popolo L, Henrissat B, Hartland RP, Latgé JP (1999) The protein encoded by *GAS1* of *Saccharomyces cerevisiae* displays the same enzymatic activity as Gel1p of *Aspergillus fumigatus* and corresponds to a β(1-3)glucanosyltransferase. Curr Genet 35:440

Mühlschlegel FA, Fonzi WA (1997) *PHR2* of *Candida albicans* encodes a functional homolog of the pH-regulated gene *PHR1* with an inverted pattern of pH-dependent expression. Mol Cell Biol 17:5960–5967

Mrsa V, Seidl M, Gentzsch M, Tanner W (1997) Specific labeling of cell wall proteins by biotinylation. Identification of four covalently linked O-mannosylated proteins of *S. cerevisiae*. Yeast 13:1145–1149

Nagahashi S, Lussier M, Bussey H (1998) Isolation of *Candida glabrata* homologs of the *Saccharomyces cerevisiae* KRE9 and KNH1 genes and their involvement in cell wall beta-1,6-glucan synthesis. J Bacteriol 180:5020–5029

Nakajima T, Yoshida M, Nakamura M, Hiura N, Matsuda K (1984) Structure of the cell wall proteogalactomannan from *Neurospora crassa*. II. Structural analysis of the polysaccharide part. J Biochem (Tokyo) 96:1013–1020

Nakazawa T, Horiuchi H, Ohta A, Takagi M (1998) Isolation and characterization of *EPD1*, an essential gene for pseudohyphal growth of a dimorphic yeast, *Candida maltosa*. J Bacteriol 180:2079–2086

Navarro-Garcia F, Alonso-Monge R, Rico H, Pla J, Sentandreu R, Nombela C (1998) A role for the MAP kinase gene *MKC1* in cell wall construction and morphological transitions in *Candida albicans*. Microbiology 144:411–424

Nguyen TH, Fleet GH, Rogers PL (1998) Composition of the cell walls of several yeast species. Appl Microbiol Biotechnol 50:206–212

Orlean P (1997) Biogenesis of yeast wall and surface components. In: Pringle JR, Broach JR, Jones EW (eds) The Molecular and Cellular Biology of the Yeast *Saccharomyces*, vol 3. Cold Spring Harbor Laboratory Press, Cold Spring Harbor, NY, pp 229–362

Orlean PAB, Krag SS (eds) (1999) Glycans of the secretory pathway in yeast. Biochim Biophys Acta 1426, no 2, Special Issue

Osmond BC, Specht CA, Robbins PW (1999) Chitin synthase. III. Synthetic lethal mutants and "stress related" chitin synthesis that bypasses the *CSD3/CHS6* localization pathway. Proc Natl Acad Sci USA 96:11206–11210

Park IC, Horiuchi H, Hwang CW, Yeh WH, Ohta A, Ryu JC, Takagi M (1999) Isolation of csm1 encoding a class V chitin synthase with a myosin motor-like domain from the rice blast fungus, *Pyricularia oryzae*. FEMS Microbiol Lett 170:131–139

Peberdy JF (1990) Fungal cell walls – a review. In: Kuhn PJ, Trinci APJ, Jung MJ, Goosey MW, Copping LG (eds) Biochemistry of cell walls and membranes in fungi. Springer, Berlin Heidelberg New York, pp 5–30

Polacheck I, Rosenberger RF (1977) *Aspergillus nidulans* mutant lacking alpha-(1,3)-glucan, melanin, and cleistothecia. J Bacteriol 132:650–656

Popolo L, Vai M (1999) The Gas1 glycoprotein, a putative wall polymer cross-linker. Biochim Biophys Acta 1426:385–400

Popolo L, Gilardelli D, Bonfante P, Vai M (1997) Increase in chitin as an essential response to defects in assembly of cell wall polymers in the *ggp1Δ* mutant of *Saccharomyces cerevisiae*. J Bacteriol 179:463–469

Qadota H, Python CP, Inoue SB, Arisawa M, Anraku Y, Zheng Y, Watanabe T, Levin DE, Ohya Y (1996) Identification of yeast Rho1p GTPase as a regulatory subunit of 1,3-beta-glucan synthase. Science 272:279–281

Ram AF, Wolters A, Ten Hoopen R, Klis FM (1994) A new approach for isolating cell wall mutants in *Saccharomyces cerevisiae* by screening for hypersensitivity to calcofluor white. Yeast 10:1019–1030

Ram AFJ, Kapteyn JC, Montijn RC, Caro LH, Douwes JE, Baginsky W, Mazur P, Van den Ende H, Klis FM (1998) Loss of the plasma membrane bound protein *Gas1p* in *Saccharomyces cerevisiae* results in the release of β1,3-glucan into the medium and induces a compensation mechanism to ensure cell wall integrity. J Bacteriol 180:1418–1424

Rico H, Herrero E, Miragall F, Sentandreu R (1991) An electron microscopy study of wall expansion during *Candida albicans* yeast and mycelial growth using concanavalin A-ferritin labelling of mannoproteins. Arch Microbiol 156:111–114

Roemer T, Bussey H (1995) Yeast Kre1p is a cell surface O-glycoprotein. Mol Gen Genet 249:209–216

Roncero C, Duran A (1985) Effect of calcofluor white and Congo red on fungal cell wall morphogenesis: in vivo activation of chitin polymerization. J Bacteriol 163:1180–1185

Roncero C, Valdivieso MH, Ribas JC, Duran A (1988) Isolation and characterization of *Saccharomyces cerevisiae* mutants resistant to calcofluor white. J Bacteriol 170:1950–1954

Ruiz-Herrera J (1992) Fungal cell wall. Structure, synthesis, and assembly. CRC Press, Boca Raton

Russo P, Kalkkinen N, Sareneva H, Paakkola J, Makarow M (1992) A heat shock gene from *Saccharomyces cerevisiae* encoding a secretory glycoprotein. Proc Natl Acad Sci USA 89:3671–3675

San-Blas G, San-Blas F (1977) *Paracoccidioides brasiliensis*: cell wall structure and virulence. A review. Mycopathology 62:77–86

Saporito-Irwin SM, Birse CE, Sypherd PS, Fonzi WA (1995) *PHR1*, a pH-regulated gene of *Candida albicans*, is required for morphogenesis. Mol Cell Biol 15:601–613

Schekman R, Brawley V (1979) Localized deposition of chitin on the yeast cell surface in response to mating pheromone. Proc Natl Acad Sci USA 76:645–649

Schoffelmeer EAM (1999) Biochemical aspects of the cell wall of *Fusarium oxysporum*. PhD thesis, University of Amsterdam, The Netherlands

Schoffelmeer EAM, Klis FM, Sietsma JGH, Cornelissen BJC (1999) The cell wall of *Fusarium oxysporum*. Fungal Genet Biol 27:275–282

Schreuder MP, Mooren ATA, Toschka HY, Verrips CT, Klis FM (1996) Immobilizing proteins on the surface of yeast cells. Trends Biotechnol 14:115–120

Sentandreu R, Mormeneo S, Ruiz-Herrera J (1994) Biogenesis of the fungal cell wall. In: Wessels JGH, Meinhardt F (eds) The Mycota, vol 1. Springer, Berlin Heidelberg New York, pp 111–124

Shaw JA, Mol PC, Bowers B, Silverman SJ, Valdivieso MH, Duran A, Cabib EJ (1991) The function of chitin synthase-2 and synthase-3 in the *Saccharomyces-cerevisiae* cell cycle. J Cell Biol 114:111–123

Shematek EM, Braatz JA, Cabib E (1980) Biosynthesis of the yeast cell wall. I. Preparation and properties of β-1,3-glucan synthetase. J Biol Chem 255:888–894

Shepherd MG (1987) Cell envelope of *Candida albicans*. Crit Rev Microbiol 15:7–25

Sietsma JH, Wessels JG (1977) Chemical analysis of the hyphal wall of *Schizophyllum commune*. Biochim Biophys Acta 496:225–239

Sietsma JH, Wessels JGH (1981) Solubility of (1→3)-beta-D/(1→6)-beta-D-glucan in fungal walls: importance of presumed linkage between glucan and chitin. J Gen Microbiol 125:209–212

Sietsma JH, Wessels JGH (1990) The occurrence of glucosaminoglycan in the wall of *Schizosaccharomyces pombe*. J Gen Microbiol 136:2261–2265

Sietsma JH, Wessels JGH (1994) Apical wall biogenesis. In: Wessels JGH, Meinhardt F (eds) The Mycota, vol 1. Springer, Berlin Heidelberg New York, pp 125–141

Sietsma JH, Beth-Din A, Ziv V, Sjollema KA, Yarden O (1996) The localization of chitin synthase in membranous vesicles (chitosomes) in *Neurospora crassa*. Microbiology 142:1591–1596

Smits G, Kapteyn JC, Van Den Ende H, Klis FM (1999) Cell wall dynamics in yeast. Curr Opin Microbiol 2:348–352

Specht CA, Liu Y, Robbins PW, Bulawa CE, Iartchouk N, Winter KR, Riggle PJ, Rhodes JC, Dodge CL, Culp DW, Borgia PT (1996) The chsD and chsE genes of *Aspergillus nidulans* and their roles in chitin synthesis. Fungal Genet Biol 20:153–167

Staab JF, Sundstrom P (1998) Genetic organization and sequence analysis of the hypha-specific cell wall protein gene *HWP1* of *Candida albicans*. Yeast 14:681–686

Staab JF, Ferrer CA, Sundstrom P (1996) Developmental expression of a tandemly repeated, proline-and glutamine-rich amino acid motif on hyphal surfaces on *Candida albicans*. J Biol Chem 271:6298–6305

Staab JF, Bradway SD, Fidel PL, Sundstrom P (1999) Adhesive and mammalian transglutaminase substrate properties of *Candida albicans* Hwp1. Science 283:1535–1538

Staebell M, Soll DR (1985) Temporal and spatial differences in cell wall expansion during bud and mycelium formation in *Candida albicans*. J Gen Microbiol 131:1467–1480

St Leger RJ, Staples RC, Roberts DW (1992) Cloning and regulatory analysis of starvation-stress gene, ssgA, encoding a hydrophobin like protein from the entomopathogenic fungus *Metarizium anisopliae*. Gene 120:119–124

Strahl-Bolsinger S, Gentzsch M, Tanner W (1999) Protein *O*-mannosylation. Biochim Biophys Acta 1426:297–308

Stringer MA, Dean RA, Sewall TC, Timberlake WE (1991) Rodletless, a new *Aspergillus* developmental mutant induced by directed gene inactivation. Genes Dev 5:1161–1171

Surarit R, Gopal PK, Shepherd MG (1988) Evidence for a glycosidic linkage between chitin and glucan in the cell wall of *Candida albicans*. J Gen Microbiol 134:1723–1730

Szaniszlo PJ, Kang MS, Cabib E (1985) Stimulation of beta(1-3)glucan synthetase of various fungi by nucle-

oside triphosphates: generalized regulatory mechanism for cell wall biosynthesis. J Bacteriol 161:1188–1194

Talbot NJ (1997) Fungal biology: growing into the air. Curr Biol 7:R78–R81

Talbot NJ, Ebbole DJ, Hamer JE (1993) Identification and characterization of *MPG1*, a gene involved in pathogenicity from the rice blast fungus *Magnaporthe grisea*. Plant Cell 5:1575–1590

Tanaka H, Watanabe T (1995) Glucanases and chitinases of *Bacillus circulans* WL-12. J Ind Microbiol 14:478–483

Tanaka K, Nambu H, Katoh Y, Kai M, Hidaka Y (1999) Molecular cloning of homologs of *RAS* and *RHO1* genes from *Cryptococcus neoformans*. Yeast 15:1133–1139

Thompson JR, Douglas CM, Li W, Jue CK, Pramanik B, Yuan X, Rude TH, Toffaletti DL, Perfect JR, Kurtz M (1999) A glucan synthase *FKS1* homolog in *Cryptococcus neoformans* is single copy and encodes an essential function. J Bacteriol 181:444–453

Toda T, Dhut S, Superti-Furga G, Gotoh Y, Nishida E, Sugiura R, Kuno T (1996) The fission yeast pmk1+ gene encodes a novel mitogen-activated protein kinase homolog which regulates cell integrity and functions coordinately with the protein kinase C pathway. Mol Cell Biol 16:6752–6764

Toh-e A, Yasunaga S, Nisogi H, Tanaka K, Oguchi T, Matsui Y (1993) Three yeast genes, *PIR1*, *PIR2* and *PIR3*, containing internal tandem repeats, are related to each other, and *PIR1* and *PIR2* are required for tolerance to heat shock. Yeast 9:481–494

Trinci APJ, Collinge AJ (1975) Hyphal wall growth in *Neurospora crassa* and *Geotrichum candidum*. J Gen Microbiol 91:355–361

Trinci APJ, Wiebe MG, Robson GD (1994) The mycelium as an integrated entity. In: Wessels JGH, Meinhardt F (eds) The Mycota, vol 1. Springer, Berlin Heidelberg New York, pp 175–193

Valdivieso MH, Mol PC, Shaw JA, Cabib E, Duran A (1991) *CAL1*, a gene required for activity of chitin synthase 3 in *Saccharomyces cerevisiae*. J Cell Biol 114:101–109

Valentin E, Herrero W, Pastor JFI, Sentandreu R (1984) Solubilization and analysis of mannoprotein molecules from the cell wall of *Saccharomyces cerevisiae*. J Gen Microbiol 130:1419–1428

Van der Valk P, Marchant R, Wessels JGH (1977) Ultrastructural localization of polysaccharides in the wall and septum of the basidiomycete *Schizophyllum commune*. Exp Mycol 1:69–82

Vermeulen CA, Wessels JGH (1986) Chitin biosynthesis by a fungal membrane preparation. Evidence for a transient non-crystalline state of chitin. Eur J Biochem 158:411–415

Villa TG, Notario V, Villanueva JR (1980) Chemical and enzymic analysis of *Pichia polymorpha* cell walls. Can J Microbiol 26:169–174

Von Sengbusch P, Hechler J, Müller U (1983) Molecular architecture of fungal cell walls. An approach by use of fluorescent markers. Eur J Cell Biol 30:305–312

Wessels JGH (1994) Developmental regulation of fungal cell wall formation. Annu Rev Phytopathol 32:413–437

Wessels JGH (1996) Fungal hydrophobins: proteins that function at an interface. Trends Plant Sciences 1:9–15

Wessels JGH (1997) Hydrophobins: proteins that change the nature of the fungal surface. Adv Microb Physiol 38:1–45

Wessels JGH, Sietsma JH (1981) Fungal cell walls: a survey. In: Tanner W, Loewus FA (eds) Encyclopedia of plant physiology, vol 13B. Plant carbohydrates II. Extracellular carbohydrates. Springer, Berlin Heidelberg New York, pp 352–394

Williamson G, Belshaw NJ, Williamson MP (1992) O-Glycosylation in Aspergillus glucoamylase. Confirmation and role in binding. Biochem J 282:423–428

Wood PJ (1980) Specificity in the interaction of direct dyes with polysaccharides. Carbohydr Res 85:271–287

Wösten HAB, De Vries OMH, Wessels JGH (1993) Interfacial self-assembly of a fungal hydrophobin into a rodlet layer. Plant Cell 5:1567–1574

Wösten HAB, Asgeirsdottir SA, Krook JH, Drenth JHH, Wessels JGH (1994) The Sc3p hydrophobin self-assembles at the surface of aerial hyphae as a protein membrane constituting the hydrophobic rodlet layer. Eur J Cell Biol 63:122–129

Wösten HAB, Richter M, Willey JM (1999) Structural proteins involved in emergence of microbial aerial hyphae. Fungal Genet Biol 27:153–160

Xoconostle-Cazares B, Specht CA, Robbins PW, Liu Y, Leon C, Ruiz-Herrera J (1997) Umchs5, a gene coding for a class IV chitin synthase in Ustilago maydis. Fungal Genet Biol 22:199–208

Yu L, Goldman R, Sullivan P, Walker GF, Fesik SW (1993) Heteronuclear NMR studies of 13C-labeled yeast cell wall beta-glucan oligosaccharides. J Biomol NMR 3:429–441

Zhu Y, Yang C, Magee DM, Cox RA (1996) Coccidioides immitis antigen 2: analysis of gene and protein. Gene 181:121–125

Ziman M, Chuang JS, Schekman RW (1996) Chs1p and Chs3p, two proteins involved in chitin synthesis, populate a compartment of the Saccharomyces cerevisiae endocytic pathway. Mol Biol Cell 7:1909–1919

Zlotnik H, Fernandez MP, Bowers B, Cabib E (1984) Saccharomyces cerevisiae mannoproteins form an external cell wall layer that determines wall porosity. J Bacteriol 159:1018–1026

Zonneveld BJ (1972) Morphogenesis in Aspergillus nidulans. The significance of α-1,3-glucan of the cell wall and α-1,3-glucanase for cleistothecium development. Biochim Biophys Acta 273:174–187

10 Bridging the Divide: Cytoskeleton-Plasma Membrane-Cell Wall Interactions in Growth and Development

I. Brent Heath

CONTENTS

I. Introduction

The fundamental difference between living and non-living materials is the organization characteristic of cells, including the plethora of structures well known in cell biology. A critical feature of this organization is the ability of cells to maintain a selective barrier between themselves and the surrounding environment. The selectivity is not absolute, it ranges from very low to very high permeability, with many intermediates, and it is subject to physiological and environmental regulation. This barrier is of course the **plasma membrane** (PM), and it is the great divide between living and non-living worlds.

While the PM is the great divide, it also transmits information and materials. This transmission is bidirectional: cellular products are secreted and environmental inputs are received. Transmission

of soluble molecules and ions is a well-understood phenomenon beyond the range of this review. However, there are also transmissions of information that depend on more long-term interactions across the PM. A dominant set of these interactions comprises those that involve the **cytoskeleton** on the inside and the environment on the outside.

The "environment" is not simply defined, nor separable from the cell. It clearly includes substrates surrounding the cell, including the soil, host tissues or adjacent cells of the same species. However, at a finer level, it is the inside of the **extracellular matrix** (ECM); perhaps more wisely referred to as the exocellular matrix (Wyatt and Carpita 1993). However, the latter term has not gained wide acceptance and the distinction need not be made from what is a well-entrenched term in the animal literature. ECM is a term derived from the animal literature, where it refers to the universal ensemble of proteins, glycoproteins and proteoglycans that surround the simplistically termed "naked" cells. However, it applies equally to the cell walls of plant and fungal cells (see Chap. 8, this Vol.). The walls are compositionally comparable to the ECM, although certainly with differences, and conceptually similar. The most prominent differences from ECMs are that walls tend to be thicker and stiffer. Nevertheless, in both animals and walled organisms, the ECM lies outside the PM and is an additional interface between the living and non-living worlds. While **cytoskeleton–PM–ECM interactions** are well recognized for their importance in multicellular animal development and physiology, their significance in plant and fungal biology is only recently beginning to emerge. It is both the presence and functions of these interactions that are the subject of this review.

Biology Department, York University, 4700 Keele Street, Toronto, Ontario M3J 1P3, Canada

The Mycota VIII
Biology of the Fungal Cell
Howard/Gow (Eds.)
© Springer-Verlag Berlin Heidelberg 2001

II. Animal Models

In a large diversity of animal cells, indeed probably all, the basic "bridge across the divide" involves three fundamental components, (1) an inner ensemble of linked proteins, the **membrane skeleton**, which is connected to, and may be considered a part of, the **cytoskeleton**, (2) integral **trans-PM** proteins that interact with the membrane skeleton via cytoplasmic domains and which may, or may not, interact with the ECM via extracellular domains, and (3) **ECM proteins**, **glycoproteins** and **proteoglycans**. The ECM molecules are highly heterogeneous, with diverse mechanical properties and capacities for mutual interactions; however, many bear binding sites for specific interactions with the trans-PM proteins. There are certainly interorganismal and intercellular variations in these components, and the list of each is still increasing. However, there are a number of well-known dominant molecules of each type that form the essential background to the analysis of the fungal systems.

A. Membrane Skeleton Molecules

The most common and quantitatively dominant membrane skeleton molecules are **F-actin** [polymerized or filamentous actin, the degree of polymerization (number of monomers and thus length of the filament) being generally unknown (12–14 in erythrocytes) and probably highly variable] and members of the **spectrin** superfamily. These two types of molecules are typically both present in the membrane skeleton of any particular cell, although the relative quantities vary. It is unclear whether there are any membrane skeletons that lack either component.

Members of the spectrin superfamily of proteins (including **fodrin** in brain and **dystrophin** and α-**actinin** in other animal cell types) are generally restricted to the membrane skeleton. They do not appear to extend into the cytoplasm, and thus do not become a component of the cytoskeleton. However, there are reports of spectrin associated with Golgi bodies (Beck et al. 1994) and other intracellular membranes (Fath et al. 1985; Zagon et al. 1986) and it has recently been suggested that spectrins may indeed play a role in intracellular membrane interactions (Matteis and Morrow 2000). Individual spectrin units are typically flexi-

ble 200-nm long tetramers that bind to a number of other proteins, including actin, to form junctions in a two-dimensional **polygonal network**. This network is attached to the PM at the junctions via direct interactions with integral PM proteins such as **glycophorin** and an **anion-exchange protein**, and indirect interactions via **ankyrin** to integral membrane proteins such as **Na⁺-K⁺-ATPases** and **Cl⁻/HCO₃⁻ exchangers** (Yeaman et al. 1999). There is substantial diversity in the spectrin superfamily of molecules, but the basic features include an actin binding site, multiple (4–26, with corresponding variations in molecular length) ~110 amino acid repeats, Ca^{2+}-binding domains and binding sites for other proteins such as ankyrin and the other integral PM proteins mentioned above (Bennett 1990).

There are a diversity of other molecules that have been described as forming part of the membrane skeleton in erythrocytes and other cell types. They are no doubt important in determining the mechanical and regulatory properties of the membrane skeleton, but since homologues have not been described in walled cells, there is little point in describing them in detail.

B. Integral Transplasma Membrane Molecules

As noted above, there are many integral trans plasma membrane (trans-PM) proteins to which membrane skeletons of diverse cells bind directly or indirectly with high affinity. For spectrins, including α-actinin, these include the Na⁺-K⁺-ATPase, the Cl⁻/HCO₃⁻ exchanger, at least three related glycophorins, a **voltage-gated Na⁺ channel**, hypothetically, **stretch-activated ion channels** and possibly **cadherins** (Yap et al. 1997; Bennett 1990). For actin, directly or indirectly via a number of **actin binding proteins** (ABPs), the integral proteins include **integrins** and **cadherins** (Yeaman et al. 1999). With respect to attaching the membrane skeleton to the PM, binding to any integral protein is significant. In the case of the channel proteins, such binding is also likely to influence their distribution and activities. However, the mechanical integrity of the membrane skeleton/PM complex is likely to be greater if it is also linked to the ECM. Furthermore, as we shall see, a significant activity at the membrane skeleton/PM may be anchorage to provide a base against which intracellular force can be exerted, a fact that clearly indicates the desirability of attachment to the ECM. Of the above molecules, the **integrins** and possibly the **glycophorins** are known to have this capability. The ion channels do not interact with the ECM and the cadherins primarily mediate cell-to-cell adhesion by homotypic interactions with each other (Yap et al. 1997), a process unlikely to be significant in membrane skeleton/PM interactions with the walls of plants and fungi. However, there are reports of heterotypic interactions between cadherins and other proteins (Takeichi 1995), so they could also have a role in wall adhesions.

The best-known animal "model interaction" involving membrane skeletons, the cytoskeleton, the PM and the ECM is the **focal adhesion** (or focal contact), in which **integrins** are the dominant linkers between the inside and outside of the cell (Jockusch et al. 1995; Burridge and Chrzanowska-Wodnicka 1996). However, lesser-known molecules, such as **syndecan 4** and **dystroglycan,** are also present that are transmembrane glycoproteins interacting with both the ECM and the cytoskeleton (Burridge and Chrzanowska-Wodnicka 1996). The existence of homologues of these molecules in walled cells is totally unknown, but their little-appreciated presence in the very well-studied focal adhesions indicates the potential for discovering molecules with similar properties in walled cells.

C. Extracellular Matrix Molecules

The best-known ECM molecules that bind to the transmembrane proteins are the proteins that interact with integrins (see Chap. 8, this Vol.). These include **collagens, fibronectin, vitronectin, laminin** and **thromobospondin** (Jockusch et al. 1995). The most common binding site for integrins on these ECM proteins is an **RGD** (arginine-glycine-aspartic acid) motif, hence the use of peptides containing this sequence as specific inhibitors of integrin-ECM interactions. However, there are a number of other motifs that have been shown to mediate integrin binding to ECM molecules (Ruoslahti 1996). Thus, while a specific cellular response to RGD peptides is an indication of integrin-ECM interactions, the absence of such a response does not indicate the absence of either integrin or other trans-PM protein to ECM interactions. Given the evolutionary distances between walled cells and the often highly specialized animal cells in which adhesion has been studied,

differences in both motifs and ECM proteins are highly probable.

III. Walled Cells: Plants Do It Too

In contrast to animal cells, which have the ability to migrate across surfaces and dramatically change their shapes as they go, the shape of walled cells is typically imposed by the cell wall. Changes in walled cell shape must involve changes in the wall, but there are likely cytoplasmic influences. Superficially, **turgor pressure** obviates the "need" for PM-ECM adhesion. Both reversible plasmolysis and the formation of viable protoplasts support this suggestion. However, in animal cells, the cytoplasm-PM-ECM connection is not only related to adhesion and motility, it also plays a major role in **signaling** between the environment and the cell (e.g. Burridge and Chrzanowska-Wodnicka 1996). Walled cells have similar requirements, as witnessed by the diversity of tropisms and responses to both pathogens and symbionts. Developmental morphogenesis also involves appropriate responses to neighboring cells. Thus there is a priori justification for expecting cytoplasm-PM-ECM connections in walled cells that are comparable to those in animal cells, as previously discussed (Wyatt and Carpita 1993; Fowler and Quatrano 1997). An accelerating accumulation of experimental data supports this expectation. Since the biology of plant and algal cells is in many respects comparable to that of the Mycota, it is worth briefly reviewing the evidence for the sort of cytoplasm-PM-ECM interactions that are emerging from plant and algal groups. The existence of such interactions in other walled cells, as well as in animal cells, supports the hypothesis that they are also present and significant in the fungi.

In essence, the evidence for plant and algal cytoplasm-PM-ECM interactions is of two types: (1) morphological and functional indications of interactions and (2) existence of molecules related to well-known (largely animal) molecules of the types mentioned previously.

A. Morphological and Functional Interactions

The archetypal morphological indication of PM-ECM interaction is the **Hechtian strand**, fine cytoplasmic extensions between plasmolyzed cytoplasm and the cell wall and known for over 100 years (Bower 1883, reviewed in Oparka 1994). While some strands are likely to result from wall adhesion mediated via **plasmodesmata**, others are certainly not because they occur on walls lacking plasmodesmata (e.g. Kropf et al. 1993; Reuzeau and Pont-Lezica 1995; Henry et al. 1996). Thus the terminations at the wall must indicate strong and localized membrane-wall adhesion. The Hechtian strand is typically a very fine strand, with a correspondingly small attachment area at the wall, but larger adhesions are also known (Kagawa et al. 1992; Henry et al. 1996). **Hydroxyproline-rich glycoproteins** may play a role in these adhesions, but **vitronectin-like** molecules are more likely responsible (Pont-Lezica et al. 1993). In a number of adhesions, the interaction appears to involve **integrins**, or integrin-like proteins, since the adhesions are disrupted by RGD peptides that compete for integrin-binding sites on ECM molecules in animal cells (e.g. Schindler et al. 1989; Wayne et al. 1992; Zhu et al. 1993; Henry et al. 1996; Ryu et al. 1997; Barthou et al. 1999). (Here, and subsequently, the terms for these proteins should more appropriately be integrin-like, spectrin-like, etc. However, the unqualified terms will be used for simplicity.)

At the ultrastructural level, the very long-standing observations of morphological interactions between microtubules and the PM (Ledbetter and Porter 1963) and both microtubules and actin filaments and the PM in cells synthesizing **cellulose cell wall fibrils** (e.g. Seagull and Heath 1980) indicate cytoskeleton-PM-ECM interactions. More generally, the well-known influence of cytoplasmic microtubules on cellulose synthesis in developing secondary plant cell walls (reviewed in Heath 1974; Heath and Seagull 1982) can only be explained by the existence of dynamic cytoskeleton-PM-ECM interactions. Also indicative of a specialized part of the cytoskeleton forming a true plant membrane skeleton is the observation of a membrane skeleton containing actin in isolated PM vesicles (Sonesson and Widell 1993).

Another type of indication of the existence of cytoskeleton-PM-ECM interactions is shown by the abilities of **extracellular molecules to influence the cytoskeleton**. For example, cell wall proteins stabilize microtubules against cold shock (Akashi and Shibaoka 1991; Akashi et al. 1990) and extracellular proteases disrupt cytoplasmic actin orga-

nization and cytoplasmic streaming (Masuda et al. 1991; Ryu et al. 1995).

B. Animal-Related Molecules

Molecules related to animal cytoskeleton-PM-ECM complexes are increasingly identified from plant materials. Examples include **spectrins** (Michaud et al. 1991; de Ruijter and Emons 1993; Faraday and Spanswick 1993; Reuzeau et al. 1997; de Ruijter et al. 1998; Holzinger et al. 1999; Braun and Wasteneys 2000), **integrins** (Quatrano et al. 1991; Nelson and Bohnert 1994; Gens et al. 1996; Katembe et al. 1997; Reuzeau et al. 1997; Falk et al. 1998; Lynch et al. 1998; Laval et al. 1999; Swatzell et al. 1999), **vitronectins** and **fibronectins** (Sanders et al. 1991; Wagner et al. 1992; Pont-Lezica et al. 1993; Zhu et al. 1993; Wang et al. 1994; Gens et al. 1996) and **cadherins** (Baluska et al. 1999). In general terms, the primary criteria for identification of these plant molecules have been immunological cross-reactivity with animal antibodies, predicted molecular weights and predicted cellular localizations (although the anti-cadherin was reported to localize to both the endoplasmic reticulum and nucleoplasm, unexpected sites). However, there have been further characterizations, with unexpected results.

One of the **vitronectins** from *Lilium* (Wang et al. 1994) turned out to be a phosphoglycerate mutase (Wang et al. 1996) and that from *Nicotiana* has strong sequence similarity to a translation elongation factor, yet still shows many functional features, including cellular localization, of animal vitronectins (Zhu et al. 1994). Thus, while initial indications of the presence of cytoskeleton-PM-ECM ensemble molecules in plants are interesting and certainly indicate the possibility of animal homologues and analogs in plants, the evidence is still far from conclusive. Nevertheless, it is very clear that there are functional interactions between the cytoskeleton, the PM and the ECM, as noted above. Thus, it must only be a matter of time before the characterization of the molecules involved is accomplished. Interestingly, it has been suggested that one of the major classes of plant cell wall proteins, the **arabinogalactan glycoproteins**, function as vitronectin equivalents (Pennell et al. 1989). The recent demonstration of the influence of a selective inhibitor of these molecules (Yariv reagent) on pollen tube tip growth (Roy et al. 1999) provides support for such a suggestion. Furthermore, Akashi and Shibaoka (1991) have suggested that the very common cell wall **hydroxyproline-rich glycoproteins** (also known as **extensins**) function as cell wall elements interacting with the cytoskeleton. Given the evolutionary distance between plants and animals, it would not be surprising if significantly different molecules with common functions have evolved in the two groups, as is apparently the case with the *Nicotiana* vitronectin (Zhu et al. 1994). Since the fungi are phylogenetically more closely related to animals than plants, their cytoskeleton-PM-ECM related molecules may be less divergent from those of animals, but that remains to be seen.

IV. Fungal Processes

As we shall see, the Mycota exhibit a range of observations similar to those discussed for plants, which indicate that they too have evolved cytoskeleton-PM-ECM interactions. In essence, the primary evidence for the existence of a membrane skeleton-PM-ECM ensemble in fungi includes

1. Descriptions of **membrane skeleton-like structures and components** such as peripheral **actin** (reviewed in Heath 1990) and **spectrin** (Kaminskyj and Heath 1995; Degousée et al. 2000; Fig. 1).
2. Descriptions of **trans-PM proteins** such as **integrins** (Marcantonio and Hynes 1988; Santoni et al. 1994; Hostetter et al. 1995; Kaminskyj and Heath 1995; Gale et al. 1996) and **cell wall proteins** with putative integrin-binding sites (Laurent et al. 1999).
3. The existence of **PM-cell wall adhesions** (Kaminskyj and Heath 1995; Bachewich and Heath 1997a).
4. Effects of **RGD peptides** on growth or differentiation (Corrêa et al. 1996).
5. Morphogenic effects of expression of a heterologous **integrin** gene in *Saccharomyces* (Gale et al. 1996).
6. Morphological observations of **cytoplasmic specialization adjacent to the PM**, in the region of a putative membrane skeleton (Howard 1981; McKerracher and Heath 1986c; Baba et al. 1989; Kaminskyj and Heath 1996; Heath and Janse van Rensburg 1996; Fig. 1).

In addition to the above evidence, there is an increasing body of literature identifying a number of proteins (e.g. Sla1p, formins, Arp2p, Arp3p) that specifically interact with, and regulate the formation of, PM-associated actin arrays in both *Saccharomyces* and *Schizosaccharomyces* (Li et al. 1995; Evangelista et al. 1997; Lechler and Li 1997; Fujiwara et al. 1998; Ayscough et al. 1999; Morrell et al. 1999; Yang et al. 1999; Tang et al. 2000). These clearly form part of the membrane skeleton and no doubt future work will reveal homologues in hyphal species as well.

However, there are many other observations that point to the functioning of membrane skeleton-PM-ECM ensembles in numerous processes found in the fungi. I shall provide a more detailed analysis of these processes.

A. Morphogenesis

Hyphal and yeast (both budding and fission) morphogenesis depends on the accurate regulation of cell **surface extensibility**, which, by definition, involves the membrane skeleton-PM-ECM ensemble. While some models of the process place extension regulation solely in the cell wall, another places more emphasis on the membrane skeleton, but also with hypothesized linkages through the PM to the cell wall. The latter is most relevant to the present topic and has been reviewed extensively (Heath 1990, 1994, 1995; Jackson and Heath 1990b; Kaminskyj and Heath 1996; Heath and Janse van Rensburg 1996; Gupta and Heath 1997). The main points made in these papers are:

1. Hyphal tips are enriched in **actin** that is associated with the apical PM.
2. The organization of the apical **actin changes** between growing and non-growing hyphae.
3. The organization of the apical **actin reflects the extensibility** of the hyphal tip, being most concentrated in the regions of greatest extensibility, at least in some species.
4. **Disruption** of the apical **actin** leads to **disruption of hyphal growth**, tip shape and or tip bursting.
5. Actin mutants have **increased osmotic sensitivity**.
6. Morphological abnormalities result from disruption of **actin-binding proteins**.
7. Initiation of **new tips** involves the assembly of new tip-specific patterns of PM-associated actin.

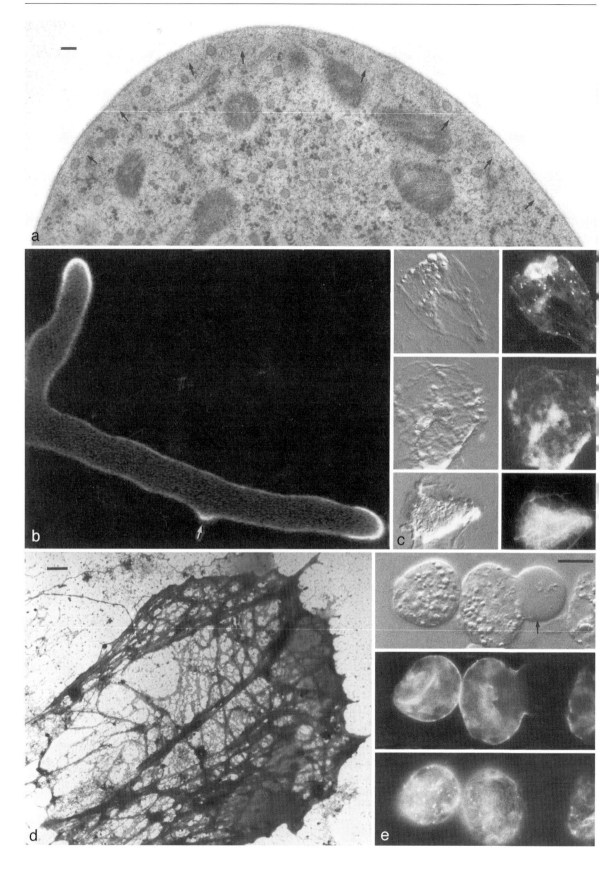

8. **Integrin** and **spectrin** concentrate in the hyphal tips, but only shown for one species.
9. Tips contain a **differentiated layer of cytoplasm** adjacent to the apical PM, in the predicted region of the membrane skeleton (Fig. 1).

There are a number of other observations that support and extend this previously reviewed evidence for the role of the membrane skeleton-PM-ECM ensemble in determining tip morphogenesis.

With respect to **spectrin**, its presence has now been extended to *Neurospora*, where it forms a very tightly concentrated "**cap**" in the tips of growing hyphae (Degousée et al. 2000; Fig. 1). This extends the reporting of spectrin to a Dikaryomycota species, from the previous single report from an Oomycota species. It also demonstrates that, in a species where the PM-associated apical actin is less obviously "cap like" and less abundant than it is in the Oomycota, spectrin is more prominent. This suggests the possibility of functional complementation (i.e. one or the other may dominate in different species) between these two membrane skeleton components. It is also notable that in *Neurospora*, the spectrin is identified on the basis of immunoreactivity, molecular mass, PM-association and actin-binding capacity, but further characterization is lacking. In *Saccharomyces*, which lacks an obvious spectrin homologue, a much smaller protein of unknown function shares some domains with spectrins (Donnelly et al. 1993). Mutation of this gene causes disruption of the normal cortical PM-associated actin and abnormal cell morphology. Any evolutionary or functional relationship between these *Neurospora* and *Saccharomyces* proteins is purely speculative,

but these apparent similarities are intriguing and suggest that functionally homologous proteins may have substantial differences.

With respect to the participation of **integrins**, there are two very interesting observations. In germinating *Uromyces* spores, the germ tubes are initially abnormally swollen when produced in the presence of **RGD peptides** (Corrêa et al. 1996), thus implicating a role for integrins in germ-tube morphogenesis. However, paradoxically, continued germ-tube growth in the same medium is morphologically normal. Subsequent induction of differentiation of the germ tubes into **appressoria**, which is also a morphogenetic process, is again inhibited by the peptide. The explanation of these results is obscure, although, as suggested for the interspecies complementation between actin and spectrin, perhaps the initial encounter with the peptide disables the normal functioning of integrin, and then another membrane skeleton-PM-ECM component substitutes for the integrin for the duration of germ-tube growth until another integrin "use" (i.e. appressorium formation signaling) is required.

In *Candida*, an **integrin** gene product seems to be involved in host adhesion (see Sect. IV.G). This gene has been expressed in *Saccharomyces*, where it induces the **formation of germ tubes** quite unlike any structure normally produced by that species (Gale et al. 1996). The mechanisms involved in the process are unclear, but the ability of an integrin to induce germ tubes in a budding yeast is certainly an indication of a role for related proteins in hyphal morphogenesis.

With respect to the participation of **connections through the PM** to the cell wall, perhaps the

Fig. 1. Illustrations of putative membrane skeletons from *Neurospora crassa* (**a, b**) and *Saprolegnia ferax* (**c–e**). **a** Near-median longitudinal section of a freeze-substituted hyphal tip showing the characteristic zone of exclusion containing only diffuse amorphous material and vesicles, presumably in transit, adjacent to the inside of the plasma membrane (*PM*) in the region occupied by a putative membrane skeleton. Approximate inner margin indicated by *arrows* (cf. Kaminskyj and Heath 1996). *Bar* 100 nm. **b** Hyphal tips immunostained with an antibody directed against animal spectrin, showing dominant PM staining which forms caps over both established and emerging new (*arrow*) tips. Details in Degousée et al. (2000), courtesy of G. Gupta and N. Degousée. **c** Images of "wet cleaved" protoplast PM patches. Protoplasts were attached to a plastic surface, cleaved with a nitrocellulose sheet (Brands and Feltkamp 1988) to remove the upper PM and cytoplasm and then stained with rhodamine-labeled phalloidin to reveal attached actin filaments. *Left panels* Nomarski differential interference contrast images; *right panels* rhodamine fluorescence images of the same fields. **d** Similar material to **c**, but cleaved by a jet of buffer across the attached protoplast and viewed in the transmission electron microscope to show attached filamentous material. *Bar* 1 μm. **e** Median (*upper two panels*) and upper surface (*bottom panel*) focal plane images of intact protoplasts stained with rhodamine-labeled phalloidin, showing cytoplasm and PM-associated actin (*lower panels*), that is fibrillar and punctate (*lowest panel*), as it is in hyphae (Heath 1987), and absence of actin associated with the artefactually protruded or "herniated" PM (*arrow, upper panel*). These protoplasts were concentrated by centrifugation in the enzyme solution that removed their walls and the fixative was immediately added to the pellet, thus rendering it unlikely that any cell wall had regenerated prior to fixation and staining. *Bar* 10 μm applies to **b, c** and **e**

most salient observations are the presence of adhesions between the apical cytoplasm (presumably including the PM) and the cell wall in growing hyphal tips (Kaminskyj and Heath 1995; Bachewich and Heath 1997a). These adhesions are enriched in actin and integrin (Kaminskyj and Heath 1995) and are clearly regulated since they are less abundant or strong in slow or non-growing hyphae (Kaminskyj and Heath 1995; Bachewich and Heath 1997a).

In addition to the above specific observations on the participation of components of the membrane skeleton-PM-ECM ensemble, there is also evidence for the predicted function of the membrane skeleton in strengthening the PM. **Protoplasts** are generally osmotically fragile (as predicted if the cell wall is playing its obviously important role in providing strength to the cell surface, with or without the membrane skeleton), but this fragility is not always as great as might be supposed. Protoplasts of *Saprolegnia* retain their actin-containing membrane skeleton and, when the osmolarity of their medium is reduced below isotonicity, some lyse while many others do not (Heath et al. 2000). This result indicates significant strength in the PM, presumably imparted by the PM-associated actin. Interestingly, under some unexplained circumstances, "tears" occur in the protoplast membrane skeletons and from these "tears" unsupported PM (lacking associated actin) protrudes (Heath, unpubl.; Fig. 1), thus indicating the normal role of the membrane skeleton in strengthening the PM.

Taken together, the above observations provide strong support for the hypothesis that a **membrane skeleton-PM-ECM ensemble plays a major role in regulating the extensibility of hyphal tips**, although the details of both the way in which the different elements interact and are regulated remain unclear, and will be very difficult to clarify in such a highly dynamic situation.

B. Organelle Motility and Mitosis

The regulated movement and positioning of fungal organelles, including nuclei, is of critical importance to many aspects of fungal cellular and developmental biology and has been reviewed extensively (e.g. McKerracher and Heath 1987; Heath 1994; Steinberg 1998). In addition to the obvious examples of motility, spindle elongation during **anaphase B** and the subsequent **telophase**

are also at least conceptually related to **nuclear motility** and are likely to employ similar motor and regulation. In general terms, there is evidence for both **microtubules** and **actin** playing direct roles in these processes, with the necessary force generation residing in mechanochemical motor molecules such as the **kinesins, dyneins** and **myosins**. Discussion of these topics is outside the scope of this review, and in many cases it is clear that it is the non-peripheral components of the cytoskeleton that are involved. However, there is increasing evidence for involvement of the peripheral membrane skeleton-PM-ECM ensemble.

On a purely theoretical basis, if **force** is applied to an organelle such that it moves, there must be an **equal and opposite force** applied to some other structure(s). In the cellular context the "other structure" can most certainly include the cytoskeleton; the cytoplasmic milieu is clearly not a simple solution. Rather it is permeated by the complex network of interlinked filamentous proteins of which microtubules and actin are the most prominent components. However, this organization presents further complexity in its own right since the organelles must move through the cytoskeletal network, that is hypothetically both weak enough to permit their passage yet strong enough to sustain the forces responsible for that movement. With respect to larger organelles such as the nuclei, this is a non-trivial problem; they displace a very large percentage of the cytoplasmic volume as they pass along the hyphae or through a yeast cell.

In basic respects, the problems of **organelle motility** are comparable to **cell motility**, where the forces for movement are transmitted to the substrate over which the cell moves. This may be a secreted ECM or an inanimate substrate such as a coverslip. This aspect of cellular behavior is comparable to cytoplasmic migration within the hypha and is discussed in more detail in Section IV.C. However, cell motility is simpler in that the load-bearing substrate does not surround the cell as it does the moving organelle in organelle motility. Conceptually, one hypothetical solution to the problem of separation of substrate and moving organelle is to use the membrane skeleton-PM-ECM ensemble as substrate. This concept would leave the central cytoplasm/cytoskeleton relatively weak and deformable, capable of sustaining motility of only small organelles while deforming for the larger organelles. There are supporting data for this hypothesis.

Organelle motility based on force application to the membrane skeleton-PM dates back to Aist and Bern's work on **anaphase B** in *Nectria* (Aist and Berns 1981). They showed pulling forces exerted on the mitotic spindle, associated with microtubules emanating from the spindle poles (**astral microtubules**) and the interaction of some of those microtubules with the PM. In a more extensive study, Aist and Bayles (1991) described the presence of **10 nm diameter filaments** associated with the sites of interaction between the astral microtubules and the PM, but only about 2% of the astral microtubules approached the PM. This led them to conclude that those microtubules may be of little importance in force generation. Nevertheless, the interactions with the filaments and the PM do indicate a potentially interesting force-generating system at the PM that requires explanation. Interestingly, the filaments were not only larger than actin filaments, but also anaphase B was insensitive to presumed actin disruption by **cytochalasin**. At face value, this suggests that the filaments are a non-actin part of a membrane skeleton. Unfortunately, further studies on the same species showing the role of **dynein in aster formation** and function (Inoue et al. 1998a,b) do not help localize the site of force generation.

In a study comparable to those above, but focused on **postmitotic nuclear migration** in *Pleurotus*, it was also observed that only 2% of astral microtubules associated with migrating nuclei terminated in the vicinity of the PM (Kaminskyj et al. 1989). However, 24% of non-astral microtubules laterally associated with the nuclear envelope had such associations and could potentially mediate nucleus to membrane skeleton-PM-ECM interactions. Nevertheless, for a number of reasons, the authors' conclusions favored nuclear interactions with the cytoskeleton, not necessarily with the membrane skeleton. A similar conclusion was reached concerning the role of microtubules adjacent to **moving nuclei in *Basidiobolus*** (McKerracher and Heath 1985, 1986b). In contrast, in *Saprolegnia*, interphase nuclei are consistently positioned with their centrioles facing, and in close proximity to, the PM, where their associated **astral microtubules** are well placed to interact with a membrane skeleton (Heath and Kaminskyj 1989). In these same hyphae, the nuclei become distorted during plasmolysis-induced cytoplasmic streaming, and the consistent pattern of distortion indicates that their **peripheral regions are more firmly anchored** than their more central

regions (Bachewich and Heath 1997a). In the only species examined with the appropriate technique, the peripheral cytoplasm in the region of the putative membrane skeleton is distinctly **enriched in filamentous material** (McKerracher and Heath 1986c), but the generality of this observation remains to be seen.

While there are numerous other studies of mitotic and interphase nuclear migration in mycelial species, none have identified the location of the force-generating machinery (e.g. review of Morris et al. 1995). In contrast, there is an emerging consensus that, in *Saccharomyces*, nuclear migration does involve interactions between **astral microtubules and the cell cortex**, mediated by a diversity of cortical proteins (recently reviewed in Reinsch and Gönczy 1998, in which it is noted that the cell periphery is involved in nuclear positioning in many cell types, see also Lee et al. 1999). Thus, while there are a priori reasons and some experimental evidence to believe that nuclear movements in both mitosis and interphase involve the membrane skeleton-PM-ECM ensemble, the generality and details of the phenomenon remain uncertain.

The relationship between astral microtubules and the membrane skeleton-PM may be complex because PMs isolated from *Neurospora* are not only enriched in actin, spectrin and integrin, but also tubulin, the organization of which is unknown (Degousée et al. 2000). Both microtubules and tubulin patches associate with *Saccharomyces* protoplast PMs (Vavricková et al. 1998). Furthermore, proteins specifically able to interact with both microtubules and actin are becoming well characterized in animal cells (Leung et al. 1999).

The role of the membrane skeleton in motility of **other organelles** is as uncertain as with the nuclei. For both **mitochondria** (Simon et al. 1995, 1997; Hermann et al. 1997; Yaffe 1999) and **vacuoles** (Hill et al. 1996) there is evidence from *Saccharomyces* that actin is important for their directed movements during budding, and, since the *Saccharomyces* actin arrays are attached to the PM, this indicates a role for the membrane skeleton-PM-ECM ensemble. Within mycelial species, there is no evidence for a direct role. Indeed, for a class of highly **motile vesicles** of unknown function, they clearly do not associate with the cell periphery (Heath 1988). Conversely, since the general cytoplasm does seem to be linked to the cell wall (see Sect. IV.C), the argument can be made that indirectly all movements related to the cytoskeleton transmit force to the membrane skeleton-PM-ECM ensemble. Perhaps the critical point is to consider the entire cytoskeleton as an **integrated ensemble** capable of both generating and sustaining forces (e.g. McKerracher and Heath 1986b) and transferring these to the cell wall, with variations in the distance and number of intermediates between the site of force generation and its transfer to the cell wall. This concept then permits incorporation of the demonstration of organelle motility in **protoplasts**, where clearly there is no cell wall (Steinberg and Schliwa 1993). The cells were attached to a glass substrate that perhaps assumed the role of the normally present cell wall, as suggested by Chiu et al. (1997).

C. Cytoplasmic Migration

Reinhardt (1892) introduced the idea that fungi are **tube-dwelling amoebae**, a concept that is still viable and has recently been reviewed (Heath and Steinberg 1999). The essential feature of the idea is that the cytoplasm migrates through the tubular hypha like an animal cell migrates over a solid substrate. The concept is best exemplified in genera such as *Basidiobolus* (Robinow 1963; McKerracher and Heath 1986a) and *Uromyces* (Heath and Heath 1978, 1979), but is more widespread. Essentially, any hypha that has subapical vacuolation in effect shows cytoplasmic migration of the apical mass of cytoplasm as it moves out of the vacuolate regions. Since the cytoplasm normally retains its association with the hyphal tip, its migrations must be net unidirectional towards the tip. Comparison with animal cell migrations suggests the presence of **PM-wall adhesions**, and its polarization towards the hyphal tips suggests that these adhesions should be restricted to, more abundant or stronger at hyphal tips relative to subapical regions.

   A number of pieces of data indicate that there are **apical adhesions** and that they are more abundant or stronger in the hyphal tips

1. Plasmolysis produces **remnants of PM and actin** attached to the apical walls. These remnants are larger at the tips than the subapical Hechtian strands (Kaminskyj and Heath 1995; Bachewich and Heath 1997a).
2. Plasmolysis **draws the cell wall inward** in a region just behind the apex, where the PM

remains attached to the wall via very broad adhesions (Bachewich and Heath 1997a).
3. Hyphal tips contain tip-high gradients of **integrins** (Kaminskyj and Heath 1995; Degousée et al. 2000).
4. Cytoplasm can be induced to **contract**, and when it does the contraction is predominantly unidirectional towards the tips (McKerracher and Heath 1986a, 1987; Heath 1990; Kaminskyj et al. 1992; Bachewich and Heath 1997a).
5. The peripheral plaques of actin, that are typically concentrated in the tips of most hyphae outside of the Oomycota, have been suggested to be **focal adhesion equivalents** (Hoch and Staples 1983; Adams and Pringle 1984; Roberson 1992). There is little direct evidence for this. Their motility in *Saccharomyces* (Doyle and Botstein 1996) argues against this interpretation. However, there is no reason to believe that all plaques have similar form or function; those in hyphae appear to be related to **filasomes** (Howard 1981; Bourett and Howard 1991; Roberson 1992), whose function is unknown but does not exclude an adhesive-type role. Thus the focal adhesion suggestion may still be valid for hyphae.
6. Fungal cell walls contain proteins with putative integrin-binding sites (RGD sequence), comparable to those on animal ECM molecules (Laurent et al. 1999).
7. At the tips of *Schizosaccharomyces*, there are specific "granules" that appear to attach actin filaments to the PM, as might be seen for a small **focal adhesion equivalent** (Kanbe et al. 1989).
8. When cytoplasmic migration into Oomycota **sporangia** ceases, after basal cross-wall insertion, the previously abundant apical PM-wall adhesions are much reduced (Bachewich and Heath 1997a).

   Together, the above data indicate that cytoplasmic migration through hyphae probably involves the functional equivalents of **apical focal adhesions**. In addition, the morphology of cytoplasm re-entering hyphal tips from which it had previously been induced to withdraw is very reminiscent of **amoeboid** animal cell behavior (Bachewich and Heath 1997b, 1999). The cytoplasmic aspects of migration are beyond the scope of this review and have been discussed previously (McKerracher and Heath 1987; Heath 1990, 1994; Heath and Steinberg 1999).

D. Ion Transporter Localizations

All critically examined growing hyphal tips contain a steep tip-high gradient of $[Ca^{2+}]_{cyt}$, reviewed in Jackson and Heath (1993) and more recently by Levina et al. (1995), Hyde and Heath (1997), and Silverman-Gavrila and Lew (2000). There are also reported gradients of $[H^+]_{cyt}$, although tip-high (Turian 1979, 1981; McGillviray and Gow 1987), tip-low (Roncal et al. 1993; Robson et al. 1996) and no gradients (Bachewich and Heath 1997b; Parton et al. 1997) have all been reported in different species (see Chap. 4, Sect. III.C, this Vol.). The presence of a dynamic gradient implies both localized input (at the tip, for a tip-high gradient) and removal (subapically, for a tip-high gradient).

With respect to Ca^{2+}, their input in *Saprolegnia* appears to involve **stretch-activated (mechanosensitive)** Ca^{2+}-transmitting channels (Garrill et al. 1993) in the PM, although such appear to be dispensable in *Neurospora* (Levina et al. 1995). Localized influx may be due to either higher channel activation levels and/or higher channel numbers. The former is likely at the tip when the channels are stretch activated since the expanding tip is likely to produce PM stretching, but such a phenomenon is almost impossible to demonstrate experimentally. However, there is also evidence for a higher **apical concentration of channels** (Garrill et al. 1993; Levina et al. 1994). Again, there are two basic mechanisms that could achieve a gradient of channel densities, apical insertion and subapical removal or some form of "**anchorage**" of the channels at the apex. Since the PM is constantly being synthesized at the hyphal tips by the **exocytosis** of the vesicles that transport cell wall material to the developing wall, apical insertion is highly likely. Presumably, general membrane turnover, or possibly **endocytosis** (Hoffmann and Mendgen 1998), would account for subapical removal. However, there is also direct evidence that, at least in *Saprolegnia*, the channel gradient is maintained by action of the actin-containing membrane skeleton (Levina et al. 1994). Such would be comparable to the known interaction between the membrane skeleton of erythrocytes and their ion channels (Bennett 1990). Furthermore, in *Saprolegnia*, the **actin-containing membrane skeleton** characteristic of the growing tips actually forms prior to the protrusion of a new tip (Bachewich and Heath 1998), at about the same time as the **trans-PM ion fluxes**

characteristic of growing tips are also initiated (Kropf et al. 1983). Thus the new membrane skeletons are formed in a time and position in which they could function to recruit channels for new tip formation, although there is not direct evidence for such a process. Furthermore, if the actin-containing membrane skeleton is prevented from forming by inhibitor treatment, polarization to form a new tip does not occur (Bachewich and Heath 1998), although it is not clear whether this is due to inability to cluster ion transporters or to assemble other elements of the tip-forming apparatus.

This suggested **localization of ion channels** by the membrane skeleton does not involve the ECM. However, it is comparable to the situation at the tips of developing algal rhizoids where tip formation involves both channel (Shaw and Quatrano 1996) and actin (Kropf et al. 1989) clustering at the tip and where Ca^{2+} currents enter the cell (Kropf 1992). Interestingly, in this situation, the "**fixation**" of this ensemble apparently requires the participation of the cell wall (Kropf et al. 1993), thus indicating involvement of tripartite membrane skeleton-PM-ECM interactions. Direct evidence for such is lacking in hyphae, but there are co-localizations of channels, membrane skeleton actin and cytoplasm-wall adhesions in *Saprolegnia* tips (Sect. IV.C). The cast may well be assembled, but the plot of the play is not yet clear.

Further data on the mechanisms responsible for localizing ion transporters is lacking for any fungi. Indeed, there are no other data that demonstrate that channels are in fact distributed (as opposed to activated) in a non-random manner, although such is certainly likely, and has been suggested, with the reports of various ion current patterns associated with hyphae (e.g. Schreurs and Harold 1988; Takeuchi et al. 1988; Lew 1999; see Harold 1994). Furthermore, it has been suggested that stretch-activated channels are involved in thigmo-sensing (Zhou et al. 1991; Watts et al. 1998; see Sect. IV.G) where they are likely to be active and probably concentrated at the contact area, but such remains interesting speculation until their distribution can be described in detail.

E. Cell Wall Synthesis

Cell walls are essentially two-phase composites with **fibrillar** elements (e.g. chitin) embedded in, and attached to, a **matrix** phase; the structure and synthesis of all of which have been reviewed by Wessels (1990) and Ruiz-Herrera (1991). There is substantial diversity in the composition of both phases, especially the matrix phase (see Sentandreu et al. 1994). However, for present purposes, the main point is that matrix phase materials are secreted primarily by **exocytosis** of vesicles derived from the endomembrane system and fibrillar components are primarily synthesized by **PM-bound enzymes.** There are post-secretory and post-synthetic changes in the polymers and their interactions (Wessels 1990), but these are beyond the scope of this review. However, since the synthesis and composition of cell walls is developmentally regulated and typically highly localized, it is likely that membrane skeleton-PM-ECM interactions are involved. There is direct evidence for roles of the ensemble in both matrix and fibril synthesis localization.

1. Exocytosis Localization

The primary site of exocytosis in hyphae is the hyphal tip. One of the features universally present in growing (and thus exocytosing) tips is a **tip-high gradient of actin**, adjacent to, and often associated with, the PM (Heath 1990). Merely by its presence and high concentration it must influence exocytosis; the vesicles must traverse the actin in order to fuse with the PM. However, the role of actin in **exocytosis localization** is probably more specific than this, although direct data on its precise functions are sparse.

One facet of exocytosis localization is the vectorial transport of the vesicles so that they achieve high concentration at their site of **exocytosis**. Analysis of both inhibitor-treated cells and mutations in both cytoskeletal proteins (actin and microtubules) and their associated motors indicate that both actin and microtubules are involved in vesicle transport, although details of the way in which the two components interact are unclear (see Heath 1994; Steinberg 1998; see also Johnston et al. 1991; Kanbe et al. 1993; Schott et al. 1999). In these studies it is unclear what population of actin may be involved. In *Saccharomyces* there are both peripheral plaques and more central filaments of actin, the latter focusing on the sites of exocytosis (e.g. Adams and Pringle 1984). Here it is most likely the central filaments that are important, but in hyphal species the dominant reported actin ensembles are the peripheral plaques that

are apparently not well positioned to function in vesicle transport. However, peripheral actin may not be the only population of actin present in the cytoplasm, as previously discussed in general (Heath 1990; Kaminskyj and Heath 1996), and demonstrated directly, in hyphae of the Oomycota (Jackson and Heath 1990a) and Dikaryomycota (Bourett and Howard 1991; Roberson 1992).

With respect to the present discussion, it is the peripheral, **PM-associated actin** that is of interest, and there is direct evidence that it does function in transport of vesicles to their sites of exocytosis. In *Saprolegnia*, the vesicles move radially from the subapical Golgi bodies and then move towards the tips only in the peripheral cytoplasm adjacent to, and aligned along, the actin filaments that are located adjacent to the PM (Heath and Kaminskyj 1989). Similarly, in the same organism, the formation of radial arrays of PM-associated actin focusing on the sites of new tip formation (Bachewich and Heath 1998) is consistent with the vesicles being transported along that actin and being brought to their site of exocytosis, although such is certainly not proven (see Sect. IV.D for another similarly consistent function).

An additional observation indicating that the PM-associated actin in *Saprolegnia* functions in the transport of vesicles to the hyphal tips is seen in the detailed observations of the way in which growing **tips swell in response to actin disruption** (Gupta and Heath 1997). As noted in Sect. IV.A, actin disruption results in initial acceleration of tip growth, followed by cessation of tip growth and subapical swelling of the cell wall that would not normally have expanded (i.e. had reached normal maximum hyphal diameter). The extent of this subapical swelling is similar to the extent of predicted tip growth that would have occurred in the same period of time and is thus most easily explained by exocytosis of the *in transit* population of vesicles at subapical sites following disruption of their normal transport to the hyphal tips. Tip swelling induced by actin disruption is a common phenomenon among hyphae of diverse species (Heath 1990, 1994), but it has not been described in sufficient detail in other species to determine the generality of the above sequence.

Further direct evidence for a role of the PM-associated actin in determining the site of exocytosis in *Saprolegnia* is seen when that actin (and also possibly the cytoplasmic actin, that was not analyzed) is disrupted for a long time with **latrunculin B** (Bachewich and Heath 1998). One result of this treatment is the delocalized production of much-thickened cell wall throughout the apical and subapical regions of the hyphae. While not analyzed, it is likely that this wall contains both matrix and fibrils and thus involved delocalized and misdirected exocytosis. A similar phenotype, including actin disorganization, was observed in *Aspergillus* when a gene involved in **nuclear migration** was disrupted (Chiu et al. 1997). The function of the gene product is unknown, but, given the possible relationship between organelle motility and the membrane skeleton (Sect. IV.B), common disruption of nuclear motility and cell wall deposition by a single gene disruption may have its basis in the membrane skeleton-PM-ECM ensemble.

Among the Dikaryomycota, tip growth and thus exocytosis involves the **Spitzenkörper** (Lopez-Franco and Bracker 1996). While it is clear that the Spitzenkörper includes a substantial fraction of the apical exocytotic vesicles (Grove and Bracker 1970), its precise composition and role in tip growth is unclear. One hypothesis (Bartnicki-Garcia et al. 1989) describes it as functioning as a vesicle supply center (VSC), with an obvious role in exocytosis. With respect to the membrane skeleton-PM-ECM ensemble, it is interesting to note that the Spitzenkörper is rich in **actin** (Bourett and Howard 1991) and highly **mobile** (Lopez-Franco and Bracker 1996; Riquelme et al. 1998). The molecules responsible for this mobility are unknown, but there must be a base against which the movements are generated and the apical membrane skeleton-PM-ECM ensemble is an obvious, although undemonstrated, candidate. Whatever the precise composition and role of the Spitzenkörper proves to be, it is associated with exocytotic vesicles, contains actin, and likely interacts with the inside of the PM in some way, thus giving it a role in regulating exocytosis.

Once the exocytotic vesicles have arrived at the PM, exocytosis depends on appropriate docking and subsequent fusion of membranes. These processes are based on a family of proteins known as **SNAREs**, one set of which is designated t-SNAREs ("t" for target). In principle, the t-SNAREs are located on the target membrane (in the case of the hyphal tip, the apical PM) where they form part of a receptor system to mediate the docking of vesicles with complementary v-SNAREs ("v" for vesicle) (Rothman 1994; Ferro-Novick and Jahn 1994; Pelham 1999). The t-SNAREs are part of the membrane skeleton-

PM-ECM ensemble, being attached to the PM. The apical gradient of exocytosis in a hyphal tip predicts a **gradient of t-SNAREs**, and Drubin and Nelson (1996) have suggested that PM-associated actin may play a role in generating or maintaining such a gradient. While much of the SNARE story has been elucidated in *Saccharomyces*, where at least one PM t-SNARE is *not* restricted to sites of exocytosis (Brennwald et al. 1994), there has been little work on these proteins in hyphal species. However, recently, a t-SNARE has been identified in *Neurospora* and shown to have a tip-high concentration at apical PMs (Gupta and Heath 2000). Thus there is reason to believe that SNAREs form part of the hyphal membrane skeleton and need to be considered in future studies on the regulation of exocytosis and thus cell wall matrix production.

2. Fibril Synthesis Localization

As mentioned in Section III.A, the formation of oriented **cell wall fibrils** (cellulose fibrils) in secondary plant cell walls certainly involves both PM-inserted enzymes and microtubules and actin filaments adjacent to the PM in what could be defined as a membrane skeleton. There is no such evidence for a role of microtubules in fungal wall fibril synthesis. However, there is evidence for a role of an actin-containing membrane skeleton in fibril synthesis. Perhaps the most comprehensive study is in *Schizosaccharomyces* where an apparent PM-located **α-1,3-glucan** (a fibrillar wall polymer; Ruiz-Herrera 1991) **synthase** co-localizes with **peripheral actin**; its distribution is dependent on that actin and it can in turn influence the distribution of the actin (Katayama et al. 1999). The nature of the interaction between the enzyme and actin is unknown, but the glucan is essential to the properties of the cell wall and the data clearly show the importance of membrane skeleton-related actin interacting with a PM enzyme and thus influencing the synthesis of the ECM. More recently, in *Saccharomyces*, Tang et al. (2000) have shown that part of a **β-1,3-glucan synthase** interacts directly with Sla1p, apparently independently of the latter's interactions with actin (referred to in Sect. IV), thus indicating that other components of the membrane skeleton can also potentially control fibril synthesis localization.

In other studies, the influence of peripheral actin or another putative membrane skeleton component in wall fibril synthesis is also very clear. For example, the previously described massive abnormal subapical wall thickenings, produced by either **actin disruption** (Bachewich and Heath 1998) or mutation of a gene involved in **nuclear migration** (Chiu et al. 1997), are very rich in fibrillar wall components, the localization of which has clearly been dramatically disturbed. These results indicate that the normal localization of the synthetic enzymes is dependent on the appropriate membrane skeleton (or likely membrane skeleton) components. Comparable results, with a strong correlation between timing and position of PM-associated actin and wall fibril synthesis – and concomitant disruption of both by an actin-disrupting drug – are seen in **regenerating protoplasts** of *Schizosaccharomyces* (Osumi et al. 1998). Similarly, chitin fibril synthesis in *Saccharomyces* is co-localized with peripheral actin and delocalized by disruption of the actin (Novick and Botstein 1985; reviewed for yeasts in general by May and Hyams 1998) or myosin (Johnston et al. 1991).

In hyphal species **septum formation** is typically accomplished with PM-associated actin (e.g. Girbardt 1979; Hoch and Howard 1980; Howard 1981; Roberson 1992; Tsukamoto et al. 1996; Momany and Hamer 1997). There are two pieces of data that suggest that the actin may be involved in guiding chitin synthases. The actin arrays are at least partially **circumferential**, as are the chitin fibrils (Scurfield 1967). This coincidence of orientation during fibril formation suggests direct actin guidance of the synthetic enzymes. Similarly, during stages of septum formation, actin fibrils radiate along the PM from the site of septum formation in a position where they are well situated for **recruiting PM-located synthetic enzymes** from adjacent regions of the PM (Butt and Heath 1988; Roberson 1992; Momany and Hamer 1997). Irrespective of these suggested direct interactions between peripheral actin and cell wall fibrils, the very consistent formation and maintenance of specific peripheral actin arrays adjacent to the PM during the formation of the fibril-rich extracellular septum are strong indications of some functional linkage between the two.

Another morphogenetic process in which the membrane skeleton-PM-ECM ensemble apparently plays a significant role is in **mushroom stipe elongation**. During this process one of the major cellular activities is extensive cell wall metabolism,

including high levels of chitin synthase activity (reviewed in Kamada 1994). As indicated above, the synthase is a PM-located enzyme, but there is also an indication for a role of peripheral actin in the process. Stipe elongation is gravitropic and the gravi-perceptive system is inhibited by **cytochalasin disruption of stipe cell actin** (Monzer 1995). Analysis of actin organization in the stipe cells was not extensive, only changes in that located around the periphery of nuclei was shown to be disrupted, but such is likely to be only part of the cellular complement. Changes in cell wall extensibility, that must explain the gravitropic response, are most likely mediated at the PM where the chitin synthase is located. Thus the actin target of the cytochalasin effect is also most likely located at the PM.

F. Environmental Sensing and Interactions

In one respect it is self-evident that interactions between the environment and the cytoplasm involve at least some aspects of the cytoskeleton-PM-ECM ensemble; bidirectional information has to traverse this interface in some form. However, whether such traversal involves more than simple trans-PM diffusion of "appropriate" molecules is less evident. Perhaps the most compelling argument for a role of the membrane skeleton-PM-ECM ensemble in environmental sensing is in the **thigmotropic interactions** characteristic of a number of fungal pathogens, perhaps best known for the rust fungi.

Rust fungal germ tubes sense the **topography** of the host leaf surface and respond by both steering their direction of **growth** and **differentiating** into infection structures. That this phenomenon is truly morphology sensing is shown by their ability to respond appropriately to purely artificial **morphological cues**, with no chemical host information present (Wynne 1976; Hoch et al. 1987a; Hoch and Staples 1987; Heath and Perumalla 1988; Collins and Read 1997; Read et al. 1997). The sensory mechanism is very subtle; it can differentiate between signals differing by only about $0.25\,\mu$m (Hoch et al. 1987a). Signal perception is dependent on the **microtubular** part of the cytoskeleton (Hoch et al. 1987b) that is concentrated into a layer of parallel microtubules adjacent to the PM, specifically in the sensory region (Bourett et al. 1987; Kwon et al. 1991a,b). These microtubules are close enough to the PM to be considered a part of the membrane skeleton. Interestingly, perception is independent of actin (Tucker et al. 1986). Perception is also dependent on a proteinaceous (probably glycoprotein) extracellular material (Epstein et al. 1987). Thus, components of the entire membrane skeleton-PM-ECM ensemble are involved. However, most interestingly, differentiation is inhibited by **RGD peptides** (Corrêa et al. 1996), indicating a role for **integrins** and the PM-ECM interface in the entire sensory system, and making it appear similar to animal sensory systems.

There are many other interactions between fungi and their plant host organisms (e.g. Read et al. 1992; Dean 1997; Howard 1997), but in none of these is there data as extensive as discussed above for the rust fungi. However, in the case of *Magnaporthe* and other species (Howard 1997), the complex **penetration structures** that form involve the production of **extracellular adhesives** (clearly a type of ECM material), unique **cell wall patterns** and intimately adjacent **actin arrays** (Bourett and Howard 1992) that must certainly involve cytoskeleton-PM-ECM interactions. The localized nature of the actin arrays, adjacent to the PM, might well qualify them as membrane skeletons, although their functions are difficult to demonstrate and remain unknown. There are also increasing reports of RGD-containing or **vitronectin-** and **fibronectin**-related extracellular proteins produced by the fungi, whose expression is consistent with a role in mediating plant-fungal interactions (van West et al. 1998; Martin et al. 1999; Dean et al. 1994). It is too early to understand the roles of these proteins, but presence of the RGD motif implicates them in interacting with integrin-like molecules of the fungal PM. Such an interaction is especially likely with the RGD-containing cell wall protein (of undetermined function) found at the PM-wall interface in *Trichoderma* conidia (Puyesky et al. 1999). Equally, they may be involved in fungus-host adhesion in similar ways to the RGD-containing proteins of the animal pathogenic *Candida* species discussed below.

Candida albicans contains a gene (*αINT 1*) encoding a protein with most of the functional **α-integrin** domains, including cross-reactivity with anti-animal-integrin antibodies, but with very little overall sequence similarity to animal integrins (Gale et al. 1996). This protein is apparently an integral PM protein with an extracellular domain that mediates cell aggregation. It is likely to be the protein that mediates **RGD-peptide-blockable adhesion** of the fungus to animal epithelial cells (Bendel and Hostetter 1993). Interestingly, the protein also contains an RGD sequence that is not present in animal integrins and the function of which is unknown. Since αINT1p binds to RGD-containing proteins and itself contains an RGD sequence, it is potentially able to bind to itself and thus possibly contribute to the integrity of the cell wall and PM-ECM interactions. However, there is no direct evidence that αINT1p does mediate PM-ECM interactions as it does in animal cells. Nevertheless, the existence of this fungal protein with integrin properties but little overall sequence similarity to other integrins reveals both the existence of an interesting class of molecules and the difficulty of seeking animal protein homologues by searching sequences of known fungal genes.

Very different types of environmental interactions are the assorted **mating interactions** of fungi (Gooday 1992; Casselton and Kües 1994; Esser and Blaich 1994; Glass and Nelson 1994; Kämper et al. 1994). All of these interactions involve the fungus sensing its environment (typically a neighboring hypha) and responding in some way via no doubt complex signaling and effector pathways. Clearly, the ECM or cell wall is a primary part of the receptor system and equally clearly the perceived signal must traverse the PM. Whether the intracellular part of the sensory system involves a membrane skeleton or other parts of the cytoskeleton remains to be seen, although, certainly after fusion, the cytoskeleton becomes reorganized (e.g. Raudaskoski 1998). All of these systems are candidates for further investigation of the cytoplasm-PM-ECM ensemble. Unfortunately, in spite of large amounts of molecular and genetic information, there is little that advances our understanding of the basic questions being addressed in this review. Perhaps the most intriguing observation is that the **fimbriae**, which are involved in mating in *Ustilago* species, are closely related to **collagen** (Celerin et al. 1996), which is a substrate for animal PM-ECM adhesions, again emphasizing the possible similarities with animal systems.

A final type of environmental interaction is shown between adjacent hyphae when they grow as cohesive tissues to produce structures such as **fruiting bodies**. The intriguing question of how normally divergent hyphae switch to parallel cohesive growth has barely been addressed. Minimally, it involves changes in hyphal tip growth steering mechanisms, coordination of growth rates and adhesive interactions between hyphae. These changes presumably involve signaling between hyphae, as well as adhesion, and obviously the signaling could relate to adhesion as found in animal cells. A possible indication of this process is the observation of the fruit body specific expression of a *Lentinus* gene encoding a **RGD-containing extracellular protein** that mediates adhesion between *Lentinus* hyphae, animals cells and *Saccharomyces* cells (Kondoh et al. 1995; Yasuda and Shishido 1997; Yasuda et al. 1997). The mechanisms by which this protein mediates adhesion between the walled cells is unclear, since in the animal model adhesion occurs between the PM-located integrins and the RGD-containing proteins. Furthermore, whether this adhesion has any role in intracellular signaling, with or without the cytoskeleton, is a totally unknown but intriguing possibility.

V. Conclusions

We have seen that, in some respects, the concept that there are cytoskeleton-PM-ECM interactions involving a specialized part of the cytoskeleton, the membrane skeleton, is intuitively obvious. All organisms must interact bidirectionally with their environment via this ensemble. We have also seen that very many of the fundamentally important processes that fungi engage in involve the membrane skeleton-PM-ECM ensemble. More importantly, we are now beginning to investigate the details of these interactions and processes. It is clear that we are a very long way from a complete understanding of any of them. However, there are sufficient data to indicate that the basic concepts that apply to the better-studied animal kingdom have their similarities among the plants and fungi. There is evidence, albeit often rather preliminary, that similar molecular components and ensembles are present and have similar functions in the fungi.

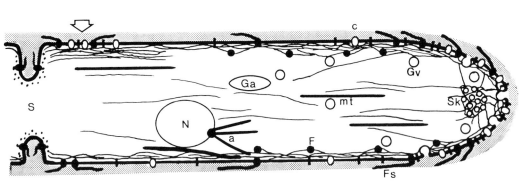

Fig. 2. Diagrammatic representation of the cytoskeleton-membrane skeleton-PM-ECM interactions discussed in this review. The *heavy line* enclosing the hypha, with vesicles fusing at the tip, indicates the PM. Cell wall fibril synthases are indicated by *black ovals* inserted in the PM, with attached extracellular fibrils. *White trans-PM ovals* represent ion channels (*c*). The membrane skeleton consists of actin and spectrin fibrils attached to the inside of the PM via the ion channels, synthases and integrins. The latter, represented by *trans-PM bars*, insert into and attach to the matrix components of the wall, indicated by *stippling*. Vesicles (*Gv*) derived from the Golgi apparatus (*Ga*) are shown in transit along both the inner surface of the membrane skeleton and microtubules (*mt*) lying deeper in the cytoplasm. Astral microtubules (*a*) emanating from the spindle pole bodies of the nucleus (*N*) are shown interacting with both cytoplasmic actin (*fine lines*) and the membrane skeleton, while microtubules laterally associated with the nuclear envelope are shown only associated with the membrane skeleton. *Broad arrow* indicates future site of branch formation, with increased abundance of adjacent membrane skeleton fibrils that are postulated to have moved channels and synthases into the branch initiation site. At the septum (*S*), circumferential membrane skeleton filaments are seen in cross section (*dots*) on either side of the developing septum, associated with the synthases, while longitudinal elements of the membrane skeleton are positioned to recruit more synthases. In the growing apex, the Spitzenkörper is shown to be rich in both vesicles and membrane skeleton filaments that extend into the membrane skeleton. Subapically, filasomes (*F*) are shown attached to and as extensions of the membrane skeleton. Actin filaments are also shown extending from the membrane skeleton and Spitzenkörper back into the cytoplasm. This diagram incorporates information from both Dikaryomycota and Oömycota. Certainly the latter lack both septa and Spitzenkörper, but there are no data that deny the postulated features as generalities among the former. Also, as indicated in the text, there are no compelling data that prove many of the illustrated features for the Dikaryomycota

Thus we have seen that there is a membrane skeleton, which contains at least actin and spectrin, that there are integrins, presumably in the PM, and that the ECM interacts with the PM and involves RGD. The postulated features of the membrane skeleton-PM-ECM ensemble are summarized in Fig. 2.

Given the amount of investigative time devoted to the membrane skeleton-PM-ECM ensemble in the fungi, and the fact that even in the best-known fungus, *Saccharomyces cerevisae*, only about half of the known genes have been assigned a function, it would be surprising if there are not many more, possibly fungal-specific, molecules involved in the fungal membrane skeleton-PM-ECM ensemble. It is hoped that additional thought directed towards these issues will result in further elucidation of the participating molecules. Only then will it become possible to more meaningfully address the questions of both functions and regulation of the ensemble. However, since so many of the basic aspects of fungal biology are dependent on the membrane skeleton-PM-ECM ensemble, it is clear that future progress along the lines of the work presented in this review is a requisite to our understanding, and possible future manipulation, of fungal biology.

Acknowledgements. Unpublished observations and the preparation of this work were supported by grants from NSERC, which are gratefully acknowledged. Former and present students and postdoctoral workers, who are recognized by the appropriate citations, substantially influenced the evolution of the ideas. Kate Clark provided valuable technical assistance in the production of the data in Fig. 1.

References

Adams AEM, Pringle JR (1984) Relationship of actin and tubulin distribution to bud growth in wild-type and morphogenetic mutant *Saccharomyces cerevisiae*. J Cell Biol 98:934–945

Aist JR, Bayles CJ (1991) Ultrastructural basis of mitosis in the fungus *Nectria haematococca* (sexual stage of *Fusarium solani*). Protoplasma 161:111–122

Aist JR, Berns MW (1981) Mechanism of chromosome separation during mitosis in *Fusarium* (Fungi imperfecti). New evidence from ultrastructural and laser microbeam experiments. J Cell Biol 91:446–458

Akashi T, Shibaoka H (1991) Involvement of transmembrane proteins in the association of cortical micro-

tubules with the plasma membrane in tobacco BY-2 cells. J Cell Sci 98:169–174

Akashi T, Kawasaki S, Shibaoka H (1990) Stabilization of cortical microtubules by the cell wall in cultured tobacco cells: effects of extensin on the cold stability of cortical microtubules. Planta 182:363–369

Ayscough KR, Eby JJ, Lila T, Dewar H, Kozminski KG, Drubin DG (1999) Sla1p is a functionally modular component of the yeast cortical actin cytoskeleton required for the correct localization of both Rho1p-GTPase and Sla2p, a protein with talin homology. Mol Biol Cell 10:1061–1075

Baba M, Baba N, Ohsumi Y, Kanaya K, Osumi M (1989) Three-dimensional analysis of morphogenesis induced by mating pheromone a factor in *Saccharomyces cerevisiae*. J Cell Sci 94:207–216

Bachewich CL, Heath IB (1997a) Differential cytoplasm-plasma membrane-cell wall adhesion patterns and their relationships to hyphal tip growth and organelle motility. Protoplasma 200:71–86

Bachewich CL, Heath IB (1997b) The cytoplasmic pH influences hyphal tip growth and the cytoskeleton-related organization. Fungal Genet Biol 21:76–91

Bachewich CL, Heath IB (1998) Radial F-actin arrays precede new hypha formation in *Saprolegnia*: implications for establishing polar growth and regulating tip morphogenesis. J Cell Sci 111:2005–2016

Bachewich CL, Heath IB (1999) Cytoplasmic migrations and vacuolation are associated with growth recovery in hyphae of *Saprolegnia*, and are dependent on the cytoskeleton. Mycol Res 103:849–858

Baluska F, Samaj J, Volkmann D (1999) Proteins reacting with cadherin and catenin antibodies are present in maize showing tissue-, domain-, and development-specific associations with endoplasmic-reticulum membranes and actin microfilaments in root cells. Protoplasma 206:174–187

Barthou H, Petitprez M, Brière C, Souvré A, Alibert G (1999) RGD-mediated membrane-matrix adhesion triggers agarose-induced embryoid formation in sunflower protoplasts. Protoplasma 206:143–151

Bartnicki-Garcia S, Hergert F, Gierz G (1989) Computer simulation of fungal morphogenesis and the mathematical basis for hyphal (tip) growth. Protoplasma 153:46–57

Beck KA, Buchanan JA, Malhotra V, Nelson WJ (1994) Golgi spectrin: identification of an erythroid β-spectrin homolog associated with the Golgi complex. J Cell Biol 127:707–723

Bendel CM, Hostetter MK (1993) Distinct mechanisms of epithelial adhesion for *Candida albicans* and *Candida tropicalis*; identification of the participating ligands and development of inhibitory peptides. J Clin Invest 92:1850–1849

Bennett V (1990) Spectrin-based membrane skeleton: a multipotential adaptor between plasma membrane and cytoplasm. Physiol Rev 70:1029–1065

Bourett TM, Howard RJ (1991) Ultrastructural immunolocalization of actin in a fungus. Protoplasma 163:199–202

Bourett TM, Howard RJ (1992) Actin in penetration pegs of the fungal rice blast pathogen, *Magnaporthe grisea*. Protoplasma 168:20–26

Bourett TM, Hoch HC, Staples RC (1987) Association of the microtubule cytoskeleton with the thigmotropic signal for appressorium formation in *Uromyces*. Mycologia 79:540–545

Bower FO (1883) On plasmolysis and its bearing upon the relationship between cell wall and protoplasm. Q J Microsc Sci 23:151–168

Brands R, Feltkamp CA (1988) Wet cleaving of cells: a method to introduce macromolecules into the cytoplasm. Exp Cell Res 176:309–318

Braun M, Wasteneys G (2000) Actin in characean rhizoids and protonemata. Tip growth, gravity sensing and photomorphogenesis. In: Staiger C, Baluska F, Volkmann P, Barlow P (eds) Actin: a dynamic framework for multiple plant cell functions. Kluwer, Dordrecht, pp 237–258

Brennwald P, Kearns B, Champion K, Keranen S, Bankaitis V, Novick P (1994) Sec9 is a SNAP-25-like component of a yeast SNARE complex that may be the effector of Sec4 function in exocytosis. Cell 79:245–258

Burridge K, Chrzanowska-Wodnicka M (1996) Focal adhesions, contractility and signaling. Annu Rev Cell Dev Biol 12:463–519

Butt TM, Heath IB (1988) The changing distribution of actin and nuclear behavior during the cell cycle of the mite-pathogenic fungus *Neozygites* sp. Eur J Cell Biol 46:499–505

Casselton LA, Kües U (1994) Mating-type genes in homobasidiomycetes. In: Wessels JGH, Meinhardt F (eds) The Mycota, vol I. Growth, differentiation and sexuality. Springer, Berlin Heidelberg New York, pp 307–322

Celerin M, Ray JM, Schisler NJ, Day AW, Stetler-Stevenson WG, Laudenbach DE (1996) Fungal fimbriae are composed of collagen. EMBO J 15:4445–4453

Chiu YH, Xiang X, Dawe AL, Morris NR (1997) Deletion of *nudC*, a nuclear migration gene of *Aspergillus nidulans*, causes morphological and cell wall abnormalities and is lethal. Mol Biol Cell 8:1735–1749

Collins TJ, Read ND (1997) Appressorium induction by topographical signals in six cereal rusts. Physiol Mol Plant Pathol 51:169–179

Corrêa A, Staples RC, Hoch HC (1996) Inhibition of thigmostimulated cell differentiation with RGD-peptides in *Uromyces* germlings. Protoplasma 194:91–102

de Ruijter N, Emons A (1993) Immunodetection of spectrin antigens in plant cells. Cell Biol Int 17:169–182

de Ruijter NC, Brook MB, Bisseling T, Emons AMC (1998) Lipochito-oligosaccharides re-initiate root hair tip growth in *Vicia sativa* with high calcium and spectrin-like antigen at the tip. Plant J 13:341–350

Dean RA (1997) Signal pathways and appressorium morphogenesis. Annu Rev Phytopathol 35:211–234

Dean RA, Lee Y, Mitchell TK, Whitehead DS (1994) Signalling systems and gene expression regulating appressorium formation in *Magnaporthe grisea*. In: Zeigler RS, Leong SA, Teng PS (eds) Rice blast disease. Cambridge Univ Press, Cambridge, pp 23–34

Degousée N, Gupta GD, Lew RR, Heath IB (2000) A putative spectrin-containing membrane skeleton in hyphal tips of *Neurospora crassa*. Fungal Genet Biol 30:33–44

Donnelly S, Pocklington M, Pallotta D, Orr E (1993) A proline-rich protein, verprolin, involved in cytoskeletal organization and cellular growth in the yeast *Saccharomyces cerevisiae*. Mol Microbiol 10:585–596

Doyle T, Botstein D (1996) Movement of yeast cortical actin cytoskeleton visualized *in vivo*. Proc Natl Acad Sci USA 93:3886–389

Drubin D, Nelson W (1996) Origins of cell polarity. Cell 84:335–344

Epstein L, Laccetti LB, Staples RC, Hoch HC (1987) Cell-substratum adhesive protein involved in surface contact responses of the bean rust fungus. Physiol Mol Plant Pathol 30:373–388

Esser K, Blaich R (1994) Heterogenic incompatibility. In: Wessels JGH, Meinhardt F (eds) The Mycota, vol I. Growth, differentiation and sexuality. Springer, Berlin Heidelberg New York, pp 211–234

Evangelista M, Blundell K, Longtine MS, Chow CJ, Adames N, Pringle JR, Peter M, Boone C (1997) Bni1p, a yeast formin linking Cdc42p and the actin cytoskeleton during polarized morphogenesis. Science 276:118–122

Falk A, Laboure AM, Gulino D, Mandaron P, Falconet D (1998) A plant surface protein sharing structural properties with animal integrins. Eur J Biochem 253:552–559

Faraday CD, Spanswick RM (1993) Evidence for a membrane skeleton in higher plants: a spectrin-like polypeptide co-isolates with rice root plasma membranes. FEBS Lett 318:313–316

Fath KR, Graham SF, Keates R (1985) Association of fodrin with brain microtubules. Can J Biochem Cell Biol 63:372–381

Ferro-Novick S, Jahn R (1994) Vesicle fusion from yeast to man. Nature 370:191–193

Fowler JE, Quatrano RS (1997) Plant cell morphogenesis: plasma membrane interactions with the cytoskeleton and cell wall. Annu Rev Cell Dev Biol 13:697–743

Fujiwara T, Tanaka K, Mino A, Kikyo M, Takahashi K, Shimizu K, Takai Y (1998) Rho1p-Bni1p-Spa1p interactions: implication in localization of Bni1p at the bud site and regulation of the actin cytoskeleton in *Saccharomyces cerevisiae*. Mol Biol Cell 9:1221–1233

Gale C, Finkel D, Tao N, Meinke M, McClellan M, Olson J, Kendrick K, Hostetter M (1996) Cloning and expression of a gene encoding an integrin-like protein in *Candida albicans*. Proc Natl Acad Sci USA 93:357–361

Garrill A, Jackson SL, Lew RR, Heath IB (1993) Ion channel activity and tip growth: tip-localized stretch-activated channels generate an essential Ca^{2+} gradient in the oomycete *Saprolegnia ferax*. Eur J Cell Biol 60:358–365

Gens JS, Reuzeau C, Doolittle KW, NcNally JG, Pickard BG (1996) Covisualization by computational optical-sectioning microscopy of integrin and associated proteins at the cell membrane of living onion protoplasts. Protoplasma 194:215–230

Girbardt M (1979) A microfilamentous septal belt (FSB) during induction of cytokinesis in *Trametes versicolor* (L. ex. Fr.). Exp Mycol 3:215–228

Glass NL, Nelson MA (1994) Mating-type genes in mycelial ascomycetes. In: Wessels JGH, Meinhardt F (eds) The Mycota, vol I. Growth, differentiation and sexuality. Springer, Berlin Heidelberg New York, pp 295–306

Gooday GW (1992) The fungal surface and its role in sexual interactions. In: Callow JA, Green JR (eds) Perspectives in plant cell recognition. SEB seminar series. Cambridge Univ Press, Cambridge, pp 33–58

Grove SN, Bracker CE (1970) Protoplasmic organization of hyphal tips among fungi: vesicles and Spitzenkörper. J Bacteriol 104:989–1009

Gupta G, Heath IB (1997) Actin disruption by latrunculin B causes turgor-related changes in tip growth of *Saprolegnia* hyphae. Fungal Genet Biol 21:64–75

Gupta G, Heath IB (2000) A tip-high gradient of a putative plasma membrane SNARE approximates the exocytotic gradient in hyphal apices of the fungus *Neurospora crassa*. Fungal Genet Biol 29:187–199

Harold FM (1994) Ionic and electrical dimensions of hyphal growth. In: Wessels JGH, Meinhardt F (eds) The Mycota, vol I. Growth, differentiation and sexuality. Springer, Berlin Heidelberg New York, pp 89–110

Heath IB (1974) A unified hypothesis for the role of membrane bound enzyme complexes and microtubules in plant cell wall synthesis. J Theor Biol 47:445–449

Heath IB (1987) Preservation of a labile cortical array of actin filaments in growing hyphal tips of the fungus *Saprolegnia ferax*. Eur J Cell Biol 44:10–16

Heath IB (1988) Evidence against a direct role for cortical actin arrays in saltatory organelle motility in hyphae of the fungus *Saprolegnia ferax*. J Cell Sci 91:41–47

Heath IB (1990) The roles of actin in tip growth of fungi. Int Rev Cytol 123:95–127

Heath IB (1994) The cytoskeleton in hyphal growth, organelle movements, and mitosis. In: Wessels JGH, Meinhardt F (eds) The Mycota, vol I. Growth, differentiation and sexuality. Springer, Berlin Heidelberg New York, pp 43–65

Heath IB (1995) Integration and regulation of hyphal tip growth. Can J Bot 73 [Suppl 1]:S131–S139

Heath IB, Heath MC (1978) Microtubules and organelle movements in the rust fungus *Uromyces phaseoli* var. *vignae*. Cytobiologie 16:393–411

Heath IB, Heath MC (1979) Structural studies of the development of infection structures of cowpea rust, *Uromyces phaseoli* var. *vignae*. II. Vacuoles. Can J Bot 57:1830–1837

Heath IB, Janse van Rensburg EJ (1996) Critical evaluation of the VSC model for tip growth. Mycoscience 37:1–10

Heath IB, Kaminskyj SGW (1989) The organization of tip-growth related organelles and microtubules revealed by quantitative analysis of freeze-substituted oomycete hyphae. J Cell Sci 93:41–52

Heath IB, Seagull RW (1982) Oriented cellulose microfibrils and the cytoskeleton: a critical comparison of models. In: Lloyd CW (ed) The cytoskeleton in plant growth and development. Academic Press, London, pp 163–182

Heath IB, Steinberg G (1999) Mechanisms of hyphal tip growth: tube dwelling amoebae revisited. Fungal Genet Biol 28:79–93

Heath IB, Gupta G, Bai S (2000) Plasma membrane-associated actin filaments, but not microtubules, are essential for both polarization and hyphal morphogenesis in *Saprolegnia ferax* and *Neurospora crassa*. Fungal Genet Biol 30:45–62

Heath MC, Perumalla CJ (1988) Haustorial mother cell development by *Uromyces vignae* on collodion membranes. Can J Bot 66:736–741

Henry CA, Jordan JR, Kropf DL (1996) Localized membrane-wall adhesions in *Pelvetia* zygotes. Protoplasma 190:39–52

Hermann GJ, King EJ, Shaw JM (1997) The yeast gene, *MDM20*, is necessary for mitochondrial inheritance and organization of the actin cytoskeleton. J Cell Biol 137:141–153

Hill KL, Catlett NL, Weisman LS (1996) Actin and myosin function in directed vacuole movement during cell division in *Saccharomyces cerevisiae*. J Cell Biol 135:1535–1549

Hoch HC, Howard RJ (1980) Ultrastructure of freeze-substituted hyphae of the basidiomycete *Laetisaria arvalis*. Protoplasma 103:281–297

Hoch HC, Staples RC (1983) Visualization of actin *in situ* by rhodamine-conjugated phalloidin in the fungus *Uromyces phaseoli*. Eur J Cell Biol 32:52–58

Hoch HC, Staples RC (1987) Structural and chemical changes among the rust fungi during appressorium development. Annu Rev Phytopathol 25:231–247

Hoch HC, Staples RC, Whitehead B, Comeau J, Wolf ED (1987a) Signaling for growth orientation and cell differentiation by surface topography in *Uromyces*. Science 235:1659–1662

Hoch HC, Tucker BE, Staples RC (1987b) An intact microtubule cytoskeleton is necessary for mediation of the signal for cell differentiation in *Uromyces*. Eur J Cell Biol 45:209–218

Hoffmann J, Mendgen K (1998) Endocytosis and membrane turnover in the germ tube of *Uromyces fabae*. Fungal Genet Biol 24:77–85

Holzinger A, de Ruijter N, Emons A, Lütz-Meindel U (1999) Spectrin-like proteins in green algae (Desmidaceae). Cell Biol Int 23:335–344

Hostetter M, Tao N-J, Gale C, Herman D, McClellan M, Sharp R, Kendrick K (1995) Antigenic and functional conservation of an integrin I-domain in *Saccharomyces cerevisiae*. Biochem Mol Med 55:122–130

Howard RJ (1981) Ultrastructural analysis of hyphal tip cell growth in fungi: Spitzenkörper, cytoskeleton and endomembranes after freeze-substitution. J Cell Sci 48:89–103

Howard RJ (1997) Breaching the outer barriers – cuticle and cell wall penetration. In: Carroll GC, Tudzynski P (eds) The Mycota, vol V. Plant relationships, part A. Springer, Berlin Heidelberg New York, pp 43–60

Hyde GJ, Heath IB (1997) Ca^{2+} gradients in hyphae and branches of *Saprolegnia ferax*. Fungal Genet Biol 21:238–251

Inoue S, Turgeon BG, Yoder OC, Aist JR (1998a) Role of fungal dynein in hyphal growth, microtubule organization, spindle pole body motility and nuclear migration. J Cell Sci 111:1555–1566

Inoue S, Yoder O, Turgeon B, Aist J (1998b) A cytoplasmic dynein required for mitotic aster formation in vivo. J Cell Sci 111:2607–2614

Jackson SL, Heath IB (1990a) Visualization of actin arrays in growing hyphae of the fungus *Saprolegnia ferax*. Protoplasma 154:66–70

Jackson SL, Heath IB (1990b) Evidence that actin reinforces the extensible hyphal apex of the oomycete *Saprolegnia ferax*. Protoplasma 157:144–153

Jackson SL, Heath IB (1993) The roles of calcium ions in hyphal tip growth. Microbiol Rev 57:367–382

Jockusch BM, Bubeck P, Giehl K, Kroemker M, Moschner J, Rothkegel M, Rüdiger M, Schlüter K, Stanke G, Winkler J (1995) The molecular architecture of focal adhesions. Annu Rev Cell Dev Biol 11:379–416

Johnston GC, Prydz K, Ryd M, van Deurs B (1991) The *Saccharomyces cerevisiae* MYO2 gene encodes an essential myosin for vectorial transport of vesicles. J Cell Biol 113:539–552

Kagawa T, Kadota A, Wada M (1992) The junction between the plasma membrane and the cell wall in fern protonemal cells, as visualized after plasmolysis, and its dependence on arrays of cortical microtubules. Protoplasma 170:186–190

Kamada T (1994) Stipe elongation in fruit bodies. In: Wessels JGH, Meinhardt F (eds) The Mycota, vol I. Growth, differentiation and sexuality. Springer, Berlin Heidelberg New York, pp 367–379

Kaminskyj SGW, Heath IB (1995) Integrin and spectrin homologues, and cytoplasm-wall adhesion in tip growth. J Cell Sci 108:849–856

Kaminskyj SGW, Heath IB (1996) Studies on *Saprolegnia ferax* suggest the general importance of the cytoplasm in determining hyphal morphology. Mycologia 88: 20–37

Kaminskyj SGW, Yoon KS, Heath IB (1989) Cytoskeletal interactions with post-mitotic migrating nuclei in the oyster mushroom fungus, *Pleurotus ostreatus*: evidence against a force-generating role for astral microtubules. J Cell Sci 94:663–674

Kaminskyj SGW, Jackson SL, Heath IB (1992) Fixation induces differential polarized translocations of organelles in hyphae of *Saprolegnia ferax*. J Microsc 167:153–168

Kämper J, Bölker M, Kahmann R (1994) Mating-type genes in heterobasidiomycetes. In: Wessels JGH, Meinhardt F (eds) The Mycota, vol I. Growth, differentiation and sexuality. Springer, Berlin Heidelberg New York, pp 323–332

Kanbe T, Kobayashi I, Tanaka K (1989) Dynamics of cytoplasmic organelles in the cell cycle of the fission yeast *Schizosaccharomyces pombe*. J Cell Sci 94:647–656

Kanbe T, Akashi T, Tanaka K (1993) Effect of cytochalasin A on actin distribution in the fission yeast *Schizosaccharomyces pombe* studied by fluorescent and electron microscopy. Protoplasma 176:24–32

Katayama S, Hirata D, Arellano M, Perez P, Toda T (1999) Fission yeast α-glucan synthase Mok1 requires the actin cytoskeleton to localize the sites of growth and plays an essential role in cell morphogenesis downstream of protein kinase C function. J Cell Biol 144: 1173–1186

Katembe WJ, Swatzell LJ, Makaroff CA, Kiss JZ (1997) Immunolocalization of integrin-like proteins in *Arabidopsis* and *Chara*. Physiol Plant 99:7–14

Kondoh O, Muto A, Kajiwara S, Takagi J, Sato Y, Shishido K (1995) A fruit body-specific cDNA, *mfbaAc*, from the mushroom *Lentinus edodes* encodes a high-molecular-weight cell-adhesion protein containing an Arg-Gly-Asp motif. Gene 154:31–37

Kropf DL (1992) Establishment and expression of cellular polarity in fucoid zygotes. Microbiol Rev 56:316–339

Kropf DL, Lupa MDA, Caldwell JH, Harold FM (1983) Cell polarity: endogenous ion currents precede and predict branching in the water mold *Achyla*. Science 220:1385–1387

Kropf DL, Berge SK, Quatrano RS (1989) Actin localization during *Fucus* embryogenesis. Plant Cell 1:191–200

Kropf DL, Coffman HR, Kloareg B, Glenn P, Allen VW (1993) Cell wall and rhizoid polarity in *Pelvetia* embryos. Dev Biol 160:303–314

Kwon YH, Hoch HC, Aist JR (1991a) Initiation of appressorium formation in *Uromyces appendiculatus*: orga-

nization of the apex, and the responses involving mictotubules and apical vesicles. Can J Bot 69:2560–2573

Kwon YH, Hoch HC, Staples RC (1991b) Cytoskeletal organization in *Uromyces* urediospore germling apices during appressorium formation. Protoplasma 165:37–50

Laurent P, Voiblet C, Tagu D, de Carvalho D, Nehls U, De Bellis R, Balestrini R, Bauw G, Bonfante P, Martin F (1999) A novel class of ectomycorrhiza-regulated cell wall polypeptides in *Pisolithus tinctorius*. Mol Plant Microbe Interact 12:862–871

Laval V, Chabannes M, Carrière M, Canut H, Barre A, Rougé P, Pont-Lezica R, Galaud J-P (1999) A family of *Arabidopsis* plasma membrane receptors presenting animal β-integrin domains. Biochim Biophys Acta 1435:61–70

Lechler T, Li R (1997) In vitro reconstitution of cortical actin assembly sites in budding yeast. J Cell Biol 138:95–103

Ledbetter MC, Porter KR (1963) A "microtubule" in plant cell fine structure. J Cell Biol 19:239–250

Lee L, Lee SK, Evangelista M, Boone C, Pelham D (1999) Control of mitotic spindle position by the *Saccharomyces cerevisiae* formin Bni1p. J Cell Biol 144: 947–961

Leung C, Sun D, Zheng M, Knowles D, Liem R (1999) Microtubule actin cross-linking factor (MACF): a hybrid of dystonin and dystrophin that can interact with the actin and microtubule cytoskeletons. J Cell Biol 147:1275–1285

Levina NN, Lew RR, Heath IB (1994) Cytoskeletal regulation of ion channel distribution in the tip-growing organism *Saprolegnia ferax*. J Cell Sci 107:127–134

Levina NN, Lew RR, Hyde GJ, Heath IB (1995) The roles of Ca²⁺ and plasma membrane ion channels in hyphal tip growth of *Neurospora crassa*. J Cell Sci 108: 3405–3417

Lew RR (1999) Comparative analysis of Ca²⁺ and H⁺ flux magnitude and location along growing hyphae of *Saprolegnia ferax* and *Neurospora crassa*. Eur J Cell Biol 78:892–902

Li R, Zheng Y, Drubin DG (1995) Regulation of cortical actin cytoskeleton assembly during polarized cell growth in budding yeast. J Cell Biol 128:599–615

Lopez-Franco R, Bracker CE (1996) Diversity and dynamics of the Spitzenkörper in growing hyphal tips of higher fungi. Protoplasma 195:90–111

Lynch T, Lintilhac P, Domozych D (1998) Mechanotransduction molecules in the plant gravisensory response: amyloplast/statolith membranes contain a β₁ integrin-like protein. Protoplasma 201:92–100

Marcantonio EE, Hynes RO (1988) Antibodies to the conserved cytoplasmic domain of the integrin β₁ subunit react with proteins in vertebrates, invertebrates and fungi. J Cell Biol 106:1765–1772

Martin F, Laurent P, de Carvalho D, Voiblet C, Balestrini R, Bonfante P, Tagu D (1999) Cell wall proteins of the ectomycorrhizal basidiomycete *Pisolithus tinctorius*: identification, function, and expression in symbiosis. Fungal Genet Biol 27:161–174

Masuda Y, Takagi S, Nagai R (1991) Protease-sensitive anchoring of microfilament bundles provides tracks for cytoplasmic streaming in *Vallisneria*. Protoplasma 162:151–159

Matteis AD, Morrow J (2000) Spectrin tethers and mesh in the biosynthetic pathway. J Cell Sci 113:2331–2343

May KM, Hyams JS (1998) The yeast cytoskeleton: the closer we look, the more we see. Fungal Genet Biol 24:110–122

McGillviray AM, Gow NAR (1987) The transhyphal electrical current of *Neurospora crassa* is carried principally by protons. J Gen Microbiol 133:2875–2881

McKerracher LJ, Heath IB (1985) Microtubules around migrating nuclei in conventionally-fixed and freeze-substituted cells. Protoplasma 125:162–172

McKerracher LJ, Heath IB (1986a) Polarized cytoplasmic movement and inhibition of saltations induced by calcium-mediated effects of microbeams in fungal hyphae. Cell Motil Cytoskeleton 6:136–145

McKerracher LJ, Heath IB (1986b) Fungal nuclear behaviour analysed by ultraviolet microbeam irradiation. Cell Motil Cytoskeleton 6:35–47

McKerracher LJ, Heath IB (1986c) Comparison of polyethylene glycol and diethylene glycol disterate embedding methods for the preservation of fungal cytoskeletons. J Electron Microsc Tech 4:347–360

McKerracher LJ, Heath IB (1987) Cytoplasmic migration and intracellular organelle movements during tip growth of fungal hyphae. Exp Mycol 11:79–100

Michaud D, Guillet G, Rogers PA, Charest PM (1991) Identification of a 200 kDa membrane-associated plant cell protein immunologically related to human β-spectrin. FEBS Lett 294:77–80

Momany M, Hamer JE (1997) Relationship of actin, microtubules, and crosswall synthesis during septation in *Aspergillus nidulans*. Cell Motil Cytoskeleton 38:373–384

Monzer J (1995) Actin filaments are involved in cellular graviperception of the basidiomycete *Flammulina velutipes*. Eur J Cell Biol 66:151–156

Morrell J, Morphew M, Gould K (1999) A mutant of Arp2p causes partial disassembly of the Arp2/3 complex and loss of cortical actin function in fission yeast. Mol Biol Cell 10:4201–4215

Morris NR, Xiang X, Beckwith SM (1995) Nuclear migration advances in fungi. Trends Cell Biol 5:278–282

Nelson D, Bohnert H (1994) Plasma membrane-cell wall adhesion in ice plant. Plant Physiol [Suppl] 105:58

Novick P, Botstein D (1985) Phenotypic analysis of temperature-sensitive yeast actin mutants. Cell 40:405–416

Oparka KJ (1994) Plasmolysis: new insights into an old problem. New Phytol 126:571–591

Osumi M, Sato M, Ishijima SA, Konomi M, Takagi T, Yaguchi H (1998) Dynamics of cell wall formation in fission yeast, *Schizosaccharomyces pombe*. Fungal Genet Biol 24:178–206

Parton RM, Fischer S, Malho R, Papasouliotis O, Jelitto TC, Leonard T, Read ND (1997) Pronounced cytoplasmic pH gradients are not required for tip growth in plant and fungal cells. J Cell Sci 110:1187–1198

Pelham H (1999) SNAREs and the secretory pathway-lessons from yeast. Exp Cell Res 247:1–8

Pennell RI, Knox JP, Scofield GN, Selvendran RR, Roberts K (1989) A family of abundant plasma membrane-associated glycoproteins related to the arabinogalactan proteins is unique to flowering plants. J Cell Biol 108:1967–1977

Pont-Lezica RF, McNally JG, Pickard BG (1993) Wall-to-membrane linkers in onion epidermis: some hypotheses. Plant Cell Environ 16:111–123

Puyesky M, Benhamou N, Ponce-Noyola P, Bauw G, Ziv T, Montagu MV, Herrera-Estrella A, Horwitz BA (1999) Developmental regulation of *comp 1*, a gene encoding a multidomain conidiospore surface protein of *Trichoderma*. Fungal Genet Biol 27:88–99

Quatrano RS, Brian L, Aldridge J, Schultz T (1991) Polar axis fixation in *Fucus* zygotes: components of the cytoskeleton and extracellular matrix. Development [Suppl] 1:11–16

Raudaskoski M (1998) The relationship between B-mating-type genes and nuclear migration in *Schizophyllum commune*. Fungal Genet Biol 24:207–227

Read ND, Kellock LJ, Knight H, Trewavas AJ (1992) Contact sensing during infection by fungal pathogens. In: Callow JA, Green JR (eds) Perspectives in plant cell recognition. SEB seminar series. Cambridge Univ Press, Cambridge, pp 137–172

Read ND, Kellock LJ, Collins TJ, Gundlach AM (1997) Role of topography sensing for infection-structure differentiation in cereal rust fungi. Planta 202:163–170

Reinhardt MO (1892) Das Wachsthum der Pilzhyphen. Jahrb Wiss Bot 23:479–566

Reinsch S, Gönczy P (1998) Mechanisms of nuclear positioning. J Cell Sci 111:2283–2295

Reuzeau C, Pont-Lezica RF (1995) Comparing plant and animal extracellular matrix-cytoskeleton connections – are they alike? Protoplasma 186:113–121

Reuzeau C, Doolittle K, McNally J, Pickard B (1997) Co-visualization in living onion cells of putative integrin, putative spectrin, actin, putative intermediate filaments, and other proteins at the cell membrane and in an endomembrane sheath. Protoplasma 199:173–197

Riquelme M, Reynaga-Peña CG, Gierz G, Bartnicki-Garcia S (1998) What determines growth direction in fungal hyphae? Fungal Genet Biol 24:101–109

Roberson RW (1992) The actin cytoskeleton in hyphal cells of *Sclerotium rolfsii*. Mycologia 84:41–51

Robinow CF (1963) Observations on cell growth, mitosis, and division in the fungus *Basidiobolus ranarum*. J Cell Biol 17:123–152

Robson GD, Prebble E, Rickers A, Hosking S, Denning DW, Trinci APJ, Robertson W (1996) Polarized growth of fungal hyphae is defined by an alkaline pH gradient. Fungal Genet Biol 20:289–298

Roncal T, Ugalde UO, Irastorza A (1993) Calcium-induced conidiation in *Penicillium cyclopium*: calcium triggers cytosolic alkalinization at the hyphal tip. J Bacteriol 175:879–886

Rothman J (1994) Mechanisms of intracellular protein transport. Nature 372:55–63

Roy SJ, Holdaway-Clarke TL, Hackett GR, Kunkel JG, Lord EM, Hepler PK (1999) Uncoupling secretion and tip growth in lily pollen tubes: evidence for the role of calcium in exocytosis. Plant J 19:379–386

Ruiz-Herrera J (1991) Fungal cell wall: structure, synthesis, and assembly. CRC Press, Boca Raton

Ruoslahti E (1996) RGD and other recognition sequences for integrins. Annu Rev Cell Dev Biol 12:697–715

Ryu JH, Takagi S, Nagai R (1995) Stationary organization of the actin cytoskeleton in *Vallisneria*: the role of stable microfilaments at the end walls. J Cell Sci 108:1531–1539

Ryu JH, Mizuno K, Takagi S, Nagai R (1997) Extracellular components implicated in the stationary organization of the actin cytoskeleton in mesophyll cells of *Vallisneria*. Plant Cell Physiol 38:420–432

Sanders LC, Wang C-S, Walling LL, Lord EM (1991) A homolog of the substrate adhesion molecule vitronectin occurs in four species of flowering plants. Plant Cell 3:629–635

Santoni G, Gismondi A, Liu JH, Punturieri A, Santoni A, Frati L, Piccoli M, Djeu JY (1994) *Candida albicans* expresses a fibronectin receptor antigenically related to alpha 5 beta 1 integrin. Microbiology 140:2971–2979

Schindler M, Meiners S, Cheresh DA (1989) RGD-dependent linkage between plant cell wall and plasma membrane: consequences for growth. J Cell Biol 108:1955–1965

Schott D, Ho J, Pruyne D, Bretscher A (1999) The COOH-terminal domain of Myo2p, a yeast myosin V, has a direct role in secretory vesicle targeting. J Cell Biol 147:791–807

Schreurs WJA, Harold FM (1988) Transcellular proton current in *Achlya ambisexualis* hyphae: relationship to polarized growth. Proc Natl Acad Sci USA 85:1534–1538

Scurfield G (1967) Apical pore in fungal hyphae. Nature 214:740–741

Seagull RW, Heath IB (1980) The organization of cortical microtubule arrays in the radish root hair. Protoplasma 103:205–229

Sentandreu R, Mormeneo S, Ruiz-Herrera J (1994) Biogenesis of the fungal cell wall. In: Wessels JGH, Meinhardt F (eds) The Mycota, vol I. Growth, differentiation and sexuality. Springer, Berlin Heidelberg New York, pp 111–124

Shaw SL, Quatrano RS (1996) Polar localization of a dihydropyridine receptor in living *Fucus* zygotes. J Cell Sci 109:335–342

Silverman-Gavrila LB, Lew RR (2000) Calcium and tip growth in *Neurospora crassa*. Protoplasma 213:203–217

Simon VR, Swayne TC, Pon LA (1995) Actin-dependent mitochondrial motility in mitotic yeast and cell-free systems: identification of a motor activity on the mitochondrial surface. J Cell Biol 130:345–354

Simon VR, Karmon SL, Pon LA (1997) Mitochondria inheritance: cell cycle and actin cable dependence of polarized mitochondrial movements in *Saccharomyces cerevisiae*. Cell Motil Cytoskeleton 37:199–210

Sonesson A, Widell S (1993) Cytoskeleton components of inside-out and right-side-out plasma membrane vesicles from plants. Protoplasma 177:45–52

Steinberg G (1998) Organelle transport and molecular motors in fungi. Fungal Genet Biol 24:161–177

Steinberg G, Schliwa M (1993) Organelle movements in the wild type and wall-less fz;sg;os-1 mutants of *Neurospora crassa* are mediated by cytoplasmic microtubules. J Cell Sci 106:555–564

Swatzell LJ, Edelmann RE, Makaroff CA, Kiss JZ (1999) Integrin-like proteins are localized to plasma membrane fractions, not plastids, in *Arabidopsis*. Plant Cell Physiol 40:173–183

Takeichi M (1995) Morphogenetic roles of classic cadherins. Curr Opin Cell Biol 7:619–627

Takeuchi Y, Schmid J, Caldwell JH, Harold FM (1988) Transcellular ion currents and extension of *Neurospora crassa* hyphae. J Membr Biol 101:33–41

Tang H-Y, Xu J, Cai M (2000) Pan1p, End3p, and Sla1p, three yeast proteins required for normal cortical actin cytoskeleton organization, associate with each other

and play essential roles in cell wall morphogenesis. Mol Cell Biol 20:12–25

Tsukamoto M, Yamamoto M, Tanabe S, Kamada T (1996) Cytochalasin-E-resistant mutants of *Coprinus cinereus*: isolation and genetic, biochemical, and cytological analyses. Fungal Genet Biol 20:52–58

Tucker BE, Hoch HC, Staples RC (1986) The involvement of F-actin in *Uromyces* cell differentiation: the effects of cytochalasin E and phalloidin. Protoplasma 135:88–101

Turian G (1979) Cytochemical gradients and mitochondrial exclusion in the apices of vegetative hyphae. Experentia 35:1164–1166

Turian G (1981) Low pH in fungal bud initials. Experentia 37:1278–1279

van West P, de Jong AJ, Judelson HS, Emons AMC, Govers F (1998) The *ipi*O gene of *Phytophthora infestans* is highly expressed in invading hyphae during infection. Fungal Genet Biol 23:126–138

Vavricková P, Kohlwein S, Dráber P, Hasek J (1998) Accumulation of β-tubulin-containing fibres in the cortical domain of *Saccharomyces cerevisiae* spheroplasts. Protoplasma 202:11–16

Wagner VT, Brian L, Quatrano RS (1992) Role of a vitronectin-like molecule in embryo adhesion of the brown alga *Fucus*. Proc Natl Acad Sci USA 90:3644–3648

Wang CS, Walling LL, Gu YQ, Ware CF, Lord EM (1994) Two classes of proteins and messenger-RNAs in *Lilium longiflorum* L. identified by human vitronectin probes. Plant Physiol 104:711–717

Wang JL, Walling LL, Jauh GY, Gu Y-Q, Lord EM (1996) Lily cofactor-independent phosphoglycerate mutase: purification, partial sequencing, and immunolocalization. Planta 200:343–352

Watts HJ, Véry A-A, Perera THS, Davies JM, Gow NAR (1998) Thigmotropism and stretch-activated channels in the pathogenic fungus *Candida albicans*. Microbiology 144:689–695

Wayne R, Staves MP, Leopold AC (1992) The contribution of the extracellular matrix to gravisensing in characean cells. J Cell Sci 101:611–623

Wessels JGH (1990) Role of cell wall architecture in fungal tip growth generation. In: Heath IB (ed) Tip growth in plant and fungal cells. Academic Press, San Diego, pp 1–29

Wyatt SE, Carpita NC (1993) The plant cytoskeleton-cell-wall continuum. Trends Cell Biol 3:413–417

Wynn WK (1976) Appressorium formation over stomates by the bean rust fungus: response to a surface contact stimulus. Phytopathology 66:136–146

Yaffe MP (1999) The machinery of mitochondrial inheritance and behaviour. Science 283:1493–1497

Yang S, Cope MJTV, Drubin DG (1999) Sla2p is associated with the yeast cortical actin cytoskeleton via redundant localization signals. Mol Biol Cell 10:2265–2283

Yap AS, Brieher WM, Gumbiner BM (1997) Molecular and functional analysis of cadherin-based adherens junctions. Annu Rev Cell Dev Biol 13:119–146

Yasuda T, Shishido K (1997) Aggregation of yeast cells induced by the Arg-Gly-Asp motif-containing fragment of high-molecular-mass cell-adhesion protein MFBA, derived from the basidiomycetous mushroom *Lentinus edodes*. FEMS Microbiol Lett 154:195–200

Yasuda T, Ishihara H, Amano H, Shishido K (1997) Generation of basidiomycetous hyphal cell-aggregates by

addition of the Arg-Gly-Asp motif-containing fragment of high-molecular-weight cell-adhesion protein
MFBA, derived from the basidiomycete *Lentinus
edodes*. Biosci Biotechnol Biochem 61:1587–1589

Yeaman C, Grindstaff KK, Nelson WJ (1999) New perspectives on mechanisms involved in generating
epithelial cell polarity. Physiol Rev 79:73–98

Zagon IS, Higbee R, Riederer BM, Goodman SR (1986)
Spectrin sub-types in mammalian brain: an immunoelectron microscopic study. J Neurosci 6:2977–2986

Zhou X-L, Stumpf MA, Hoch HC, Kung C (1991) A
mechanosensitive channel in whole cells and membrane patches of the fungus *Uromyces*. Science 253:
1415–1417

Zhu J-K, Shi J, Singh U, Wyatt SE, Bressan RA, Hasegawa
PM, Carpita NC (1993) Enrichment of vitronectin-
and fibronectin-like proteins in NaCl-adapted plant
cells and evidence for their involvement in plasma
membrane-cell wall adhesion. Plant J 3:637–646

Zhu J-K, Damsz B, Kononowicz AK, Bressan RA,
Hasegawa PM (1994) A higher plant extracellular
vitronectin-like adhesion protein is related to the
translational elongation factor-1α. Plant Cell 6:393–
404

11 Microtubules and Molecular Motors

In Hyung Lee and Michael Plamann

CONTENTS

I. Introduction

The principal components of the cytoskeleton are microtubules and actin microfilaments. These cytoskeletal elements are required for a variety of cellular processes including polarized growth, organelle movement and positioning, secretion, endocytosis, cell division, and chromosome segregation. In addition to the cytoskeleton, proper organelle transport and distribution also require the mechanochemical enzymes myosin, for actin-dependent transport, and kinesin and cytoplasmic dynein, for microtubule-dependent transport (reviewed in Mooseker and Cheney 1995; Hirokawa 1998). Each of these molecular motors generates movement by converting the energy of ATP hydrolysis into movement relative to the surface of actin filaments or microtubules.

Both yeasts and mycelial fungi have provided sophisticated experimental systems for the identification and characterization of microtubule- and actin microfilament-dependent processes. The yeast *Saccharomyces cerevisiae* has proven particularly well suited to study the cytoskeleton. The availability of a complete genome sequence and a broad variety of biochemical, genetic, and protein localization methodologies has led to the identification of many genes and proteins that are required for organization and function of the cytoskeleton (reviewed in Winsor and Schiebel 1997). A limitation of cytoskeletal studies in yeast is that, while microtubule organization is similar to that of other eukaryotes, microtubule-dependent transport of vesicular cargoes is absent (Huffaker et al. 1988; Jacobs et al. 1988). In contrast, mycelial fungi have both an elaborate cytoplasmic microtubule cytoskeleton and clearly demonstrable microtubule-dependent organelle transport (for earlier reviews see Heath 1994a,b; McKerracher and Heath 1987). In addition, recent studies have identified both kinesins and cytoplasmic dyneins in mycelial fungi and determined that they are significantly conserved both in structure and function to the respective motors of higher eukaryotes. These findings indicate that mycelial fungi provide good microbial systems for the analysis of microtubule-based motors and transport. In this chapter, we will focus on recent advances in the study of microtubule organization and function in mycelial fungi. Particular attention will be given to the progress that has been made in the identification and characterization of the fungal kinesins and cytoplasmic dynein.

School of Biological Sciences, University of Missouri-Kansas City, Kansas City, Missouri 64110-2499, USA

The Mycota VIII
Biology of the Fungal Cell
Howard/Gow (Eds.)
© Springer-Verlag Berlin Heidelberg 2001

II. Microtubules

A. Functions

One of the most noted functions of microtubules is in the formation of the mitotic spindle; however, cytoplasmic microtubules have also been found to play a central role in hyphal growth and intracellular transport of organelles to hyphal tips. A common cytological feature of hyphal tips of septate fungi is the Spitzenkörper or "apical body". The Spitzenkörper was discovered over 70 years ago as a densely staining or refractive spheroid body located at tips of actively growing hyphae (Brunswik 1924). The Spitzenkörper consists of a cluster of apical (secretory) vesicles (Grove and Bracker 1970; Howard 1981), and has been proposed to act as a vesicle supply center for hyphal growth (Bartnicki-Garcia et al. 1995). Examination of the cytoskeleton of fungal hyphae has revealed a network of microfilaments at hyphal tips and cytoplasmic microtubule tracks in all areas of hyphae including tip regions (Howard 1981). The observation that cytoplasmic vesicles are in close association with cytoplasmic microtubules contributed to the hypothesis that cytoplasmic microtubules function as a transport system for the movement of secretory vesicles to hyphal tips (Howard and Aist 1977, 1980; Howard 1981). In support of this model, treatment of hyphae with the anti-microtubule agent benomyl (i.e., methyl benzimidazole-2-ylcarbamate) resulted in disassembly of microtubules, disappearance of the Spitzenkörper, and a uniform distribution of vesicles (Howard and Aist 1977, 1980).

More recent studies have used video-enhanced contrast microscopy to study microtubule involvement in positioning of the Spitzenkörper at the hyphal tip and organelle movements within the hypha. The position of the Spitzenkörper has been shown to determine the shape of the fungal hypha (Bartnicki-Garcia et al. 1995). Inhibitors of microtubules, but not of actin microfilaments, resulted in loss of hyphal growth directionality, suggesting that cytoplasmic microtubules control the position of the Spitzenkörper at hyphal tips (Riquel et al. 1998).

The movement of nuclei, mitochondria, and vesicles were examined in *Neurospora crassa* and found to be dependent on cytoplasmic microtubules (Steinberg and Schliwa 1993).

All organelle movements observable by video-enhanced contrast microscopy correlated completely with the localization of cytoplasmic microtubule tracks. In addition, treatment of hyphae or protoplasts with anti-microtubule agents resulted in the cessation of all observable organelle movement, while treatment with anti-actin agents had no effect (Steinberg and Schliwa 1993). Recently, vesicle movement was examined in kinesin and cytoplasmic dynein mutants of *N. crassa* (see below; Seiler et al. 1999). Loss of kinesin function resulted in defective transport to the Spitzenkörper, while loss of cytoplasmic dynein resulted in complete cessation of retrograde transport (i.e., transport from hyphal tips to distal regions). Therefore, cytoplasmic microtubules of mycelial fungi, like those of higher eukaryotes, function along with microtubule-associated motors to drive organelle transport within the cytoplasm (cf. Chap. 2, Sect. VII.A,B, this Vol.).

B. Microtubule Organization

Microtubules are primarily composed of two related proteins, α- and β-tubulin, that share ~40% identity (Little and Seehaus 1988). Microtubule organization in fungi and other eukaryotes is controlled in large part by the microtubule organizing centers (MTOCs). These are the sites at which microtubule polymerization initiates (Picket-Heaps 1969). In fungi, MTOCs are called spindle pole bodies (SPBs) and are found at the poles of the mitotic spindle apparatus (reviewed in Heath 1994b). For an SPB-associated microtubule, the microtubule minus end is located at the SPB while the distal end represents the plus end. One of the central constituents of MTOCs is γ-tubulin, which has ~30% identity with α- and β-tubulins (Oakley and Oakley 1989; Oakley et al. 1990). γ-Tubulin was discovered in *A. nidulans* as a conformational suppressor of a benomyl-resistant mutant, *benA*, that has a mutation in one of the two known *A. nidulans* β-tubulin-encoding genes (Oakley and Oakley 1989). Localization studies have shown that in vegetative hyphae of *A. nidulans* and *N. crassa*, γ-tubulin is found only at SPBs, suggesting that SPBs represent the only MTOCs in vegetative hyphae (Oakley et al. 1990; Minke et al. 1999a). The significance of this finding is that MTOCs are a major determinant of microtubule polarity, and therefore establish the direction that kinesins and dyneins will transport cargo in the cell. The lack of γ-tubulin-containing MTOCs at sites other than SPBs suggests that all microtubules formed in *A. nidulans* and *N. crassa* initiate at the nuclear-associated SPBs. However, there does not appear to be a consistent pattern to microtubule organization, as not all fungal MTOCs appear to be nuclear associated. In *Schizosaccharomyces pombe*, MTOCs can be identified at the ends of cells (Hagan and Hyams 1988; Horio et al. 1991). MTOCs can also be found at hyphal tips during sexual development of *Sordaria macrospora* (Thompson-Coffe and Zickler 1992). In *Uromyces phaseoli*, repolymerization of cytoplasmic microtubules, following depolymerization with anti-tubulin agents, occurs first at the hyphal apex and not near the nuclear SPBs, suggesting that a MTOC is located in the apical region of these hyphae (Hoch and Staples 1985). In contrast, repolymerization of microtubules in *A. nidulans*, following treatment with an antimicrotubule agent, occurs at random sites along hyphae (de Andrade-Monteiro and Martinez-Rossi 1999). These sites of repolymerization coincide with bright immunofluorescent dots observed following localization with an anti-tubulin antibody. Small, punctate dots containing β-tubulin are also observed in hyphae of *Trichoderma viride* (Czymmek et al. 1996). Therefore, in some fungi, there may be non-nuclear-associated MTOCs that do not contain γ-tubulin.

There are three classes of fungal microtubules: intranuclear spindle microtubules; astral microtubules that extend off the cytoplasmic face of the SPBs; and cytoplasmic microtubules (reviewed in McKerracher and Heath 1987; Heath 1994a,b). Electron microscopy and indirect immunofluorescence have been used to examine the organization of cytoplasmic microtubules. Some cytoplasmic microtubules are found in close proximity to nuclear-associated SPBs, while other cytoplasmic microtubules are not associated with SPBs (McKerracher and Heath 1985a,b). Cytoplasmic microtubule tracks are primarily associated with the cell cortex, can be >50 μm in length, and can extend from distal regions of hyphae to hyphal tips (Czymmek et al. 1996; Tinsley et al. 1996; Bourett et al. 1999; Minke et al. 1999a). This pattern is consistent with the proposed role for cytoplasmic microtubules in long-range transport.

There is considerable variability in microtubule organization in fungi. In most eukaryotes, cytoplasmic microtubules disassemble during mitosis. This is also true for many fungi (Salo et al. 1989). In contrast, *N. crassa* is syncytial with asynchronous nuclear division, and cytoplasmic microtubules appear stable throughout the nuclear division cycle (Minke et al. 1999a). It has been proposed that the continuous presence of cytoplasmic microtubules throughout the nuclear division cycle allows for the very fast hyphal extension rates (>1.5 μm/s) of *N. crassa* (Minke et al. 1999a). In *N. crassa*, some cytoplasmic microtubules have also been observed to connect SPBs of adjacent nuclei (Minke et al. 1999a). It is likely that these microtubules play a role in the movement and distribution of nuclei in hyphae.

III. Kinesins in Mycelial Fungi

A. Identification and Structure

Since the kinesin class of microtubule-associated motors was first discovered in 1985 (Brady 1985; Scholey et al. 1985; Vale et al. 1985), more than 100 kinesin proteins have been identified in eukaryotes (reviewed in Hirokawa 1998; Kirchner et al. 1999). The kinesin superfamily is divided into various subfamilies according to the sequence homology and position of motor domain, the oligomerization state of the motor, and the presence of homologous sequences within the neck regions (Cole and Scholey 1995; Vale and Fletterick 1997; Hirokawa 1998). The proteins of one subfamily, that includes the founding members of the kinesin superfamily, share very high sequence homology in the motor domain and have been classified as conventional kinesins. The other members of the kinesin superfamily are called kinesin-like proteins (KLPs) due to reduced sequence homology in the motor domain relative to the conventional kinesins (Bloom and Endow 1995; Moore and Endow 1996).

At present, four conventional kinesins and three KLPs have been identified in the mycelial fungi (Table 1). Biochemical approaches used in the purification of kinesin from higher eukaryotes have been successfully applied to isolate conventional fungal kinesins from *N. crassa* and *Syncephalastrum racemosum* (Steinberg and Schliwa 1995; Steinberg 1997). Molecular searches using polymer chain reaction (PCR)-based strategies based on conserved residues in the motor domain have led to the identification of additional con-

Table 1. Kinesins in mycelial fungi

Motor	Organism	Amino acids	Functions or mutant phenotypes	Reference(s)
Conventional kinesins				
Nkin	*N. crassa*	926	Spitzenkörper establishment; hyphal growth and morphology; nuclear spacing	Steinberg and Schliwa (1995, 1996); Seiler et al. (1997, 1999)
Kin2	*U. maydis*	968	Dikaryotic hyphal growth; pathogenicity, microvesicle establishment; vacuole formation	Lehlmer et al. (1997); Steinberg et al. (1998)
NhKIN1	*N. hematococca*	929	Spitzenkörper formation; hyphal growth and morphology; mitochondrial positioning	Wu et al. (1998)
Synkin	*S. racemosum*	935	Not determined	Steinberg (1997); Grummt et al. (1998)
Kinesin-like proteins				
BIMC	*A. nidulans*	1184	Spindle pole body separation; bipolar spindle formation	Enos and Morris (1990)
KLPA	*A. nidulans*	770	Spindle separation	O'Conell et al. (1993)
Kin1	*U. maydeis*	1459[a]	Not known	Lehlmer et al. (1997)

[a] Predicted size using the first start codon of two putative in-frame start codons.

ventional kinesins from *Ustilago maydis* (Lehmler et al. 1997) and *Nectria haematococca* (Wu et al. 1998) and a KLP from *U. maydis* (Lehmler et al. 1997) and *A. nidulans* (O'Connell et al. 1993). An additional KLP has been identified from *A. nidulans* using a genetic approach (Enos and Morris 1990).

Conventional kinesins from mycelial fungi show about 50% amino acid identity with the motor domain of conventional kinesin from neuronal tissue, and 70–90% identity with one another (Steinberg 1998). The overall structure of conventional kinesins of mycelial fungi is very similar to that of higher eukaryotes except that fungal conventional kinesins lack associated light chains (Fig. 1). In higher eukaryotes, conventional kinesins typically consist of two heavy (~120 kDa) and two light (~60 kDa) chains (Bloom et al. 1988; Kuznetsov et al. 1988). Three distinct domains are found within the heavy chain of conventional kinesins of higher eukaryotes, and these domains are conserved in the fungal conventional kinesins (Fig. 1; Bloom and Endow 1995; Steinberg 1998). The N-terminal globular motor domain contains a microtubule-binding site and a conserved P-loop motif required for ATP binding and hydrolysis. An α-helical coiled-coil stalk domain located in the central region is important for dimerization. The C-terminal globular tail domain, which interacts with kinesin light chains, is required to mediate interaction of kinesin with organelles or other cargoes. Kinesin light chains of higher eukaryotes

Fungal conventional kinesin

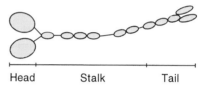

Head   Stalk   Tail

Fig. 1. Schematic drawing of fungal conventional kinesin. In higher eukaryotes, conventional kinesin is a rod-shaped molecule consisting of two heavy chains and two light chains. Conventional kinesins of fungi have the two heavy chains, but lack light chains. The N-terminus of the heavy chain forms a globular head domain that has ATPase- and microtubule-binding activity. The C-terminus can be divided into a long stalk domain, consisting of extensive coiled-coil structure (*small, shaded ovals*), and a small globular tail domain that is believed to be important for cargo binding. Kinesin heavy chains appear to dimerize by interaction between the coiled-coil domains of the stalk

contain a long series of N-terminal heptad repeats and several imperfect tandem repeats closer to the C-terminus. The absence of a light chain in all fungal conventional kinesins characterized so far is surprising as the kinesin light chain has an important role in kinesin-membrane interaction in higher eukaryotes. In addition, kinesin light chains have been proposed to regulate the ATPase activity of kinesin heavy chains. The ATPase activity of a kinesin heavy chain alone is greater than that of a kinesin heavy chain complexed with a kinesin light chain (Hackney et al. 1991, 1992). In the absence of cargo, kinesin may be folded by bringing the kinesin heavy chain motor domain and the kinesin light chain close to each other and possibly allowing kinesin light chains to regulate the ATPase activity of kinesin heavy chains in vivo. When binding to membranous organelles, the kinesin becomes active as the head domain unfolds from the tail (Hackney et al. 1991, 1992). It is possible that mycelial fungi contain a factor, as yet unidentified, that has a function comparable to the kinesin light chains of higher eukaryotes.

Another interesting aspect of the structure of fungal kinesins is that they have additional P-loop motifs. All characterized fungal conventional kinesins contain a second P-loop motif in the motor domain (Steinberg and Schliwa 1995; Lehmler et al. 1997; Grummt et al. 1998). *N. haematococca* conventional kinesin (NhKIN1) contains a third P-loop motif toward the end of the presumptive stalk domain (Wu et al. 1998). An additional interesting feature of NhKIN1 is that it contains a leucine zipper motif that overlaps with a region in the C-terminus domain (Wu et al. 1998). The biological importance of these additional P-loop motifs and the leucine zipper motif is unknown.

B. Functions of Fungal Kinesins

In higher eukaryotes, localization studies and functional assays indicate that conventional kinesin acts as a plus-end-directed microtubule motor involved in anterograde membrane transport (Pfister et al. 1989; Brady et al. 1990; Hirokawa et al. 1991). In various types of cells, kinesin associates with the ER, Golgi complex, mitochondria, endosomes and lysosomes, suggesting involvement of kinesin motors in the intracellular transport of membranous organelles (Bloom and Endow 1995; Barton and Goldstein 1996; Hackney 1996; Moore and Endow 1996; Vale and Fletterick 1997). Like conventional kinesins from higher eukaryotes, purified fungal conventional kinesins are plus-end-directed motors that are able to power microtubule gliding and the transport of membranous organelles along microtubules in vitro (Steinberg and Schliwa 1995, 1996; Steinberg 1997; Steinberg et al. 1998; see also Chap. 12, this Vol.). Although the specific cargoes for conventional kinesin motors of mycelial fungi are still not known, all fungal conventional kinesins seem to be involved in vesicle transport

toward hyphal tips. Surprisingly, conventional kinesins are not essential for viability, suggesting the presence of functionally redundant microtubule-associated motor proteins or redundant transport mechanisms involving different cytoskeletal components such as actin and myosin (Lilli and Brown 1992; Barton and Goldstein 1996). Several lines of evidence indicate that actinased transport of membranous organelles does occur in mycelial fungi (Heath 1994a,b; McGoldrick et al. 1995). Therefore, in the absence of a conventional kinesin, the fungal cell may continue to grow, albeit slowly and abnormally, by utilizing an actin-based transport system.

The role of conventional kinesins in the transport of mitochondria is not consistent among mycelial fungi. Mitochondrial positioning is affected in the conventional kinesin deletion mutants of *N. haematococca* (Wu et al. 1998), but not in the conventional kinesin mutants of *N. crassa* and *U. maydis* (Lehmler et al. 1997; Seiler et al. 1997). Mitochondrial movement in *N. crassa* is known to be microtubule-dependent; therefore, it is very likely that other microtubule-associated motors function to move and position mitochondria in *N. crassa* (Steinberg 1998). Conventional kinesins may also be involved in nuclear migration in mycelial fungi. In conventional kinesin-deficient cells of *N. crassa*, the spacing of nuclei is lost and nuclei are often clustered in small groups separated by large gaps (Seiler et al. 1997).

While conventional kinesins are involved in vesicular trafficking and organelle transport, many of the KLPs function primarily in nuclear division (Brady 1995; Barton and Goldstein 1996; Vallee and Sheetz 1996). At present, only three KLPs have been identified from mycelial fungi. Functional analysis of two KLPs identified in *A. nidulans* indicates that these fungal KLPs are involved in mitosis. A description of each of the conventional kinesins and KLPs isolated from mycelial fungi is presented below.

1. *Neurospora crassa*

The first conventional kinesin from mycelial fungi was identified in *N. crassa* by a biochemical approach (Steinberg and Schliwa 1995). Steinberg and Schliwa (1995) adopted an isolation protocol commonly used in the purification of microtubule-dependent motor proteins from higher eukaryotes. Two polypeptides (105/108 kDa) were copurified and shown to produce vigorous microtubule gliding activity with four- to eightfold faster velocity than conventional kinesins of higher eukaryotes. Molecular cloning showed that the 105- and 108-kDa polypeptides are homologues of conventional kinesin and are isoforms of the same protein, possibly from differential phosphorylation. *N. crassa* conventional kinesin (Nkin) consists of 926 residues and its motor domains shows 53–57% amino acid identity with that of other conventional kinesins. Like other members of the kinesin superfamily, Nkin consists of an N-terminal motor domain, a central stalk-like domain and a small globular tail domain.

A Nkin-deficient mutant was constructed and found to be viable, but slow growing. The mutant is characterized by hyphae that have an increased diameter and a high incidence of branching. In addition, nuclei often clustered in small groups resulting in long anucleate regions of hyphae. Although the movement of organelles that can be visualized by video microscopy is hardly affected, the Spitzenkörper, an accumulation of vesicles located at the hyphal apex and linked to hyphal growth, was absent or much reduced (Seiler et al. 1997). This observation implies that kinesin is essential for the transport of small secretory vesicles that are required for normal hyphal growth and secretion. Indeed, protein secretion is substantially reduced in the Nkin-deficient cells (Seiler et al. 1999).

2. *Ustilago maydis*

Two kinesin-encoding genes, *kin1* and *kin2*, have been identified from the phytopathogenic fungus *U. maydis* (Lehmler et al. 1997). The motor domain of Kin1 is most similar to human CENP-E protein with 40% identity. CENP-E is a centromere-binding protein implicated in chromosome movement during prometaphase (Yen et al. 1991); however, there is no evidence that Kin1 has a comparable function in *U. maydis*. The *kin2* gene encodes a 968 amino acid polypeptide that has 52% amino acid sequence identify with *N. crassa* conventional kinesin, Nkin. Recently, Kin2 was partially purified and showed similar motility properties to Nkin of *N. crassa* (Steinberg et al. 1998).

The deletion of *kin1* has no significant effect on cell morphology and growth of yeast-like haploid cells, formation of dikaryons or elongation of dikaryotic hyphae. A *kin2* deletion has no obvious effects on viability and morphology of

vegetatively growing yeast-like haploid cells. However, the deletion mutation of *kin2* severely affects the extension of dikaryotic hyphae, the migration of the cytoplasm into the tip cell component, and pathogenicity. Therefore, conventional kinesin of *U. maydis* appears to be required for the long-range transport of vesicles during hyphal growth.

Green fluorescent protein (GFP) fusion proteins were used to localize Kin2 (Lehmler et al. 1997). GFP-Kin2 was evenly distributed in the cytoplasm of haploid cells and dikaryotic hyphae, suggesting that kinesin is bound to submicroscopic vesicles. In *U. maydis*, microvesicles accumulate in the hyphal apex in the wild type and are required for hyphal growth. In dikaryotic hyphae of Δ*kin2*, microvesicles were observed, but they were not clustered at hyphal tips as in the wild type (Lehmler et al. 1997). Therefore, Kin2 is probably required for the vectorial transport of submicroscopic vesicles to tip regions. In addition, the large basal vacuoles that are typically observed in the wild type were absent in the Δ*kin2* mutant. Only scattered, small vacuoles (200–400 nm) were observed in dikaryotic hyphae of the mutant, suggesting that Kin2 is involved in the formation of vacuoles (Steinberg 1998; see also Chap. 12, this Vol.).

3. *Nectria haematococca*

Nectria haematococca conventional kinesin (NhKIN1) was identified by a PCR-based approach (Wu et al. 1998). NhKIN1 encodes a 929 amino acid polypeptide with 77% amino acid sequence identity to Nkin. Like Nkin and Kin2, NhKIN1 contains a second P-loop motif in the motor domain. A third P-loop motif is located at the end of the presumptive stalk domain. In addition, NhKIN1 contains a leucine zipper motif in the C-terminal region.

Deletion of the NhKIN1 gene results in helical or wavy hyphal growth, reduced hyphal diameter, a reduction in the size of the Spitzenkörper, and an ~50% reduction in colony growth rate. Unlike conventional kinesins from other mycelial fungi (Lehmler et al. 1997; Seiler et al. 1997), NhKIN1 appears to be involved in the positioning and transport of mitochondria. In deletion mutants, mitochondria did not occupy their normal subapical position behind the Spitzenkörper, and there were long regions of hyphae without mitochondria. The deletion mutation had no effect on mitosis or microtubule organization.

4. *Syncephalastrum racemosum*

The conventional kinesin from *S. racemosum* (Synkin) has been identified by purification (Steinberg 1997) and its gene cloned by PCR amplification (Grummt et al. 1998). Two polypeptides (112/115 kDa) copurified, with no accompanying light chains. Like other fungal kinesins, Synkin contains a second P-loop motif in the motor domain. Biochemical characterization suggests that Synkin is a dimer with a more globular shape than conventional kinesin of higher eukaryotes. In vitro, the enzyme was able to drive the microtubule-dependent movement of vesicles isolated from *S. racemosum, N. crassa,* and *A. nidulans.* Vesicles from *S. racemosum* were also transported in vitro by *N. crassa* kinesin. However, vesicles from pig brain were not transported by the *S. racemosum* or *N. crassa* kinesin motors, suggesting that vesicles of mycelial fungi contain specific proteins required for kinesin-vesicle interaction.

5. *Aspergillus nidulans*

In *Aspergillus nidulans*, conventional kinesin has not yet been identified; however, two KLPs have been isolated. BIMC, the first kinesin homologue identified in the mycelial fungi, is a polypeptide of 1184 amino acids with greatest homology to *Drosophila melanogaster* kinesin heavy chain and the *S. cerevisiae* Cin8p KLP (Enos and Morris 1990; Hoyt et al. 1992). In *bimC* mutant cells, spindle pole body separation is impaired, suggesting that KLP is required for proper mitosis in *A. nidulans*. However, nuclear migration is not affected in the *bimC* mutation. In addition, the *bimC* mutant grows in typical polar fashion, suggesting that vesicle transport to hyphal tips is not affected.

The second *A. nidulans* KLP, KLPA, was identified by a PCR strategy based on the conserved sequences of the kinesin heavy chain (O'Connell et al. 1993). Unlike conventional kinesins, and most KLPs that have N-terminal motor domains, KLPA contains a C-terminal motor domain like KAR3 of *S. cerevisiae* and NCD of *D. melanogaster* (McDonald et al. 1990; Meluh and Rose 1990; Walker et al. 1990). The deletion of *klpA* causes no observable mutant phenotype on

its own. However, the *klpA* deletion does suppress the temperature-sensitive mitotic spindle defect observed with the *bimC* mutation (O'Connell et al. 1993). This genetic suppression provides evidence for a functional role for KLPA in the mitotic spindle.

IV. Cytoplasmic Dynein and Dynactin in Mycelial Fungi

A. Identification and Structure

Dynein was first discovered in eukaryotes as an ATPase that powers the movement of cilia and flagella (Gibbons and Rowe 1965). Approximately 20 years later, a cytoplasmic form of dynein required to transport membranous cargo to the minus ends of microtubules was discovered in brain tissue (Lye et al. 1987; Paschal et al. 1987; Paschal and Vallee 1987). Additional molecular searches for cytoplasmic dynein revealed that it is ubiquitous among eukaryotes (reviewed in Hirokawa 1998; Milisav 1998). Cytoplasmic dynein function requires an additional multi-subunit complex known as **dynactin (dynein activator)** (Gill et al. 1991; Schroer and Sheetz 1991). Together, cytoplasmic dynein/dynactin represent the most complex of the molecular motors operating in the cytoplasm.

In higher eukaryotes, cytoplasmic dynein is a 20S protein complex composed of two heavy chains (~500 kDa), three intermediate chains (~70 kDa), four light intermediate chains (55–60 kDa), and light chains (8 and 23 kDa) (Fig. 2; reviewed in Holzbaur and Vallee 1994; Allan 1996). The cytoplasmic heavy chain contains four P-loop motifs in its central region. The central and C-terminal regions of the cytoplasmic dynein heavy chain form a large globular domain that interacts with microtubules and has motor activity, while the N-terminal region forms a thin stalk that interacts with other dynein subunits and is thought to be the site of cargo binding (reviewed in Hirokawa 1998).

Dynactin is also a 20S multi-subunit complex composed of at least ten subunits including p150[Glued], p62, p50/dynamitin, Arp1 (*actin-related protein*), Arp11, actin-capping protein a and b subunits, p27, p25, and p24 (Fig. 2; reviewed in Allan 1996; Schroer 1996; Schroer et al. 1996; Holleran et al. 1998). Ultrastructural analysis of purified

A. Cytoplasmic dynein

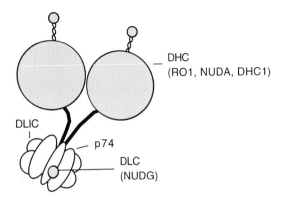

B. Dynactin

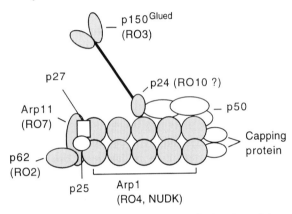

Fig. 2. Schematic drawing of cytoplasmic dynein and dynactin. **A** Fungal cytoplasmic dynein and **B** dynactin are predicted to form molecular motors similar to those of higher eukaryotes. The identities and proposed interaction of the cytoplasmic dynein and dynactin subunits are indicated. The subunits whose genes have been identified in mycelial fungi are indicated in *gray*. See the text for a detailed description of these complex structures. *DHC* Dynein heavy chain; *DLIC* dynein light intermediate chain; *DLC* dynein light chain

with dynein, as the p150^{Glued} subunit has been shown to interact with the dynein intermediate chain (Karki and Holzbaur 1995; Vaughan et al. 1995; Waterman-Storer et al. 1995; Echeverri et al. 1996; Burkhardt et al. 1997).

Cytoplasmic dynein has not been purified from a mycelial fungus. However, genetic analyses have been used in *N. crassa* and *A. nidulans* to identify genes that encode polypeptides homologous to mammalian cytoplasmic dynein/dynactin subunits (Table 2, Fig. 2). Interestingly, additional fungal genes have been identified that encode polypeptides that are not formal cytoplasmic dynein or dynactin subunits, but are apparently required for cytoplasmic dynein/dynactin function. These genes have been proposed to act as regulators of specific cytoplasmic dynein/dynactin functions. The fungal genes identified as required for cytoplasmic dynein/dynactin function are summarized below.

1. *Neurospora crassa*

Genetic and molecular analysis of a set of genes required for normal hyphal growth led to the identification of genes required for cytoplasmic dynein and dynactin function in *N. crassa*. *N. crassa ro* py mutants were isolated as partial suppressors of a *cot-1* mutation that affects hyphal elongation (Yarden et al. 1992; Plamann et al. 1994; Bruno et al. 1996). These mutants display curled hyphal growth and abnormal nuclear distribution (Garnjobst and Tatum 1967; Perkins et al. 1982). Molecular characterization revealed that *ro-1*, *ro-3*, and *ro-4* encode cytoplasmic dynein heavy chain, p150^{Glued}, and Arp1, respectively (Table 2; Plamann et al. 1994; Robb et al. 1995; Tinsley et al. 1996). These findings suggested that other *ro* mutants may be defective in cytoplasmic dynein/dynactin function and have led to the isolation of more than 1000 *ro* mutants. The *N. crassa ro* mutants define 23 complementation groups and represent the largest collection of cytoplasmic dynein/dynactin mutants available for any organism (Bruno et al. 1996). Molecular characterization of additional *ro* genes revealed that *ro-2* and *ro-7* encode proteins homologous to the p62 and Arp11 subunits of mammalian dynactin complex (Lee and Plamann, unpubl.; Minke et al., unpubl.). The *ro-10* and *ro-11* genes were found to encode novel 24 and 75-kDa proteins, respectively (Minke et al. 1999b). In the absence of RO10, p150^{Glued} is degraded suggesting that RO10 may be a subunit

dynactin reveals that Arp1 exists as a 37-nm actin-like filament consisting of 8–10 Arp1 monomers (Schafer et al. 1994). The α and β subunits of capping protein $\beta2$ associate with the barbed end of the Arp1 filament (Schafer et al. 1994), while the pointed-ends of the Arp1 filament are bound by a complex consisting of p62, p27, p25 and Arp11 (Schafer et al. 1994; Eckley et al. 1999). The p150^{Glued} subunit contains a microtubule-binding site at its N-terminus and, together with the p50/dynamitin and p24 subunits, forms a projecting sidearm from the Arp1 filament. This sidearm is important for interaction of dynactin

Table 2. Dynein and dynactin subunits in mycelial fungi

Subunit	Organism	Amino acids	Mammalian homologue	Functions or mutant phenotypes	Ref(s)
Dynein					
RO1	*N. crassa*	4367	Heavy chain	Nuclear distribution; hyphal growth and morphology; vacuole formation	Plamann et al. (1994); Seiler et al. (1999)
NUDA	*A. nidulans*	4344	Heavy chain	Nuclear migration; hyphal growth; asexual and sexual development	Xiang et al. (1994, 1995c)
DHC1	*N. hematococca*	4349	Heavy chain	Nuclear distribution; microtubule organization; spindle pole separation	Inoue et al. (1998a)
NUDG	*A. nidulans*	94	8-kDa light chain	Nuclear migration	Beckwith et al. (1998)
Dynactin					
RO3	*N. crassa*	1300	p150[Glued]	Same as RO1	Tinsley et al. (1996)
RO4	*N. crassa*	380	Arp1	Same as RO1	Plamann et al. (1994)
RO2	*N. crassa*	634	p62	Same as RO1	Lee and Plamann (unpubl.)
RO7	*N. crassa*	647	Arp11	Same as RO1	Minke et al. (unpubl.)
NUDK	*A. nidulans*	380	Arp1	Nuclear migration	Xiang et al. (1999)
Additional proteins likely involved in dynein function					
RO10	*N. crassa*	217	Dynactin subunit?	Nuclear distribution; hyphal growth and morphology; RO3 stability	Minke et al. (1999b)
RO11	*N. crassa*	697	Not known	Nuclear distribution; hyphal growth and morphology	Minke et al. (1999b)
NUDC	*A. nidulans*	198	Not known	Nuclear migration; regulator of dynein function	Osmani et al. (1990)
NUDF	*A. nidulans*	444	LIS-1	Same as NUDC	Xiang et al. (1995b)

of dynactin that is required for incorporation of p150^{Glued} into the dynactin complex. RO11 is interesting in that the first 200 residues are predicted to comprise an α-helical coiled-coil region, and the last ~500 residues have a predicted pI of 12.0. Loss of RO11 results in about a twofold increase in the levels of cytoplasmic dynein heavy chain and p150^{Glued}, and a slight alteration in dynein/dynactin localization (Minke et al. 1999b).

2. *Aspergillus nidulans*

Molecular characterization of **nud (nuclear distribution) mutants** of *A. nidulans* revealed that some *nud* genes encode subunits of cytoplasmic dynein or dynactin. The *nudA* gene encodes the dynein heavy chain (Xiang et al. 1994). The *nudG* gene encodes an 8-kDa protein that is homologous to the 8-kDa cytoplasmic dynein light chain (Beckwith et al. 1998). As expected, NUDG co-immunoprecipitates with NUDA, and *nudG*; $\Delta nudA$ double mutants show similar growth defects to each single mutant. Recently, the *nudK* gene has been shown to encode a 380 amino acid polypeptide that is homologous to Arp1 (Xiang et al. 1999). The *nudC* and *nudF* proteins are not homologous to any subunit of dynein or dynactin, but are proposed as potential regulators that control cytoplasmic dynein function (Willins et al. 1997; Xiang et al. 1995b). NUDF contains six WD-40 repeats that are found in regulatory proteins suggesting that NUDF may act as a regulator (Xiang et al. 1995b). Multiple copies of the *nudF* gene can suppress the *nudC3* mutation, suggesting a similar regulatory role for NUDC. Additionally, several suppressors of *nudF* were found to be mutant alleles of *nudA*, and *nudF* was found to be epistatic to *nudA*, implying that NUDA and NUDF function in the same nuclear migration pathway (Willins et al. 1997). Although these results suggest that NUDF and NUDC function as regulators, how they control cytoplasmic dynein function is unknown.

3. *Nectria haematococca*

Recently, a cytoplasmic dynein heavy chain was isolated from *N. haematococca* using PCR techniques (Inoue et al. 1998a). The *CDHC1* gene encodes a 4349 amino acid polypeptide that shows >70% identity to cytoplasmic dynein heavy chains of *N. crassa* and *A. nidulans*. The deletion mutants are viable but grow very slowly. In cytoplasmic

dynein-deficient mutants, astral microtubules are absent and post-mitotic nuclear migration is limited, resulting in nonuniform distribution of interphase nuclei. In addition, the Spitzenkörper is significantly reduced in the dynein mutant, suggesting a role for CDHC1 in secretory vesicle transport.

B. Functions of Fungal Cytoplasmic Dynein

Cytoplasmic dynein has been implicated in a variety of cellular functions involving vesicle trafficking and mitosis (reviewed in Vallee and Sheetz 1996; Hirokawa 1998; Karki and Holzbaur 1999). These functions include retrograde axonal transport, vesicle movement within the endocytic pathway, formation of endoplasmic reticulum networks, organization of Golgi, and formation of the mitotic spindle.

Dynactin interacts physically with cytoplasmic dynein, through contacts between p150^{Glued} of dynactin and a dynein intermediate chain (Karki and Holzbaur 1995; Vaughan and Valee 1995; Burkhardt et al. 1997), and is required for efficient translocation of vesicles in vitro (Gill et al. 1991; Schroer and Sheetz 1991). Genetic analyses in fungi revealed that cytoplasmic dynein and dynactin function within the same cellular pathway. *S. cerevisiae* mutants defective in cytoplasmic dynein heavy chains exhibit defects in spindle orientation and distribution of nuclei into the daughter cells (Eshel et al. 1993; Li et al. 1993). Similar phenotypes are found in null mutants of yeast Arp1 (Clark and Meyer 1994; Muhua et al. 1994), and Arp1/dynein heavy chain double mutants show no additive effects (Muhua et al. 1994). Null mutants of *Neurospora* p150^{Glued} (*ro-3*) and Arp1 (*ro-4*) are phenotypically identical to cytoplasmic dynein heavy chain (*ro-1*) mutants (Plamann et al. 1994; Robb et al. 1995; Tinsley et al. 1996). In addition, immunolocalization studies revealed dynein heavy chains and p150^{Glued} are similarly localized to hyphal tips in *N. crassa* (Minke et al. 1999b).

One proposed role for dynactin is that it serves as a cargo binding site. The interaction between cytoplasmic dynein and the Golgi membrane has been shown to occur through the interaction of Arp1 with spectrin (Holleran et al. 1996). However, dynactin seems to be more than simply a linker between dynein and membranous cargo. Dynein, by itself, can bind vesicles and cross-link them to microtubules, but it is unable to translocate along the microtubule without dynactin. Dynactin may regulate the motor activity of cytoplasmic dynein, as has been proposed for the conventional kinesin light chains (Hackney et al. 1991, 1992).

1. Dynein-Dependent Vesicle Trafficking in Fungi

Dynein and dynactin are concentrated at growing hyphal tips in *N. crassa* and *A. nidulans* (Xiang et al. 1995c; Minke et al. 1999b), suggesting that cytoplasmic dynein may be involved in transport of vesicles found within the Spitzenkörper. In *N. crassa*, dynein and dynactin mutants have a prominent Spitzenkörper and show vesicle transport toward hyphal tips, demonstrating that tip-directed transport is still active. However, transport from hyphal tips to distal regions is eliminated, suggesting that, like higher eukaryotes, dynein/dynactin of fungi are required for retrograde vesicle transport (Seiler et al. 1999).

Vacuole accumulation near tips in dynein/dynactin mutants of *N. crassa* (Seiler et al. 1999; Lee and Plamann, unpubl. obs.) may be due to impaired retrograde transport of organelles targeted to the vacuole system. Some vacuoles in fungi are comparable to lysosomes in vertebrate cells (see Chap. 12, this Vol.), and dynein is known to be involved in movements of lysosomal compartments in vertebrate cells (Aniento et al. 1993; Harada et al. 1998).

2. Fungal Dynein and the Mitotic Spindle

The role of cytoplasmic dynein in mitosis is well demonstrated in mammalian cells (reviewed in Merdes and Cleveland 1997; Karki and Holzbaur 1999). Immunolocalization studies have shown that cytoplasmic dynein is localized to kinetochores and spindle microtubules (Pfarr et al. 1990; Steuer et al. 1990). Injection of anti-dynein antibodies prevents centrosome separation during mitosis (Vaisberg et al. 1993), and dynein is required for mitotic spindle assembly in vitro (reviewed in Merdes and Cleveland 1997).

Initially, cytoplasmic dynein was not thought to be involved in mitosis in mycelial fungi. In dynein mutants of *N. crassa* and *A. nidulans*, nuclei divide with no obvious defects in spindle formation (Plamann et al. 1994; Xiang et al. 1994). However, recent evidence suggests that cytoplasmic dynein of mycelial fungi is also involved in mitosis. In dynein-deficient cells of *N. haemato-*

cocca, mitosis occurs normally from prophase to anaphase A, but post-mitotic nuclear migration is limited. This can be explained by the observation that the aster-like arrays of cytoplasmic microtubules, normally focused at the spindle pole bodies, are absent in these mutants (Inoue et al. 1998a). In the wild type, spindle pole bodies are separated by both a pushing force, generated as the spindle elongates, and a pulling force operating through the astral microtubules (Inoue et al. 1998b). When the anaphase B spindle in the dynein mutant was cut by a laser microbeam, spindle pole body separation was almost stopped suggesting that the pulling force contributed by astral microtubules is lacking (Inoue et al. 1998b). These results show that cytoplasmic dynein is important for the formation and maintenance of astral microtubules that exert the pulling force for spindle elongation. In contrast to *N. haematococca*, astral microtubules are not affected in dynein-deficient cells in *S. cerevisiae* where the Cin8p and Kip1p kinesins contribute to the anaphase movement of chromosomes by interacting with astral microtubules (Yeh et al. 1995; Carminati and Stearns 1997). Spindle elongation occurs in the dynein mutants of *N. haematococca* and *S. cerevisiae* (Hoyt and Geiser 1996; Inoue et al. 1998b). Other microtubule-associated motors, probably kinesin-like motors, may be involved in generating the pushing force within the spindle. Kinesin-like proteins BIMC and KLPA are involved in spindle elongation in *A. nidulans* (Enos and Morris 1990; O'Connell et al. 1993).

Genetic analysis of *A. nidulans* cytoplasmic dynein also reveals an interaction between dynein and genes required for mitotic functions. Synthetically lethal mutants were isolated for a null mutant of dynein heavy chain (***sld* for synthetic lethal without dynein**). Two of the *sld* mutants, *sldA* and *sldB*, also show a strong synthetic lethal interaction with a mutation in the mitotic kinesin *bimC* (Efimov and Morris 1998). The SLDA and SLDB proteins show sequence homology to the spindle assembly checkpoint proteins BUB1 and BUB3 from *S. cerevisiae* (Hoyt et al. 1991; Roberts et al. 1994). These results suggest that cytoplasmic dynein plays a role in spindle assembly or chromosome separation (Efimov and Morris 1998).

3. Fungal Dynein and Nuclear Migration

The role of cytoplasmic dynein in nuclear migration is demonstrated in all three mycelial fungi in

which cytoplasmic dyneins have been characterized (reviewed in Fischer 1999). In mycelial fungi,
nuclei are more or less evenly distributed within
long hyphae. However, in fungal dynein mutants,
nuclear clustering is a common phenotype. *N.
crassa* dynein mutants show highly asymmetric
nuclear distribution with long (>200 μm) anucleate
regions in hyphae (Robb et al. 1995; Bruno et al.
1996; Tinsley et al. 1996; Minke et al. 1999a,b). In
A. nidulans temperature-sensitive *nud* mutants,
nuclei fail to move into germ tubes when conidia
are germinated at restrictive temperature (Morris
1976; Xiang et al. 1994, 1999; Beckwith et al. 1998).
In *N. haematococca* dynein-deficient cells, two to
ten nuclei are clustered near septa (Inoue et al.
1998b).

Nuclear migration is not completely blocked
in the dynein-deficient mutants, suggesting that
there are additional motor proteins that are
involved in nuclear migration in mycelial fungi.
The characteristics of bypass suppressors of *nudA*,
identified in *A. nidulans*, point to a dynein-
independent system for nuclear migration
(Goldman and Morris 1995). Two dynein bypass
mutations, *sldD* and *sldE*, cause an abnormal
nuclear morphology and a more severe nuclear
migration defect than *nudA* mutants. It has been
proposed that these mutants may identify a
dynein-independent system for nuclear migration
(Efimov and Morris 1998).

In *A. nidulans nud* mutants formation of both
conidia and ascospores is severely suppressed
(Xiang et al. 1994; Beckwith et al. 1998). Because
nuclear migration is required for developmental
structures, this may be an indirect result of nuclear
migration defects. However, *N. haematococca*
dynein-deficient cells form conidia although they
are abnormally shaped and less abundant relative
to the wild type (Inoue et al. 1998b), and *N. crassa
ro* mutants produce abundant conidia (Plamann et
al. 1994). Therefore, the role of cytoplasmic dynein
in asexual reproduction is not clear. Cytoplasmic
dynein might be involved to different extents in
different mycelial fungi.

A number of models have been proposed to
explain how cytoplasmic dynein participates in the
movement of nuclei within growing hyphae (Fig.
3). In one model (Fig. 3A), cytoplasmic dynein
binds to the nuclear envelop and transports nuclei
as it translocates along microtubules (Xiang et al.
1995a). In support of this model, nuclei have been
observed to move in one direction along microtubules in wall-less mutants and protoplasts of

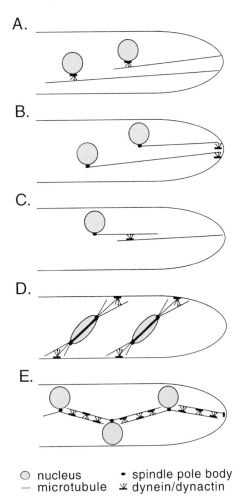

○ nucleus   • spindle pole body
— microtubule   ⊻ dynein/dynactin

Fig. 3. Models proposed to explain the possible mechanisms by which cytoplasmic dynein is involved in nuclear
migration in mycelial fungi. *Symbols* representing nuclei,
microtubules, SPBs, and cytoplasmic dynein/dynactin complexes are indicated. Details concerning each model are
explained in the text. The models are modified from those
presented previously. (Plamann et al. 1994; Xiang et al.
1994; Efimov and Morris 1998)

N. crassa (Steinberg and Schliwa 1993). Also, in
Xenopus, isolated nuclei are able to move along
microtubules (Reinsch and Karsenti 1997). In a
second model (Fig. 3B), cytoplasmic dynein is
anchored at hyphal tips and, by depolymerizing
microtubules, it reels in nuclei that are attached to
the microtubules through their spindle pole bodies
(Xiang et al. 1995a). In support of this model,
cytoplasmic dynein is concentrated at the apex
of hyphae rather than on the nuclear envelope
(Xiang et al. 1994; Minke et al. 1999b). In a third
model (Fig. 3C), cytoplasmic dynein is anchored
to microtubules that are attached to the hyphal
tip. Dynein then translocates along astral micro-

tubules toward the spindle pole body, resulting in movement of a nucleus to the tip (Xiang et al. 1995a). In a fourth model (Fig. 3D), cytoplasmic dynein pulls astral microtubules emanating from spindle pole bodies as in the second model, but cytoplasmic dynein binds to the cell cortex instead of hyphal tips. In this model, the positions of the dynein motors along the cortex determine the position of daughter nuclei within the hypha (Efimov and Morris 1998). In a fifth model (Fig. 3E), cytoplasmic dynein crossbridges antiparallel microtubules attached to spindle pole bodies of adjacent nuclei and exerts force to slide microtubules across each other. As a result, adjacent nuclei pull each other and even distribution of nuclei is achieved by balanced force. In this model the tip-proximal nucleus is anchored at the hyphal tip (Plamann et al. 1994; Plamann 1996).

V. Conclusions

During the past 5 years, great progress has been made in determining the roles of cytoplasmic microtubules and microtubule-associated motors in organelle movement and positioning in mycelial fungi. Current studies have led to the identification of conventional kinesins and cytoplasmic dynein, and shown that they are required for multiple processes including transport of membranous cargoes to and from hyphal tips, vacuole formation and nuclear movement and distribution. Most interesting are the distinctive differences in the structure and function of microtubule-associated motors of mycelial fungi and those of yeasts. Many of the structural features of conventional kinesins and cytoplasmic dynein of mycelial fungi are conserved relative to other eukaryotes (Holzbaur and Vallee 1994; Kirchner et al. 1999). However, in yeast, conventional kinesins are not present and cytoplasmic dynein is significantly diverged relative to that of mycelial fungi and other eukaryotes. These findings are consistent and supportive of earlier studies indicating that microtubules of yeast are required for spindle formation and nuclear movement, but are not involved in organelle transport as they are in other eukaryotes (Huffaker et al. 1988; Jacobs et al. 1988).

While conventional kinesins and dynein are well conserved structurally within the higher fungi, there are significant differences in the func-

tions of these motors. For example, a null mutant of conventional kinesin results in defective mitochondrial movement in *N. haematococca* but in the related fungus, *N. crassa*, mitochondrial movement appears unaffected (Seiler et al. 1997; Wu et al. 1998). This observation suggests that we will fail to define all the functions of conventional kinesins, and perhaps dyneins, in fungi by a detailed characterization of only one species.

There are also conflicting observations regarding the organization of microtubules in fungi. One of the major unanswered questions concerning microtubule-based transport is the polarity of cytoplasmic microtubules within hyphae. However, the answer to this question is likely to be complex, as variation is observed in both the localization of MTOCs between fungi (Hagan and Hyams 1988; Oakley et al. 1990), and the localization of MTOCs during vegetative growth versus sexual development (Thompson-Coffe and Zickler 1992; Minke et al. 1999a). Additional variation is observed in the stability of cytoplasmic microtubules across the nuclear division cycle of different fungi (Raudaskoski et al. 1991; Minke et al. 1999a). These results indicate that microtubule organization is dynamic and is likely to vary considerably between fungi and at different stages of the fungal life cycle.

Finally, we see the diversity of microtubule organization and motor function observed in mycelial fungi as a source of strength. *S. cerevisiae* has provided a powerful molecular genetic system to study the cytoskeleton, and has allowed us to identify many of the genes and proteins required for cytoskeleton organization. However, one of the most interesting properties of the cytoskeleton is its evolutionary and developmental plasticity. With the mycelial fungi, we have molecular genetic systems that will allow us to define not only the molecular constituents of the cytoskeleton, but also the mechanisms by which the cytoskeleton can be altered to meet the needs of the organism.

Acknowledgements. This work was supported by NIH grant GM51217 to M.P.

References

Allan V (1996) Motor proteins: a dynamic duo. Curr Biol 6:630–633

Aniento F, Emans N, Griffiths G, Gruenberg J (1993) Cytoplasmic dynein-dependent vesicular transport from early to late endosomes. J Cell Biol 123:1373–1387

Bartnicki-Garcia S, Bartnicki DD, Gierz G, Lòpez-Franco R, Bracker CE (1995) Evidence that Spitzenkörper behavior determines the shape of a fungal hypha: a test of the hyphoid model. Exp Mycol 19:153–159

Barton N, Goldstein L (1996) Going mobile: microtubule motors and chromosome segregation. Proc Natl Acad Sci USA 93:1735–1742

Beckwith SM, Roghi CH, Liu B, Morris NR (1998) The "8-kD" cytoplasmic dynein light chain is required for nuclear migration and for dynein heavy chain localization in Aspergillus nidulans. J Cell Biol 143:1239–1247

Bloom G, Endow S (1995) Motor proteins. 1. Kinesins. Protein Profile 1:1059–1116

Bloom G, Wagner M, Pfister K, Brady S (1988) Native structure and physical properties of bovine brain kinesin and identification of the ATP-binding subunit polypeptide. Biochemistry 27:3409–3416

Bourett T, Czymmek KJ, Howard RJ (1999) An improved method for affinity probe localization in whole cells of filamentous fungi. Fungal Genet Biol 24:3–13

Brady S (1985) A novel brain ATPase with properties expected for the fast axonal transport motor. Nature 317:73–75

Brady S (1995) A kinesin medley: biochemical and functional heterogeneity. Trends Cell Biol 5:159–164

Brady S, Pfister K, Bloom G (1990) A monoclonal antibody against kinesin inhibits both anterograde and retrograde fast axonal transport in squid axoplasm. Proc Natl Acad Sci USA 87:1061–1065

Bruno KS, Tinsley JH, Minke PF, Plamann M (1996) Genetic interactions among cytoplasmic dynein, dynactin, and nuclear distribution mutants of Neurospora crassa. Proc Natl Acad Sci USA 93:4775–4780

Brunswik H (1924) Untersuchungen über die Geschlecht und Kernverhaltnisse bei den Hymenomycetengattung Coprinus. In: Goebel K (ed) Botanische Abhandlung. Gustav Fischer Verlag, Jena, pp 1–152

Burkhardt JK, Echeverri CJ, Nilsson T, Vallee RB (1997) Overexpression of the dynamitin (p50) subunit of the dynactin complex disrupts dynein-dependent maintenance of membrane organelle distribution. J Cell Biol 139:469–484

Carminati JL, Stearns T (1997) Microtubules orient the mitotic spindle in yeast through dynein-dependent interactions with the cell cortex. J Cell Biol 138:629–641

Clark SW, Meyer DI (1994) ACT3: a putative centractin homologue in S. cerevisiae is required for proper spindle orientation of the mitotic spindle. J Cell Biol 127:129–138

Cole DC, Scholey JM (1995) Structural variations among kinesins. Trends Cell Biol 5:259–262

Czymmek KJ, Bourett TM, Howard RJ (1996) Immunolocalization of tubulin and actin in thick-sectioned fungal hyphae after freeze-substitution fixation and methacrylate de-embedment. J Microsc 181:153–161

de Andrade-Monteiro C, Martinez-Rossi NM (1999) The nucleation of microtubules in Aspergillus nidulans germlings. Genet Mol Biol 22:309–313

Echeverri CJ, Paschal BM, Vaughan KT, Vallee RB (1996) Molecular characterization of the 50-kD subunit of dynactin reveals function for the complex in chromosome alignment and spindle organization during mitosis. J Cell Biol 132:617–633

Eckley DM, Gill SR, Melkonian KA, Bingham JB, Goodson HV, Heuser JE, Schroer TA (1999) Analysis of dynactin subcomplexes reveals a novel actin-related protein associated with the Arp1 minifilament pointed end. J Cell Biol 147:307–320

Efimov VP, Morris NR (1998) A screen for dynein synthetic lethals in Aspergillus nidulans identifies spindle assembly checkpoint genes and other genes involved in mitosis. Genetics 149:101–116

Enos AP, Morris NR (1990) Mutation of a gene that encodes a kinesin-like protein blocks nuclear division in A. nidulans. Cell 60:1019–1027

Eshel D, Urrestarazu LA, Vissers S, Jauniaux J-C, Vliet-Reedijk JCv, Planta RJ, Gibbons IR (1993) Cytoplasmic dynein is required for normal nuclear segregation in yeast. Proc Natl Acad Sci USA 90:11172–11176

Fischer R (1999) Nuclear movement in filamentous fungi. FEMS Microbiol Rev 23:39–68

Garnjobst L, Tatum EL (1967) A survey of new morphological mutants in Neurospora crassa. Genetics 57:579–604

Gibbons IR, Rowe A (1965) Dynein: a protein with adenosine triphosphatase activity from cilia. Science 149:424

Gill SR, Schroer TA, Szilak I, Steuer ER, Sheetz MP, Cleveland DW (1991) Dynactin, a conserved, ubiquitously expressed component of an activator of vesicle motility mediated by cytoplasmic dynein. J Cell Biol 115:1639–1650

Goldman GH, Morris NR (1995) Extragenic suppressors of a dynein mutation that blocks nuclear migration in Aspergillus nidulans. Genetics 139:1223–1232

Goldstein L (1993) Functional redundancy in mitotic force generation. J Cell Biol 120:1–3

Grove SN, Bracker CE (1970) Protoplasmic organization of hyphal tips among fungi: vesicles and Spitzenkörper. J Bacteriol 104:989–1009

Grummt M, Pistor S, Lottspeich F, Schliwa M (1998) Cloning and functional expression of a "fast" fungal kinesin. FEBS Lett 427:79–84

Hackney DD (1996) The kinetic cycles of myosin, kinesin, and dynein. Annu Rev Physiol 58:731–750

Hackney DD, Levitt JD, Wagner DD (1991) Characterization of alpha 2 and beta 2 forms of kinesin. Biochem Biophys Res Commun 174:810–815

Hackney DD, Levitt JD, Suhan J (1992) Kinesin undergoes a 9S to 6S conformational transition. J Biol Chem 267:8696–8701

Hagan IM, Hyams JS (1988) The use of cell division cycle mutants to investigate the control of microtubule distribution in the fission yeast Schizosaccharomyces pombe. J Cell Sci 89:343–357

Harada A, Takei Y, Tanaka Y, Nonaka S, Hirokawa N (1998) Golgi vesiculation and lysosome dispersion in cells lacking cytoplasmic dynein. J Cell Biol 141:51–59

Heath IB (1994a) The cytoskeleton in hyphal growth, organelle movements, and mitosis. In: Wessels JGH, Meinhardt F (eds) The Mycota, vol I. Growth, differentiation and sexuality. Springer, Berlin Heidelberg New York, pp 43–65

Heath IB (1994b) The cytoskeleton. In: Gow NAR, Gadd GM (eds) The growing fungus, Chapman and Hall, London, pp 99–134

Hirokawa N (1998) Kinesin and dynein superfamily proteins and the mechanism of organelle transport. Science 279:519–526

Hirokawa N, Sato-Yoshitake R, Kobayashi N, Pfister K, Bloom G, Brady S (1991) Kinesin associates with anterogradely transported membranous organelles in vivo. J Cell Biol 114:295–302

Hoch HC, Staples RC (1985) The microtubule cytoskeleton in hyphae of *Uromyces phaseoli* germlings: its relationship to the region of nucleation and to the F-actin cytoskeleton. Protoplasma 124:112–122

Holleran EA, Tokito MK, Karki S, Holzbaur ELF (1996) Centractin (ARP1) associates with spectrin revealing a potential mechanism to link dynactin to intracellular organelles. J Cell Biol 135:1815–1829

Holleran EA, Karki S, Holzbaur ELF (1998) The role of the dynactin complex in intracellular motility. Int Rev Cytol 182:69–109

Holzbaur ELF, Vallee RB (1994) Dyneins: molecular structure and cellular function. Annu Rev Cell Biol 10:339–372

Horio T, Uzawa S, Jung MK, Oakley BR, Tanaka K, Yanagida M (1991) The fission yeast γ-tubulin is essential for mitosis and is localized at microtubule organizing centers. J Cell Sci 99:693–700

Howard RJ (1981) Ultrastructural analysis of hyphal tip cell growth in fungi: Spitzenkörper, cytoskeleton and endomembranes after freeze-substitution. J Cell Sci 48:89–103

Howard RJ, Aist JR (1977) Effects of MBC on hyphal tip organization, growth and mitosis of *Fusarium acuminatum*, and their antagonism by D$_2$O. Protoplasma 92:195–210

Howard RJ, Aist JR (1980) Cytoplasmic microtubules and fungal morphogenesis: ultrastructural effects of methyl benzimidazole-2-ylcarbamate determined by freeze-substitution of hyphal tip cells. J Cell Biol 87:55–64

Hoyt MA, Geiser JR (1996) Genetic analysis of the mitotic spindle. Annu Rev Genet 30:7–33

Hoyt MA, Totis L, Roberts BT (1991) *Saccharomyces cerevisiae* genes required for cell cycle arrest in response to loss of microtubule function. Cell 66:507–517

Hoyt MA, He L, Loo KK, Saunders WS (1992) Two *Saccharomyces cerevisiae* kinesin-related gene products required for mitotic spindle assembly. J Cell Biol 118:109–120

Huffaker TC, Thomas JH, Botstein D (1988) Diverse effects of β-tubulin mutations on microtubule formation and function. J Cell Biol 106:1997–2010

Inoue S, Yoder OC, Turgeon BG, Aist JR (1998a) A cytoplasmic dynein required for mitotic aster formation in vivo. J Cell Sci 111:2607–2614

Inoue S, Yoder OC, Turgeon BG, Aist JR (1998b) Role of fungal dynein in hyphal growth, microtubule organization, spindle pole body motility and nuclear migration. J Cell Sci 111:1555–1566

Jacobs CW, Adams AEM, Szaniszlo PJ, Pringle JR (1988) Functions of microtubules in the *Saccharomyces cerevisiae* cell cycle. J Cell Biol 107:1409–1426

Karki S, Holzbaur ELF (1995) Affinity chromatography demonstrates a direct binding between cytoplasmic dynein and the dynactin complex. J Biol Chem 131:385–397

Karki S, Holzbaur ELF (1999) Cytoplasmic dynein and dynactin in cell division and intracellular transport. Curr Opin Cell Biol 11:45–53

Kirchner J, Woehke G, Schliwa M (1999) Universal and unique features of kinesin motors: insights from a comparison of fungal and animal conventional kinesins. Biol Chem 380:915–921

Kuznetsov S, Vaisberg E, Shanina N, Magretova N, Chernyak V, Gelfand VI (1988) The quaternary structure of bovine brain kinesin. EMBO J 7:353–356

Lehmler C, Steinberg G, Snetselaar K, Schliwa M, Kahmann R, Bölker M (1997) Identification of a motor protein required for filamentous growth in *Ustilage maydis*. EMBO J 16:3464–3473

Li Y-Y, Yeh E, Hays T, Bloom K (1993) Disruption of mitotic spindle orientation in a yeast dynein mutant. Proc Natl Acad Sci USA 90:10096–10100

Lilli SH, Brown SS (1992) Suppression of a myosin defect by a kinesin-related gene. Nature 356:358–361

Little M, Seehaus T (1988) Comparative analysis of tubulin sequences. Comp Biochem Physiol B Biochem Mol Biol 90:655–670

Lye RJ, Porter ME, Scholey JM, McIntosh JR (1987) Identification of a microtubule-based cytoplasmic motor in the nematode *C. elegans*. Cell 51:309–318

McDonald HB, Stewart RJ, Goldstein LS (1990) The kinesin-like ncd protein of *Drosophila* is a minus end-directed microtubule motor. Cell 63:1159–1165

McGoldrick CA, Gruver C, May GS (1995) *myoA* of *Aspergillus nidulans* encodes an essential myosin I required for secretion and polarized growth. J Cell Biol 128:577–587

McKerracher LJ, Heath IB (1985a) Microtubules around migrating nuclei in conventionally-fixed and freeze substituted cells. Protoplasma 125:162–172

McKerracher LJ, Heath IB (1985b) The structure and cycle of the nucleus-associated organelle in two species of *Basidiobolus*. Mycologia 77:412–417

McKerracher LJ, Heath IB (1987) Cytoplasmic migration and intracellular organelle movements during tip growth of fungal hyphae. Exp Mycol 11:79–100

Meluh PB, Rose MD (1990) *KAR3*, a kinesin-related gene required for yeast nuclear fusion. Cell 60:1029–1041

Merdes A, Cleveland DW (1997) Pathways of spindle pole formation: different mechanisms; conserved components. J Cell Biol 138:953–956

Milisav I (1998) Dynein and dynein-related genes. Cell Motil Cytoskeleton 39:261–272

Minke PF, Lee IH, Plamann M (1999a) Microscopic analysis of *Neurospora* ropy mutants defective in nuclear distribution. Fungal Genet Biol 28:55–67

Minke PF, Lee IH, Tinsley JH, Bruno KS, Plamann M (1999b) *Neurospora crassa ro-10* and *ro-11* genes encode novel proteins required for nuclear distribution. Mol Microbiol 32:1065–1076

Moore JD, Endow SA (1996) Kinesin proteins: a phylum of motors for microtubule-based motility. Bioessays 18:207–219

Mooseker MS, Cheney RE (1995) Unconventional myosins. Annu Rev Cell Dev Biol 11:633–675

Morris NR (1976) Mitotic mutants of *Aspergillus nidulans*. Genet Res 26:237–254

Muhua L, Karpova TS, Cooper JA (1994) A yeast actin-related protein homologous to that in vertebrate dynactin complex is important for spindle orientation and nuclear migration. Cell 78:669–679

O'Connell MJ, Meluh PB, Rose M, Morris NR (1993) Suppression of the *bimC* mitotic spindle defect by deletion of *kplA*, a gene encoding a KAR3-related kinesin-like protein in *Aspergillus nidulans*. J Cell Biol 120:153–162

Oakley BR, Oakley CE, Yoon Y, Jung MK (1990) γ-Tubulin is a component of the spindle pole body that is essential for microtubule function in *Aspergillus nidulans.* Cell 61:1289–1301

Oakley CE, Oakley BR (1989) Identification of gamma-tubulin, a new member of the tubulin superfamily encoded by the *mipA* gene of *Aspergillus nidulans.* Nature 338:662–664

Osmani AH, Osmani SA, Morris NR (1990) The molecular cloning and identification of a gene product specifically required for nuclear movement in *Aspergillus nidulans.* J Cell Biol 111:543–551

Paschal BM, Vallee RB (1987) Retrograde transport by the microtubule-associated protein MAP1 C. Nature 330:181–183

Paschal BM, Shpetner HS, Vallee RB (1987) MAP1 C is a microtubule-activated ATPase which translocates microtubules in vitro and has dynein-like properties. J Cell Biol 105:1273–1282

Perkins DD, Radford A, Newmyer D, Björkman M (1982) Chromosomal loci of *Neurospora crassa.* Microbiol Rev 46:426–570

Pfarr CM, Coue M, Grissom PM, Hays TS, Porter ME, McIntosh JR (1990) Cytoplasmic dynein is localized to kinetochores during mitosis. Nature 345:263–265

Pfister KK, Wagner MC, Stenoien DL, Brady ST, Bloom GS (1989) Monoclonal antibodies to kinesin heavy and light chains stain vesicle like structure, but not microtubules, in cultured cells. J Cell Biol 108:1453–1463

Picket-Heaps JD (1969) The evolution of the mitotic apparatus: an attempt at comparative ultrastructural cytology in dividing plants. Cytobios 3:257–280

Plamann M (1996) Nuclear division, nuclear distribution and cytokinesis in filamentous fungi. J Genet 75:351–360

Plamann M, Minke PF, Tinsley JH, Bruno KS (1994) Cytoplasmic dynein and actin-related protein Arp1 are required for normal nuclear distribution in filamentous fungi. J Cell Biol 127:139–149

Raudaskoski M, Rupes I, Timonen S (1991) Immunofluorescence microscopy of the cytoskeleton in filamentous fungi after quick-freezing and low-temperature fixation. Exp Mycol 15:167–173

Reinsch S, Karsenti E (1997) Movement of nuclei along microtubules in *Xenopus* egg extracts. Curr Biol 7:211–214

Riquel M, Reynaga-Peña CG, Gierz G, Bartnicki-Garcia S (1998) What determines growth direction in fungal hyphae? Fungal Genet Biol 24:101–109

Robb MJ, Wilson MA, Vierula PJ (1995) A fungal actin-related protein involved in nuclear migration. Mol Gen Genet 247:583–590

Roberts BT, Farr KA, Hoyt MA (1994) The *Saccharomyces cerevisiae* checkpoint gene *BUB1* encodes a novel protein kinase. Mol Cell Biol 14:8282–8291

Salo V, Niini SS, Virtanen I, Raudaskoski M (1989) Comparative immunocytochemistry of the cytoskeleton in filamentous fungi with dikaryotic and multinucleate hyphae. J Cell Sci 94:11–24

Schafer DA, Gill SR, Cooper JA, Heuser JE, Schroer TA (1994) Ultrastructural analysis of the dynactin complex: an actin-related protein is a component of a filament that resembles F-actin. J Cell Biol 126:403–412

Scholey JM, Porter ME, Grissom PM, McIntosh JR (1985) Identification of kinesin in sea urchin eggs, and evidence for its localization in the mitotic spindle. Nature 318:483–486

Schroer TA (1996) Structure and function of dynactin. Semin Cell Biol 7:321–328

Schroer TA, Sheetz MP (1991) Two activators of microtubule-based vesicle transport. J Cell Biol 115:1309–1318

Schroer TA, Bingham JB, Gill ST (1996) Actin-related protein 1 and cytoplasmic dynein-based motility – what's the connection? Trends Cell Biol 6:212–215

Seiler S, Nargang FE, Steinberg G, Schilwa M (1997) Kinesin is essential for cell morphogenesis and polarized secretion in *Neurospora crassa.* EMBO J 16:3025–3035

Seiler S, Plamann M, Schliwa M (1999) Kinesin and dynein mutants provide novel insights into the roles of vesicle traffic during cell morphogenesis in *Neurospora.* Curr Biol 9:779–785

Steinberg G (1997) A kinesin-like mechanoenzyme from the zygomycete *Syncephalastrum racemosum* shares biochemical similarities with conventional kinesin from *Neurospora crassa.* Eur J Cell Biol 73:124–131

Steinberg G (1998) Organelle transport and molecular motors in fungi. Fungal Genet Biol 24:161–177

Steinberg G, Schliwa M (1993) Organelle movements in the wild type and wall-less *fz;sg;os-1* mutants of *Neurospora crassa* are mediated by cytoplasmic microtubules. J Cell Sci 106:555–564

Steinberg G, Schliwa M (1995) The *Neurospora* organellar motor: a distant relative of conventional kinesin with unconventional properties. Mol Biol Cell 6:1605–1618

Steinberg G, Schliwa M (1996) Characterization of the biophysical and motility properties of kinesin from the fungus *Neurospora crassa.* J Biol Chem 271:7516–7521

Steinberg G, Schliwa M, Lehmler C, Bolker M, Kahmann R, McIntosh J (1998) Kinesin from the plant pathogenic fungus *Ustilago maydis* is involved in vacuole formation and cytoplasmic migration. J Cell Sci 111:2235–2246

Steuer ER, Wordeman L, Schroer TA, Sheetz MP (1990) Localization of cytoplasmic dynein to mitotic spindles and kinetochores. Nature 345:266–268

Thompson-Coffe C, Zickler D (1992) Three microtubule-organizing centers are required for ascus growth and sporulation in the fungus *Sordaria macrospora.* Cell Motil Cytoskeleton 22:257–273

Tinsley JH, Minke PF, Bruno KS, Plamann M (1996) p150[Glued], the largest subunit of the dynactin complex, is nonessential in *Neurospora* but required for nuclear distribution. Mol Biol Cell 7:731–742

Vaisberg EA, Koonce MP, McIntosh JR (1993) Cytoplasmic dynein plays a role in mammalian spindle formation. J Cell Biol 123:849–858

Vaisberg EA, Grissom PM, McIntosh JR (1996) Mammalian cells express three distinct dynein heavy chains that are localized to different cytoplasmic organelles. J Cell Biol 133:831–842

Vale RD, Fletterick R (1997) The design plan of kinesin motors. Annu Rev Cell Dev Biol 13:745–777

Vale RD, Reese TS, Sheetz MP (1985) Identification of a novel force-generating protein, kinesin, involved in microtubule-based motility. Cell 42:39–50

Vallee RB, Sheetz MP (1996) Targeting of motor proteins. Science 271:1539–1544

Vaughan KT, Valee RB (1995) Cytoplasmic dynein binds dynactin through a direct interaction between the intermediate chains and p150[Glued]. J Cell Biol 131:1507–1516

Vaughan KT, Holzbaur ELF, Vallee RB (1995) Subcellular targeting of the retrograde motor cytoplasmic dynein. Biochem Soc Trans 23:50–54

Walker RA, Salmon ED, Endow SA (1990) The *Drosophila* claret segregation protein is a minus-end directed motor molecule. Nature 347:780–782

Waterman-Storer CM, Karki S, Holzbaur ELF (1995) The p150^{Glued} component of the dynactin complex binds to both microtubules and the actin-related protein centractin (Arp-1). Proc Natl Acad Sci USA 92:1634–1638

Willins DA, Liu B, Xiang X, Morris NR (1997) Mutations in the heavy chain of cytoplasmic dynein suppress the *nudF* nuclear migration mutation of *Aspergillus nidulans*. Mol Gen Genet 255:194–200

Winsor B, Schiebel E (1997) Review: an overview of the *Saccharomyces cerevisiae* microtubule and microfilament cytoskeleton. Yeast 13:399–434

Wu Q, Sandrock TM, Turgeon BG, Yoder OC, Wirsel SG, Aist JR (1998) A fungal kinesin required for organelle motility, hyphal growth, and morphogenesis. Mol Biol Cell 9:89–101

Xiang X, Beckwith SM, Morris NR (1994) Cytoplasmic dynein is involved in nuclear migration in *Aspergillus nidulans*. Proc Natl Acad Sci USA 91:2100–2104

Xiang X, Osmani AH, Osmani SA, Roghi CH, Willins DA, Beckwith S, Goldman G, Chiu Y, Xin M, Liu B, Morris NR (1995a) Analysis of nuclear migration in *Aspergillus nidulans*. Cold Spring Harb Symp Quant Biol 60:813–819

Xiang X, Osmani AH, Osmani SA, Xin M, Morris NR (1995b) *nudF*, a nuclear migration gene in *Aspergillus nidulans*, is similar to the human *LIS-1* gene required for neuronal migration. Mol Biol Cell 6:297–310

Xiang X, Roghi C, Morris NR (1995c) Characterization and localization of the cytoplasmic dynein heavy chain in *Aspergillus nidulans*. Proc Natl Acad Sci USA 92:9890–9894

Xiang X, Zuo WQ, Efimov VP, Morris NR (1999) Isolation of a new set of *Aspergiluus nidulans* mutants defective in nuclear migration. Curr Genet 35:626–630

Yarden O, Plamann M, Ebbole DJ, Yanofsky C (1992) *cot-1*, a gene required for hyphal elongation in *Neurospora crassa*, encodes a protein kinase. EMBO J 11:2159–2166

Yeh E, Skibbens RV, Cheng JW, Salmon ED, Bloom K (1995) Spindle dynamics and cell cycle regulation of dynein in the budding yeast, *Saccharomyces cerevisiae*. J Cell Biol 130:687–700

Yen TJ, Compton DA, Wise D, Zinkowski RP, Brinkley BR, Earnshaw WC, Cleveland DW (1991) CEN-P, a novel human centromere-associated protein required for progression from metaphase to anaphase. EMBO J 10:1245–1254

12 Motile Tubular Vacuole Systems

ANNE E. ASHFORD, LOUISE COLE, and GEOFFREY J. HYDE

CONTENTS

I. Introduction

The tips of actively growing fungal hyphae contain an extensive reticulum of motile and interconnected tubules and spherical vacuoles (Shepherd et al. 1993a,b; Ashford 1998; Cole et al. 1998). The system as a whole shows a range of morphologies, and the tubules exhibit several types of motility including tip extension and retraction, peristalsis-like motion and movement of varicosities. Tubular vacuoles are found in representatives of all major fungal groups (Rees et al. 1994). They were previously unrecognised because the tubules do not survive chemical fixation and round up irreversibly into strings of vesicles (Orlovich and Ashford 1993). Tubular vacuoles are preserved by freeze-substitution and elongate vacuolar profiles are described in virtually all fungal species that have been freeze-substituted (see Ashford 1998).

Motile tubular vacuole systems are best visualised in living cells after loading with fluorescent probes (Fig. 1a). Only then does their true nature become apparent. However, they are not an artefact of fluorochrome loading and can also be demonstrated in unlabelled cells by differential interference contrast microscopy (Hyde and Ashford 1997; Hyde et al. 1997). An approach combining observations in living cells with data from freeze-substituted sections has been very powerful in obtaining information about motile vacuole systems in the basidiomycete *Pisolithus tinctorius* (cf. Fig. 1a,b) and in determining what they contain. This approach has shown that the vacuole system in tip cells of this species comprises a series of relatively immobile rounded vacuoles attached to the cell membrane, from which vacuolar tubules extend along microtubule tracks to interact and fuse with other tubules and immobile vacuoles (Cole et al. 1998). The relative abundance of the tubular and rounded components varies according to cell age, position and external conditions, and in the tip cells a tubular reticulum may predominate. Although they appear as many distinct vesicles, vacuoles in *Saccharomyces cerevisiae* may also comprise a continuous reticulum, where vesicular components are interconnected by fine tubules (Jones et al. 1997).

Vacuoles have long been known as multifunctional organelles involved in storage, lysis and homeostasis (Boller and Wiemken 1986; Klionsky et al. 1990). They are now shown to be dynamic organelles that are involved in molecular exchanges and transport and their attributes are indicative of functional differentiation. The evidence upon which this is based is evaluated below.

School of Biological Science, The University of New South Wales, Sydney 2052, Australia

The Mycota VIII
Biology of the Fungal Cell
Howard/Gow (Eds.)
© Springer-Verlag Berlin Heidelberg 2001

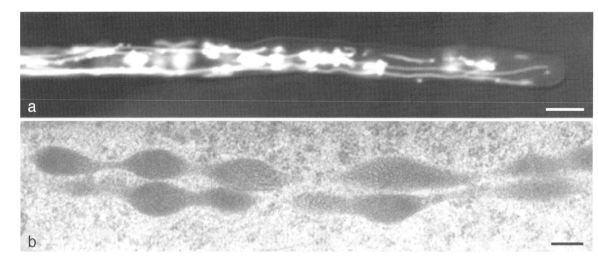

Fig. 1. Pleiomorphic tubular vacuole system in tip cells of *Pisolithus tinctorius.* **a** Labelled with carboxy-DFFDA; *bar* 5 µm; **b** in a freeze-substituted hypha; *bar* 200 nm

II. Vacuole Motility

A. Visualisation of Vacuoles in Living Fungal Cells

Vacuoles are usually visualised after staining with fluorescent probes by fluorescence or confocal laser scanning microscopy (Weisman et al. 1987; Butt et al. 1989; Shepherd et al. 1993a,b; Cole et al. 1997; Hyde and Ashford 1997). In some cells, high levels of autofluorescence in vacuoles permit their visualisation without the need for fluorescent stains; for example, *ade2* vacuoles in *S. cerevisiae* accumulate an endogenous fluorophore (Weisman et al. 1987). In most cases staining is required, and examination of unstained cells by phase contrast or differential interference contrast microscopy is also necessary to confirm that the dye has not affected vacuole integrity.

Staining of vacuoles with fluorescent probes is a useful technique for identifying specific vacuole compartments, differentiating them from other organelles and for studying vacuole biogenesis and motility (Weisman et al. 1987; Weisman and Wickner 1988; Butt et al. 1989; Raymond et al. 1990; Cole et al. 1997, 1998, 2000a,b). Many stains, such as neutral red, quinacrine and acridine orange, accumulate in acidic compartments. Like animal lysosomes, fungal vacuoles are acidic and are readily labelled by these dyes. Fluorescent probes offer several advantages over histochemical stains. They have high quantum efficiencies and can be detected at very low concentrations, min-

imising cell damage. In addition, they exhibit a range of excitation and emission spectra enabling dual- and even triple-labelling of different organelles, and their fluorescence may reflect membrane potential, intracellular pH and ions such as Ca^{2+} (Rost et al. 1995).

Many fluorescent probes that label fungal vacuoles are available commercially. Those we have used successfully include 6-carboxyfluorescein diacetate (6-CFDA), Oregon Green 488 carboxylic acid diacetate (carboxy-DFFDA), the thiol-reactive Cell Tracker reagents 7-amino-4-chloromethylcoumarin (CMAC) and 5-chloromethylfluorescein diacetate (CMFDA), and the styryl dyes N-(3-triethylammoniumpropyl)-4-(p-diethylaminophenylheximenyl)pyridinium dibromide (FM4-64) and MDY-64 (Shepherd et al. 1993a; Cole et al. 1997, 1998). The styryl dyes label membranes and the others accumulate in the vacuole lumen (Vida and Emr 1995; Cole et al. 1997, 1998; Zheng et al. 1998). The mechanism of vacuolar accumulation of the diacetate and chloromethyl probes is as follows. Both 6-CFDA and carboxy-DFFDA are non-fluorescent and readily transported across the plasma membrane. Once in the cytoplasm, the acetate is removed by cytoplasmic esterases, converting each into the anionic form which is strongly fluorescent. The resulting anions (6-CF$^-$ and carboxy-DFF$^-$) are membrane-impermeant but are sequestered rapidly into the vacuoles by organic anion transporters. This is indicated by experiments with probenecid, a drug demonstrated to be an organic anion transport inhibitor in animal cells (see Cole et al. 1997). Probenecid prevents accumulation of 6-CF and carboxy-DFF in the vacuole and they accumulate in the cytoplasm instead; however, as soon as the drug is removed, fluorescence rapidly disappears from the cytoplasm into the vacuole. The chloromethyl dyes CMAC and CMFDA are thought to react with the endogenous tripeptide glutathione via a glutathione-*S*-transferase mediated reaction to form a thiol-ether adduct. Glutathione transporters may facilitate transport of glutathione conjugates into the vacuole lumen. Reactions are discussed further in Sect. V.C.

A number of membrane-impermeant fluorescent probes, such as Lucifer Yellow CH (LY-CH) and high molecular weight fluorescent dextrans, initially developed to trace the endocytic pathway in animal cells (Swanson 1989), are internalised into vacuoles of yeasts (see Basrai et al. 1990), but not necessarily by endocytosis. *S. cerevisiae*, and both the yeast and "mould" phase of *Candida albicans*, take up LY-CH. In contrast, *P. tinctorius* does not internalise detectable amounts of either LY-CH or fluorescent dextrans, or accumulate these probes in vacuoles (Cole et al. 1997). In any case, LY-CH is not a reliable indicator of endocytosis, since it is accumulated into higher plant vacuoles by a tonoplast transport mechanism (Cole 1989; Cole et al. 1991; Oparka et al. 1991; Oparka and Hawes 1992).

Yeasts and mycelial fungi are reported also to internalise FM4-64, a vital dye for endocytosis, and the tonoplast ultimately becomes labelled (Vida and Emr 1995; Rieder et al. 1996; Hoffmann and Mendgen 1998; Steinberg et al. 1998; Zheng et al. 1998). In *S. cerevisiae*, small punctate structures that become labelled before the vacuole have been interpreted as endosomes (Vida and Emr 1995). A major fraction of internalised FM4-64 is rapidly secreted back to the medium as membrane is recycled to the cell surface, and secretion of previously internalised FM4-64 can be used as a recycling assay (Wiederkehr et al. 2000). Zheng et al. (1998) have used dual localisation of FM4-64 and carboxydichlorofluorescein diacetate (CDCFDA) with flow cytometry to isolate *S. cerevisiae* mutants that mislocalise vacuolar dyes. FM4-64 has also been very effective for investigating the effects of specific mutations on the morphology of endosomes and vacuoles in this yeast (Gaynor et al. 1998). In *Neurospora* sp., FM4-64 labelled the plasma membrane and the Spitzenkörper followed by putative early endosomes after only a few minutes (Fricker and Oparka 1999). After

20 min vacuoles were also labelled, as were all internal membranes following prolonged exposure (over 1 h). In *P. tinctorius*, the plasma membrane, tonoplast and punctate structures also became labelled with FM4-64 after prolonged treatment, but not apparently by endocytosis, since labelling was found mainly in damaged cells (Cole et al. 1998).

Fluorescent probes designed to label specific organelles are appearing at a rapid rate (Haugland 1999) and it is now possible to label several organelle types, including vacuoles, simultaneously in the same cell. For example, using carboxy-DFFDA and an ER-Tracker dye the tubular-vacuole system of *P. tinctorius* can be clearly distinguished from the endoplasmic reticulum (ER) in living cells (Cole et al. 2000a,b). Results with these probes indicate lack of any direct structural continuity between the vacuole system and the ER, and vacuoles are not formed directly from the ER, at least in this fungus (see references in Rogers 1998). The latest *Molecular Probes Handbook* is an invaluable resource for the latest information on a very wide range of probes (Haugland 1999).

Several precautions must be taken when using fluorescent probes as vacuole tracers. To avoid damage, high concentrations of the probe and prolonged exposure of the cells should be avoided (Cole et al. 1997, 1998). For similar reasons irradiation should be minimised. Although fluorochromes more stable to excitation are becoming available, such as the Alexa dyes, fading still poses a problem during image capture and z-sectioning by confocal laser scanning microscopy, and limits movie-making of vacuole motility. Success in labelling membranes and small organelle compartments (e.g. vacuoles and vesicles) depends on the continued development of very sensitive detection and imaging techniques. New developments using two-photon-optic confocal laser scanning microscopy offer high sensitivity with less photobleaching.

There is increasing use of green fluorescent protein (GFP) from the jellyfish *Aequoria victoria* (Cubitt et al. 1995) as a probe for a variety of cell functions. GFP is used as a marker for gene expression and therefore has the potential for labelling any protein for which a gene construct can be made. It has also provided a means for visualising organelles in living cells (Boevink et al. 1996, 1998; Chalfie et al. 1994). The GFP chromophore does not require any exogenous factors in order to fluoresce, merely appropriate irradiation. Thus, it is an ideal probe for monitoring intracellular processes with minimal disturbance to the living cell. Unfortunately, the wild-type GFP gene does not function in many organisms, including mycelial fungi (Fernández-Ábalos et al. 1998). Despite this, GFP has been successfully expressed in some fungi using modified GFP genes (Spellig et al. 1996; Cormack et al. 1997). In *Aspergillus nidulans*, GFP has been used to localise proteins in the ER and nucleus and has allowed the stages of cell division to be closely monitored for the first time in a living mycelial fungus (Fernández-Ábalos et al. 1998). In *S. cerevisiae*, a vacuole-targeted GFP-alkaline phosphatase fusion protein is exclusively localised to the vacuolar membrane in wild-type cells (Cowles et al. 1997). Various constructs have been made of fusion proteins between GFP and an *S. cerevisiae* protein (Btn1p) that shows a high degree of homology with Cln3p, the defective protein found in the juvenile form of the human neurodegenerative Batten disease. One *S. cerevisiae* construct is localised to both the membrane and lumen of the vacuole, while the other localises to the lumen only. Accumulation of the constructs in the vacuoles parallels the accumulation of Cln3p in the human cell lysosomes (Croopnick et al. 1998). Visualisation of a chimera of GFP and an *S. cerevisiae* SNARE protein (Snc1p) that normally mediates exocytosis has enabled demonstration of a distinct recycling pathway between the plasma membrane, Golgi and a compartment identified as an early endosome (Lewis et al. 2000). This discovery is based on redistribution of the fluorescent chimera and its co-localisation with other compartment-specific proteins in mutants where Golgi function or exocytosis is blocked. Similarly, another fusion protein between GFP and an *S. cerevisiae* protein (Yeb3p) is localised to vacuolar membranes, but it is concentrated in bands located between clustered vacuoles. The protein Yeb3p contains an "Armadillo" domain, with tandem arrays of 42 amino acid repeats, proposed to act as a scaffold for assembly of multiprotein complexes. This domain is also found on proteins that mediate the docking of cargo to membrane-localised cytoskeletal elements and it is proposed that Yeb3P links vacuoles with the actin cytoskeleton (Pan and Goldfarb 1998). Defects in the gene interfere with yeast vacuole inheritance and may also disrupt normal reticular vacuole morphology, suggesting an involvement of actin in vacuole interactions.

B. Probes and Vacuole pH

Many fluorescent probes can be used to indicate the pH of vacuoles. These include 5- or 6-CF, CDCF, 2′,7′-bis-(2-carboxymethyl)-5(and-6) carboxyfluorescein (BCECF) and the new LysoSensor probes (Preston et al. 1989; Davies et al. 1990; Haugland 1999). The emission spectrum of these probes changes with small pH changes within the physiological range, and the pH of the compartment containing the probe can be determined by the ratio of emission at two wavelengths (Tsien 1989). For example, using ratio imaging and flow cytometry, wild-type *S. cerevisiae* cells loaded with 6-CF were shown to have a vacuolar pH of 6.1, while a mutant deficient in vacuolar H^+-ATPase had an average vacuolar pH of 7.1, i.e. close to that estimated for the cytoplasm by ^{31}P nuclear magnetic resonance spectroscopy (^{31}P NMR) (Yamashiro et al. 1990). Dye ratio imaging can also be used to analyse the pH of individual vacuoles (Rost et al. 1995). This approach has shown that the pH of vacuoles is not uniform, a question that cannot be resolved by flow cytometry and NMR as these techniques only provide average values for vacuole populations. For example, the average pH for phosphate-storing vacuoles in ectomycorrhizal fungi as determined by ^{31}P NMR was pH 5.0 with a cytoplasm at pH 6.0–6.5 (Martin 1991). However, ratio imaging of 6-CF-loaded *P. tinctorius* vacuoles showed that the pH of the large vacuoles in tip cells ranged from 4.3 to 7.5, and in the penultimate cells from 4.8 to 7.2. The modal pH values suggested a tendency for the vacuoles to become more acid as they aged (Rost et al. 1995).

Uptake of probes into compartments is also pH-sensitive since it depends on whether the probe is ionised or not and whether there is an ion transporter. Changes in uptake at the plasma membrane or redistribution between internal compartments can indicate the pH at the membrane uptake site (Calahorra et al. 1998). Probes with different pK_a values have been developed for use at near neutral (e.g. fluorescein derivatives, SNARF and SNAFL pH indicators) and acidic pH (e.g. CDCF, LysoSensor probes). Fluorescent pH indicators have proved to be successful in determining the kinetics of change in pH of vacuole compartments in plant cells during development and in response to environmental conditions, e.g. light and cold treatment (Yin et al. 1990; Yoshida 1995; Davies et al. 1996). Similarly, they have been successfully applied to studies of vacuolar pH in some fungal cells. For example, using 6-CF, Abe and Horikoshi (1998) showed that elevation of hydrostatic pressure by 40–60 MPa transiently reduced the vacuolar pH by about 0.33. This effect is believed to be brought about by vacuolar H^+-ATPase.

All techniques indicate that vacuoles are more acidic than the cytoplasm. Vacuole acidification is known to play a key role in protein sorting, proteolytic processing and regulation of cytoplasmic pH (Yamashiro et al. 1990). The vacuolar H^+-ATPase (V-ATPase) is responsible for acidification of the vacuole interior and for releasing energy for transport across the vacuole membrane in fungal cells (Klionsky et al. 1990; Nelson and Nelson 1990; Yamashiro et al. 1990; Raymond et al. 1992a; Nelson and Klionsky 1996). The V-ATPase of *S. cerevisiae* is the best-characterised enzyme and the *N. crassa* enzyme appears to have a similar subunit composition (Kane et al. 1989; Bowman et al. 1997). The enzyme is a multi-subunit complex, with a ball-and-stalk structure similar to mitochondrial ATPase, except that it is larger. Hydrolysis of ATP by the V_1 domain, a cluster of eight subunits arranged into a peripheral, hydrophilic, dissociable group, drives protons through the hydrophobic region – an integral membrane domain, V_0. Reversible dissociation of the V_1 and V_0 domains is a possible mechanism for regulating V-ATPase activity and hence regulating vacuolar acidification (for review see Forgac 1998). Antibodies to various *S. cerevisiae* V-ATPase subunits are commercially available (Haugland 1999). They have the potential to label vacuolar membranes in mycelial fungi. Specific, potent V-ATPase inhibitors such as bafilomycin A and concanamycin A, both macrolide antibiotics, have been instrumental in defining the relative roles of vacuolar and plasma membrane H^+-ATPases in cytoplasmic regulation in wild-type and various mutants of *N. crassa* (Bowman et al. 1997).

C. Vacuole Probes and Electron Microscopy

One of the major challenges in vacuole labelling is the correlation of fluorescent images in living cells with ultrastructure. This is required to confirm the specificity of organelle labelling. Many probes are not specific to a particular compartment; for example, $DiOC_6(3)$ may label the ER or mitochondria (Terasaki and Reese 1992; Haugland 1999). Markers are needed to identify tubular structures at the electron microscopy (EM) level in order to differentiate tubular vacuoles from other tubules such as smooth ER. They are also needed to explore biochemical partitioning and differentiation within vacuoles. This is not so much of a problem if specific immunocytochemical markers are used, since antibodies can be conjugated with fluorophores for light microscopy and electron-opaque markers for EM. It is, however, a problem if water-soluble probes are to be retained in vacuoles. Several fluorochromes such as LY-CH are reported to be aldehyde-fixable and have been used to some advantage (see Oparka 1991). However, care must be taken, since interactions between aldehydes and the components of whole cells are complex and it is not always clear that the fluorochrome has been adequately captured in situ. Primary aldehyde fixation is not useful for vacuoles since chemical fixation destroys the integrity of the vacuole system, causing vesiculation of tubular components (Wilson et al. 1990; Orlovich and Ashford 1993).

Observations of continued membrane flow and other cellular changes during chemical fixation also cast doubt on the suitability of this approach for investigation of interactions between organelle compartments (see Mersey and McCully 1978; Hoch 1990). Methods such as low-temperature freeze-drying or anhydrous freeze-substitution are more effective for retaining water-soluble fluorescent probes in situ (Canny and McCully 1986; Linner et al. 1986a,b). It may be possible to immobilise "fixable" probes and/or better stabilise antigens as well as achieve better cell preservation by reaction with aldehydes during the substitution phase of freeze-substitution (Czymmek et al. 1996; see also Skepper 2000). Low-temperature methods are also very useful in optimising antigen survival for immunocytochemistry (Monaghan et al. 1998; Skepper 2000). For compartment labelling, the future lies with immunocytochemistry of proteins (and other substances with reactive epitopes) specific to particular membrane domains or organelles, using freeze-substituted material. Some excellent results have been obtained: for example, the localisation of clathrin in *S. cerevisiae* (Mendgen et al. 1995). It is difficult to find appropriate probes to label vacuoles in mycelial fungi, but staining sections of freeze-substituted hyphae with the lectin concanavalin A, complexed with gold, has been very successful (Bourett et al. 1993; Bourett and Howard 1994). Large vacuoles and two different types of smooth cisternae are well labelled. The content of some vacuoles including tubular vacuoles is also labelled, as well as the membranes and content of multivesicular bodies. A similar approach with more specific probes, such as those now available against *S. cerevisiae* vacuole and endosome proteins (Haugland 1999), will allow heterogeneity in function and content of the vacuole system to be systematically explored.

Various techniques involving freeze-substitution, followed by rehydration and treatment with wall-lysing enzymes, or butyl methyl methacrylate embedment and de-embedment, have been used for localisation of affinity probes in whole fungal cells by immunofluorescence (Czymmek et al. 1996; Hyde et al. 1999). The latter has been especially successful in labelling components of the cytoskeleton, and also of the endomembrane system (Czymmek et al. 1996; Bourett et al. 1998). This approach will be very valuable for visualisation of organelle systems in three dimensions by confocal microscopy, but it must be realised that there is a compromise between structural preservation and accessibility of probes to their reactive sites. This must be carefully considered in any interpretation (Bourett et al. 1998; Skepper 2000).

D. Changes in the Vacuole System with Development

Extensively developed tubular reticulate vacuole systems are commonly found in the apical and penultimate cells of actively growing hyphae. Tip cells of mycelial fungi are highly polarised, exhibiting polarised growth and a polarised organelle distribution. Polarisation of the vacuole system is especially obvious among the Oomycota where an apical filamentous vacuole network is contiguous with a large subapical vacuole that fills most of the cell volume (Allaway et al. 1997). In *P. tinctorius*, several zones can be identified in the tip cell according to vacuole morphology, motility and distribution. Tubular vacuoles extend into the extreme hyphal tip as it grows, so that the zonation remains constant and the position of the vacuole system in relation to the extending apex is maintained.

Vacuoles become larger and more stationary with increasing distance from the hyphal tip both within the tip cell and, on a larger scale, within the mycelium as a whole. Superimposed on this developmental pattern is a plasticity modulated by the environment. Non-motile vacuoles can become more motile, more tubular and more interconnected with changes in both external and internal conditions. Control of these processes is not well understood. We do know for *P. tinctorius* that fresh medium promotes tubulation and motility (Hyde and Ashford 1997) and that cultures grown between cellophane have a more extensive reticulum than those grown on a surface. Cellular damage readily converts a vacuolar reticulum into a series of larger non-motile vacuoles in both mycelial fungi and the Oomycota (Allaway et al. 1997). Treatment with anti-microtubule drugs also suppresses vacuole tubulation and motility (Hyde et al. 1999). In the Oomycota, recovery from a range of growth-inhibiting conditions is accompanied by a transient migration of vacuoles into hyphal tips, displacing tip cytoplasm. Only when cytoplasm has subsequently migrated back into the tip and displaced these vacuoles does the hypha resume growth (Bachewich and Heath 1999). It is clear that vacuole movements are independent of cytoplasmic streaming, suggesting that a different regulatory mechanism must be in operation.

In *S. cerevisiae*, vacuole tubulation and motility mediate extension of the vacuole system into developing buds or zygotes, thus ensuring that new buds receive their vacuole complement from the parent cell. Tubules, or a stream of aligned vesicles termed **segregation structures**, extend into young buds at a precise stage in their development, transferring both membrane and content. Subsequently, these new bud vacuoles rapidly separate from those of the parent cell (Weisman and Wickner 1988; Gomes de Mesquita et al. 1991; Jones et al. 1993). It is likely that an equivalent process occurs in mycelial fungi ensuring that the extending tip is constantly supplied with vacuole, and that the vacuole system as a whole moves forward to keep pace with growth. It perhaps also explains why the *kin2*-null mutants of *Ustilago maydis*, deficient in the microtubule motor kinesin, do not form large basal vacuoles and fail to exhibit the highly vacuolated cells of empty appearance, normally found distal to the growing zone in wild-type cells (Steinberg et al. 1998). Microtubules, and probably also their associated motor proteins, are therefore implicated in vacuolar tubulation and motility (see Sect. III; also see Chap. 11, this Vol.).

E. Motile Vacuoles are a General Phenomenon in Eukaryotic Cells

Vacuole motility and tubulation have been widely reported in plant and animal cells. Tubular vacuoles are found, for example, in developing guard cells (Palevitz and O'Kane 1981; Palevitz et al. 1981), root meristem cells (Davies et al. 1992; Rogers 1998), tomato petiolar hairs and other plant tissues (McCully and Canny 1985 and references therein), cultured cells and protoplasts (Hillmer et al. 1989; Cole et al. 1990; Oparka and Hawes 1992), cells involved in transport and secretion (Lazzaro and Thomson 1996), pollen tubes (Tiwari and Polito 1988) and embryos (Dr. Candida Briggs, School of Biological Science, UNSW, Sydney 2052, pers. comm.). Tubular vacuoles are also found in conducting cells of bryoid mosses (Ligrone and Duckett 1994, 1998). In higher plants, as in fungi, vacuole tubulation can be induced by changes in the external environment. Tubules in cultured tobacco cells are shown to be ripples of the large central vacuole (Verbelen and Tao 1998). As in fungi, plant tubular vacuoles are very susceptible to chemical fixation. This causes them to round up to form clusters of vesicles (Mersey and McCully 1978; Wilson et al. 1990), and so they are not likely to be detected from most electron micrographs. Animal endosomes and lysosomes also form tubules, which vary in their stability during chemical fixation (Swanson et al. 1987a,b; Hopkins et al. 1990; Robinson and Karnovsky 1991; Tooze and Hollinshead 1991). Tubulation of vacuoles is so widespread that it must surely be a fundamental property of eukaryotic cells.

III. Vacuole Motility and the Cytoskeleton

The rapid extension of vacuolar tubules, their form and the precision of their movements all point to interactions with a cytoskeletal framework. This idea is supported by experimental evidence for a functional dependence of vacuolar motility on the integrity of microtubules in mycelial fungi. Hyphae of *P. tinctorius*, like all other fungal hyphae, have longitudinally oriented networks of microtubules that are configured ideally to guide tubular extension (Allaway 1994; Czymmek et al. 1996; Hyde et al. 1999). While individual microtubules may be much shorter (a maximum of 4 μm in *P. tinctorius*) than extended vacuolar tubules (60 μm or more), adjacent microtubules overlap and may function together as much longer tracks. The type of guidance provided by microtubules in *P. tinctorius* appears to be a kinetic interaction involving motor proteins, since if microtubules are depolymerised, vacuolar tubules disappear and only spherical vacuoles are evident (Hyde et al. 1999). This precludes the alternative model whereby microtubules serve only as physical barriers that restrict tubule extension to the longitudinal axis, and are not directly involved in providing the motive force for extension. Other evidence is also consistent with a microtubule-motor protein-based mechanism. Electron microscopy of freeze-substituted *P. tinctorius* hyphae shows that vacuoles, both spherical and tubular, are always found adjacent to microtubules (Fig. 2). Fine vacuolar tubules are typically about one microtubule diameter away from neighbouring microtubules, a distance short enough to be bridged by motor proteins (Hyde et al. 1999). In vivo, a short tubule fragment can sometimes be seen to break off from the end of an extending tubular vacuole. When this occurs, the maverick fragment continues its forward journey, while the remainder of the tubule recoils in the opposite direction. This observation suggests that the

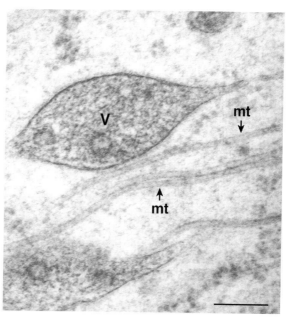

Fig. 2. Freeze-substituted *P. tinctorius* hypha showing close association between a tubular vacuole (*v*) and microtubules (*mt*); *bar* 200 nm

tubule is under tension and is being pulled along, rather than "pushed" from behind, as we would expect if, for example, hydrostatic pressure within the parent vacuole were the force responsible.

Vacuoles of other mycelial organisms also show a dependence on microtubules and associated motor proteins. Co-localisation of vacuoles and microtubules has been reported in *Saprolegnia ferax* (Allaway et al. 1997), and microtubule depolymerisation in this species suppresses vacuolation in hyphae recovering from various types of perturbation (Bachewich and Heath 1999). It has been reported that vacuole formation and movement in other mycelial fungi are also affected by microtubule depolymerisation (Herr and Heath 1982; Steinberg and Schliwa 1993). Likewise, vacuoles do not form properly in hyphae of the *kin2*-null mutant of *Ustilago maydis* that lacks normal function of the microtubule motor protein kinesin (Steinberg et al. 1998).

A dependence on an intact microtubule network is also typical of tubule formation by endomembrane compartments such as lysosomes, endosomes, the Golgi apparatus and the plasma membrane in animal cells (see Swanson et al. 1987b, 1992; van Deurs et al. 1996; Allan and Schroer 1999). In none of the animal systems, nor in any mycelial fungus, is there evidence of actin involvement in tubule formation, although there are reports that actin filaments are involved in fusion between mature endosomes and pre-existing lysosomes (van Deurs et al. 1995) and possibly also between *S. cerevisiae* vacuoles (Pan and Goldfarb 1998). The tubular ripples of the large central vacuole in tobacco protoplasts also appear to be actin-dependent (Verbelen and Tao 1998). Actin-disrupting agents, such as latrunculin and cytochalasin, do not suppress vacuolar tubules in *P. tinctorius* (Hyde et al. 1999), nor plasma membrane tubule formation in animal Hep-2 cells (van Deurs et al. 1996). In *S. cerevisiae*, migration of the vacuole into the bud does depend on actin and myosin, and involves tubule extension (Hill et al. 1996; Catlett and Weisman 1998). However, normal morphology of the *S. cerevisiae* vacuole system requires microtubule integrity (Guthrie and Wickner 1988).

While the case for microtubule involvement in hyphal vacuolar tubulation is strong, there is no evidence as to whether motor proteins are involved. Steinberg et al. (1998) provide data supporting an involvement of kinesin in vacuole formation in *Ustilago maydis*. Both kinesin and dynein are involved in the movement of endosomal and lysosomal tubules (Hollenbeck and Swanson 1990; Harada et al. 1998; Sonee et al. 1998; also see Chap. 11, this Vol.), and tubulation by the Golgi apparatus requires kinesin (see Allan and Schroer 1999). If motor proteins are involved in *P. tinctorius* vacuole tubulation, then the orientation of the microtubule tracks along which tubules extend becomes relevant, especially since tubules can extend both towards and away from the hyphal apex at the same time. Until recently, microtubules of fungal hyphae were thought to be all oriented with their plus ends away from the apex (Hoch and Staples 1985; Thompson-Coffe and Zickler 1992). However, recent work in *U. maydis* has shown that some microtubules with the reverse orientation are present also (Steinberg et al. 1998). In this case only one of the two main microtubule-associated motor proteins of eukaryotic cells would be needed for bidirectional extension of tubules – either kinesin, the plus-end-directed motor, or dynein, the minus-end-directed motor (Yamashita and May 1998). This is not meant to imply that only one type of motor is used. The vacuolar tubule retractions, often seen after extension in *P. tinctorius*, would suggest otherwise. Indeed, in some other systems, organelles appear to be endowed with a complex assemblage of motor proteins that even allows a shift from microtubule- to actin-based motility (reviewed in Allan and Schroer 1999). On the other hand, motor proteins may not be involved at all. In *Xenopus* egg extracts, microtubules attach directly to membranes at their plus ends and pull them into thin tubules as they extend at rates of $20\,\mu m\,min^{-1}$ or more (Watermanstorer et al. 1995).

Another aspect of tubule formation is whether coat proteins are involved in the initiation and/or extension of tubules in mycelial fungi. Whether proteins such as those involved in vesicle formation (e.g. clathrin, COP proteins) also encase tubules is itself a contentious point. On the one hand it has been proposed that both buds and tubules form by default from the Golgi, if coat proteins or compounds that assist coat-protein binding are either unavailable or functionally blocked (Klausner et al. 1992). Conversely, other data from the same system indicate that buds and tubules both require available coat protein (Orci et al. 1993). Three-dimensional electron microscopy indicates that Golgi buds and tubules both have coatings of some sort, but that those of buds are much more elaborate (Weidman et al. 1993). Artificially induced tubulation from the plasma membrane of nerve terminals is the most unambiguous example (Takei et al. 1995, 1998). In this system, vesicle budding requires both clathrin and the GTPase dynamin that is believed to affect pinching off of vesicles via hydrolysis of GTP. When non-hydrolysable GTPγS is added to the system, however, buds continue to grow as tubules from the membrane, encased in rings of dynamin (Takei et al. 1995). In *P. tinctorius*, GTPγS also induces massive tubulation of the vacuole system (D. Davies, School of Biological Science, UNSW, Sydney 2052, unpubl. data). This result is still too general to allow parallels to be drawn with the dynamin model, but it does distinguish tubulation of vacuoles from that of most animal endomembrane compartments that typically produce fewer tubules in response to GTPγS. Hyphae of *P. tinctorius* treated with brefeldin A also show the reverse response to that reported for animal endomembrane elements – tubulation is strongly inhibited, not promoted (Cole et al. 2000a,b).

The behaviour of tubule tips after first contact with larger vacuoles suggests a repeated exploration of the vacuole surface until exactly the right position is found for fusion to occur. Such behaviour indicates some kind of recognition event and is suggestive of an array of molecules at the tubule tips equivalent to those that mediate vesicle docking (e.g. see Waters and Pfeffer 1999). Much more work is needed to clarify this topic, but already it is clear that a variety of mechanisms operate even within individual cells.

IV. Probing the Content of Individual Vacuoles

Information on the average content of **populations of vacuoles** has been obtained from analysis of isolated vacuoles (Martinoia et al. 1979; Vaughn and Davis 1981) and from NMR spectroscopy of whole tissue. NMR spectroscopy can differentiate between different pools of a substance in different subcellular locations by small differences in the resonance frequency of the element nuclides being analysed, according to their chemical environment. For example, ^{31}P NMR has been used to measure relative changes in inorganic phosphate partitioned between vacuoles and the cytoplasm of higher plant cells during phosphorus depriva-

tion (Lee and Ratcliffe 1983). Chemical shifts of the two major pools of inorganic phosphate allocated to the vacuole and cytoplasm have been used to predict the pH of these two compartments in a number of organisms (see Ratcliffe 1994), including fungi (see Sect. II.B). The application of ^{31}P NMR has been important in confirming that fungal vacuoles contain polyphosphate as well as orthophosphate and in determining the chain length of that polyphosphate (Ashford et al. 1994; Gerlitz and Werk 1994; Gerlitz and Gerlitz 1997; see Martin et al. 1994). Similarly ^{27}Al NMR has been used to show that Al-polyphosphate complexes form rapidly when mycelium of Al-resistant *Laccaria bicolor* is grown in media containing Al. Interestingly, the chain length of polyphosphate in these complexes is much shorter (about 5 phosphate residues) than that detected by NMR in the vacuoles of non-Al treated cells (usually about 10–15 phosphate residues). All the ^{27}Al resonance is from the complexed form, suggesting that the total amount of cellular Al is sequestered rapidly in association with polyphosphate (Martin et al. 1994). It is important to realise that insoluble polyphosphate is not detected by ^{31}P NMR, as it does not resonate (Gadian 1982). The technique of NMR, as applied to ectomycorrhizal fungi, has been reviewed by Martin (1991).

Several techniques for content analysis of **individual vacuoles** are also available. These include use of microelectrodes, X-ray microanalysis and various cytochemical methods, including immunocytochemistry. Ion-selective microelectrodes have been widely used to measure ion activities in plant cells and their vacuoles (Miller 1994; Bethmann et al. 1995; Leigh 1997; Hinde et al. 1998). Often there is uncertainty about the subcellular compartment sampled especially where a cell contains a relatively high proportion of cytoplasm, but in cases where there is a large central vacuole, good results have been obtained (Hinde et al. 1998). There are also concerns about the effects of the procedure on the analysis. Probing fungal hyphae with microelectrodes is difficult because of their small size. The technique has been used mostly for microinjection (Read et al. 1992). Ion-selective electrodes are too large for probing small subcellular compartments such as motile fungal vacuoles. However, where vacuoles are large, as in the oomycete *Achlya*, it is difficult not to penetrate them (Dr. Sandra Jackson, Dept of Plant and Microbial Sciences, University of Canterbury, Christchurch, New Zealand, pers. comm.).

X-ray microanalysis provides information about the relative abundance of different elements in individual vacuoles. The spatial resolution can be very high and the technique may be used to collect data quantitatively (Hinde et al. 1998; Marshall and Xu 1998). However, because of the difficulties in controlling conditions and in use of reference standards, most analyses are only semiquantitative and only relatively high concentrations of elements can be detected. Lower limits of detection are reported to be in the range of 20–40 mM (Lazof and Laüchli 1991). It is of the utmost importance that tissues are prepared in such a way that ions are retained in situ. If a positive signal is obtained it is tempting to assume that the element has always been in that location. However, unless great care has been exercised, this is not likely to be the case. With chemical fixation membrane permeability is changed and, in an aqueous environment, ion loss, redistribution and exchange all occur, according to basic rules of chemistry. Localisations obtained after such preparative procedures may bear little relationship to those in living cells. This applies to tissue after glutaraldehyde/osmium fixation followed by ethanol dehydration and conventional embedding and it also applies to precipitates.

Rapid freezing followed either by analysis of the material while still in a frozen-hydrated state at low temperature, or after an embedding procedure where the ice has been removed without melting, can avoid these problems. Use of frozen-hydrated specimens is the preferred method to avoid element redistribution, but frozen sections are not very stable in the electron beam. In bulk specimens analytical resolution is only of the order of about 2 μm, and this applies also to depth. This resolution is insufficient to differentiate small vacuoles from surrounding cytoplasm. Etching the surface by subliming the ice, which is often done to visualise morphology, has unpredictable effects on element stability at the specimen surface. Analysis of planed frozen-hydrated bulk samples, using the carbon and oxygen X-ray images to determine morphology, avoids this problem and can yield very good results (Marshall and Xu 1998). However, the limitations of low resolution remain for small organelles.

Two methods for removal of ice, which avoid redistribution of ions and have been used successfully to analyse fungal vacuoles, are freeze-drying and freeze-substitution (Orlovich and Ashford 1993; Frey et al. 1997; Bücking et al. 1998;

Ashford et al. 1999). In both cases the specimen is embedded and sectioned, allowing better resolution. For freeze-substitution it is essential that strictly anhydrous reagents be used and that a water-attracting molecular sieve is added to the substitution fluid and resin at all steps (Canny and McCully 1986). Using such methods, mobile ions such as Na^+, Cl^-, K^+, Ca^{2+} and PO_4^{3-} can all be retained within salt-dextran droplets and localised with great precision in dry-cut sections (Orlovich and Ashford 1995). Resolution is limited by the ice crystal size and is demonstrably less than 1 μm laterally. Non-polar substitution fluids such as ether or tetrahydrofuran must be used to dissolve the ice. Although osmium/acetone is the fluid of choice for structural work (Monaghan et al. 1998), it is not a suitable substituent for localisation of diffusible substances. This is not an issue since the resolution is limited by thickness of the dry-cut sections.

X-ray microanalysis in conjunction with anhydrous methods of freeze-substitution has shown that the composition of individual vacuoles is quite distinct and differs from cell to cell. This is illustrated in a section of the Hartig net region of an ectomycorrhiza (Fig. 3). The fungal vacuoles in the Hartig net and sheath store large amounts of phosphorus, while this element cannot be detected in the root vacuoles. The fungal vacuoles also store large amounts of sulfur (Ashford et al. 1999; Vesk et al. 2000), in contrast to vacuoles in the tips of actively growing mycelium of the same fungus (Orlovich and Ashford 1993; Cole et al. 1998). The result is not surprising in view of the observation that genes coding for hydrophobins (see Chap. 7, this Vol.), which contain large amounts of cysteine, are upregulated early in mycorrhizal sheath formation (Tagu and Martin 1996).

Histochemical and ^{31}P NMR data indicate that much of the phosphorus stored in vacuoles of actively growing mycelium and mycorrhizae occurs as polyphosphate, although the relative amounts of polyphosphate and orthophosphate vary according to growth conditions, the state of the hyphae and probably the fungal species (Ashford et al. 1986, 1994; Grellier et al. 1989; Gerlitz and Werk 1994; Gerlitz and Gerlitz 1997). In freeze-substituted hyphae of P. tinctorius, phosphorus is co-localised in the vacuoles with potassium, but not with calcium. This applies to hyphae from both mycorrhizae (Fig. 3) and actively growing pure cultures (Orlovich and Ashford

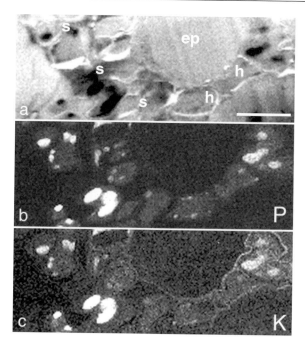

Fig. 3. Part of a *Eucalyptus pilularis/Pisolithus tinctorius* ectomycorrhiza. The specimen was plunge-frozen, freeze-substituted in anhydrous tetrahydrofuran, and dry-sectioned transversely. **a** Bright-field STEM image. *ep* Epidermal cell; *h* hyphae of the Hartig net; *s* fungal sheath. The exterior of the mycorrhiza is to the *left*; *bar* 5 μm. **b** Phosphorus X-ray image shows a strong and uniform P signal (*white pixels*) in the fungal vacuoles of both the sheath and the Hartig net, and a weaker signal in the fungal cytoplasm. **c** Potassium X-ray image. The strongest K signal is in the fungal vacuoles, in correlation with P. K is also present throughout the fungal and root cells, and in the apoplast of the Hartig net. Micrographs by courtesy of P.A. Vesk and W.G. Allaway; preparation of mycorrhiza and details of microscopy as in Ashford et al. (1999)

1993; Cole et al. 1998; Ashford et al. 1999). An association between K and P is also found in freeze-substituted protein bodies in sclerotia of *Sclerotinia minor*, while K and some P are lost from chemically fixed material (Young et al. 1993). Freeze-substituted protein bodies gave the characteristic metachromatic staining for polyphosphate showing that polyphosphate, protein, K and Mg were all present in the same storage vacuoles. However, Mg was not concentrated in the vacuoles like the other elements (Young et al. 1993). The data contrast with most of the earlier reports based on the analysis of chemically fixed specimens where polyphosphate was invariably found precipitated as granules in vacuoles and associated with Ca (see Ashford et al. 1986; Grellier et al. 1989; Orlovich and Ashford 1993; Young et al. 1993 for earlier references).

Precipitation of polyphosphate as granules in vacuoles of *P. tinctorius* was shown to occur during early stages of ethanol dehydration and interpreted as an artefact of tissue processing (Orlovich and Ashford 1993). Vacuoles of living cells of the same material grown under the same conditions also did not contain granules (see Ashford et al. 1999). This is not to say that polyphosphate-precipitating ions and hence polyphosphate granules will *never* occur in vacuoles. Vacuole cation composition in vivo will vary according to the ion supply, external conditions, fungal species and metabolic state of the cells involved (Klionsky et al. 1990). Bücking and colleagues have found globules, which they consider to be polyphosphate, in the vacuoles of living hyphae of several species in the Basidiomycotina, including an isolate of *P. tinctorius* (Bücking and Heyser 1997, 1999; Bücking et al. 1998). Their analyses also show that in freeze-dried material, Mg and K, and not Ca, are co-localised with P, while, in chemically fixed hyphae, Ca and Na replaced much of the K and Mg. Fungal vacuoles are reported to be major storage sites for calcium and other divalent cations (Klionsky 1998) and wherever these or other precipitating ions occur together with polyphosphate, precipitation might be expected. However most X-ray microanalyses of vacuoles have been carried out after conventional processing and it is not possible to draw any conclusions about their elemental composition. To clarify further the apparent protective role of vacuolar polyphosphate in precipitating toxic ions, it will be necessary to analyse vacuoles using anhydrous freezing methods, following application of such toxic ions to growing mycelium.

While only high levels of the element Ca may be detected by X-ray microanalysis, much lower amounts of Ca^{2+} ions in living cells may be measured using fluorescent probes. Ca^{2+} plays a central role in a wide range of cellular processes, including signal transduction, and it is of great interest to determine how Ca^{2+} is partitioned between the cytoplasm and organelles. Attempts to determine whether vacuoles sequester Ca^{2+} using Ca^{2+}-reactive fluorescent probes have been confounded by the tendency of these probes to accumulate in vacuoles by a general detoxifying mechanism (see Sect. V.C). Calcium green and a number of other Ca^{2+} and pH probes are sequestered into a reticulate network in hyphal tips and the larger vacuoles in the oomycete *Saprolegnia ferax*, probably by this mechanism (Hyde 1998). Reticulate tubular networks in higher fungi also accumulate Ca dyes (Read et al. 1992), and these may be tubular vacuoles.

Vacuoles are considered to be major Ca^{2+}-sequestering organelles in fungal cells (see Klionsky et al. 1990; Jackson and Heath 1993). Recent work in this area has concentrated on the activity of the Ca^{2+} pumps and transporters in the *S. cerevisiae* vacuole membrane, a topic reviewed elsewhere (see Dunn et al. 1994; Strayle et al. 1999). Vacuolar Ca^{2+} levels in *S. cerevisiae* have been determined by extrapolation from the activity of the uptake systems, using $^{45}Ca^{2+}$ radiolabelling and spectrophotometry of released contents, measured as the arsenazo(III)-Ca^{2+} complex (Dunn et al. 1994). These approaches have indicated that total Ca^{2+} concentration in vacuoles is about $2\,mM$, most bound to inorganic polyphosphates, leaving only about $30\,\mu M$ of free Ca^{2+}. This level is well above the upper limit of the sensitivity range of most commonly used fluorescent Ca^{2+}-sensitive dyes (e.g. fura-2, indo-1), and no studies have reported use of the more appropriate lower affinity Ca^{2+} dyes such as mag-fura-2 and mag-indo-1. Loading of hyphae and vacuoles with Ca^{2+} dyes is difficult (Hyde 1998; Parton and Read 1998) but it is likely that vacuoles could be loaded with low-affinity dyes using the same procedures by which vacuoles have been, inadvertently, loaded with Fluo-3 (Hyde and Heath 1997). However, potential differential uptake of dyes within the vacuole system will make it difficult to interpret results. A novel approach has used recombinant aequorin targeted to the membrane of the *S. cerevisiae* ER (Strayle et al. 1999). We believe that this approach could be modified for vacuoles.

Nitrogen compounds are stored together with P in vacuoles. Association of polyphosphate with basic proteins and amino acids is well established in yeasts and mycelial fungi (Cramer et al. 1980; Kanamori et al. 1982; Legerton et al. 1983; Kitamoto et al. 1988a,b; Young et al. 1993). Both *S. cerevisiae* and *N. crassa* store basic amino acids in the vacuole and other amino acids in the cytosol (see refs in Keenan et al. 1998). This selective uptake of basic amino acids into vacuoles depends on activity of a V-ATPase in the vacuolar membrane (see Zerez et al. 1986; Bowman et al. 1997). Proteins accumulated in vacuoles include hydrolytic enzymes. A large body of literature exists on protein targeting to vacuoles, especially in *S. cerevisiae*, showing that it occurs by more than one pathway (see Sect. V.B).

V. The Functions of Motile Vacuole Systems

Of the many diverse roles of vacuole systems in fungal cells we focus here on some aspects of function relevant to vacuole motility and differentiation.

A. Membrane Fusion and Vacuole Transport

Motile tubular vacuole systems are obvious candidates to function as a conduit for longitudinal transport along hyphae. Movement of fluorochrome pulses along vacuolar tubules is commonly observed. The hypothesis that the vacuole compartment has a role to play in the transport of substances normally stored there, in particular P and N compounds, has been argued in two recent reviews (Ashford and Orlovich 1994; Ashford 1998) and will not be covered here. If tubular vacuole systems operate as a transport conduit they have a role to play in ectomycorrhizal associations, in moving nutrients from uptake sites at the growing hyphal front along the soil mycelium to the fungus/root interface. Tubular vacuole systems are present in ectomycorrhizal mycelium but there is no evidence that they can transport molecules from one point to another (Allaway and Ashford 2000). Transport of vacuolar content and membrane into new buds and zygotes by means of tubulo-vesicular processes extending from parent vacuoles is well established in *S. cerevisiae* (Gomes de Mesquita et al. 1991; Jones et al. 1993; Raymond et al. 1990, 1992b; Weisman and Wickner 1988; Weisman et al. 1990). An in vitro assay has allowed characterisation of the process and determination of the factors required.

Molecular mechanisms of transport have hitherto been considered generally in the context that transport between intracellular compartments will be effected primarily by vesicles (e.g. see Pryer et al. 1992; Gerhardt et al. 1998; Blatt et al. 1999; Gerst 1999). This is thought to be a three-step process involving vesicle priming, docking and finally fusion. There is growing evidence that the specificity underlying vesicle docking and fusion resides in the specificity of an array of receptor proteins, on both target and vesicle membranes, called **SNAREs** (Rothman 1994), though these are not the only determinants (Götte and Lazar 1999;

Mayer 1999; Pelham 1999a). SNAREs are membrane proteins with large cytosolic domains that bind the soluble factor, *N*-ethylmaleimide-sensitive fusion protein (**NSF**), and its membrane attachment proteins called **SNAPs**. Vesicle-associated (or donor) SNAREs (**v-SNAREs**) are thought to interact specifically with their counterparts on the target membrane (**t-SNAREs**) to enable vesicle docking. Other proteins involved in docking and fusion include Rab proteins that are GTPases and also compartment-specific, NSF and various **SNAPs (soluble NSF attachment proteins)**. SNAPs bind to both v- and t-SNARES and their involvement, together with that of NSF, is a prerequisite to vesicle fusion, as distinct from docking. The topic has been extensively reviewed and there is more than one working model (Mayer 1999). SNAREs and associated proteins are highly conserved in eukaryotic cells and various *S. cerevisiae* gene products have been identified as homologues of mammalian SNAREs that were first identified in synaptic vesicle fusion during nerve transmission (Bennett 1995; reviewed Ferro-Novick and Jahn 1994; Rothman 1994; Südhof 1995; Pfeffer 1996; Gerst 1999). The *S. cerevisiae* gene product Vam3p is a t-SNARE. Together with the associated SNAP-25 homologue Vam7p, necessary for the stability of the vacuolar SNARE complex, it is localised to vacuoles. Another t-SNARE, Pep12p, is localised to the pre-vacuolar compartment (Darsow et al. 1997; von Mollard et al. 1997; Sato et al. 1998; Ungermann and Wickner 1998). However, two related t-SNARE proteins from *Arabidopsis thaliana*, AtVam3p and AtPep12p, which show a high degree of homology, are both localised to the pre-vacuolar compartment (Sanderfoot et al. 1998, 1999). Although AtVam3p replaces *S. cerevisiae* Vam3p in *vam*Δ mutants, allowing suppression of defects (Sato et al. 1997), it does not appear to be truly homologous to the yeast gene product (Sanderfoot et al. 1999). The localisation of AtVam3p to the vacuole in *A. thaliana* by Sato et al. (1997), rather than to the pre-vacuolar compartment as reported by Sanderfoot et al. (1999), might be due to vacuole rearrangements during primary aldehyde fixation in the former. Results indicate that there is more than one t-SNARE per compartment in *A. thaliana*, a contrast with the situation in *S. cerevisiae*. An *Arabidopsis* v-SNARE (AtVTI1a) reacts and co-localises with AtPep12p at the pre-vacuolar compartment and is thought to mediate docking and fusion of trans-Golgi network-derived vesicles to this compartment (Zheng et al. 1999).

The assay developed to study homotypic fusions between vacuoles from *S. cerevisiae* in vitro indicates that mature vacuoles interact and fuse by mechanisms which parallel those that mediate vesicle priming, targeting and docking (Nichols et al. 1997; Ungermann et al. 1998). The vacuole fusion assay uses two sets of vacuoles from mutant strains, one deficient in protease but containing the precursor to alkaline phosphatase, and the other alkaline phosphatase-deficient but protease-active (Haas 1995). Alkaline phosphatase activity is the measure of fusion. This elegant work suggests that *cis*-complexes of both v- and t-SNAREs exist on both the donor and recipient vesicle membranes and must first be dissociated to prime the vesicles for subsequent docking and fusion (Ungermann et al. 1999). The ATP hydrolysis-mediated priming step involves the action of the *S. cerevisiae* gene products Sec17p (an α-SNAP), Sec18p (NSF protein, an ATP-driven chaperone) and LMA1 (its co-chaperone), which are all **SNARE regulators**. Priming disassembles the *cis*-SNARE complex, releasing Sec17p and activating the t-SNARE. The co-chaperone LMA1, a novel protein containing thioredoxin, stabilises the primed t-SNARE (Xu and Wickner 1996; Xu et al. 1997, 1998). Primed vacuoles that come into contact are initially tethered (the first stage of docking) by a reversible reaction independent of SNAREs, but which requires YPt7p, a small GTP-binding protein of the Rab/Ypt family of GTPases. The interaction becomes irreversible through association of the SNAREs of the apposed membranes, forming a *trans*-v-t-SNARE complex (the second stage of docking). *trans*-SNARE pairing triggers release of luminal calcium to interact with calmodulin and mediate events leading to fusion (see Price et al. 2000a,b and references therein). The *cis*-SNARE complexes on the same membrane contain five SNAREs, together with the two chaperones and LMA1, and are fully functional in mediating fusion (Sato and Wickner 1998; Ungermann et al. 1999). Two of the SNAREs in this complex, Vam2p and Vam6p, are thought to mediate the link between priming and docking (Price et al. 2000a). They are released as a Vam2/6p complex upon Sec18p-driven priming and can then bind to Ypt7p to initiate docking. Vesicle targeting and membrane fusion are areas of intensive research and further information can be found in a number of recent reviews (Bryant and Stevens 1998; Blatt et al. 1999; Clague 1999; Gerst 1999; Götte and Lazar 1999; Jahn and Sudhof 1999; Mayer 1999; Pelham 1999a,b; Sanderfoot and Raikhel 1999). It will be interesting to see where these SNARE complexes occur in the context of the whole vacuole surface in vivo and how they relate to sites of vacuole membrane budding and tubule initiation in yeasts and mycelial fungi.

Two of the *S. cerevisiae* SNAREs that participate in *cis*-SNARE complexes, Vam3p and Vam6p, have unusual bipolar patch locations on the vacuoles, suggesting that specialised domains of the vacuolar membrane may be involved in fusion (Nakamura et al. 1997; Wada et al. 1997). SNARE-type mechanisms may also mediate specificity of targeting and docking at the tips of tubules. If this is the case, organelle linkages via tubules have the potential for delivery of specific content to specific compartments, depending on control of fusion and fission in time and space. This seems likely. Retrograde transport between organelles is known to operate via tubules, and there is evidence that some retrograde transport (though not involving vacuoles) is mediated by SNARE-type reactions (see Pelham 1999a).

B. Vacuoles and the Endocytic Pathway

Fungal vacuoles, like mammalian lysosomes, are lytic organelles and contain hydrolases (Klionsky et al. 1990). All three major compartments of the endosomal/lysosomal system in animal cells may form tubules, and transport is not necessarily unidirectional or mediated by vesicles (Robinson et al. 1986; Heuser 1989; Hopkins et al. 1990; Knapp and Swanson 1990; Deng et al. 1991; Tooze and Hollinshead 1991; Dunn and Maxfield 1992; Parton et al. 1992; Jarhaus et al. 1994; Thilo et al. 1995). It is not clear how motile vacuole systems in fungi, as demonstrated by fluorochromes, relate to these animal cell compartments. Whether the tubular vacuoles labelled by fluorochromes are equivalent only to tubular lysosomes, or whether they represent a greater part of the endosomal pathway in fungi, is a matter of conjecture (Ashford and Orlovich 1994; Ashford 1998).

In *S. cerevisiae* it has been shown that some hydrolases synthesised on the ER pass from Golgi bodies to the vacuoles through a pre-vacuolar compartment (Raymond et al. 1992a; Vida et al. 1993). This compartment is distal to the Golgi and

is considered equivalent to the late-endosomal compartment of animal cells and intermediate between the biosynthetic pathway leading to the vacuole and the endosomal pathway (Rieder et al. 1996). Protein targeting to the *S. cerevisiae* vacuole has been the subject of intensive research involving a wide variety of mutants, including mutants deficient in transport to the vacuole and proteolytic processing. Mutants of **vacuolar protein sorting (VPS) genes** result in hydrolase missorting and secretion as well as abnormal vacuole morphologies (Raymond et al. 1990, 1992b). Over 40 *VPS* genes have been identified and these have been differentiated into six major groups according to their cellular effects (see Wurmser and Emr 1998). In contrast, mutations of *PEP* genes lack the vacuolar protease activity necessary to render hydrolases active and those of *VAM* genes have abnormal vacuole morphologies. There are at least nine *VAM* genes, subdivided into two classes based on mutant phenotype (Wada et al. 1992), and many are identified as SNAREs and associated proteins (see Sect. V.A).

Several pathways of protein transport to the vacuole have been identified in *S. cerevisiae*. Two occur via the Golgi, but only one of them involves the pre-vacuolar compartment, and some proteins including enzymes are taken up directly from the cytoplasm (see reviews of Horazdovsky et al. 1995; Stack et al. 1995; Conibear and Stevens 1998; Klionsky 1998; Odorizzi et al. 1998). There are also reports that *S. cerevisiae* vacuoles degrade entire organelles and portions of the cytosol via autophagy, as in mammalian cells. The process involves surrounding the structure with a membrane to produce an autophagosome that is then targeted to the vacuole (see, for example, Campbell and Thorsness 1998; Chiang and Chiang 1998; Kametaka et al. 1998; Lang et al. 1998).

The endosomal pathway in animal cells is not only a pathway for autolysis of unwanted cell components. It is also an endocytic pathway, sorting and processing molecules entering from the cell surface or external solution, and recycling plasma membrane constituents (Gruenberg and Maxfield 1995). There is much molecular and genetic evidence for an endocytic pathway in *S. cerevisiae* similar to that in animal cells (Raths et al. 1993; see Roth et al. 1998). However, the pathway is not well characterised morphologically. Two intermediate compartments, in which the pheromone or mating factor, α-factor, sequentially accumulates after internalisation, have been characterised by cell fractionation and are considered equivalent to early and late endosomes (Hicke et al. 1997; Riezman 1993; Singer-Krüger et al. 1993). In a vacuolar protein-sorting mutant, *vps28*, both anterograde and retrograde transport are impaired and markers from both the secretory and endocytic pathway accumulate in a compartment thought to be the late endosome, also called pre-vacuolar, compartment (Rieder et al. 1996). The mutant is one of a group (class E *vps* mutants) in which this compartment is always enlarged (Raymond et al. 1992a). In this mutant, FM4-64 stains the compartment intensely, showing that it is located near the large vacuole. Localisation of vacuolar H+-ATPase, one of the blocked proteins, shows that in chemically fixed cells the compartment consists of stacks of curved cisternae (Rieder et al. 1996).

Internalisation of positively charged nanogold particles by wild-type *S. cerevisiae* protoplasts has also been used to target compartments of the endocytic pathway (Prescianotto-Baschong and Riezman 1998). Gold particles are restricted to the plasma membrane at $0\,^{\circ}$C, but, on raising the temperature to $15\,^{\circ}$C, they are rapidly internalised first to small vesicles, then to a tubulo-vesicular compartment suggestive of early endosomes, later to multivesicular bodies, and finally to the vacuoles. Multivesicular bodies are considered to be the transporting compartment carrying molecules from early to late endosomes in animal cells (see van Deurs et al. 1993; Futter et al. 1996) and are equated with late endosomes in *S. cerevisiae*. Further confirmation that there are distinct early and late endosomal compartments in *S. cerevisiae* has come from a study of a mutant with the *YJL204c* (renamed *RCY1*, or recycling 1) gene disrupted (Wiederkehr et al. 2000). In this mutant there is an early block in the endocytic pathway prior to the intersection with the vacuole-sorting pathway. Mutation of *RCY1* leads to accumulation of an enlarged compartment that contains the t-SNARE Tlg1p and lies close to areas of cell expansion. After their internalisation, the endocytic markers Lucifer Yellow, FM4-64 and Ste2p are all found in a similar enlarged compartment identified as early endosomal. This compartment recycles membrane rapidly to the cell surface in *S. cerevisiae*, thus balancing high rates of membrane internalisation via endocytosis.

Multivesicular bodies are very common in tip cells of mycelial fungi, supporting the idea of an endocytic pathway; however, their structural relationships with vacuoles have not been clearly

established. There is also molecular evidence for retrograde transport between vacuoles, the pre-vacuolar compartment and the Golgi in *S. cerevisiae*. Proteins resident in the trans-Golgi network are retrieved from the distal pre-vacuolar compartment via an aromatic amino acid sorting motif (Bryant et al. 1998). A hybrid protein based on alkaline phosphatase containing this retrieval motif reaches the pre-vacuolar compartment, not by following the classical secretory carboxypeptidase pathway, but rather by signal-dependent retrograde transport from the vacuole, indicating that in *S. cerevisiae* the vacuole is not a terminal compartment. This step is blocked by mutation of the *VAC7* gene, normally required for vacuole inheritance. Vti1p, a v-SNARE required for delivery of both carboxypeptidase Y and alkaline phosphatase to the vacuole, is also known to undergo retrograde transport out of the vacuole as part of its normal cellular itinerary (Bryant et al. 1998). Data of this nature suggest that exchanges between the vacuole *sensu stricto* and earlier compartments of the endosomal pathway are complex, and that traffic is not unidirectional. Much of this retrograde traffic may well take place through tubules.

C. The Vacuole as a Detoxifying Compartment

Mycelial fungi unavoidably take up potentially toxic substances from their natural environment. These may be animal or plant products, or synthetic compounds released into the environment from industrial pollution or agrochemical applications. Two possible mechanisms to avoid a build-up of such potentially harmful compounds inside the cytoplasm are (1) chemical modification or (2) compartmentation within a specific organelle. The sequestration of substances potentially toxic to the cytoplasm is a function of vacuoles and explains the vacuolar accumulation of many fluorescent probes (Oparka and Hawes 1992; Wink 1993; Coleman et al. 1997).

The drug **probenecid** has contributed enormously to our understanding of the role of vacuoles as detoxifying compartments. Probenecid is a known inhibitor of anion transporters in animal cells and prevents sequestration of organic anions including LY-CH from the cytosol into the endosomal-lysosomal system (Steinberg et al. 1987, 1988). Similarly, probenecid prevents uptake of fluorescein isothiocyanate across the tonoplast into plant vacuoles (Cole et al. 1990). In *P. tinctorius*, probenecid inhibits accumulation of 6-CF and carboxy-DFF in the tubular vacuole system (see Sect. II.A), demonstrating that the esterases that cleave 6-CFDA and carboxy-DFFDA reside in the cytoplasm, and that fungal vacuole membranes carry probenecid-inhibitable organic anion transporters as in animal and plant cells (Cole et al. 1997).

Further evidence for a role of fungal vacuoles in detoxification by mechanisms similar to those in other eukaryotic cells comes from use of the **chloromethyl dyes** CMAC and CMFDA. These dyes are highly lipophilic. They are readily internalised by fungal cells and accumulate in the vacuoles, indicating the presence both of intracellular thiols and a glutathione-mediated transport pathway (Cole et al. 1997; Haugland 1999). There is considerable evidence supporting a glutathione-dependent detoxification pathway in plant and animal cells (see Coleman et al. 1997). It is conceivable that endogenous glutathione is required to protect the fungal cell from chemical toxicity as in other eukaryotic cells (Coleman et al. 1997). It is, however, unlikely that probenecid-inhibitable organic anion transporters are also responsible for transport of glutathione conjugates across the tonoplast, since probenecid does not completely inhibit the sequestration of CMAC or CMF into vacuoles of *P. tinctorius*. Nonetheless, probenecid-sensitive and glutathione-conjugate transporters no doubt play a similar detoxifying role in fungal cells. In plants, both types of transporters have been shown to be ATP-dependent (Cole et al. 1990; Martinoia et al. 1993). ATP-dependent vacuole transporters such as the **multi-drug resistance-associated proteins** (MRPs), a subfamily of the ATP-binding cassette (ABC) transporters, have also been characterised in *S. cerevisiae* cells (reviewed by Rea 1999). Little is known about these transporters in mycelial fungi.

D. Vacuole Differentiation

In fungi there has been a tendency to use language such as "the vacuole" and to generate measurements that are averages of vacuole populations, giving the impression of uniformity. This leaves unexplained the paradox of how storage products, and the hydrolases that degrade them, apparently coexist in the same compartments. There is very little information about precisely what individual vacuoles in fungi contain, but there is now evidence from higher plants that cells maintain several different functionally distinct vacuolar compartments. Three distinct types of vacuole have been recognised, all of which may occur in the same cell (see Di Sansebastiano et al. 1998 and references therein; Rogers 1998). Two contain stored protein while the third is lytic. These vacuoles differ in both soluble and membrane proteins and they undergo dynamic fusion and exchanges during plant growth and development. There are also several pathways in vacuole targeting. The view has been expressed that, compared with animal and plant vacuoles, *S. cerevisiae* vacuoles are both morphologically and biochemically less diverse and may be less sophisticated in their organisation (Sato et al. 1997). The same may be true of mycelial fungi, but the morphology and dynamic exchanges of the *P. tinctorius* vacuole system would suggest otherwise.

VI. Conclusions

Tubular vacuoles are much more common than our understanding from the ultrastructure of chemically fixed cells would lead us to believe. Though they have the potential to act as a continuum via fusions and fissions over time, tubular vacuole systems are morphologically differentiated. Whether this indicates functional differentiation or bears biological significance regarding vacuole motility and behaviour remains to be seen. These questions will only be elucidated when we can specifically label and identify the content and activities of specific subcompartments, and determine what exchanges occur and how they are controlled. New molecular probes and cytological approaches will allow us to tackle these problems.

There have been significant advances in our understanding of the molecular pathways of vacuole targeting and endomembrane compartmentation in the last decade. These confirm the views that fundamental cellular processes in fungi parallel those in other eukaryotic cells, that there is an endocytic pathway operating at least in some fungi, and that vacuoles and lysosomes are equivalent. This progress has not been matched with growth in understanding of how endomembrane compartments interact spatially to exchange material. More specifically, the role of transient tubular connections in such exchanges is not clear. Conventional electron microscopy with chemical fixation does not allow us to deal with these problems adequately. Many vacuole-specific probes are becoming available as a consequence of the advances in yeast genetics and post-genomics. It will be important to apply these to mycelial fungi that may have added layers of control relevant to longitudinal transport and communication in an elongate, multicellular system. Solutions will be based on a combination of targeting such specific molecules, and advanced methods for their visualisation, including cryofixation and observations on living cells.

Acknowledgements. We would like to thank Bill Allaway and Danielle Davies for their help in the preparation of this manuscript. This work was funded by ARC grant A19701621.

References

Abe F, Horikoshi K (1998) Analysis of intracellular pH in the yeast *Saccharomyces cerevisiae* under elevated hydrostatic pressure: a study in baro-(piezo) physiology. Extremophiles 2:223–228

Allan VJ, Schroer TA (1999) Membrane motors. Curr Opin Cell Biol 11:476–482

Allaway WG (1994) Microtubules and the cytoplasmic motile tubular vacuole system in *Pisolithus tinctorius* hyphae. In: Stephenson AG, Kao T-H (eds) Current topics in plant physiology, vol 12. Pollen-pistil interactions and pollen tube growth. American Society of Plant Physiology, MA, pp 265–267

Allaway WG, Ashford AE (2001) Motile tubular vacuoles in extramatrical mycelium and sheath hyphae of ectomycorrhizal systems. Protoplasma 215:218–225

Allaway WG, Ashford AE, Heath IB, Hardham AE (1997) Vacuolar reticulum in oomycete hyphal tips; an additional component of the Ca^{2+} regulatory system? Fungal Genet Biol 22:209–220

Ashford AE (1998) Dynamic pleiomorphic vacuoles: are they endosomes and transport compartments in fungal hyphae? Adv Bot Res 28:119–159

Ashford AE, Orlovich DA (1994) Vacuole transport, phosphorus, and endosomes in the growing tips of fungal hyphae. In: Stephenson AG, Kao T-H (eds) Current topics in plant physiology, vol 12. Pollen-pistil interactions and pollen tube growth. American Society of Plant Physiology, MA, pp 135–149

Ashford AE, Peterson RL, Dwarte D, Chilvers GA (1986) Polyphosphate granules in eucalypt mycorrhizas: determination by energy dispersive X-ray microanalysis. Can J Bot 64:677–687

Ashford AE, Ryde S, Barrow KD (1994) Demonstration of a short chain polyphosphate in *Pisolithus tinctorius* and the implications for phosphorus transport. New Phytol 126:239–247

Ashford AE, Vesk PA, Orlovich DA, Markovina A-L, Allaway WG (1999) Dispersed polyphosphate in fungal vacuoles of *Eucalyptus pilularis/Pisolithus tinctorius* ectomycorrhizas. Fungal Genet Biol 28:21–33

Bachewich C, Heath IB (1999) Cytoplasmic migrations and vacuolation are associated with growth recovery in hyphae of *Saprolegnia*, and are dependent on the cytoskeleton. Mycol Res 103:849–858

Basrai MA, Naider F, Becker JM (1990) Internalization of lucifer yellow in *Candida albicans* by fluid phase endocytosis. J Gen Microbiol 136:1059–1065

Bennett MK (1995) SNAREs and the specificity of transport vesicle targeting. Curr Opin Cell Biol 7:581–586

Bethmann B, Thaler M, Simonis W, Schonknecht G (1995) Electrochemical potential gradients of H^+, K^+, Ca^{2+} and Cl^- across the tonoplast of the green alga *Eremosphaera viridis*. Plant Physiol 109:1317–1326

Blatt MR, Leyman B, Geelen D (1999) Molecular events of vesicle trafficking and control by SNARE proteins in plants. New Phytol 144:389–418

Boevink P, Santa Cruz S, Hawes C, Harris N, Oparka KJ (1996) Virus-mediated delivery of the green fluorescent protein to the endoplasmic reticulum of plant cells. Plant J 10:935–941

Boevink P, Oparka K, Santa Cruz S, Martin B, Betteridge A, Hawes C (1998) Stacks on tracks: the plant Golgi apparatus traffics on an actin/ER network. Plant J 15:441–447

Boller T, Wiemken A (1986) Dynamics of vacuolar compartmentation. Ann Rev Plant Physiol 37:137–164

Bourett TM, Howard RJ (1994) Enhanced labelling of concanavalin A binding sites in fungal membranes using a double-sided, indirect method. Mycol Res 98:769–775

Bourett TM, Picollelli MA, Howard RJ (1993) Postembedment labelling of intracellular concanavalin A binding sites in freeze-substituted fungal cells. Exp Mycol 17:223–235

Bourett TM, Czymmek KJ, Howard RJ (1998) An improved method for affinity probe localization in whole cells of filamentous fungi. Fungal Genet Biol 24:3–13

Bowman EJ, O'Neill FJ, Bowman BJ (1997) Mutations of *pma-1*, the gene encoding the plasma membrane H^+-ATPase of *Neurospora crassa*, suppress inhibition of growth by Concanamycin A, a specific inhibitor of vacuolar ATPases. J Biol Chem 272:14776–14786

Bryant NJ, Stevens TH (1998) Vacuole biogenesis in *Saccharomyces cerevisiae*: protein transport pathways to the yeast vacuole. Microbiol Mol Biol Rev 62:230–247

Bryant NJ, Piper RC, Weisman LS, Stevens TH (1998) Retrograde traffic out of the yeast vacuole to the TGN occurs via the prevacuolar/endosomal compartment. J Cell Biol 142:651–663

Bücking H, Heyser W (1997) Intracellular compartmentation of phosphorus in roots of *Pinus sylvestris* L. and the implications for transfer processes in ectomycorrhizae. In: Rennenberg H, Eschrich W, Zeiger H (eds) Trees – contributions to modern tree physiology. Backhuys Publishers, Leiden, pp 377–391

Bücking H, Heyser W (1999) Elemental composition and function of polyphosphates in ectomycorrhizal fungi – an X-ray microanalytical study. New Phytol 103:31–39

Bücking H, Beckmann S, Heyser W, Kottke I (1998) Elemental contents in vacuolar granules of ectomycorrhizal fungi measured by EELS and EDXS. A comparison of different methods and preparation techniques. Micron 29:53–61

Butt TM, Hoch HC, Staples RC, St Leger RJ (1989) Use of fluorochromes in the study of fungal cytology and differentiation. Exp Mycol 13:303–320

Calahorra M, Martínez GA, Hernández-Cruz A, Peña A (1998) Influence of monovalent cations on yeast cytoplasmic and vacuolar pH. Yeast 14:501–515

Campbell CL, Thorsness PE (1998) Escape of mitochondrial DNA to the nucleus in *yme1* yeast is mediated by vacuolar-dependent turnover of abnormal mitochondrial compartments. J Cell Sci 111:2455–2464

Canny MJ, McCully ME (1986) Locating water-soluble vital stains in plant tissues by freeze-substitution and resin-embedding. J Microsc 142:63–70

Catlett NL, Weisman LS (1998) The terminal tail region of a yeast myosin-V mediates its attachment to vacuole membranes and sites of polarized growth. Proc Natl Acad Sci USA 95:14799–14804

Chalfie M, Tu Y, Euskirchen G, Ward WW, Prasher DC (1994) Green fluorescent protein as a marker for gene expression. Science 263:802–805

Chiang M-C, Chiang H-L (1998) Vid24p, a novel protein localized to the fructose-1,6-bisphosphatase-containing vesicles, regulates targeting of fructose-1,6-bisphosphatase from the vesicles to the vacuole for degradation. J Cell Biol 140:1347–1356

Clague MJ (1999) Membrane transport: take your fusion partners. Curr Biol 9:R258–260

Cole L (1989) Endocytosis and the transport of fluorescent probes in suspension-cultured plant cells. MSc thesis, Oxford Univ, Oxford

Cole L, Coleman J, Evans D, Hawes C (1990) Internalisation of fluorescein isothiocyanate and fluorescein isothiocyanate-dextran by suspension-cultured plant cells. J Cell Sci 96:721–730

Cole L, Coleman J, Kearns A, Morgan G, Hawes C (1991) The organic anion transport inhibitor, probenecid, inhibits the transport of Lucifer Yellow at the plasma membrane and the tonoplast in suspension-cultured plant cells. J Cell Sci 99:545–555

Cole L, Hyde G, Ashford AE (1997) Uptake and compartmentalisation of fluorescent probes by *Pisolithus tinctorius* hyphae: evidence for an anion transport mechanism at the tonoplast but not for fluid-phase endocytosis. Protoplasma 199:18–29

Cole L, Orlovich DA, Ashford AE (1998) Structure, function and motility of vacuoles in filamentous fungi. Fungal Genet Biol 24:86–100

Cole L, Davies D, Hyde GJ, Ashford AE (2000a) ER-Tracker dye and BODIPY-brefeldin A differentiate the endoplasmic reticulum and Golgi bodies from the tubular vacuole system in living hyphae of *Pisolithus tinctorius*. J Microsc 197:239–248

Cole L, Davies D, Hyde GJ, Ashford AE (2000b) Brefeldin A affects radial growth, endoplasmic reticulum, Golgi bodies, tubular vacuole system and secretory pathway in *Pisolithus tinctorius*. Fungal Genet Biol 29:95–106

Coleman JOD, Blake-Kalff MMA, Davies TGE (1997) Detoxification of xenobiotics by plants: chemical modification and vacuolar compartmentation. Trends Plant Sci 2:144–151

Conibear E, Stevens TH (1998) Multiple sorting pathways between late Golgi and the vacuole in yeast. Biochim Biophys Acta 1404:211–230

Cormack BP, Bertram G, Egerton M, Gow NAR, Falkow S, Brown AJP (1997) Yeast-enhanced green fluorescent protein (yEGFP): a reporter of gene expression in *Candida albicans*. Microbiology 143:303–311

Cowles CR, Odorizzi G, Payne GS, Emr SD (1997) The AP-3 adaptor complex is essential for cargo-selective transport to the yeast vacuole. Cell 91:109–118

Cramer CL, Vaughn LE, Davis RH (1980) Basic amino acids and inorganic polyphosphates in *Neurospora crassa*: independent regulation of vacuolar pools. J Bacteriol 142:945–952

Croopnick JB, Choi HC, Mueller DM (1998) The subcellular location of the yeast *Saccharomyces cerevisiae* homologue of the protein defective in the juvenile form of Batten disease. Biochem Biophys Res Commun 250:335–341

Cubitt AB, Heim R, Adams SR, Boyd AE, Gross LA, Tsien RY (1995) Understanding, improving and using green fluorescent proteins. TIBS 20:448–455

Czymmek KJ, Bourett TM, Howard RJ (1996) Immunolocalization of tubulin and actin in thick-sectioned fungal hyphae after freeze-substitution fixation and methacrylate de-embedment. J Microsc 181:153–161

Darsow T, Rieder SE, Emr SD (1997) A multispecificity syntaxin homologue, Vam3p, essential for autophagic and biosynthetic protein transport to the vacuole. J Cell Biol 138:517–529

Davies JM, Brownlee C, Jennings DH (1990) Measurement of intracellular pH in fungal hyphae using BCECF and digital image microscopy. Evidence for a primary proton pump in the plasmalemma of a marine fungus. J Cell Sci 96:731–736

Davies KL, Davies MS, Francis D (1992) Vacuolar development in the root meristem of *Festuca rubra* L. New Phytol 121:581–585

Davies TGE, Steele SH, Walker DJ, Leigh R (1996) An analysis of vacuole development in oat aleurone protoplasts. Planta 198:356–364

Deng Y, Griffiths G, Storrie B (1991) Comparative behaviour of lysosomes and the pre-lysosome compartment (PLC) in in vivo cell fusion experiments. J Cell Sci 99:571–582

Di Sansebastiano G-P, Paris N, Marc-Martin S, Neuhaus J-M (1998) Specific accumulation of GFP in a non-acidic vacuolar compartment via a C-terminal propeptide-mediated sorting pathway. Plant J 15:449–457

Dunn KW, Maxfield FR (1992) Delivery of ligands from sorting endosomes to late endosomes occurs by maturation of sorting endosomes. J Cell Biol 117:301–310

Dunn T, Gable K, Beeler T (1994) Regulation of cellular Ca^{2+} by yeast vacuoles. J Biol Chem 269:7273–7278

Fernández-Ábalos JM, Fox H, Pitt C, Wells B, Doonan JH (1998) Plant-adapted green fluorescent protein is a versatile vital reporter for gene expression, protein localization and mitosis in the filamentous fungus, *Aspergillus nidulans*. Mol Microbiol 27:121–130

Ferro-Novick S, Jahn R (1994) Vesicle fusion from yeast to man. Nature 370:191–193

Forgac M (1998) Structure and function of the vacuolar (H^+)-ATPases. FEBS Lett 440:258–263

Frey B, Brunner I, Walther P, Scheidegger C, Zierold L (1997) Element localization in ultrathin cryosections of high-pressure frozen ectomycorrhizal spruce roots. Plant Cell Environ 20:929–937

Fricker MD, Oparka KJ (1999) Imaging techniques in plant transport: meeting review. J Exp Bot 50:1089–1100

Futter CE, Pearse A, Hewlett LJ, Hopkins CR (1996) Multivesicular endosomes containing internalised EGF-EGF receptor complexes mature and then fuse directly with lysosomes. J Cell Biol 132:1011–1023

Gadian DG (1982) Nuclear magnetic resonance and its applications to living systems. Oxford Univ Press, New York

Gaynor EC, Chen C-Y, Emr SD, Graham TR (1998) ARF is required for maintenance of yeast Golgi and endosome structure and function. Mol Biol Cell 9:653–670

Gerhardt B, Kordas TJ, Thompson CM, Patel P, Vida T (1998) The vesicle transport protein Vps33p is an ATP-binding protein that localizes to the cytosol in an energy-dependent manner. J Biol Chem 273:15818–15829

Gerlitz TGM, Gerlitz A (1997) Phosphate uptake and polyphosphate metabolism of mycorrhizal and non-mycorrhizal roots of pine and of Suillus bovinus at varying external pH measured by in vivo ^{31}P-NMR. Mycorrhiza 7:101–106

Gerlitz TGM, Werk WB (1994) Investigations on phosphate uptake and polyphosphate metabolism by mycorrhized and nonmycorrhized roots of beech and pine as investigated by in vivo ^{31}P-NMR. Mycorrhiza 4:207–214

Gerst JE (1999) SNAREs and SNARE regulators in membrane fusion and exocytosis. Cell Mol Life Sci 55:707–734

Gomes de Mesquita DS, Ten Hoopen R, Woldringh CL (1991) Vacuolar segregation to the bud of Saccharomyces cerevisiae: an analysis of morphology and timing in the cell cycle. J Gen Microbiol 137:2447–2454

Götte M, Lazar T (1999) The ins and outs of yeast vacuole trafficking. Protoplasma 209:9–18

Grellier B, Strullu DG, Martin F, Renaudin S (1989) Synthesis in vitro, microanalysis and ^{31}P NMR study of metachromatic granules in birch mycorrhizas. New Phytol 112:49–54

Gruenberg J, Maxfield FR (1995) Membrane transport in the endocytic pathway. Curr Opin Cell Biol 7:552–563

Guthrie BA, Wickner W (1988) Yeast vacuoles fragment when microtubules are disrupted. J Cell Biol 107:115–120

Haas A (1995) A quantitative assay to measure homotypic vacuole fusion in vitro. Methods Cell Sci 17:283–294

Harada A, Takei Y, Kanai Y, Tanaka Y, Nonaka S, Hirokawa N (1998) Golgi vesiculation and lysosome dispersion in cells lacking cytoplasmic dynein. J Cell Biol 141:51–59

Haugland RP (1999) Handbook of fluorescent probes and research chemicals, 7th edn. Molecular Probes, Eugene, OR (http://www.PROBES.com)

Herr FB, Heath MC (1982) The effects of antimicrotubule agents on organelle positioning in the cowpea rust fungus Uromyces phaseoli var. vignae. Exp Mycol 6:15–24

Heuser J (1989) Changes in lysosome shape and distribution correlated with changes in cytoplasmic pH. J Cell Biol 108:855–864

Hicke L, Zanolari B, Pypaert M, Rohrer J, Riezman H (1997) Transport through the yeast endocytic pathway occurs through morphologically distinct compartments and requires an active secretory pathway and Sec18p/n-ethylmaleimide-sensitive fusion protein. Mol Biol Cell 8:13–31

Hill KL, Catlett NL, Weisman LS (1996) Actin and myosin function in directed vacuole movement during cell division in Saccharomyces cerevisiae. J Cell Biol 135:1535–1549

Hillmer SL, Quader H, Robert-Nicoud M, Robinson DG (1989) Lucifer yellow uptake in cells and protoplasts of Daucas carota visualized by laser scanning microscopy. J Exp Biol 40:417–423

Hinde P, Richardson P, Koyro H-W, Tomos AD (1998) Quantitative X-ray microanalysis of solutes in individual plant cells: a comparison of microdroplet and in situ frozen-hydrated data. J Microsc 191:303–310

Hoch HC (1990) Preservation of cell ultrastructure by freeze-substitution. In: Mendgen K, Lesemann D-E (eds) Electron microscopy of plant pathogens. Springer, Berlin Heidelberg New York, pp 1–16

Hoch HC, Staples RC (1985) The microtubule cytoskeleton in hyphae of Uromyces phaseoli germlings: its relationship to the region of nucleation and to the F-actin cytoskeleton. Protoplasma 124:112–122

Hoffmann J, Mendgen K (1998) Endocytosis and membrane turnover in the germ tube of Uromyces fabae. Fungal Genet Biol 24:77–85

Hollenbeck PJ, Swanson JA (1990) Radial extension of macrophage tubular lysosomes supported by kinesin. Nature 346:864–866

Hopkins CR, Gibson A, Shipman M, Miller K (1990) Movement of internalized ligand-receptor complexes along a continuous endosomal reticulum. Nature 346:335–339

Horazdovsky BF, DeWald DB, Emr SD (1995) Protein transport to the yeast vacuole. Curr Opin Cell Biol 7:544–551

Hyde GJ (1998) Calcium imaging: a primer for mycologists. Fungal Genet Biol 24:14–23

Hyde GJ, Ashford AE (1997) Vacuole motility and tubule-forming activity in Pisolithus tinctorius hyphae are modified by environmental conditions. Protoplasma 198:85–92

Hyde GJ, Heath IB (1997) Ca^{2+} gradients in hyphae and branches of Saprolegnia ferax. Fungal Genet Biol 21:238–251

Hyde GJ, Cole L, Ashford AE (1997) Mycorrhiza movies. Mycorrhiza 7:167–169

Hyde GJ, Davies D, Perasso L, Cole L, Ashford AE (1999) Microtubules but not actin microfilaments regulate vacuole motility and morphology in hyphae of Pisolithus tinctorius. Cell Motil Cytoskeleton 42:114–124

Jackson SL, Heath IB (1993) Roles of calcium ions in hyphal tip growth. Microbiol Rev 57:367–382

Jahn R, Sudhof TC (1999) Membrane fusion and exocytosis. Annu Rev Biochem 68:863–911

Jahraus A, Storrie B, Griffiths G, Desjardins M (1994) Evidence for retrograde traffic between terminal lysosomes and the prelysosomal/late endosomal compartment. J Cell Sci 107:145–157

Jones EW, Webb GC, Hiller MA (1997) Biogenesis and function of the yeast vacuole. In: Pringle JR, Broach JR, Jones EW (eds) The molecular and cellular biology of the yeast Saccharomyces, vol 3. Cell cycle and cell biology. Cold Spring Harbor Laboratory Press, Cold Spring Harbor, pp 363–470

Jones HD, Schliwa M, Drubin DG (1993) Video microscopy of organelle inheritance and motility in budding yeast. Cell Motil Cytoskeleton 25:129–142

Kametaka S, Okano T, Ohsumi M, Ohsumi Y (1998) Apg14p and Apg6/Vps30p form a protein complex essential for autophagy in the yeast, *Saccharomyces cerevisiae*. J Biol Chem 273:22284–22291

Kanamori K, Legerton TL, Weiss RL, Roberts JD (1982) Nitrogen-15 spin-lattice relaxation times of amino acids in *Neurospora crassa* as a probe of intracellular environment. Biochemistry 21:4916–4920

Kane PM, Yamashiro CT, Stevens TH (1989) Biochemical characterisation of the yeast vacuolar H^+-ATPase. J Biol Chem 264: 19236–19244

Keenan KA, Utzat CD, Zielinski TK (1998) Isolation and characterization of strains defective in vacuolar ornithine permease in *Neurospora crassa*. Fungal Genet Biol 23:237–247

Kitamoto K, Yoshizawa K, Ohsumi Y, Anraku Y (1988a) Dynamic aspects of vacuolar and cytosolic amino acid pools of *Saccharomyces cerevisiae*. J Bacteriol 170:2683–2686

Kitamoto K, Yoshizawa K, Ohsumi Y, Anraku Y (1988b) Mutants of *Saccharomyces cerevisiae* with defective vacuolar function. J Bacteriol 170:2687–2691

Klausner RD, Donaldson JG, Lippincott-Schwartz J (1992) Brefeldin A: insights into the control of membrane traffic and organelle structure. J Cell Biol 116: 1071–1080

Klionsky DJ (1998) Nonclassical protein sorting to the yeast vacuole. J Biol Chem 273:10807–10810

Klionsky DJ, Herman PK, Emr SD (1990) The fungal vacuole: composition, function, and biogenesis. Microbiol Rev 54:266–292

Knapp PE, Swanson JA (1990) Plasticity of the tubular lysosome compartment in macrophages. J Cell Sci 95:433–439

Lang T, Schaeffeler E, Bernreuther D, Bredschneider M, Wolf DH, Thumm M (1998) Aut2p and Aut7p, two novel microtubule-associated proteins, are essential for delivery of autophagic vesicles to the vacuole. EMBO J 17:3579–3607

Lazof D, Laüchli A (1991) Complementary analysis of freeze-dried and frozen-hydrated plant tissue by electron-probe X-ray microanalysis: spectral resolution and analysis of calcium. Planta 184:327–333

Lazzaro MD, Thomson WW (1996) The vacuolar-tubular continuum in living trichomes of chickpea (*Cicer arietinum*) provides a rapid means of solute delivery from base to tip. Protoplasma 193:181–190

Lee RB, Ratcliffe RG (1983) Phosphorus nutrition and the intracellular distribution of inorganic phosphate in pea root tips: a quantitative study using ^{31}P NMR. J Exp Bot 34:1222–1244

Legerton TL, Kanamori K, Weiss RL, Roberts JD (1983) Measurements of cytoplasmic and vacuolar pH in *Neurospora* using nitrogen-15 nuclear magnetic resonance spectroscopy. Biochemistry 22:899–903

Leigh RA (1997) Solute composition of vacuoles. In: Leigh RA, Sanders D (eds) The plant vacuole. Adv Bot Res, vol 25. Academic Press, New York, pp 171–194

Lewis MJ, Nichols BJ, Prescianotto-Baschong C, Riezman H, Pelham HRB (2000) Specific retrieval of the exocytic SNARE Snc1p from yeast early endosomes. Mol Biol Cell 11:23–38

Ligrone R, Duckett JG (1994) Cytoplasmic polarity and endoplasmic microtubules associated with the nucleus and organelles are ubiquitous features of food-conducting cells in bryoid mosses (Bryophyta). New Phytol 127:601–614

Ligrone R, Duckett JG (1998) The leafy stems of *Sphagnum* (Bryophyta) contain highly differentiated polarized cells with axial arrays of endoplasmic microtubules. New Phytol 140:567–579

Linner JG, Bennett SC, Harrison DS, Steiner AL (1986a) Cryopreparation of tissue for electron microscopy. In: Muller M, Becker RP, Boyde A, Wolosewicke JJ (eds) Science of biological specimen preparation. SEM Inc, AMF O'Hare, Chicago, pp 165–174

Linner JG, Livesey SA, Harrison DS, Steiner AL (1986b) A new technique for removal of amorphous phase tissue water without ice crystal damage: a preparative method for ultrastructure analysis and immunoelectron microscopy. J Histochem Cytochem 34:1123–1135

Marshall AT, Xu W (1998) Quantitative elemental X-ray imaging of frozen-hydrated biological samples. J Microsc 190:305–316

Martin F (1991) Nuclear magnetic resonance studies in ectomycorrhizal fungi. Methods Microbiol 23:121–148

Martin F, Rubini P, Côté R, Kottke I (1994) Aluminium polyphosphate complexes in the mycorrhizal basidiomycete Laccaria bicolor: a ^{27}Al-nuclear magnetic resonance study. Planta 194:241–246

Martinoia E, Heck U, Boller TH, Wiemken A, Matile PH (1979) Some properties of vacuoles isolated from *Neurospora crassa* slime variant. Arch Microbiol 120: 31–34

Martinoia E, Grill E, Tommasini R, Kreuz K, Amrhein N (1993) An ATP-dependent glutathione S-conjugate 'export' pump in the vacuolar membrane of plants. Nature 364:247–249

Mayer A (1999) Intracellular membrane fusion: SNARES only? Curr Opin Cell Biol 11:447–452

McCully ME, Canny MJ (1985) The stabilization of labile configurations of plant cytoplasm by freeze-substitution. J Microsc 139:27–33

Mendgen K, Bachem U, Stark-Urnau M, Xu H (1995) Secretion and endocytosis at the interface of plants and fungi. Can J Bot 73:S640–648

Mersey B, McCully ME (1978) Monitoring the course of fixation of plant cells. J Microsc 114:49–76

Miller AJ (1994) Ion-selective microelectrodes. In: Harris NJ, Oparka K (eds) Plant cell biology – a practical approach. IRL Press, Oxford, pp 283–296

Monaghan P, Perusinghe N, Müller M (1998) High-pressure freezing for immunocytochemistry. J Microsc 192:248–258

Nakamura N, Hirata A, Ohsumi Y, Wada Y (1997) Vam2/Vps41p and Vam6/Vps39p are components of a protein complex on the vacuolar membranes and involved in the vacuolar assembly in the yeast *Saccharomyces cerevisiae*. J Biol Chem 272:11344–11349

Nelson N, Klionsky DJ (1996) Vacuolar H^+-ATPase: from mammals to yeast and back. Experientia 52:1101–1110

Nelson H, Nelson N (1990) Disruption of genes encoding subunits of yeast vacuolar H^+-ATPase causes conditional lethality. Proc Natl Acad Sci USA 87:3503–3507

Nichols BJ, Ungermann C, Pelham HR, Wickner WT, Haas A (1997) Homotypic fusion mediated by t- and v-SNAREs. Nature 387:199–202

Odorizzi G, Cowles CR, Emr SD (1998) The AP-3 complex: a coat of many colours. Trends Cell Biol 8:282–288

Oparka KJ (1991) Uptake and compartmentation of fluorescent probes by plant cells. J Exp Bot 42:565–579

Oparka KJ, Hawes CJ (1992) Vacuolar sequestration of fluorescent probes in plant cells. J Microsc 166:15–27

Oparka KJ, Cole L, Wright KM, Hawes CJ, Coleman JOD (1991) Fluid-phase endocytosis and the subcellular distribution of fluorescent probes in plant cells. In: Hawes CR, Coleman JOD, Evans DE (eds) Endocytosis, exocytosis and vesicle traffic in plants. Cambridge Univ Press, Cambridge, pp 81–102

Orci L, Palmer DJ, Ravazzola M, Perrelet A, Amherdt M, Rothman JE (1993) Budding from Golgi membranes requires the coatomer complex of non-clathrin coat proteins. Nature 362:648–652

Orlovich DA, Ashford AE (1993) Polyphosphate granules are an artefact of specimen preparation in the ectomycorrhizal fungus *Pisolithus tinctorius*. Protoplasma 173:91–105

Orlovich DA, Ashford AE (1995) X-ray microanalysis of ion distribution in frozen salt/dextran droplets after freeze-substitution and embedding in anhydrous conditions. J Microsc 180:117–126

Palevitz BA, O'Kane DJ (1981) Epifluorescence and video analysis of vacuole motility and development in stomatal cells of *Allium*. Science 214:443–445

Palevitz BA, O'Kane DJ, Kobres RE, Raikhel NV (1981) The vacuole system in stomatal cells of *Allium*. Vacuole movements and changes in morphology in differentiating cells as revealed by epifluorescence, video and electron microscopy. Protoplasma 109: 23–55

Pan X, Goldfarb DS (1998) *YEB3/VAC8* encodes a myristylated armadillo protein of the *Saccharomyces cerevisiae* vacuolar membrane that functions in vacuole fusion and inheritance. J Cell Sci 111:2137–2147

Parton RG, Schrotz P, Bucci C, Gruenberg J (1992) Plasticity of early endosomes. J Cell Sci 103:335–348

Parton RM, Read ND (1998) Calcium and pH imaging in living cells. In: Lacey AJ (ed) Light microscopy in biology – a practical approach. IRL Press, Oxford

Pelham HRB (1999a) SNAREs and the secretory pathway – lessons from yeast. Exp Cell Res 247:1–8

Pelham HRB (1999b) The Croonian lecture 1999. Intracellular membrane traffic: getting proteins sorted. Philos Trans R Soc Lond B Biol Sci 354:1471–1478

Pfeffer SR (1996) Transport vesicle docking: SNAREs and associates. Annu Rev Cell Biol 12:441–461

Prescianotto-Baschong C, Riezman H (1998) Morphology of the yeast endocytic pathway. Mol Biol Cell 9:173–189

Preston RA, Murphy RF, Jones EW (1989) Assay of vacuolar pH in yeast and identification of acidification-defective mutants. Proc Natl Acad Sci USA 86: 7027–7032

Price A, Wickner W, Ungermann C (2000a) Proteins needed for vesicle budding from the Golgi complex are also required for the docking step of homotypic vacuole fusion. J Cell Biol 148:1223–1229

Price A, Seals D, Wickner W, Ungermann C (2000b) The docking stage of yeast vacuole fusion requires the transfer of proteins from a *cis*-SNARE complex to a Rab/Ypt protein. J Cell Biol 148:1231–1238

Pryer NK, Wuestehube LJ, Schekman R (1992) Vesicle-mediated sorting. Annu Rev Biochem 61:471–516

Ratcliffe RG (1994) In vivo NMR studies of higher plants and algae. Adv Bot Res 20:43–123

Raths S, Rohrer J, Crausaz F, Riezman H (1993) *end*3 and *end*4: two mutants defective in receptor-mediated and fluid-phase endocytosis in *Saccharomyces cerevisiae*. J Cell Biol 120:55–65

Raymond CK, O'Hara P, Eichinger G, Rothman JH, Stevens TH (1990) Molecular analysis of the yeast *VPS3* gene and the role of its product in vacuolar protein sorting and vacuolar segregation during the cell cycle. J Cell Biol 111:877–892

Raymond CK, Howald-Stevenson I, Vater C, Stevens TH (1992a) Morphological classification of the yeast vacuolar protein sorting mutants: evidence for a prevacuolar compartment in class E *vps* mutants. Mol Biol Cell 3:1389–1402

Raymond CK, Roberts CJ, Moore KE, Howald I, Stevens TH (1992b) Biogenesis of the vacuole in *Saccharomyces cerevisiae*. Int Rev Cytol 139:59–120

Rea PA (1999) MRP subfamily ABC transporters from plants and yeast. J Exp Bot 50:895–913

Read ND, Allan WTG, Knight H, Knight MR, Malhó R, Russel A, Shacklock PS, Trewavas AJ (1992) Imaging and measurement of cytosolic free calcium in plant and fungal cells. J Microsc 166:57–86

Rees B, Shepherd VA, Ashford AE (1994) Presence of a motile tubular vacuole system in different phyla of fungi. Mycol Res 98:985–992

Rieder SE, Banta LM, Köhrer K, McCaffery JM, Emr SD (1996) Multilamellar endosome-like compartment accumulates in yeast *vps28* vacuolar protein sorting mutant. Mol Biol Cell 7:985–999

Riezman H (1993) Yeast endocytosis. Trends Cell Biol 3:273–277

Robinson JM, Karnovsky MJ (1991) Rapid-freezing cytochemistry: preservation of tubular lysosomes and enzyme activity. J Histochem Cytochem 39:787–792

Robinson JM, Okada T, Castellot JJ Jr, Karnovsky MJ (1986) Unusual lysosomes in aortic smooth muscle cells: presence in living and rapidly frozen cells. J Cell Biol 102:1615–1622

Rogers JC (1998) Compartmentation of plant cell proteins in separate lytic and protein storage vacuoles. J Plant Physiol 152:653–658

Rost FWD, Shepherd VA, Ashford AE (1995) Estimation of vacuolar pH in actively growing hyphae of the fungus *Pisolithus tinctorius*. Mycol Res 99:549–553

Roth AM, Sullivan DM, Davis NG (1998) A large PEST-like sequence directs the ubiquitination, endocytosis and vacuolar degradation of the yeast a-factor receptor. J Cell Biol 142:949–961

Rothman JE (1994) Mechanisms of intracellular protein transport. Nature 372:55–63

Sanderfoot AA, Raikhel NV (1999) The specificity of vesicle trafficking: coat proteins and SNAREs. Plant Cell 11:629–641

Sanderfoot AA, Ahmed SU, Marty-Mazars D, Rapoport I, Kirchhausen T, Marty F, Raikhel NV (1998) A putative vacuolar cargo receptor partially colocalizes with AtPEP12 on a prevacuolar compartment in *Arabidopsis* roots. Proc Natl Acad Sci USA 95:9920–9925

Sanderfoot AA, Kovaleva V, Zheng H, Rapoport I, Raikhel NV (1999) The t-SNARE AtVAM3p resides on the prevacuolar compartment in Arabidopsis root cells. Plant Physiol 121:929–938

Sato K, Wickner W (1998) Functional reconstitution of Ypt7p GTPase and a purified vacuole SNARE complex. Science 281:700–702

Sato MH, Nakamura N, Ohsumi Y, Kouchi H, Kondo M, Hara-Nishimura I, Nishimura M, Wada Y (1997) The *AtVAM3* encodes a syntaxin-related molecule implicated in the vacuolar assembly in *Arabidopsis thaliana*. J Biol Chem 272:24530–24535

Shepherd VA, Orlovich DA, Ashford AE (1993a) A dynamic continuum of pleiomorphic tubules and vacuoles in growing hyphae of a fungus. J Cell Sci 104:495–507

Shepherd VA, Orlovich DA, Ashford AE (1993b) Cell-to-cell transport via motile tubules in growing hyphae of a fungus. J Cell Sci 105:1173–1178

Singer-Krüger B, Rainer F, Crausaz F, Riezman H (1993) Partial purification and characterization of early and late endosomes from yeast. J Biol Chem 268:14376–14386

Skepper J (2000) Immunocytochemical strategies for electron microscopy: choice or compromise. J Microsc 199:1–36

Sonee M, Barrón E, Yarber FA, Hamm-Alvarez SF (1998) Taxol inhibits endosomal-lysosomal membrane trafficking at two distinct steps in CV-1 cells. Am J Physiol Cell Physiol 275:C1630–C1639

Spellig T, Bottin A, Kahmann R (1996) Green fluorescent protein (GFP) as a vital new marker in the phytopathogenic fungus Ustilago maydis. Mol Gen Genet 252:503–509

Stack JH, Horazdovsky BF, Emr SD (1995) Receptor-mediated protein sorting to the vacuole in yeast: roles for a protein kinase, a lipid kinase and GTP binding proteins. Annu Rev Cell Dev Biol 11:1–33

Steinberg G, Schliwa M (1993) Organelle movements in the wild type and wall-less fz;sg;os-1 mutants of Neurospora crassa are mediated by cytoplasmic microtubules. J Cell Sci 106:555–564

Steinberg G, Schliwa M, Lehmler C, Bölker M, Kahmann R, McIntosh JR (1998) Kinesin from the plant pathogenic fungus Ustilago Maydis is involved in vacuole formation and cytoplasmic migration. J Cell Sci 111:2235–2246

Steinberg TM, Newman AS, Swanson JA, Silverstein SC (1987) Macrophages possess probenecid-inhibitable organic anion transporters that remove fluorescent dyes from the cytoplasmic matrix. J Cell Biol 105:2695–2702

Steinberg TM, Swanson JA, Silverstein SC (1988) A prelysosomal compartment sequesters membrane-impermeant fluorescent dyes from the cytoplasmic matrix of J774 macrophages. J Cell Biol 107:887–896

Strayle J, Pozzan T, Rudolph HK (1999) Steady-state free Ca^{2+} in the yeast endoplasmic reticulum reaches only $10\,\mu M$ and is mainly controlled by the secretory pathway pump Pmr1. EMBO J 18:4733–4743

Südhof TC (1995) The synaptic vesicle cycle: a cascade of protein-protein interactions. Nature 375:645–653

Swanson J (1989) Fluorescent labelling of endocytic compartments. Methods Cell Biol 29:137–151

Swanson J, Burke E, Silverstein SC (1987a) Tubular lysosomes accompany stimulated pinocytosis in macrophages. J Cell Biol 104:1217–1222

Swanson J, Bushnell A, Silverstein SC (1987b) Tubular lysosomes morphology and distribution within macrophages depend on the integrity of cytoplasmic microtubules. Proc Natl Acad Sci USA 84:1921–1925

Swanson J, Locke A, Ansel P, Hollenbeck PJ (1992) Radial movement of lysosomes along microtubules in permeabilized macrophages. J Cell Sci 103:201–209

Tagu D, Martin F (1996) Molecular analysis of cell wall proteins expressed during the early steps of ectomycorrhizal development. New Phytol 133:73–85

Takei K, McPherson PS, Schmid SL, De Camilli P (1995) Tubular membrane invaginations coated by dynamin rings are induced by GTP-γS in nerve terminals. Nature 374:186–192

Takei K, Haucke V, Slepnev V, Farsad K, Salazar M, Chen H, De Camilli P (1998) Generation of coated intermediates of clathrin-mediated endocytosis on protein-free liposomes. Cell 94:131–141

Terasaki M, Reese TS (1992) Characterization of endoplasmic reticulum by co-localization of BiP and dicarbocyanine dyes. J Cell Sci 101:315–322

Thompson-Coffe C, Zickler D (1992) Three microtubule-organizing centers are required for ascus growth and sporulation in the fungus Sordaria macrospora. Cell Motil Cytoskeleton 22:257–273

Thilo L, Stroud E, Haylett T (1995) Maturation of early endosomes and vesicular traffic to lysosomes in relation to membrane recycling. J Cell Sci 108:1791–1803

Tiwari SC, Polito VS (1988) Organisation of the cytoskeleton in pollen tubes of Pyrus communis: a study employing conventional and freeze-substitution electron microscopy, immunofluorescence, and rhodamine-phalloidin. Protoplasma 147:100–112

Tooze J, Hollinshead M (1991) Tubular early endosomal networks in AtT20 and other cells. J Cell Biol 115:635–653

Trey K, Sato TK, Darsow T, Emr SD (1998) Vam7p, a SNAP-25-like molecule, and Vam3p, a syntaxin homolog, function together in yeast vacuolar protein trafficking. Mol Cell Biol 18:5308–5319

Tsien RY (1989) Fluorescent indicators of ion concentrations. In: Taylor DL, Wang YL (eds) Fluorescence microscopy of living cells in culture, part B. Quantitative fluorescence microscopy – imaging and spectroscopy. Methods Cell Biol 30X:127–156

Ungermann C, Wickner W (1998) Vam7p, a vacuolar SNAP-25 homolog, is required for SNARE complex integrity and vacuole docking and fusion. EMBO J 17:3269–3276

Ungermann C, Nichols BJ, Pelham HRB, Wickner W (1998) A vacuolar v-t-SNARE complex, the predominant form in vivo and on isolated vacuoles, is disassembled and activated for docking and fusion. J Cell Biol 140:61–69

Ungermann C, Fischer von Mollard G, Jensen ON, Margolis N, Stevens TH, Wickner W (1999) Three v-SNAREs and two t-SNARES, present in a pentameric cis-SNARE complex on isolated vacuoles, are essential for homotypic fusion. J Cell Biol 145:1453–1442

van Deurs B, Holm PK, Kayser L, Sandvig K, Hansen SH (1993) Multivesicular bodies in HEp-2 cells are maturing endosomes. Eur J Cell Biol 61:208–224

van Deurs B, Holm PK, Kayser L, Sandvig K (1995) Delivery to lysosomes in the human carcinoma cell line HEp-2 involves an actin filament-facilitated fusion between mature endosomes and preexisting lysosomes. Eur J Cell Biol 66:309–323

van Deurs B, von Bulow F, Vilhardt F, Holm PK, Sandvig K (1996) Destabilization of plasma membrane structure by prevention of actin polymerization – microtubule-dependent tubulation of the plasma membrane. J Cell Sci 109:1655–1665

Vaughn LE, Davis RH (1981) Purification of vacuoles from Neurospora crassa. Mol Cell Biol 1:797–806

Verbelen JP, Tao W (1998) Mobile arrays of vacuole ripples are common in plant cells. Plant Cell Rep 17:917–920

Vesk P, Ashford AE, Markovina A-L, Allaway WG (2000) Apoplasmic barriers and their significance in the exodermis and sheath of *Eucalyptus pilularis/ Pisolithus tinctorius* ectomycorrhizas. New Phytol 145: 333–346

Vida TA, Emr SD (1995) A new vital stain for visualizing vacuolar membrane dynamics and endocytosis in yeast. J Cell Biol 128:779–792

Vida TA, Huyer G, Emr SD (1993) Yeast vacuolar proenzymes are sorted in the late Golgi complex and transported to the vacuole via a prevacuolar endosome-like compartment. J Cell Biol 121:1245–1256

von Mollard GF, Nothwehr SF, Stevens TH (1997) The yeast v-SNARE Vti1p mediates two vesicle transport pathways through interactions with the t-SNAREs Sed5p and Pep12p. J Cell Biol 137:1511–1524

Wada Y, Ohsumi Y, Anraku Y (1992) Genes for directing vacuolar morphogenesis in *Saccharomyces cerevisiae*. I. Isolation and characterization of two classes of *vam* mutants. J Biol Chem 267:18665–18670

Wada Y, Nakamura N, Ohsumi Y, Hirata A (1997) Vam3p, a new member of syntaxin related protein, is required for vacuolar assembly in the yeast *Saccharomyces cerevisiae*. J Cell Sci 110:1299–1306

Waterman-Storer CM, Gregory J, Parsons SF, Salmon ED (1995) Membrane/microtubule tip attachment complexes (TACs) allow the assembly dynamics of plus ends to push and pull membranes into tubulovesicular networks in interphase *Xenopus* egg extracts. J Cell Biol 130:1161–1169

Waters MG, Pfeffer SR (1999) Membrane tethering in intracellular transport. Curr Opin Cell Biol 11:440–446

Weidman P, Roth R, Heuser J (1993) Golgi membrane dynamics imaged by freeze-etch electron microscopy – views of different membrane coatings involved in tubulation versus vesiculation. Cell 75:123–133

Weisman LS, Wickner W (1988) Intervacuole exchange in the yeast zygote: a new pathway in organelle communication. Science 241:589–591

Weisman LS, Bacallao R, Wickner W (1987) Multiple methods of visualizing the yeast vacuole permit evaluation of its morphology and inheritance during the cell cycle. J Cell Biol 105:1539–1547

Weisman LS, Emr SD, Wickner WT (1990) Mutants of *Saccharomyces cerevisiae* that block intervacuole vesicular traffic and vacuole division and segregation. Proc Natl Acad Sci USA 87:1076–1080

Wiederkehr A, Avaro S, Prescianotto-Baschong C, Haguenauer-Tsapis R, Riezman H (2000) The F-box protein Rcy1p is involved in endocytic membrane traffic and recycling out of an early endosome in *Saccharomyces cerevisiae*. J Cell Biol 149:397–410

Wilson TP, Canny MJ, McCully ME, Lefkovitch LP (1990) Breakdown of cytoplasmic vacuoles: a model of endomembrane rearrangement. Protoplasma 155:144–152

Wink M (1993) The plant vacuole: a multifunctional compartment. J Exp Bot 44:231–246

Wurmser AE, Emr SE (1998) Phosphoinositide signaling and turnover: PtdIns(3)P, a regulator of membrane traffic, is transported to the vacuole and degraded by a process that requires lumenal vacuolar hydrolase activities. EMBO J 17:4930–4942

Xu Z, Wickner W (1996) Thioredoxin is required for vacuole inheritance in *Saccharomyces cerevisiae*. J Cell Biol 132:787–794

Xu Z, Mayer A, Muller E, Wickner W (1997) A heterodimer of thioredoxin and I^B_2 cooperates with Sec18p (NSF) to promote yeast vacuole inheritance. J Cell Biol 136:299–306

Xu Z, Sato K, Wickner W (1998) LMA1 binds to vacuoles at Sec18p (NSF), transfers upon ATP hydrolysis to a t-SNARE (Vam3p) complex, and is released during fusion. Cell 93:1125–1134

Yamashiro CT, Kane PM, Wolczyk DF, Preston RA, Stevens TH (1990) Role of vacuolar acidification in protein sorting and zymogen activation: a genetic analysis of the yeast vacuolar proton-translocating ATPase. Mol Cell Biol 7:3737–3749

Yamashita RA, May GS (1998) Motoring along the hyphae – molecular motors and the fungal cytoskeleton. Curr Opin Cell Biol 10:74–79

Yin Z-H, Neimanis S, Wagner U, Heber U (1990) Light-dependent pH changes in leaves of C_3 plants. I. Recording pH changes in various cellular compartments by fluorescent probes. Planta 182:244–252

Yoshida S (1995) Low-temperature-induced alkalization of vacuoles in suspension-cultured cells of mung bean (*Vigna radiata* [L.] Wilczek). Plant Cell Physiol 36:1075–1079

Young N, Bullock S, Orlovich DA, Ashford AE (1993) Association of polyphosphate with protein in freeze-substituted sclerotia of *Sclerotinia minor*. Protoplasma 174:134–141

Zerez CR, Weiss RL, Franklin C, Bowman BJ (1986) The properties of arginine transport in vacuolar membrane vesicles of *Neurospora crassa*. J Biol Chem 261:8877–8882

Zheng B, Wu JN, Schober W, Lewis DE, Vida T (1998) Isolation of yeast mutants defective for localization of vacuolar vital dyes. Proc Natl Acad Sci USA 95: 11721–11726

Zheng H, von Mollard GF, Kovaleva V, Stevens TH, Raikhel NV (1999) The plant vesicle associated SNARE AtVTI1a likely mediates vesicle transport from the TGN to the prevacuolar compartment. Mol Biol Cell 10:2251–2264

13 Genomics for Fungi

J.W. Bennett[1] and Jonathan Arnold[2]

CONTENTS

I. Introduction

Since 1995, the genomes of dozens of bacteria and several model eukaryotes have been completely sequenced and these data have been deposited in public databases. With the exception of the yeast *Saccharomyces cerevisiae*, fungal genomes were not among these early genomics accomplishments. Nevertheless, despite a slow start, fungal genomics is now poised to reveal a comprehensive picture of fungal gene structure and function, to resolve questions about evolutionary relationships among mycelial fungi, and to decipher the biochemical and cellular relationships between fungi and other eukaryotic life forms. Genomics databases and their accompanying bioinformatics tools are among the most powerful scientific resources available to contemporary fungal cell biologists.

Molecular biology first extended the reach and power of genetic analysis to species that were once considered genetically intractable. In the old genetics, genes were recognized only when a mutant allele was available. With genetic engineering approaches (sometimes called "reverse genetics") we can find genes without mutants. Genomics has greatly accelerated the shift from mutation-based gene discovery to sequence-based gene discovery. Indeed, a major justification for completely sequencing a genome is that it can identify all the genes in an organism. Further, genomics has turned biology into an information science. Advances in genomics are inseparable from advances in computer technology and the expansion of the Internet. Individual scientists have unprecedented access to immense resources of data distributed across international networks with online data retrieval and literature searching via the World Wide Web (Schatz 1997).

Currently, many fungal genome sequencing projects are driven by anticipated medical benefits projected from the study of human pathogens. Pharmaceutical companies hope to define new targets for drug discovery, identify immunogenic targets for vaccine development, as well as to isolate novel natural products from difficult-to-cultivate species. Other economic justifications for fungal genomics come from agriculture and indus-

---

[1] Department of Cell and Molecular Biology, Tulane University, New Orleans, Louisiana 70118, USA
[2] Department of Genetics, University of Georgia, Athens, Georgia 30602, USA

The Mycota VIII
Biology of the Fungal Cell
Howard/Gow (Eds.)
© Springer-Verlag Berlin Heidelberg 2001

try, with the promise of new approaches to plant pathogen control and improved yields for industrial fermentation. However, perhaps the greatest promise of fungal genomics lies in fundamental biology. Classical genetic workhorses such as *Aspergillus nidulans* and *Neurospora crassa* can be re-harnessed to extract the meaning of many of the recently discovered genes that are of unknown function.

The ability to analyze entire genomes accelerates the pace of gene discovery and alters the kinds of questions that can be addressed in model organisms. Considerable new jargon and an inundation of acronyms have been added to the biological lexicon. We now have "*in silico* experiments"; "point and click biology"; "BAC libraries" ; "BLAST searches" ; and so forth. See Tables 1 and 2 for a listing of some of the more commonly encountered genomics acronyms and terminology.

Even more than in most fields of contemporary science, a review of genomics is condemned to almost instant obsolescence. The best we can do is present a "snap shot" in time. Our goals are to give a brief introduction to the general history and practice of genomics research; to review the properties of fungi that make them desirable targets for

Table 1. Primer of acronyms regularly encountered in genomics research

BAC	Bacterial artificial chromosome
EMBL	European Molecular Biology Laboratory (now EBI)
EBI	European Bioinformatics Institute
BLAST	Basic local alignment search tool
EST	Expressed sequence tags
EUROFAN	European Functional Analysis Network (for *S. cerevisiae*)
FGDB	Fungal Genome Data Base
GRAIL	Gene Recognition and Analysis Link
ORF	Open reading frame
PAC	P1 (phage) artificial chromosome
PCR	Polymerase chain reaction
HGP	Human Genome Project
MIPS	Martinsried Institute for Protein Sequences (now Munich Information Center for Protein Sequences)
NCBI	National Center for Biotechnology Information (GenBank)
RFLP	Restriction fragment length polymorphism
STS	Sequence tagged sites
TIGR	The Institute for Genome Research (Rockville, Maryland)
UTR	Untranslated regions
XML	eXtensible markup language
YAC	Yeast (*S. cerevisiae*) artificial chromosomes
YPD	Yeast (*S. cerevisiae*) proteome data base

genomics research; to describe the status of several fungal genome projects at the start of the year 2001; to point out some growing problems with traditional conventions for gene nomenclature; and to comment on the probable impact of fungal genomics research on future developments.

II. History

A. Definitions

Genomics has its antecedents in classical genetics. The word "genome" was originally used by cytogeneticists to refer to a single, complete set of chromosomes. A haploid (1n) organism has one genome, a diploid (2n) has two, a triploid (3n) has three, and so on, with higher levels of ploidy such as tetraploid (4n) and hexaploid (6n) (Sybenga 1972). The formation of polyploid series is common in the evolution of higher plants (Stebbins 1966). The hybridization of two closely related species followed by chromosome doubling results in the formation of allotetraploids (also called amphidiploids) in which the hybrid contains two complete diploid genomes. Both cotton and tobacco are allotetraploids, while modern cultivated wheat is of allohexaploid origin (Garber 1972; Sybenga 1972). For the classical cytogeneticist, genome analysis consisted of the identification of the diploid ancestors that contributed their chromosomes to the formation of the extant allopolyploid. Cytological examination of meiotic figures was used to determine the degree of homology and nonhomology between chromosome sets. Using the parasexual cycle, fungal geneticists performed analysis of certain mold genomes through the formation of mitotic diploids and their regression to haploidy through various aneuploid states (Pontecorvo 1956).

The modern sense of a genome is different. The new era of **genome science** or **genomics** is tied to the advent of large-scale DNA sequencing. The emphasis has shifted from the way chromosomes look under the microscope to the actual DNA sequence within chromosomes. DNA sequences are determined by automated sequencing protocols. The term genomics was coined in 1986 by Thomas Roderick to describe the mapping, sequencing and analysis of genomes, as well as to provide a name for the new journal *Genomics* (Hieter and Boguski 1997). A genome is now defined as "the totality of a cell's genetic infor-

Table 2. Primer of genomics jargon

BAC (bacterial artificial chromosome)	Vector used to clone DNA fragments in *E. coli* with inserts ranging from approx. 100–300 kb
GenBank	Public database operated by the National Center for Biotechnology Information (National Institutes of Health)
Contig	Group of cloned DNAs representing overlapping segments of a particular chromosome region and providing unbroken coverage of that region; the continuous DNA sequence generated from these DNA clones. A contig contains no gaps
Contig map	Map depicting the relative order of a linked library of overlapping clones
Depth of coverage	Number of times a particular DNA is sequenced (1× means that on average a bp has been sampled once; 8× means on average a particular base has been sequenced eight times
Functional genomics	Determining the function of genes through the use of microarrays and other methods that can study the function of many genes simultaneously
Genomic library	Collection of clones containing the entire genome of an organism cut up into many pieces, e.g., a BAC library
Homologous	In evolutionary biology, refers to genes that descend from a common ancestral gene; in genomics homology is used to describe DNA that has the same or nearly the same nucleotide sequence
ORF (open reading frame)	Series of triplets coding for amino acids without any stop codons; these sequences have the potential to be translated into polypeptides
Orthologous	Homologous sequences that descend from a single ancestral gene
Paralogous	Homologous sequences that arise through gene duplication
Proteome	Complete set of proteins that a living cell can synthesize
Proteomics	Identification and characterization of each protein, its structure and its interactions with other proteins
Structural genomics	Mapping and sequencing stages of genome analysis; also used to describe projects that aim to solve the structure of all possible proteins (Skolnick et al. 2000)
Synteny	"On the same thread"; the presence of sets of genes showing the same order in different species, often used as short hand for saying that a group of genes shows conservation of linkage

mation, including both genes and other DNA sequences" (Berg and Singer 1992). However, the determination of the DNA sequence of an organism is just the beginning of genomics research. A complete DNA sequence can be annotated to find all of the predicted genes in an organism. The resultant gene catalogue can be used to compare any given genome with all other sequenced DNA that exists in data banks. The ultimate goal is to understand the structure, function, interrelationships, and evolution of genes within each genome. Although many genomics projects target the sequencing of entire genomes, less ambitious (and more affordable) projects that target portions of genomes can also be of great utility. For example, EST-based projects for *Aspergillus nidulans* (Kupfer et al. 1997) and *Neurospora crassa* (Nelson et al. 1997) revealed that nearly half of the predicted genes had no homologies in known databases.

B. Human Genome Project

Genomics has become such an integral part of modern biology that it is easy to forget how young a science it is. Moreover, the successful sequencing of a large number of bacterial and eukaryotic model genomes tends to overshadow the fact that the early vision of genomic science focused on the human genome.

The Human Genome Project (HGP) was first proposed in the USA in the mid-1980s. The concept immediately attracted considerable debate and dissent from within the scientific community. Many biologists feared that this kind of "mindless big biology" would have an adverse impact on hypothesis-driven research. In response to community concerns, the National Research Council organized a committee and commissioned a report (Alberts et al. 1988). The committee not only supported the concept of a human genome project, it also clarified the way in which such a project should be structured and went on to endorse a number of parallel projects. For example, the report included recommendations to map first and sequence later. Other important commitments were the development of improved vector systems for cloning large DNA fragments; cheaper, automated technologies of DNA sequencing; as well as the development of better bioinformatics tools for the storage, annotation, and analysis of data. A further recommendation was the inclusion of a set of model organisms in both the mapping and sequencing efforts. The models originally chosen included the fly, *Drosophila melanogaster*; the nematode, *Caenorhabditis elegans*; the mouse, *Mus musculus*; and the bacterium, *Escherichia coli*. The mustard plant, *Arabidopsis thaliana*, and the yeast, *Saccharomyces cerevisiae*, were added later. Additional components of the HGP focused on ethical and social issues; on education and training; and the variations in human DNA that underlie disease susceptibility (Collins and Gallas 1993).

The American HGP was formally initiated on 1 October 1990, and continues to be a "work in progress." The HGP has regularly re-organized and re-shaped itself as research has progressed. Notably, the project has become increasingly international, with major collaborations provided by Japanese partners and the Sanger Centre in the United Kingdom. Simultaneously, the goals have become more ambitious. Progress in genomics has regularly outstripped predictions, often through competition from unexpected quarters. For example, the first rough map of the human genome was produced not by the American HGP but by the French group Genethon (Evry, France) with funding from the French Muscular Dystrophy Association (Hodgson 1994). Even more unexpected was the 1995 announcement that The Institute for Genome Research (TIGR), a nonprofit organization in Rockville, Maryland, founded by Craig Venter, had completed the first genome sequence of a cellular life form (see Sect. II.C). More recently, the 5-year plan for the HGP for 1998–2003 included stated goals to complete the *Drosophila* genome by 2002 (Collins et al. 1998). In actuality, the *Drosophila* genome was nearly completed in 2000 (Adams et al. 2000), and the human genome itself was announced in February, 2001 (International Human Genome Sequencing Consortium 2001; Venter et al. 2001).

C. Bacteria and Archaea

Bacterial and archaeal genomes are small in size (mostly in the range of 0.5–7 Mb), usually consisting of a single chromosome, and have a high ratio of coding to noncoding regions. The modest size of these genomes, and the relative absence of repetitive DNA, makes practicable a whole genome random shotgun approach to sequencing.

Total DNA from an organism is isolated and then broken into smaller fragments. In the random sequencing or shotgun phase of the project, the individual fragments are sequenced and then assembled together by looking for overlapping regions. Computer assembly is followed by a closure phase during which errors are resolved and gaps between overlapping regions are closed by directed sequencing of the gaps. Using this non-hierarchical approach, The Institute of Genomics Research successfully sequenced the first two bacterial genomes: *Haemophilus influenzae* (Fleischmann et al. 1995) and *Mycoplasma genitalium* (Fraser et al. 1995). At the time, many people thought that the 1.8-Mb *H. influenzae* genome was too large to be sequenced by the whole genome shotgun strategy. The indisputable success of this tactic, however, generated immediate excitement and imitation across the biological community. The following year, two more bacterial species (*Mycoplasma pneumoniae* and *Synechocystis* PCC) and one archaeal species (*Methanococcus jannaschii*) were also completed using the whole shotgun approach (Koonin et al. 1996). *E. coli*, the model bacterium targeted by the HGP, and sequenced with a more hierarchical strategy in a project that extended over 12 years, was not completed until 1997 (Blattner et al. 1997).

Since then, the number of known genome sequences for bacterial and archaeal species has been growing almost exponentially. The majority of the first ones completed were human pathogens (e.g. *Borrellia burgdorferi*, *Helicobacter pylori*, *Mycobacterium tuberculosis*, *Rickettsia prowazekii*, *Vibrio cholerae*, etc.), but species of ecological, industrial, and agricultural importance are becoming increasingly common in the databases. As sequencing technology has become faster and more automated, the pace of discovery has increased so much that it is difficult to keep track of it. In May 2000, for instance, the US Department of Energy (DOE) announced that *Enterococcus faecium* had been sequenced in a single day. In reality, only the actual electrophoresis of the random sequencing phase of the project was accomplished that day; closure was left for others in the scientific community to perform. Later that year, the DOE extended this publicity tactic by running a "Microbial Marathon" and performing the shotgun phase on the genomes of 15 biodegradative species in a single month (Amber 2000). Since the number of species with either completed or near-completed "draft" form

Table 3. Uniform resource locators (URLs) for selected major databases on the World Wide Web

Name of data base	URL
DNA Databank of Japan	http://ddbj.nig.ac.jp
European Bioinformatics Institute	http://www.ebi.ac.uk
GenBank	http://www.ncbi.nlm.nih.gov
Genoscope	http://www.genoscope.fr
Genome Sequence Database	http://www.ncgr.org/gsdb/
Human Genome Database	http://www.gdb.org/
MIPS	http://www.mips.biochem.mpg.de/
NCBI	http://www.ncbi/nlm
Proteome, Inc.	http://www.proteome.com/databases
REBASE	http://www.nbeb.com/rebase
SWISS-PROT	http://expasy.ch

Table 4. Uniform resource locators (URLs) for fungal genomics on the World Wide Web

Information source	URL
DOE Microbial Genome Program	http://www.ornl.gov/microbialgenomes
Fungal Genetics Stock Center	http://www.kumc.edu
Fungal Genome Resource Center	http://www.gene.genetics.uga.edu
National Center for Genomics Research	http://www.ncgr.org
Neurospora Genome Assembly	http://www-genome.wi.mit.edu/annota tion/fungi/neurospora/
Institute for Genomics Research (TIGR)	http://www.tigr.org
Sanger Centre	http://www.sanger.ac.uk
Stanford University	http://sequence-www.stanford.edu
University of Georgia	http://gene.genetics.uga.edu
University of Kentucky	http://biology.uky.edu
University of Manchester	http://www.aspergillus.man.ac.uk
University of Oklahoma	http://www.genome.ou.edu
Whitehead Institute	http://www.genome.wi.mit.edu

genomes changes almost weekly, the reader should access the Internet for up-to-the-minute information. Tables 3 and 4 provide some major addresses; each of these sites provides links to many other information sources (and so on ad infinitum).

III. Tools of Genomic Analysis

A. Databases

Databases include a diversity of genome data, mostly available free of charge via the Internet. They are the repository of primary nucleotide and amino acid sequence data, a place for storing sequence data derived from the experimental characterization of DNA and proteins, and for the results of analysis derived from the primary sequence data. Each database is managed by a government, academic or industrial sponsor. For example, in the USA, the National Center for Biotechnology Information (NCBI) manages GenBank, which is linked to mirror sites in England and Japan (Table 3). In Europe, the Munich Information Center for Protein Sequences (MIPS), Martinsried near Munich, Germany, develops and maintains a related database for DNA and protein sequences (Mewes et al. 1999). In addition to providing data access, public database groups develop analysis tools; furnish links with other databases and libraries; and provide other services such as download software, on-line tutorials, and glossaries of genetics terms. These databases make it possible to retrieve sequence data of interest and to query new sequences against known sequence records. Benson et al. (1998) have provided a brief introduction to GenBank tools such as BankIt, Sequin, *Entrez*, BLAST, and so forth. There are several introductory books about database access and use;

see, for example, *Internet for the Molecular Biololgist* (Swindell et al. 1995) and *The Internet and the New Biology* (Peruski and Peruski 1997). Because of the rapid speed with which the field is growing and evolving, the most up-to-date information is available directly from the Internet by logging into the home pages of the respective sites. Selected databases and their Internet addresses are listed in Table 3. All of them provide links to other databases. There are thousands of potentially relevant sites on the Internet, and the field changes fast. Very fast.

The use of mathematics, statistics, and computer science to model, analyze, store, retrieve, and distribute biological data is often called **bioinformatics**, an umbrella term that is difficult to define precisely. It has come to cover a broad variety of approaches to capturing, managing, organizing, storing, modeling, and accessing data about DNA sequences, proteins, expression arrays, and the other output of genomics research (Bishop 1999; Thompson 1999). Bioinformatics tools enable scientists to extract information about potential genes from databases, compare genes from different organisms, develop hypotheses about structure-function relationships, and query patterns and motifs derived from sequence homologues.

Biology is increasingly dependent on results available only in electronic form. Most of the DNA sequence, protein sequences, genome annotation, gene expression patterns and other genomics data are published exclusively on the World Wide Web. These data are unsuited to print media because of their immensity and because this kind of information can only be assimilated and used with the aid of computers (Botstein and Cherry 1997). A consequence of this electronic revolution is that chapters such as this one are not so much reviews (tours of published territory) as they are preludes (starting points for Internet exploration).

Another practical consequence is that as increasing amounts of sequence data have been deposited in databanks, it has become possible to do research "*in silico.*" With each passing month, such database searching or "mining" becomes more productive (Boguski 1995). This new paradigm was forecast by Walter Gilbert: "... all the 'genes' will be known (in the sense of being resident in databases available electronically), and that the starting point of a biological investigation will be theoretical. An individual scientist will begin with a theoretical conjecture, only then turning to experiment to follow or test that hypothesis. The actual biology will continue to be done as 'small science' – depending on individual insight and inspiration to produce new knowledge – but the reagents that the scientist uses will include a knowledge of the primary sequence of the organism, together with a list of all previous deductions from that sequence" (Gilbert 1991).

B. Mapping

In mapping, genes are assigned to particular chromosome regions. Classical genetic linkage maps were created using recombinational analysis; map distances were represented in units based on percent recombination between mutant markers, sometimes called the centiMorgan. Accurate genetic maps based on tetrad analysis are available for several model ascomycetes (Davis 2000). In laboratory strains of *Aspergillus nidulans*, where recombination rates are high, linkage maps can be quite long, with 350–750 additive units per chromosome (Clutterbuck 1997). The edited series on *Genetic Maps, Vol. 3. Lower Eukaryotes* by O'Brien (1993) has classical maps for the following fungi and fungal relatives: *Aspergillus nidulans, Aspergillus niger, Cochliobolus heterostrophus, Coprinus cinereus, Dictyostelium discoideum, Magnaporthe grisea, Neurospora crassa, Phycomyces blakesleeanus, Saccharomyces cerevisiae, Schizophyllum commune, Schizosaccharomyces pombe, Sordaria macrospora, Ustilago maydis*, and *U. violacea*.

Fungal chromosomes are much smaller than most mammalian and plant chromosomes. This small size, traditionally a disadvantage for classical cytogenetic analysis, has become an advantage for performing electrophoretic karyotyping by means of pulsed-field gel electrophoresis (Walz 1995; Zolan 1995; Kelkar et al. 2001). This technique resolves the genome into a series of chromosomal bands on an electrophoretic gel, allowing researchers to by-pass conventional cytological maps. When an electrophoretic karyotype is available, molecular probes can be used to map genes directly to chromosomes (Skinner et al. 1991).

Molecular approaches have freed geneticists from the need to have mutants to construct a map. Physical maps can be obtained by breaking chromosomes with restriction enzymes and then identifying the points of breakage. This kind of physical map is also called a restriction map. Physical distances are measured directly in nucleotide length, using the number of base pairs (bp) usually expressed in kilobases ($1\,kb$ = $10^3\,bp$) or megabases ($1\,Mb = 10^6\,bp$). The use of restriction fragment length polymorphisms (RFLPs) detected by hybridization of genomic DNA with anonymous DNA probes was introduced by Botstein et al. in 1980. In conjunction with the polymerase chain reaction (PCR), complex fingerprinting patterns can be produced using arbitrarily selected PCR primers that anneal to multiple loci. Physical maps can be constructed in this way for cosmids, bacterial artificial chromosomes (BACs) or other large clones of DNA, as well as for whole chromosomes (see, e.g., Tait et al. 1997).

To facilitate interpretation of physical maps, the position of landmark markers is often added. A sequence tagged site (STS) is a unique pair of primers on opposite strands. STSs can be detected by PCR amplification by means of the defining primers. High-resolution physical maps can be used to create a scaffold for organizing large-scale sequencing. In the HGP, physical maps based on STS landmarks have been used to develop so-called sequence-ready clones that consist of overlapping cosmids or BACs (Benson et al. 1998). YAC clones and other kinds of clones can be assigned a physical location by in situ hybridization (Burke et al. 1987). During the mapping stage of the HGP, YAC clones covering most of the human genome were interconnected by about 30,000 STS markers. Finally, expressed sequence tags (ESTs) can be used to derive a transcript map, i.e., a map of functional genes (Adams et al. 1991), as well as serve as gene-based STSs.

In the early years of eukaryotic genome research, mapping preceded sequencing and fungi were used as models for developing several physical mapping algorithms (Cuticchia et al. 1992; Xiong et al. 1996; Prade et al. 1997; Bhandarkar et al. 2001). The first stages of the *Neurospora crassa* genome were also chromosome based (Aign et al. 2001; Bean et al. 2001; Kelkar et al. 2001). Mapping will always be an important goal of genetic analysis for fungi and other organisms. Maps are essential for positional cloning of genes that are linked to particular phenotypes. Many contemporary projects, especially for economically important eukaryotes with enormous genomes (e.g., cat, cow, chicken, corn) seek mostly to map genes to chromosomes, limiting sequencing to biologically interesting regions or to ESTs.

C. Sequencing, Assembly, Closure, and Data Release

Sequencing DNA entails determining the exact order of nucleotides in a segment of DNA. For the human genome, that means determining the sequence of 3 billion bp (3000 Mb) of DNA. Fungal genomes are two orders of magnitude smaller, e.g., *A. nidulans* is about 30 Mb and *N. crassa* is approximately 40 Mb. There are several useful introductions to the principles of genome analysis (see, for example, Primrose 1995; Birren et al. 1997).

Nowadays, most sequencing centers use some version of the shotgun sequencing method. In this approach, a random sample of sequences is collected from a population of DNAs obtained from a larger DNA segment. Assembly involves putting together sequenced stretches of DNA into contiguous fragments ("assemblies"). Computer programs compare individual sequences to find overlaps and then use clone constraints (the sequences obtained from the two ends of a single clone of known size to reflect the distance between the two sequences) to properly place sequences in assemblies.

In a typical project, DNA is extracted, broken into pieces, and then cloned into YACS, BACs, PACs or other vectors to construct a library (Kupfer et al. 1997; Bean et al. 2001; Kelkar et al. 2001). Because of their better stability and ease of handling in *E. coli*, BACs have become the preferred vector. The DNA from each "large insert" clone is again broken into smaller pieces and subcloned. These subclones are used for the sequencing reactions followed by loading onto automated sequencing machines. The sequencing machines produce "reads" of 500–800 DNA bases as the emission of fluorescent dyes (red, green, yellow, and blue) that are recorded and deciphered automatically.

The next stage is assembly. The sequenced fragments overlap, so some DNA is sequenced many times. The overlapping segments are used to arrange the fragments in order. Special assembler software is used to assemble the sequencing data into contigs representing the DNA in the clone. These data are assembled into a working draft that covers most of a region but may still contain gaps and ambiguities. Assembling contigs becomes more difficult when there are frequent repeat sequences and high G+C content.

In most eukaryotic projects to date, genomic DNA has been first sorted into chromosome size populations (e.g., by pulsed field gel electrophoresis) before library construction. Tiling paths of ordered clones can be assembled and optimized to produce a system of minimal clone overlap. Clones can be anchored back to chromosome maps in order to build a sequence-ready map. Then individual clones are sequenced through a combination of shotgun and directed approaches. The *Caenorhabditis elegans* genome project used this strategy (*C. elegans* Sequencing Consortium, 1998). Recently, *Drosophila*, with its enormous 180-Mb genome, has been used as a test system to explore the applicability of the whole genome shotgun method to circumvent the need for a sequence-ready map prior to sequencing. The end sequences of BAC clones were used in the assembly process to provide a framework for linking contigs over larger regions (Venter et al. 1996, 1998). The completed BAC map and other hierarchical resources developed early in the *Drosophila* project served as an independent confirmation of the assembly of data from the shotgun strategy. The *Drosophila* whole genome shotgun project yielded a high quality draft, and this success is once again changing scientific attitudes about the necessity of more ordered approaches to sequencing (Adams et al. 2000). Since fungal genomes are significantly smaller than *Drosophila*, they are prime targets for whole genome random shotgun projects. The *Neurospora crassa* genome project, for example, began with a chromosome-by-chromosome approach but was completed with the whole shotgun method.

The final part of the assembly process (closure or "finishing") checks for errors, provides additional coverage to low-quality regions of the genome sequence, closes gaps and reassembles the sequences to generate single contiguous pieces. Since sequencing data are not of consistent quality, incorrect base calls, extra base calls, and other errors must be corrected during this last phase. The goal is to join sequences to contigs and contigs together. Gaps are filled and the ambiguities are resolved. This process is accomplished using an extensive combination of automated editing and analysis tools to check assemblies and to identify directed sequencing reactions for generating supplemental sequence data across the gaps (Ewing and Green 1998; Ewing et al. 1998;

Rieder et al. 1998). In the final stages, skilled individuals identify errors and then, where appropriate, collect more data. This phase is highly labor-intensive. Thus, the last stages of producing a completed sequence consume a disproportionate amount of time and money compared to the earlier, highly automated phases. What is more, completion is a moving end point. Different standards for what is variously called "closure" or "finishing" are applied at different centers.

At a meeting held in Bermuda in 1996, representatives of publicly funded sequencing centers agreed to release partially assembled, unfinished sequences onto server sites every day or "as soon as possible", without seeking patents. Although there are arguments for waiting until errors in sequenced data are cleaned up, the immediate data release policy has been widely accepted. Private companies do not have to conform to this policy nor do they have uniform alternative policies. Many companies keep their data secret while seeking patent rights on interesting sequences. At the other end of the corporate spectrum is the Merck Gene Index that has deposited all of its data in GenBank. In between are the many companies that keep data for a specified period of time, e.g., 6 months or a year, and then release their data to public data banks.

The quasi-mandatory public release policies have raised new ethical issues. Although these posted sequences are public, they have not been subjected to peer review, are often incomplete, and may contain errors and contamination. Further, these unpublished sequence data should be treated in the same way as other unpublished data are treated; an outside individual or group who wants to publish an analysis should obtain the written consent of the scientists who produced the data. Without written consent, independently published analysis can be defined as "misappropriation of data" (Hyman 2001). Unfortunately, the distinction between "public" and "in the public domain" has not been clarified sufficiently, and there have been several prominent publications in which bioinformatics groups preempted the scientists who have done the sequence work (Kennedy 2001).

The volume of new submissions is staggering. In January 2001, data from the HGP alone were piling into public databases at a rate of 10,000 bases/min (www.nhgm.nih.gov). Obviously, most of these newly collected data are far from "finished". Nevertheless, there seems to be a consensus that the benefits of immediate access outweigh the disadvantages of having to go back and check for revisions and updates.

Finally, for eukaryotes, cDNA sequencing is a valuable complement to complete genome sequencing. Whereas complete sequencing yields information on the structure of the genome, cDNA sequencing give information on which genes are expressed in a particular cell type or tissue type at a given time (Primrose 1995). Expressed sequence tags (ESTs) from cDNA sequencing projects also assist in genome sequence annotation.

D. Annotation

According to the Oxford English Dictionary, annotation is the act of making notes; an annotation is a note added to anything written; an annotator is one who makes notes to a text. In genomics, the text is the sequence of A's, T's, G's and C's in a given organism. The first goal of genomics annotation is to identify potential genes. Thus, after a segment of a genome has been sequenced, or in a whole shotgun approach, after an entire genome sequence has been determined, the data are analyzed to reveal possible protein and transfer RNA genes, similarities to known ESTs and other features of the DNA. The goal is to identify all the potential genes, pseudogenes, transposons, and other features, and to categorize them in structural and functional groups.

Computational tools do most of the work. Programs such as Genefinder, Artemis, Grail FFG, and Genmark are used to predict open reading frames (ORFs), making use of known information about species-specific codon usage and splice junctions. Currently, most of the algorithms used to scan filamentous fungal genomes are derived from nonfilamentous sources, and there is a need to develop and train programs especially for gene finding in molds (Kraemer et al. 2001).

Once extracted, gene products are analyzed in two basic ways: (1) similarity (homology), and (2) intrinsic sequence properties (Burset and Guigo 1996; Fickett 1996). Searching for a homology is the oldest means of identifying new genes. Most molecular biologists are familiar with the BLAST algorithm that was designed for finding ungapped, locally optimal sequence alignments. The BLAST family of programs can compare a nucleotide query sequence against a nucleotide sequence database, an amino acid query sequence against a protein sequence database, as well as other combinations of protein and nucleic acid comparison. BLAST algorithms find sequence similarities, i.e., they identify homologues to a query sequence (Altschul et al. 1990). The resulting alignments can be refined by other tools like FASTA. Predicted proteins are classified according to their similarity to entries in databases, using FASTA scores to quantify similarity. Pairs with high identity are scored as "strong similarity" (=highly significant). Matches can be divided into highly significant, moderately significant and weakly significant categories (Lipman and Pearson 1985; Pearson 1990, 1991; Pearson and Miller 1992). Unidentified open reading frames (URFs) are ORFs to which no function can be assigned. It is common practice to annotate these potential gene products as "hypothetical proteins". Within the yeast community they are called "orphan genes" (Dujon 1996), i.e., genes with no similarity to sequences in the public databases.

Intrinsic sequence analysis is broader in scope and ranges from predicting exons on the basis of statistical properties of sequence composition to additional computational approaches that detect protein motifs, predict secondary structure and other protein feature, as well as folding patterns. For example, there are several pattern databases such as PROSITE and PRINTS that can be joined to BLAST and FASTA searches (Pearson and Lipman 1988; Attwood et al. 2000; Skolnick et al. 2000).

During functional annotation, proteins are automatically attributed to predefined functional and/or structural categories. The first predictions of gene function are tentatively established based on DNA and amino acid sequence similarities to genes in other organisms, a method based on the assumption of an underlying evolutionary relationships between the two sequences. In evolutionary biology, homology means relationship due to descent from a common ancestral sequence. Similar sequences are homologous if they are derived through shared evolutionary history. It is also possible for sequence similarity to arise by convergent evolution. In the usage of genomics annotation, however, homology is usually used synonymously with "strong similarity".

Homologous genes are orthologous or paralogous (Fitch 1970). When a query by a new sequence elicits "hits" (a highly significant match) from another organism, the gene pairs are orthologous. Orthologs are genes in different species that evolve from a common ancestral gene. Orthologs evolve through speciation and normally retain the same function in the course of their evolution. When the "hit" is from within the query organism itself, the gene pairs are paralogous. Paralogs derive from a common ancestor through gene duplication and subsequent divergence. Paralogs evolve new functions, even if related to the original one (Tatusov et al. 1997). Correct identification of orthologs is critical for reliable prediction of gene functions in newly sequenced genomes.

After putative functions have been assigned based on sequence homologies and structural analysis, genes can be grouped into categories. The model bacterium *Escherichia coli* has one of the oldest systems for classifying genes. In the early years, functional classification used an arbitrary "one-dimensional" scheme consisting of six groups: intermediary metabolism, biosynthesis of small molecules, macromolecular metabolism, cell structure, cellular processes and other functions (Riley 1993). Some of these categories emphasized structure, and others function. This scheme did not properly encompass the multidimensional nature of "function". In a more recent classification, *E. coli* gene products were grouped into six "level 1" categories: metabolism, information transfer, regulation, transport, other processes, and cell structure. Each level was then broken down into sublevels. For example, within metabolism, "biosynthesis of building blocks" is a level 2 category. Level 2 is further subdivided into level 3 categories, i.e., the biosynthesis of amino acids, or the biosynthesis of nucleotides (Riley and Serres 2000).

Proteins can be assigned to putative functional categories using a variety of other schemes. Nelson et al. (1997) categorized *Neurospora crassa* expressed sequences according to the Expressed Gene Anatomy Database (EGAD) developed at TIGR. The categories used by Nelson et al. (1997) for *Neurospora* were: cell division cycle; cell signaling and communication; cell structure defense and detoxification; metabolism; protein synthesis; RNA synthesis; and "unclassified" (e.g., clock-controlled genes). In the MIPS functional classification used for *Saccharomyces cerevisiae*, there are 11 categories (Mewes et al. 1997). Main categories (e.g., energy, metabolism, transcription) are subdivided into subcategories (e.g., amino acid metabolism, translation) and then further divided

into sub-sub categories (e.g., electron transport proteins). Whatever the scheme, functional categories can overlap, so a given protein may be assigned to more than one category. They are fuzzy sets. Alternatively, proteins themselves can be "multifunctional", so a robust classification system must avoid simplistic one-protein to one-function categories.

There is a need for a largely or entirely computational system for comparing or transferring annotation among different species. To this end, the databases for three model organisms, FlyBase, Mouse Genome Informatics (MGI) and the *Saccharomyces* Genome Database (SGD), have started a joint project called the Gene Ontology Consortium. The goal of the consortium is to produce "a structured, precisely defined, common, controlled vocabulary for describing the roles of genes and gene products in any organism" (Gene Ontology Consortium 2000, p. 26). Three independent ontologies are being constructed: biological process, molecular function, and cellular component. Each "node" within an ontology is linked to other kinds of information including the many gene and protein databases such as SwissProt, GenBank, MIPS, etc. It is hoped that this dynamic but controlled vocabulary can be extended to all eukaryotes (Gene Ontology Consortium 2000).

How accurate is current functional annotation? Without experimental verification of every gene in every genome, it is impossible to know. However, Brenner (1999) compared three groups' annotation of the *Mycoplasma genitalium* genome, scoring for cases in which two group's descriptions were completely incompatible. The error rate was 8% for the 340 genes annotated by two of three groups; for various reasons, this is almost certainly an underestimation. It is worth quoting Brenner's commentary at length: "The annotation problem escalates dramatically beyond the single genome, for genes with incorrect functions are entered into public databases. Subsequent searches against these databases then cause errors to propagate to future functional assignments. The procedure need cycle only a few times without corrections before the resources that made computational function determination possible – the annotation databases – are so polluted as to be almost useless. To prevent errors from spreading out of control, database curation by the scientific community will be essential" (Brenner 1999, p. 132). Since fungal biologists are still in the early stages of genome research, it is important for the community to establish general awareness of the possible pitfalls of automated annotation. A subsidiary issue related to annotation involves genetic nomenclature (see Sect. V). It is hoped that the mycological community, perhaps through the Fungal Genetics Stock Center, will be able to establish some kind of standardized conventions and develop a uniform system for naming newly discovered genes.

Analysis of genomic sequence data is necessarily an on-going process and grows in an iterative fashion as additional data are added to databases and as more sophisticated algorithms are brought to bear on these data. Annotation lags behind DNA sequencing and will go on for many years into the foreseeable future, long after the DNA sequencing is finished. In the next few years, even more genomes will be completed and even more vast quantities of sequence data will be generated. The magnitude of DNA sequence data is already straining data base resources, and there is a need for more accessible ways in which to display and manipulate these data. Functional genomics (see Sect. III.E) has exacerbated the problem and created a need for better algorithms for pattern recognition, for modeling metabolic networks, and for analyzing microarrays. It has become a cliché to conclude such "needs lists" with a piety about how genomics will no doubt stimulate inventive theories in information science and computational biology.

E. Functional Genomics

It has been said that in the past we had functions in search of a sequence; now we have sequences in search of a function. In most species sequenced to date, a large number of ORFs predict proteins with no similarities to sequences in published databases. Linking unknown genes to a corresponding biological function is often called "post genomics" or "functional genomics". In the short run, the goal is to assign a putative function to each of the predicted genes in a genome. In the long run, the "functional genomics approach is to expand the scope of biological investigation from studying single genes or proteins to studying all genes or proteins at once in a systematic fashion" (Hieter and Boguski 1997). In order to start this extremely formidable project of studying the function of "all genes or proteins at once", it is useful to create working definitions of what is meant by

"function". In one useful overview, Oliver (2000) divides gene-function analysis into four tiers: the genome (the DNA sequence), the "transcriptosome" (the complete set of messenger RNAs), the "proteome" (the complete set of proteins), and the "metabolome" (the complete set of metabolites). While the DNA sequence is considered stable, the transcriptosome, the proteome and the metabolome are context dependent. Gene expression not only varies between different cell types, but also depends on the stage of the cell cycle, developmental status, and the environmental milieu of each cell.

Function may also be thought of from the perspective of biological hierarchies. At the biochemical level, function may refer to the catalytic activity of an enzyme. At the cellular level, function may refer to the physiological role of a particular protein in a particular metabolic pathway. At the phenotypic level, function is the role played by the protein in the observable appearance of the whole organism, a role traditionally detected by "loss of function" mutations. Finally, many proteins are multi-functional.

The most basic function-prediction method is annotation of sequence, based either on sequence alignment or sequence-motif methods such as Prosite or Blocks annotation. A suite of adjunct computer techniques may assist in finding proteins that are linked physiologically (in a pathway or organelle) or phylogenetically (through gene fusion) (Enright et al. 2000). These *in silico* analyses serve as a prelude to physiological studies and provide a starting point for further analysis. One of the most widely applied tools for the study of gene expression ("the transcriptosome") is the microarray (Schena et al. 1995; Shoemaker et al. 1996; DeRisi et al. 1997). DNA arrays can be made by binding DNA from genomic or cDNA libraries, or PCR products from either, onto a solid support such as a nitrocellulose filter, glass or silicon. Alternatively, oligomer probes can be synthesized directly. These arrays expedite the study of the relative abundance of thousands of individual mRNAs obtained from tissues and, moreover, they can be monitored simultaneously. Microarrays can be used to analyze gene expression patterns and define networks of genes controlled by single transcription factors (Blanchard and Hood 1996). They also can provide valuable information on comparative cellular physiology at different stages of the cell cycle, in changing environmental circumstances or in the presence or absence of drugs, toxins, and other chemicals (Young 2000).

When all is said and done, however, having the complete DNA sequence of an organism, even when supplemented with microarray expression data, is not sufficient to elucidate biological function at all its many levels. Proteomics, structural genomics and many other developments well beyond the scope of this review will take on increasing importance during the years to come (Gygi et al. 2000; Pandey and Mann 2000; Skolnick et al. 2000).

From the standpoint of fungal genomics, *Saccharomyces cerevisiae* has already become a living reagent. Other fungi have similar potential. Because of their larger genomes, and the wealth of data gleaned from decades of biochemical genetics, both *Aspergillus nidulans* and *Neurospora crassa* can provide powerful experimental systems for linking genes to function. Many filamentous fungal transcriptosome projects are in progress. For example, at Oklahoma State University, Rolf Prade is being supported by Genecor to produce microarrays containing DNA targets of all known *Aspergillus nidulans* genes. A similar project at North Carolina State University under the supervision of Ralph Dean is underway for *Magnaporthe grisea*.

IV. Fungal Genomics

A. Justifications

Plants, animals and fungi constitute the three major eukaryotic life forms. Distinguished by their absorptive mode of nutrition, efficient secretion of enzymes, and diverse modes of spore production, it has been estimated that there are over a million species of fungi, of which less than 10% are known to science (Hawksworth 1991). Reasons to sequence fungal genomes run the gamut of basic to applied with a fair amount of "organismal chauvinism" thrown in. Several early articles presented justifications for establishing whole genome projects for filamentous fungi (Arnold 1997; Bennett 1997b; May and Adams 1997). To recapitulate in brief: as "lower eukaryotes" fungi have relatively small genomes and simple life cycles. They are biochemically versatile, secreting a wide array of acids, degradative enzymes and secondary metabolites to support their absorptive mode of nutrition and to enhance their various ecological

niches. Fungi play a vital role in global ecosystems in recycling carbon and nitrogen in the environment; in the maintenance of healthy plant populations through mycorrhizal associations; and in numerous other environmental roles. The fungal kingdom encompasses both filamentous and yeast-like forms of which the budding yeast *Saccharomyces cerevisiae* is the best studied. Essential to a variety of fermentation products (wine, beer and bread), budding yeast is also the premier eukaryotic model organism. It was the first eukaryote to have its genome completely sequenced (see Sect. IV.B). In addition, four other ascomycetes are excellent genetic models: the mycelial fungi *Aspergillus nidulans*, *Podospora anserine* and *Neurospora crassa* and the fission yeast *Schizosaccharomyces pombe*. They can be grown on simple, defined media, giving exquisite experimental control over environmental parameters. Procedures for mutagenesis, crossing, transformation, gene knockouts, physical maps, and other genetic manipulations are readily available as are extensive collections of mutant stocks and accompanying genetic linkage map data. In the past, research with fungal models has provided valuable insights into many aspects of modern biology. These models are also ideally suited for functional genomics research.

The economic importance of fungi is difficult to exaggerate. In addition to their roles as major biodeteriogens and plant pathogens (see Chap 5, this Vol.), certain of their metabolites are the bedrock of the pharmaceutical industry. Famous fungal metabolites include antibiotics such as penicillin and cephalosporin; immunsuppressants such as cyclosporin and mycophenolic acid; blood pressure lowering agents such as mevalonin; and toxins such as aflatoxin. The industrial production of citric acid and industrial enzymes (e.g., glucoamylase, phytase) forms the basis of a multibillion dollar business. Fungi are also of growing concern as human pathogens, especially in developed countries. Immunocompromised patients suffering from HIV infections, taking cancer chemotherapy, or after organ transplantation, are particularly susceptible to life-threatening systemic mycoses.

B. *Saccharomyces cerevisiae*

The first eukaryotic genome to be sequenced was the yeast *Saccharomyces cerevisiae*. Supported largely by the European Union, in an unusual collaboration of more than 90 participating laboratories, the entire sequence was completed early in 1996 and released to public databases in April 1996 (Dujon 1996; Goffeau et al. 1996). The work was an international project involving over 600 scientists in Europe, North America, and Japan.

Saccharomyces cerevisiae has 16 chromosomes and a genome size of just over 12,000 kb. The composition of its DNA is moderately biased with an average G+C content of 39%. Compared with most multicellular organisms, the *S. cerevisisae* genome is compact with very few repeated sequences and introns. Gene density is not uniform along chromosomes, with some regions significantly below average. On average there is one gene per 2 kb. As predicted from classical genetics, regions around centromeres and telomeres display lower gene density (Dujon 1996). The annotated sequence consisted of approximately 6000 genes with an average size of 1450 bp (483 codons). About half of the ORFs were classified on the basis of their similarity with other proteins of known function and could be ascribed to functional categories (Goffeau et al. 1996). Nearly one-third of *S. cerevisiae* ORFs have a mammalian homologue, and several genes are useful in modeling human disease. For example, there are good matches to the human genes that cause hereditary nonpolyploid colon cancer (*MSH2* and *MLH1* in *S. cerevisiae*), and Werner's syndrome (*SGS1* in *S. cerevisiae*). Werner's syndrome exhibits several phenotypes associated with premature aging in humans; the cellular phenotype in *S. cerevisiae* includes reduced life span (Botstein et al. 1997).

The final sequence was assembled from roughly 300,000 independent sequence reads. In 1997, when a gazette of *S. cerevisiae* genes was published as a Supplement to Volume 387 of *Nature* (Goffeau et al. 1997), only 43.3% of the genes were classified as "functionally characterized," i.e. having experimentally elucidated properties, being members of defined protein families, or displaying strong homology to proteins with described biochemical function (Mewes et al. 1997). In fact, from the time the first *S. cerevisiae* chromosome was sequenced in 1992, one of the most striking results was that about half of the ORFs had no clear-cut homologues in any databases, i.e., these genes had no predicted function. With each new genome sequenced this pattern has been repeated. For organism after organism it has been found that a high proportion of predicted ORFs have no counterparts in the public databases. Dujon's (1996) words were prescient: "Thus, with the sequence of the first eukaryotic chromosome, it was the discovery of the extent of our ignorance, rather than the discovery of many new genes, that was the most conspicuous finding."

Subsequent to the completion of the sequencing and initial annotation, EUROFAN was established for the systematic analysis of the function of orphan *S. cerevisiae* genes. In one approach, systematic knock outs are used to associate orphan genes with function. In another, *S. cerevisiae* DNA microarrays enable rapid screening for genes that are induced under different physiological conditions. Out of hundreds of possible examples, here is one. Microarray technology was used to look for genes induced by copper stress. Five transcripts with increased abundance were identified in cells challenged with copper sulfate: *SOD1*, *CUP1*, *CRS5*, *FET3*, and *FTR1*. The latter two genes, which function in high affinity iron uptake, had not been previously known to be induced by copper (Gross et al. 2000).

It would take an entire separate chapter to describe the role that *S. cerevisiae* has played in the development of tools for functional genomics. See, for example, Martzen et al. (1999) and Carlson (2000) for a strategy that detects "any" biochemical activity; Hegyi and Gerstein (1999) for a comprehensive survey of the relationship between protein structure and function; Marcotte et al. (1999) for a combined algorithm for *in silico* prediction of gene function; and Uetz et al. (2000) for a review of two-hybrid screens to identify protein-protein interactions between full length ORFs. Hughes et al. (2000) have generated a compendium (a reference database) of *S. cerevisiae* expression profiles corresponding to 300 diverse mutation and chemical treatments and show that affected cellular pathways can be determined by pattern searching.

Finally, in addition to the science itself, the sociology of the *S. cerevisiae* genomics project deserves commentary. The work has been described as the "largest decentralized experiment in modern molecular biology," (Mewes et al. 1997), because the scientists who deciphered the *S. cerevisiae* genome used a chromosome-by-chromosome "network" method and displayed a high level of cooperation. The shared sense of purpose and enthusiasm invigorated the community and will be difficult to replicate in future fungal genomics ventures; nevertheless, it is a worthy goal to emulate, even on a smaller scale.

C. Filamentous Fungi: Early Initiatives

The fungal research community is an international and diverse group representing numerous disciplines (e.g., genetics, biochemistry, plant pathology, industrial microbiology, medical mycology, systematics, etc.) collectively working on hundreds of different yeasts and mycelial species. Molecular mycologists constitute a more heterogeneous community than do yeast, fly, worm, or *Arabidopsis* biologists.

A major milestone in the genomics of mycelial fungi was the First International Conference on Fungal Genomics held at Oklahoma State University, Stillwater, Oklahoma, in 1996. Organized by Rolf Prade, and timed shortly after the completion of the *Saccharomyces cerevisiae* genome, the workshop addressed the question: Which mycelial fungus to sequence first? *Aspergillus nidulans* was chosen, a steering committee was established, and negotiations were begun to organize a pilot project supported by an industrial consortium. A few months after the Oklahoma conference, a second, smaller meeting was held in New Orleans, Louisiana, where it was agreed that *Neurospora crassa* should also receive priority in mycelial fungal genomics and that the respective communities should independently seek funding for major genome projects (Aramayo and Bennett 1997; Bennett 1997a; Hamer 1997).

In March 1998, the Second International Symposium on Fungal Genomics was organized by Jonathan Arnold and held at the University of Georgia, Athens, Georgia. At the time, robust pilot projects were underway for both *Aspergillus nidulans* and *Neurospora crassa*. Progress was reported on physical mapping, EST-based analyses, the development of informatics tools, and the status of funding. Mapping projects were also reported for *Candida albicans*, *Magnaporthe grisea*, *Phytophthora infestans* and *Pneumocystis carinii*. Other species on the program included *Ashbya gossypii*, *Histoplasma capsulatum*, and *Saccharomyces cerevisiae* (Prade 1998). Subsequent to the meeting, while enthusiasm for genomics approaches continued to grow among fungal biologists, it did not do so within funding agencies. Several coalitions were unsuccessful in raising the money to support whole genome projects for either *Aspergillus nidulans* or *Neurospora crassa*.

By the time the Third International Symposium on Fungal Genomics was held at the University of Georgia in July 2000, no effort was made to unify efforts around a few model systems. Instead there were a diversity of topics and species. Biological clocks, mating factors, pathogenicity, gene silencing by mutation (RIP), gene silencing by methylation, development and other biological features of fungal systems received considerable attention, as did the development of computer models for keeping databases up to date, performing comparative genomic analyses, and using proteome databases as a resource (Anonymous 2000). In the four short years since the Oklahoma meeting, it had become apparent that molecular mycologists wanted multiple projects encompassing both mycelial and yeast genomes. Then, just a few months after the Third International Symposium, good news started to pour in. Major funding had been secured for public genome projects on *Aspergillus fumigatus*, *Neurospora crassa*, *Pneumocystis carinii* and *Phanerochaete chrysosporium*. Sequencing projects for the yeasts *Candida albicans* and *Schizosaccharomyces pombe* were nearing completion; faster than expected, progress had been made on the *Cryptococcus neoformans* genome. Moreover, robust EST projects were underway for numerous other species, within both the private and the public sectors, and several proprietary, whole genome sequencing projects were being sponsored by industry.

D. A Survey of Fungal Genomics Initiatives

This section is meant to be illustrative rather than exhaustive. Selected fungal genome projects are listed in alphabetical order by species name. [Note: *Saccharomyces cerevisiae* was covered in Section IV.B] Each entry briefly summarizes the biological and/or economic significance of the species in question and, where relevant, cites one or more illustrative publications. Since data accumulation is so rapid, the reader is advised to consult the World Wide Web (Table 4) to obtain up-to-date information on these projects. In some cases, where genome research is being conducted by an industrial entity, data may not be publicly available.

1. *Ashbya gossypii*

One of the first applications of a genomic approach to gene discovery in mycelial fungi was for *Ashbya gossypii*, a species known both as a cotton pathogen and for its ability to overproduce riboflavin for industry (Demain 1972). Taking advantage of the availability of nearly completed genome data for *S. cerevisiae*, single-read sequence analysis of the termini of eight randomly picked clones of *Ashbya* were compared with published databases. Seven sequences with homology to *S. cerevisiae* genes were discovered; one of these genes appeared to code for the carboxy terminus of the *S. cerevisiae* threonine synthase (THR4). The putative AgTHR4 gene was subsequently cloned and sequenced, revealing an overall identity with the *S. cerevisiae* gene of 67.4% at the amino acid level. Surprisingly, a conservation of gene order was discovered around the THR4 locus in *A. gossypii* and *S. cerevisiae* (Altmann-Johl and Philippsen 1996). On-going functional genomics projects have included development of a gene replacement strategy (Wendland et al. 2000).

2. *Aspergillus flavus* Complex

The *Aspergillus flavus* complex encompasses a cluster of anamorphic, haploid species that run the economic gamut from mycotoxin producers such as *A. flavus* and *A. parasiticus* to the koji molds *A. oryzae* and *A. sojae*. Aflatoxins are among the most toxic and carcinogenic of natural products whereas koji products such as soy sauce, miso, and sake are used extensively in Asian cuisine. Species in the *A. flavus* complex have a high degree of DNA homology and are closely related phylogenetically (Geiser et al. 1998).

Aflatoxin biosynthesis is one of the best studied of the fungal polyketide pathways. The pathway genes are clustered in both *A. flavus* and *A. parasiticus*; most of them have been cloned and sequenced (Bennett et al. 1997; Minto and Townsend 1997; Woloshuk and Prieto 1998). Homologous toxin genes are present in both *A. oryzae* and *A. sojae* (Klich et al. 1997; Watson et al. 1999) and there is evidence for regulation of some of them (Klich et al. 1997), although none of the koji molds has ever been shown to produce aflatoxin. At least five of the aflatoxin pathway genes are clustered the same way in *A. parasiticus* and *A. oryzae*; however, certain deletions within the cluster in *A. oryzae* apparently prevent the koji molds from producing toxins (Kusumoto et al. 2000).

Since *A. oryzae* and *A. sojae* have GRAS status (Generally Regarded As Safe) with the US Food and Drug Administration, economic and public health considerations make it important to maintain a separate classification of the molds in this group. They can be distinguished by RFLP analysis (Klich et al. 1987) or random amplified polymorphic DNA analysis (Yuan et al. 1995).

The economic importance of the *A. flavus* group is being used to justify genomics efforts with hopes that there will be improvements in commercial scale fermentations of the koji molds, as well as in methods of aflatoxin control. Masayuki Machida's group at the National Institute of Bioscience and Human Technology in Tsukuba City, in collaboration with several other Japanese laboratories, has assembled cDNA libraries of *A. oryzae* cultures grown under different liquid and solid state conditions. Approximately 13,000 ESTs fell into approximately 5000 nonredundant sequence clusters; a BLAST search showed that about 40% of the nonredundant sequences had no similarity to any genes or proteins registered in databases (Machida et al. 2000). EST projects are also underway for *A. flavus* in a number of substitute "university" for American laboratories, e.g., the University of Oklahoma, the University of Georgia. A karyotype of seven linkage groups has been identified and a physical map has been partly constructed (Kelkar and Arnold, unpubl.). Several companies are said to support proprietary projects and the US Department of Agriculture is expanding research on genomics of the aflatoxigenic species.

3. *Aspergillus fumigatus*

Aspergillus fumigatus is a well-known vertebrate pathogen first isolated in 1863 from avian lungs. It is an even more common saprophyte found in decaying organic matter, especially associated with compost and manure undergoing thermophilic decompositions. It produces prodigious numbers of small conidia and is almost universally isolated from air samples. Medically, *A. fumigatus* is the most common cause of both invasive and noninvasive human aspergillosis, and has caused concern in recent years as diagnosis of the invasive form has skyrocketed in immunocompromised patients, with accompanying high mortality (Latge 1999). *A. fumigatus* is also implicated in human allergies, as well as disease states in many vertebrate and insect species (Smith 1989).

It is an anamorphic, haploid fungus with an estimated genome size of 30 Mb. About 50 genes have been cloned; a transformation system has been developed, and the parasexual cycle has been elucidated (Weidner et al. 1998; Latge 1999), but our overall understanding of *A. fumigatus* genetics is rudimentary. Because it is the most pathogenic *Aspergillus*, the bulk of the molecular research on this species has been focused on identifying virulence genes. By analogy to other aspergilli, it is believed that *A. fumigatus* has eight chromosomes; however, electrophoretic karyotype studies by Amar and Moore (1998) resolved only four bands, and another study by Tobin et al. (1997) resolved only five bands. Attempts to produce a better-resolved electrophoretic karyotype are underway at the University of Minnesota in the laboratory of Jo-Anne VanBurik.

A BAC library has been constructed from a clinical isolate of *A. fumigatus* (AF293) recovered from a patient who had received long-term steroid therapy for rheumatoid arthritis and who succumbed to systemic aspergillosis. In a pilot project, ten physically linked BAC clones (~1 Mb) were sequenced at the Sanger Centre. An international genome project under the leadership of David Denning, University of Manchester, was initiated in the fall of 2000. A whole genome shotgun approach is being used, with sequencing performed at TIGR, the Sanger Centre, and a group at Salamanca University in Spain. At TIGR, three libraries will be constructed with 2, 10, and 15–20 kb inserts. There will be 100,000 reads from the 2-kb library, 40,000 from the 10-kb library, and 20,000 from the BAC library to give, respectively, 5X, 2X, and 1X coverage. Annotation will be performed at both TIGR and the Sanger Centre. In parallel, a transcriptome study is underway at the Institute Pasteur. It is hoped that this publicly funded project will stimulate the release of already existing *A. fumigatus* genome data from private companies.

4. *Aspergillus nidulans* (teleomorph: *Emericella nidulans*)

Since the discovery and elucidation of the parasexual cycle, *Aspergillus nidulans* has been a popular genetic model (Pontecorvo et al. 1952; Pontecorvo 1956), especially for developmental biology (Timberlake 1990; Adams and Yu 1998). Large numbers of mutants have been characterized, and extensive genetic linkage maps are available (Clutterbuck 1994). In addition to mitotic crossing over and haploidization, the construction of the *A. nidulans* genetic map has been aided by the use of translocation stocks (Kafer 1977). *A.*

nidulans is well amenable to molecular manipulations (Timberlake 1991).

The genome of *A. nidulans* is approximately 28 kb, divided into eight chromosomes numbered in order of discovery. This is somewhat confusing because the convention across most of eukaryotic genetics is for chromosomes to be numbered in decreasing size order. In *A. nidulans*, however, the smallest chromosome is called IV, not VIII. Whatever their numbered label, these linkage groups provide excellent anchorage for physical map data (Prade et al. 1997). To date, with few exceptions, a gratifying self-consistency has been found between the genetic and contig maps (Xiong et al. 1996; Clutterbuck 1997; Prade 2000). Nevertheless, given the growing popularity of whole genome shotgun approaches, Clutterbuck's (1997) cautionary reminder is worth repeating: DNA-based maps are determined by contiguity only, with the result that one false linkage can displace a whole segment, i.e., "the map is as reliable as its weakest link" (Clutterbuck 1997).

An EST study conducted at the University of Oklahoma has estimated that *A. nidulans* has 8000 genes, yielding a significantly lower gene density than *S. cerevisiae*, with an average prediction of one gene every 3.3 kb. A homology based assignment of ORFs in cosmid SW06E08 revealed 12 potential genes and one transposable element. Seven of the genes showed similarity to those found in other organisms (Kupfer et al. 1997). Annotation of the overlapping cosmids W30B01 and W02H02 covering a 45-kb region of chromosome VIII revealed 17 genes and a gene density of one gene per 2.6 kb, yielding a revised prediction of one gene every 3 kb. Comparison of the chromosome VIII genome region with the available ESTs also showed a second transposable element and two genes with no known homologues. Additionally, the genes for spermidine synthase and transketolase were shown to have 3′ overlapping convergent transcription units. The EST database has also revealed a set of six paralogous heat shock 30 genes, only one of which had previously been detected (Kupfer, personal commun.).

A multicompany industrial consortium, under the leadership of Nigel Dunn-Coleman (Genencor Corp.), supported the sequencing of chromosome IV of *A. nidulans* (Dunn-Coleman and Prade 1998). This 2.9-Mb chromosome was targeted not only because of its small size but also because it contains the sterigmatocystin gene cluster (Brown et al. 1996). Chromosome IV was cloned into 147 cosmids and then subcloned and sequenced at an average coverage of 4.5X. The sequencing work was done at Texas A&M University under the leadership of Nancy Keller; the annotation work was done at Oklahoma State University under the leadership of Rolf Prade (Prade et al. 2001a). Genencor is also sponsoring the development of a ~1500 EST microarray resource through Dr. Prade's laboratory (Prade et al. 2001b). Transcriptome and proteome studies are underway in Steve Oliver's group in Manchester and Geoff Turner's group at the University of Sheffield.

A large shotgun sequence for *A. nidulans* has been generated by Cereon Corp. (formerly Millenium) of Cambridge, Massachusetts. These data can be obtained through special agreements with the company.

5. *Aspergillus niger*

Aspergillus niger is an anamorphic species of great industrial importance. Classified as GRAS by the US Food and Drug Administration, it is widely used in the industrial production of citric acid and a variety of industrial enzymes such as amylases, pectinases, and proteases (Godfrey and West 1996). Eight linkage groups have been defined and rudimentary genetic maps are available (Debets et al. 1993) as well as an electrophoretic karyotype (Verdoes et al. 1994). The genome is estimated to be about 30 Mb. In September 2000, DSM N.V (Amsterdam, Netherlands) announced that it had commissioned the German genomics consortium Gene Alliance to determine the entire DNA sequence of *A. niger* using a BAC by BAC sequencing approach. It is believed to be the largest industrial genome contract in Europe to date and is scheduled for completion in September 2001. In the USA, Genencor is sponsoring a cDNA sequencing project for *A. niger* at North Carolina State University in the laboratory of Ralph Dean.

6. *Candida albicans*

Candida albicans is a diploid fungus with a genome size of approximately 16 Mb that grows in both yeast and hyphal modes. Each cell contains around 20 mitochondria, equivalent to about 1 Mb of DNA. As a yeast, it is a common commensal of the human vagina and gastrointestinal tract. *C. albicans* is also an opportunistic pathogen. Mucosal candidiasis is the most prevalent infection in HIV-positive individuals; it is also one of the most common causes of bloodstream infections in hospitals (DeBacker et al. 2000). Funding for genome projects has been based on the hope that whole genome sequence will identify genes involved in pathogenesis. Since the laboratory manipulation of *C. albicans* is easier than other fungal pathogens, it has become a model for studying dimorphism and fungal virulence factors (Kron and Gow 1995; Brown and Gow 1999). A number of genetic and genomic tools have been developed, e.g., transformation, gene reporter systems, gene inactivation, cosmid-based contigs (Scherer and Magee 1990; Pla et al. 1996; Tait et al 1997), and both a public and a proprietary genome project have produced draft versions. The public project is organized through the University of Minnesota; sequencing is done at the Stanford Genome Center using shotgun sequencing and assembly. The propriety database for *Candida albicans* is available from Incyte Genomics, Palo Alto, California, for a fee. Preliminary estimates suggest that there are about 7000 ORFs in the 16-Mb genome, yielding a similar gene density to *S. cerevisiae*, which has about 6000 genes in its 12-Mb genome. A high number of the *Candida*-specific genes seem to be involved in catabolic pathways (DeBacker et al. 2000).

C. albicans has traditionally been classified as asexual since neither mating nor meiosis has ever been observed. Recently, with the sequencing of the *C. albicans* genome, orthologs have been discovered for *S. cerevisiae* mating (*CPH1*, *STE20*, and *STE6*) and meiotic (*DMC1*) genes, suggesting that *C. albicans* may have a cryptic sexual cycle (Hull et al. 2000; Magee and Magee 2000). By making slight genetic alterations at the mating type-like locus (*MTL*), Hull et al. (2000) were able to generate *C. albicans* derivatives that were capable of mating in mouse, but not on the laboratory media tested. In a parallel approach, Magee and Magee (2000) took advantage of the fact that the *MTL* locus was located on chromosome 5. When sorbose is the sole carbon source, only cells monosomic for chromosome 5 grow. Appropriately paired auxotrophic monosomic strains formed stable prototrophs on laboratory media that were tetraploid in DNA content. These findings have implications for other species in which cryptic recombinational mechanisms are suspected (e.g., Geiser et al. 1998), and demonstrate the power of known gene inventories to permit manipulations that can expand our view of fungal life cycles.

7. *Cryptococcus neoformans* (Teleomorph: *Filobasidiella neoformans*)

Cryptococcosis, also known as "European blastomycosis", is caused by the basidiomycetous pathogen, *Cryptococcus neoformans*, and occurs mostly in immunocompromised patients. Primary infections are usually pulmonary and asymptomatic; meningitis is a common consequence following dissemination from the lungs, particularly in AIDS patients. The central nervous system disease is the form most frequently diagnosed and unless treated immediately has a grave prognosis. Prior to the HIV epidemic, cryptococcosis was diagnosed only sporadically. Currently, it is the fourth most commonly recognized cause of life-threatening infection among AIDS patients, occurring in about 8% of patients in the USA (Kwon-Chung and Bennett 1992; Casadevall and Perfect 1998).

The most important natural source of the anamorphic *C. neoformans* state is soil contaminated with avian droppings, especially pigeons. Clinical isolates are usually haploid. Heterothallism was discovered in *Cryptococcus neoformans* in 1975, but the natural reservoir of the teleomorphic *Filobasidiella* state is unknown. Basidiospores are pathogenic when injected into mice (Kwon-Chung and Bennett 1992).

Classical genetic studies have been conduced with serotype D isolates (Whelan and Kwon-Chung 1986; Whelan 1987). Animal models for the study of virulence have been developed, and molecular tools such as transformation and gene disruption by homologous recombination are well developed, as is signature tagged mutagenesis (Nelson et al. 2001). These and other justifications for a *C. neoformans* genome sequencing project have been summarized by Heitman et al. (2000). Currently, a *Cryptococcus neoformans* (strain JEC21 = B-4500, serotype D) genome project is underway at the Stanford Genome Technology Center and the Nagasaki University School of Medicine with funding from the National Institutes of Health. Four-fold shotgun coverage of the 21-Mb (haploid) genome has been accomplished. The mitochondrial sequence has been completed.

Raw shotgun data are being released daily; contigs-in-process are released as rapidly as possible. These partially annotated data can be accessed through the Stanford web site. In addition, an EST database of serotype D strain JEC21 is under construction at the University of Oklahoma as part of a collaboration between Juneann Murphy and Bruce Roe.

8. *Histoplasma capsulatum*

Histoplasma capsulatum is the cause of histoplasmosis, the most common pulmonary mycosis of humans. *H. capsulatum* thrives in humid soils associated with bird and bat droppings. Most cases of histoplasmosis are asymptomatic or mild, but a small number progress to a systemic disease, yielding a rapidly fatal infection. There is an increasing incidence of opportunist histoplasmosis in patients with compromised immune systems (Rippon 1988; Levitz 1991). *H. capsulatum* is dimorphic, growing as a yeast at 35 °C and as mycelium at lower temperatures. The yeast form is pathogenic whereas the mycelial form is saprobic. Infections result predominantly from inhalation of conidia produced by the mycelial form. The so-called dimorphic switch is also important for a number of other systemic fungal pathogens in which the yeast is the pathogenic state, the mycelium is saprobic, and dimorphism is regulated by temperature. For example, *Blastomyces dermatitidis* causes blastomycosis, a disease similar to histoplasmosis. Both *Blastomyces* and *Histoplasma* are anamorphs of the same teleomorph genus, *Ajellomyces*. In *H. capsulatum*, using a targeted gene replacement technique, Sebghati et al. (2000) have implicated a calcium-binding protein in virulence. The gene is inactive in the mycelial form but switches on when the fungus converts to the yeast form of growth. Mutants with the inactivated gene were a 1000-fold less virulent in mice (Sebghati et al. 2000).

A project at the University of California, Berkeley, and the University of California Medical School, San Francisco, has developed microarrays of random genomic fragments from *H. capsulatum*. These arrays are being used to identify yeast-specific and mycelia-specific genes. Incyte Genomics is sponsoring the partial sequencing of genes with provocative transcriptional profiles. Details are available through the Stanford Microarray Database.

9. *Magnaporthe grisea*

Magnaporthe grisea (anamorph: *Pyricularia grisea* = [*P. oryzae*]) is the cause of rice blast disease. It also can attack more than 50 other species of grasses and sedges, and is one of the most economically significant of the plant pathogens (see Chap 5, this Vol.).

M. grisea is haploid and heterothallic. Tools for molecular manipulations are well developed (Skinner et al. 1993); several genetic maps have been constructed (Romao and Hamer 1992; Sweigard et al. 1993); and the genome size is estimated to be 38 Mb (Hamer et al. 1989). Despite the broad host range of the species, individual strains have a narrow host range, and the genetics of virulence has received considerable attention (Sweigard et al. 1995; Leung et al. 1998). Physical analysis of the *M. grisea* genome has been facilitated through the construction of a BAC library (Zhu et al. 1997). BAC clones can be transformed back into *M. grisea* for complementation studies (Diaz-Perez et al. 1996). In the study of avirulence genes, a number of laboratories have made estimates of the relationship between genetic and physical map distances. In a chromosome walking study of the AVR1-CO39 avirulence gene, for example, meiotic cross-over points were unevenly distributed and a 14-fold variation in the relationship between genetic and physical distances was measured. Both *A. nidulans* and *N. crassa* show a better correlation between genetic and physical distances than does *M. grisea* (Farman and Leong 1998). An EST project for *M. grisea* is underway at the North Carolina State University.

10. *Neurospora crassa*

Neurospora crassa is a haploid, orange-colored ascomycete found on recently burned fields in tropical and subtropical areas. The premier genetic model for mycelial fungi, it was instrumental in the research leading to the one gene-one enzyme hypothesis (Beadle and Tatum 1941), and is nowadays widely used in both research and education. *N. crassa* has seven chromosomes, first elucidated cytologically by McClintock (1945). The chromosomes can be separated by pulsed field gel electrophoresis and, using this technique, the genome size is estimated to be 42.9 M (Orbach et al. 1988). Over 800 characterized genes and a detailed genetic map of the seven linkage groups are available (Perkins et al. 1982; Davis 2000; Perkins et al.

2000). A low amount of repetitive sequence and a GC-content of 54% make *N. crassa* amenable for large-scale sequencing (Radford and Parish 1997). A pilot project at the University of New Mexico used expressed sequences from conidial, mycelial and perithecial stages to conduct a preliminary analysis of the *N. crassa* genome. Considerably more ESTs with significant matches in the databases (56%) were present in germinating conidia than in mycelial or perithecial samples. Overall, over 50% of ESTs showed no similarity to previously identified genes (Nelson et al. 1997; Braun et al. 2000). Another cDNA sequencing project examining clock-controlled genes has been reported (Zhu et al. 2001). A physical map is also well developed (Aign et al. 2001; Hall et al. 2001). The mapping approach used nonoverlapping cosmids that were linked by adjacent "tiles" through hybridization to a common cosmid (Bhandarkar et al. 2001; Kelkar et al. 2001). Similar physical mapping strategies had been used previously to develop physical maps for both *Aspergilllus nidulans* (Prade et al. 1997) and *Schizosaccaromyces pombe* (Hoheisel et al. 1993). Using both cosmid and BAC libraries, chromosome specific subsets of clones have been hybridized to assign chromosomal locations to over 75% of cosmid clones (Kelkar et al. 2001; Aign et al. 2001). In the first years of *Neurospora* genomics efforts, a German group sequenced and analyzed about 10 Mb of nonredundant regions from chromosomes II and V. These data are available in the MIPS databases (Schulte et al. 2001). It is estimated that *Neurospora* has ~11,000 genes, making it almost as gene rich as *Drosophila* with ~13,600 genes (Adams et al. 2000; Bean et al. 2001; Kelkar et al. 2001). In the fall of 2000, the US National Science Foundation funded a proposal to complete the *N. crassa* genome using a whole shotgun approach. Sequencing was performed at the Whitehead Institute at the Massachusetts Institute of Technology and the completed sequence was released to the Whitehead Institute web site on 14 February 2001. The strong foundation of biochemical and molecular research will render the *Neurospora* genome an especially accessible system for future functional genomics studies.

11. *Phanerochaete chrysosporium*

Phanerochaete chrysosporium, a basidiomycete sometimes called the "model white rot", is used extensively for research on fungal lignin degradation and bioremediation. Under conditions of nitrogen, carbon or sulfur deprivation, it produces families of ligninolytic enzymes, including lignin peroxidase and manganese-dependent peroxidase. These enzymes use hydrogen peroxide to promote oxidation of lignin to free radicals that then undergo spontaneous rearrangements (Kirk and Farrell 1987). The peroxidase genes are encoded in structurally related families that have been cloned and sequenced (Broda et al. 1996). In addition to lignin degradation, *P. chrysosporium* is capable of breaking down many environmental pollutants, sometimes in complex mixtures. It is believed that the capacity for xenobiotic degradation is associated with the metabolic apparatus of lignin degradation (Barr and Aust 1994). A restriction fragment length polymorphism (RFLP)-based map is available (Raeder et al. 1989), and a number of genetic studies have been directed toward the study of the lignin-degrading system (Alec and Gold 1991). In the hope of learning more about its capacity to degrade toxic wastes, the US Department of Energy (DOE) has sequenced the 40-Mb *P. chrysosporium genome* (see the DOE web site), unexpectedly making this basidiomycete one of the first mycelial fungi with a draft genomic sequence in the public databases.

12. *Phytophthora infestans* and *P. sojae*

Phytophthora infestans is the etiological agent of potato late blight, perhaps the most notorious of plant pathogens because of the Irish potato famine. *Phytophthora sojae* is the causal agent of a major root and stem blight of soybeans. *Phytophthora* species are members of the Oomycota, a group that shares a number of morphological and ecological similarities with fungi but, based on analysis of rDNA sequences, is no longer classified in the Kingdom Fungi (Forster et al. 1990). Because they are vegetatively diploid, genetic studies based on mutant phenotypes are difficult in *P. infestans* and other members of the genus (Judelson 1996). Using the AFLP fingerprint protocol, Van der Lee et al. (1997) constructed a linkage map with 190 polymorphic DNA markers over ten linkage groups [note: AFLP is a DNA fingerprint technique that is not an acronym (Vos et al. 1995)]. More recently, Van der Lee et al. (2001) have used this resource to locate avirulence genes. The exact chromosome number of *P. infestans* is

unknown; resolution of a karyotype by pulsed field gel electrophoresis was unsuccessful (Tooley and Carras 1992). However, photometric studies indicate that it has an enormous genome of ~240 Mb, four times the size of the 62-Mb genome of *P. sojae* (Tooley and Therrein 1987). A BAC library of *P. sojae* has been developed in Brett Tyler's laboratory at the University of California, Davis; sequencing and analysis are underway at the National Center for Genome Research, Santa Fe, New Mexico. Comparative analysis of ESTs from mycelia, infected tissue, and zoospores of *P. sojae* and mycelia of *P. infestans* have been published (Kamoun et al. 1999; Qutob et al. 2000). A functional classification scheme was devised for oomycete genes using ten categories based on plant and EST functional catalogues, plus two new categories especially for oomycetes: category 11 comprises elicitors, avirulence factors and pathogenicity factors; category 12 involves cell defense (Kamoun et al. 1999). The novel members of the elicitin protein family revealed by sequencing are of particular interest to plant pathologists. These data have been complied in searchable databases (Waugh et al. 2000). A BAC library has been constructed for *P. infestans* (Randall and Judelson 1999). Recently, the US Department of Agriculture has funded the sequencing of a series of additional ESTs of in vitro and potato infection stages for *P. infestans*. Another proprietary EST project has been funded by Syngenta (Novartis) and will be released publicly at the end of 2002.

13. *Pneumocystis carinii*

Pneumocystis carinii is a pulmonary pathogen; one variety (*P. carinii*. f. sp. *carinii*) causes pneumonia in rats; another variety (*P. carinii* f. sp. *hominis*) causes pneumonia in humans, especially those with AIDS. It is an unusual fungus classified in the Archaeascomycetes, that has little or no ergosterol, and that cannot be cultivated on artificial media (Edman et al. 1988). In addition, it has a complex antigenic expression system consisting of a family of mannosylated surface antigens (Sunkin and Stringer 1996). With a genome smaller than *S. cerevisiae* (7.7 Mb), and a well-resolved (albeit variable) karyotype (Cushion et al. 1993), it is believed that DNA sequencing will shed light on its pathogenicity, unique antigenic system, and place in fungal evolution. Approximately 24 *P. carinii* genes have been cloned and sequenced, with 60–65% A+T, and up to nine introns per gene.

A proposal for a *P. carinii* genome project was unveiled in 1997 at the Fifth International Workshop on *Pneumocystis* (Lille, France). This 5-year plan proposes to construct cosmid libraries of the rat *P. carinii* genome in two vector systems; create ESTs; build physical maps by assigning clones to chromosomes; create chromosome-specific libraries; order the clones to chromosomes; and then subclone ordered cosmids for sequencing. The rat *P. carinii* genome would be used to guide the construction and characterization of human *P. carinii* cosmid libraries. Annotation would be followed by concurrent functional genomic studies (Cushion and Arnold 1998). This project is scheduled for completion in 2004. Analysis of ~5000 sequence runs in the EST database indicates that 50% do not have homologues to any known gene sequences; of the remaining, the most similarity is to genes from *Schizosaccharomyces pombe*. It is predicted that *P. carinii* will have about 4080 genes. Currently at least 1.25 Mb of sequence has been accumulated on the Web, and an early report on this genome project is given in Smulian et al. (2001). Details are available through the University of Kentucky genome projects web site.

14. *Schizosaccharomyces pombe*

Schizosaccharomyces pombe, "the fission yeast", was first isolated from an African millet beer called pombe. It is heterothallic, dimorphic (i.e. capable of switching from yeast to pseudo-hyphal morphology) and can grow vegetatively as both a haploid and a diploid. Haploid *S. pombe* has three chromosomes and a genome size of 14 Mb. It has been touted as a good model for mammalian cells, and has figured prominently in research on the cell division cycle (Forsburg and Nurse 1991; Nurse 1994). It has also been used to study signal transduction pathways, recombination and the meiotic cycle, and cellular differentiation (Hayles and Nurse 1992). There is an extensive gene database (Lennon and Lehrach 1992).

Because of its relatively close evolutionary relationship *to Saccharomyces cerevisiae*, its inherent biological interest, and its compact genome with low repetitive DNA, *Schizosaccharomyces pombe* was an early target for genomic research and was used to develop physical mapping strategies using YACs (Maier et al. 1992). Hoheisel et al. (1993) produced high-resolution physical maps using cosmids and PACs. A pilot genome project

at the Sanger Centre, under the leadership of Bart Barrell with funding from the Wellcome Trust, grew into an international sequencing consortium with funding from the European Commission and Japan. This *S. pombe* project has used a "top-down" clone-by-clone approach, and several intermediate reports have appeared (Lucas et al. 2000; Xiang et al. 2000). There is an average gene density of one gene every 2.4 kb. The genome has 1.2 Mb of rDNA and 0.4 Mb devoted to telomeres and centromeres. The status of sequence data is posted at the Sanger Centre web site. Although the final publication describing the completed analysis of the *Schizosaccharomyces pombe* genome has not yet appeared, one group of informatics researchers used data already deposited in databanks to perform a comparison with *Saccharomyces cerevisiae*. They found that *Saccharomyces cerevisiae* has lost approximately 300 genes since divergence from the common ancestor with *S. pombe*, notably genes for spliceosome and signalosome components (Aravind et al. 2000).

15. Others

The mounting incidence of nosocomial fungal infections and the shortage of good drugs to treat them means that many pharmaceutical companies are pinning their hopes on genomics approaches to find new targets for drug screening programs. Agricultural companies are also seeking new ways to control plant pathogens. Manufacturers of commercial fungal products hope to improve product yields. When these economic justifications for fungal genomics are coupled with the sequencing capacity that will become available in major sequencing centers upon the completion of the human and mouse genomes, it is obvious that fungal genomics is about to enter a period of rapid growth. When EST projects are also considered, it is even difficult to estimate the number of projects already underway. It should be noted that NCBI (Genbank) has a database specifically for ESTs.

The smut basidiomycete *Ustilago maydis* is being sequenced at Lion Bioscience, Heidelberg, Germany. Syngenta (Novartis) and several other industrial giants are producing low coverage shotgun data for a number of species such as *Nectria haematococca* and *Trichoderma reeseii*. Paradigm Genetics, Research Triangle Park, North Carolina, is supporting work on several species including *Mycosphaerella graminicola* (anamorph:

Septoria tritici), the cause of wheat leaf blotch; and *Botrytis cinerea*, a cosmopolitan leaf pathogen sometimes called "the common gray mold".

Species that are high on the mycology "wish list" include: *Claviceps purpurea* (ergot alkaloid producer), *Cochliobolus heterostrophus* (maize pathogen), *Penicillium chrysogenum* (penicillin producer) and *Ophistoma ulmi* (cause of Dutch Elm Disease). The zygomycetes, a group characterized by fusing gametangia that form thick-walled zygospores and nonseptate hyphae, include a number of research-ready genera: *Absidia*, *Mucor*, *Phycomyces*, and *Rhizopus*. Basidiomycetes such as *Agaricus bisporous* (cultivated button mushroom), *Coprinus cinereus* (meiotic model), *Lentinus edodes* (shitake) and *Pleurotus ostreatus* (oyster mushroom), *Schizophyllum commune* (a model for mating type studies) are also appealing targets.

V. Gene Nomenclature

The need for standardized naming systems is well known in taxonomy, information theory, and in many other intellectual endeavors. Standardization of terms is an important and frequently cantankerous issue in science, one that has a long history of successful resolution in organic chemistry and biological systematics. Chemists, for example, have established international groups to regularize the naming of compounds. The Nomenclature Committee of the International Union of Biochemistry and Molecular Biology manages the well-known numbering system for enzymes (the "Enzyme Commission" or "EC number"). International taxonomic commissions similarly oversee animal, plant and bacterial nomenclature. (Note: for historical reasons, fungi are governed by the rules of botanical nomenclature.) Massive genomics databases have now focused attention on the issue of standardization for genetic nomenclature.

The primary purpose of a gene symbol is to provide a unique identifier for a specific locus. In general, gene names are brief and try to convey the character by which the gene is recognized. In the early years, genes were named rather haphazardly but, as genetics matured as a discipline, and the number of known mutants grew larger, uniform systems for naming genes evolved around model organisms. These rules and conventions of

nomenclature developed out of different traditions, and are not consistent among model systems. They usually try to embed information about phenotype, allelism/nonallelism, and dominance. For most fungi, three letter symbols are used, but older symbols are retained and may consist of one to five letters. In *Aspergillus nidulans*, *Neurospora crassa*, and *Schizosaccharomyces pombe*, gene names are given in lower case italics (*trp*); in *Saccharomyces cerevisiae* they are presented as italic capital letters (*CDC*). In some systems, letters of the alphabet are used to distinguish alleles (*trpC*); in other cases numbers are used (CDC28). The use of hyphens, superscripts and other nonnumerical or alphabetic symbols is quite idiosyncratic across systems. Elaborate rules concerning designation of dominance relations, distinguishing alleles from one another and so on, have also been developed for different models.

The gene nomenclature conventions for mycelial fungi have been summarized in detail by Bennett and Lasure (1985). More recently, in 1998, *Trends in Genetics* published a genetic nomenclature guide as a supplement to volume 13. It covers bacteria, protozoa, yeast, mycelial fungi, plants, invertebrates and vertebrates. In addition, many other nomenclature resources are available on the World Wide Web.

Aspergillus nidulans follows bacterial conventions (Demerec et al. 1966; Clutterbuck 1973). Names for genes are in lowercase italics (*trp*), with a capital letter of the alphabet to designate a specific gene in a pathway (*trpC*). Names of proteins are written in regular type, beginning with a capital letter (TrpC). Alleles for a given gene are designated with numbers (*trpC1*, *trpC2*, etc.); when allelic relationships of a mutant have not yet been determined, the capital letter is replaced by a hyphenated number (*trp-58*). The superscript "plus" (*trpC*[+]) indicates wild-type alleles. In general, for *Aspergillus nidulans*, dominance is not part of the primary gene symbol.

The conventions for *Neurospora crassa* resemble those developed for *Drosophila* and have been summarized by Perkins et al. (1982). Most genes are named with three-letter symbols (*tub*), but two-letter symbols are also common (*ad*). Recessive genes are written entirely in lowercase italics; when a mutant allele is known to be dominant, the first letter is capitalized (*Sk*, spore killer). The superscript "plus" designates wild type (*ad*[+]). To distinguish nonallelic loci, numbers separated by a hyphen are used (*ad-1*, *ad-2*). The use of the hyphen is similar to *Caenorhabditis elegans* and *Drosophila*, but differs from bacteria, *Aspergillus nidulans*, *Saccharomyces cerevisiae*, and *Schizosaccharomyces pombe*.

The original conventions used for *Saccharomyces cerevisiae* have been summarized by Sherman (1981). Three-letter symbols are used, with dominant alleles denoted by uppercase italics (*URA3*), and the recessive allele (*ura3*) by lowercase italics. For most structural genes, the nonmutant (wild-type) allele is dominant to the mutant form, and the convention in publishing linkage maps has been to use the nonmutant symbol (i.e., *URA3*). The protein product of the gene is written without italics (Ura3). The *Saccharomyces* Genome Database (SGD) has established guidelines for naming genes and maintains a "Global Gene Hunter" to help with the choice of unique gene names. Formal definitions are stipulated. "A reserved gene name is a unique identifier for a locus that will be published shortly in a scientific journal. This name has been submitted to SGD along with at least one piece of scientific information regarding the gene or its product (i.e., description, phenotype, gene product, ORF name, mapping information). A standard gene name is a unique gene name that is published in a scientific journal and is the primary accepted name for that gene. Other published names for that gene are 'Not Standard' names, or 'aliases'." (http://genome-www.stanford.edu/Saccharomyces/gene_guidelines.html).

In *Schizosaccharomyces pombe*, genes are designated by three lowercase italics, immediately followed by an unhyphenated number (*cdc01*, *cdc15*). A gene database has been published that lists gene names and also includes a description of the gene product or phenotype, map location, and possible synonyms (Lennon and Lehrach 1992). Other information is available through the Sanger Centre web site.

As classical and molecular genetic approaches have been applied to larger and larger numbers of fungi, scientists have adopted one or more of these nomenclature conventions to particular organisms. All too often, authors have made up their own system, or used gene names from inappropriate models. For example, in many of the papers on *Aspergillus fumigatus*, genes have been named following *S. cerevisiae* conventions rather than *A. nidulans* conventions (e.g., Tobin et al. 1997). Based on an Internet survey conducted by Michael Anderson at the University of Manchester, and after considerable discussion, the Steering Committee for the *A. fumigatus* genome project recommended that *A. fumigatus* should follow *A. nidulans* conventions.

The burgeoning numbers of new putative genes produced by genomic research have put new focus on old conventions of gene nomenclature. For the unwary, minor differences in notations can lead to erroneous assumptions. For example, the symbol *lys-1* would designate an uncharacterized lysine-requiring mutant in *A. nidulans* but would designate a characterized allele in *N. crassa*. Far more problematic is the lack of standardization for the three-letter epithets. The use of the same name for different genes causes one kind of problem; the use of different names for the same gene causes another kind of problem. When genes are clearly orthologous, it would be desirable to use the same or similar gene name across all species. Since there is a core of common eukaryotic genes, one could envision a system in which gene names for these central housekeeping genes could be unified (Rubin et al. 2000). A more modest goal would be to unify nomenclature within respective plant, animal, and fungal communities. Because budding yeast is such a magnificent model, it has been suggested that it would be appropriate to adopt *S. cerevisiae* names to all other fungi, a suggestion that no doubt will generate considerable heat among those who have grown up with allegiance to mycelial models.

As the magnitude of data increases, there is an increasing urgency about the need to impose defined, rigid rules. Community standards for gene nomenclature need to be identified and community consensus to follow these rules needs to be garnered.

These are not trivial issues. Wholesale changing of traditional names will cause problems with respect to accessing the print literature. Not unifying nomenclature will cause problems with database annotation. Currently, bar-code-like clone numbers are performing yeoman's service as surrogates for more mnemonically satisfactory names. The next few years will, of necessity, see an expansion of the Gene Ontology efforts.

VI. Applications

A. Biotechnology

The word genome is in danger of linguistic overuse, and has replaced "biotechnology" as the chic noun to be coupled with a variety of adjectives to describe contemporary disciplines: e.g., agricultural genomics, ecosystem genomics, nutritional genomics. The commercial hype surrounding genomics is an extension of the market-driven excitement elicited by genetic engineering and biotechnology since the late 1970s.

Genetic engineering brought a renaissance to drug discovery; genomics has further enriched this new era. Genomics research promises an understanding of the fitness of strains in industrial settings, unique ways to control plant and animal pathogens, the isolation of metabolic pathways from difficult-to-cultivate species and other potentially profitable outcomes. Traditional pharmaceutical companies have recognized the power of genomics to accelerate drug discovery through "data mining". Public and proprietary gene sequence databases are scrutinized to identify sequences of potential interest (Brocklehurst et al. 1999). By isolating genes that are unique to pathogens, new targets are identified, against which libraries of drug candidates can be screened, in so-called "genotypic drug discovery". A useful table of public databases relating genomic and proteomic information available on the Internet is provided by Bailey et al. (1999).

The commercial possibilities have attracted venture capitalists and several major companies have been founded including Affymetrix (Santa Clara, California), Celera (Rockville, Maryland), Genome Therapeutics (Waltham, Massachusetts), and Incyte Genomics (Palo Alto, California). Unlike the biotechnology start-ups of the 1980s, the new genomics companies have succeeded in generating immediate revenue by selling information (Cohen 1997). They have been influential in developing many of the tools that have sped the pace and the accuracy of genomics research. In addition, they have spawned another wave of start-up companies in proteomics such as Proteome, Inc. (Beverly, Massachusetts), Oritaba (Odense, Denmark), and Genomic Solutions (Ann Arbor, Michigan) (Service 2000). Large companies such as Celera and Incyte combine both genomic and proteomic approaches.

Several of the comprehensively curated protein databases maintained by Proteome, Inc. are of particular interest to fungal biologists. These include the Yeast Proteome Database (YPD), and the Fission Yeast Proteome Database (PombePD). The most recent addition, Myco-PathPD, is a database devoted to pathogenic fungi based on data from *Aspergillus flavus*, *A. fumigatus*, and *A. niger*; *Blastomyces dermatitidis*; *Candida albicans*, *C. dubliniensis*, *C. glabrata*, *C. guilliermondii*, *C. krusei*, *C. lusitaniae*, *C. parapsilosis*, *C. pseudotropicalis* and *C. tropicalis*; *Coccidioides immitis*; *Cryptococcus neoformans*; *Histoplasma capsulatum*; and *Pneumocystis carinii*.

There is a seductive aspect to genomics and proteomic projects, alluring some into the belief that once the complete DNA sequence is deciphered, or the complete expression pattern is worked out, an almost magical kind of biological comprehension will inevitably follow. Philosophical issues aside, it is quite apparent that neither meaning nor profits automatically leap out of computerized sequence analysis. These data are best exploited in the context of good experimental science.

In some countries, public fears of biotechnology, especially of genetically engineered foods, have extended to genomics research. Public distrust is greater in Europe than in the USA, and seems to be more pronounced for research on higher plants and animals than for bacteria and fungi (Gaskell et al. 1999). Luckily for molecular mycologists, the manipulation of fungal genetic material poses fewer ethical problems than the alteration of animal genomes. Similarly, the use of genetically engineered enzymes in food preparation is less problematic to consumers than the direct consumption of modified grains, fruits, and vegetables.

B. Comparative Genomics

Sequence data are becoming the common language for the integration of mycology and all of biology. The wealth of data deposited into public databases provides opportunities for whole genome comparisons. Sometimes relationships between extant species and their ancestors can be deduced from DNA sequence organization, thus providing insights into genome evolution. The International Sequence Database Collaboration maintains a taxonomy database with the names of all organisms that are represented in primary sequence databases with at least one nucleotide or protein sequence (http://www.ncbi.nlm.nih.gov/Taxonomy/taxonomyhome.html).

Genomes not only allow us to read the evolutionary history of organisms, they also allow us to identify those sequences that are crucial to basic life processes such as DNA replication, protein synthesis, gene regulation, and intermediary metabolism and, consequently, have been retained unaltered over long periods of evolutionary time. There is a remarkable amount of gene sequence homology between humans, worms, mice, *Drosophila* and *S. cerevisiae*. Such compelling evidence of cross phylum sequence conservation gives new meaning to the concept of "unity of biochemistry".

Comparisons between the *S. cerevisiae* and the *N. crassa* genomes have been informative. Both are members of the Ascomycota, estimated to have diverged from each other 300–400 million years ago (Berbee and Taylor 1993; Taylor et al. 1999). Compared to yeast, *N. crassa* has more morphological complexity and the genome is about three times larger. Braun et al. (2000) conducted a large-scale homology search using BLAST to query sequence to three databases: the set of translated ORFs from *S. cerevisiae*; a set of translated nonfungal ORFs in public databases; and the human and mouse EST database. Only a third of the predicted genes from *N. crassa* had clear homologies in the database queries, while 57% of the predicted genes from *S. cerevisiae* had clear homologues in the same databases. These data extended and supported earlier EST studies that had shown that *N. crassa* had more orphan genes than *S. cerevisiae* (Nelson 1997). Most of the *N. crassa* genes with identifiable homologues were common to both *S. cerevisiae* and nonfungal species. Only nine genes had homologues in *S. cerevisiae* but not in the other nonfungal databases. These were candidates for fungal-specific proteins; three were cell wall components (e.g. Gas1p), two candidates were transcription factors (Emc22p and Sok2p). Additional searches of these nine cases were conducted against sequence sets from other fungi; homologues of all nine were found in *Candida albicans* and seven of the nine were found in *Schizosaccharomyces pombe* (Braun et al. 2000).

Comparative studies also allow us to tabulate known genes by class, and organize the inventory of known genes. When it is known how a gene product functions in a model system, sequence homology can be used to extrapolate from models to other species. We can study families of genes rather than single mutants; whole biochemical pathways, not single reactions. Many essential genes remain grouped together in the same posi-

tion; such synteny can be useful when model genomes are used to understand relevant processes in pathogens and industrial species. *S. cerevisiae* genes shed light on human disease; mouse genes can inform mold metabolism. No doubt, as discussed earlier, *Aspergillus nidulans* and *Neurospora crassa* sequences will be used to extrapolate to their counterparts in genetically less tractable fungal species.

VII. Conclusions

Genome science has become an essential part of the scientific infrastructure and is a critical component of 21st century biology. Metaphors used to describe the various large-scale genome projects include "The Holy Grail", "The Rosetta Stone" and the "Biological Periodic Table". As the final stages of the human, *Drosophila*, and mouse genome projects come to a close, excess sequencing capacity will be freed up in all the major sequencing centers. It is safe to say that the genomes of hundreds of species spanning all domains of life will soon be sequenced. Genomics approaches are scientifically appealing, especially for organisms that are difficult to culture and/or are experimentally recalcitrant. Other strong candidates for sequencing are organisms with economic importance. The relatively small size of fungal genomes, combined with the industrial, agricultural and medical significance of many species, ensures that the genome bandwagon will soon be full of molds, rusts, smuts, mildews, mushrooms, truffles and such.

The glut of data coming out of genomics projects is often described as a "flood", "deluge", "avalanche", or "explosion". The choice of these disaster metaphors reflects the way in which the overwhelming amount of sequence information makes many biologists feel as if they are "drowning" in data, or are being "buried" in data. The information is so voluminous that it is difficult to comprehend. Most biologists lack the skills in computer science to take full advantage of *in silico* research. Further, the growing abstraction of genomics research is a marked departure from traditional laboratory- and field-based biology. Nevertheless, genomics marks a shift away from the extreme reductionism of molecular biology, and allows insights into biological complexity never before possible. The whole is, indeed, greater than the sum of its parts. An organism is more than the sum of its genes. Genomics shifts emphasis away from single genes to whole genomes. Post-genomic analysis seeks to decipher functional networks and analyze complexity in a more profound way. Ideally, after a sequence is determined and annotated, additional genomic analysis will precede the shift to "small science" experiments. These additional analyses will include genome-scale expression analysis at the RNA, protein, and pathway levels. The wealth of data produced in the genome-scale studies will provide an informed starting point for designing "wet" laboratory experiments.

The sequencing of a genome demonstrates the powers as well as the limits of modern molecular biology. Current genomics data provide an algorithm for talking about fundamental questions but do not provide answers to these questions.

Acknowledgements. Many colleagues have provided help. Special thanks go to Maria Costanzo, Rhian Gwilliam, Richard Hyman, Sophien Kamoun, Jeff Shuster, Chuck Staben, and Brett Tyler for help in tracking down information. Useful input was also provided by Bruce Birren. Maria Costanzo, Doris Kupfer and Bill Nierman reviewed the manuscript. Sue Boucher, Karl Esser, Neil Gow and Rick Howard gave outstanding editorial support. The Burroughs-Wellcome Foundation has provided travel funds to a number of conferences. We are also grateful for support by MCB-9630910.

References

Adams MD, Kelley JM, Gocayne JD et al. (1991) Complementary DNA sequencing: expressed sequence tags and human genome project. Science 252:1651–1656

Adams MD, Celniker SE, Holt RA, Evans CA, Gocayne JD et al. (2000) The genome sequence of *Drosophila melanogaster*. Science 287:2185–2195

Adams TH, Yu JH (1998) Coordinate control of secondary metabolite production and asexual sporulation in *Aspergillus nidulans*. Curr Opin Microbiol 1:674–677

Aign V, Schulte U, Hoheisel JD (2001) Hybridization mapping of *Neurospora crassa* linkage groups II and V. Genetics 157:1015–1020

Alberts BM, Botstein D, Brenner S (1988) Report of the Committee on Mapping and Sequencing the Human Genome, National Academy of Sciences, Washington, DC

Alec M, Gold MH (1991) Genetics and molecular biology of the lignin-degrading basidiomycete *Phanerochaete chrysosporium*, In: Bennett JW, Lasure LL (eds) More

Gene Manipulations in Fungi. Academic Press, New York, pp 320–341

Altmann-Johl R, Philippsen P (1996) AgTHR4, a new selection marker for transformation of the filamentous fungus *Ashbya gossypii*, maps in a four-gene cluster that is conserved between *A. gossypii* and *Saccharomyces cerevisiae*. Mol Gen Genet 250:69–80

Altschul FS, Gish W, Miller W, Myers EW, Lipman DJ (1990) Basic local alignment search tool. J Mol Biol 215:403–410

Amar YG, Moore MM (1998) Mapping of the nitrate-assimilation gene cluster (*crnA-niiA-niaD*) and characterization of the nitrite reductase gene (*niiA*) in the opportunistic fungal pathogen *Aspergillus fumigatus*. Curr Genet 33:206–215

Amber D (2000) Production genomics. The Scientist 14:1, 14, 16, 19

Anonymous (2000) Abstracts Third International Symposium on Fungal Genomics, International Business Communications, Southborough, Massachusetts

Aramayo R, Bennett JW (1997) The importance of fungal genomes. Am Soc Microbiol News 63:176–177

Aravind L, Watanabe H, Lipman DJ, Koonin EV (2000) Lineage-specific loss and divergence of functionally linked genes in eukaryotes. Proc Natl Acad Sci USA 97:11319–11324

Arnold J (1997) Editorial. Fungal Genet Biol 21:254–257

Attwood TK, Croning MDR, Flower DR, Lewis AP, Maybe JE (2000) PRINTS-S: the database formerly known as PRINTS. Nucleic Acids Res 28:225–227

Bailey DS, Furness LM, Dean PM (1999) New tools for quantifying molecular diversity. PharmaInformatics. Trends Suppl 6–9

Barr DP, Aust SD (1994) Mechanisms white rot fungi use to degrade pollutants. Environ Sci Technol 28:79A–87A

Beadle GW, Tatum EL (1941) Genetic control of biochemical reaction in *Neurospora*. Proc Natl Acad Sci USA 27:499–506

Bean LE, Dvoracheck WH, Braun EL, Errett A, Saenz GS et al. (2001) Analysis of the *pdx-1* (*snz-1/sno-1*) region of the *Neuorspora crassa* genome: correlation of pyridoxine-requiring phenotypes with mutations in two structural genes. Genetics 157:1067–1075

Bennett JW (1997a) Open letter to fungal researchers. Fungal Genet Biol 21:2

Bennett JW (1997b) White paper: genomics for filamentous fungi. Fungal Genet Biol 21:3–7

Bennett JW, Chang P-K, Bhatnagar D (1997) One gene to whole pathway: the role of norsolorinic acid in aflatoxin research. Adv Appl Microbiol 45:1–15

Bennett JW, Lasure L (1985) Conventions for gene symbols. In: Bennett JW, Lasure L (eds) Gene manipulations in fungi. Academic Press, San Diego, pp 537–544

Benson DA, Boguski MS, Lipman DJ, Ostell J, Ouellette BFF (1998) GenBank. Nucleic Acids Res 26:1–7

Berbee ML, Taylor JW (1993) Dating the evolutionary radiations of the true fungi. Can J Bot 71:1114–1127

Berg P, Singer M (1992) Dealing with genes. The language of heredity. University Science Books, Mill Valley, California, p 247

Bhandarkar SM, Machaka SA, Shete SS, Kota RN (2001) Parallel computation of a maximum likelihood estimator of a physical map. Genetics 157:1021–1043

Birren B, Green ED, Klaholz S, Myers RM, Roskams J (1997) Genome Analysis, a Laboratory Manual, vol. 1.

Analyzing DNA. Cold Spring Harbor Press, Cold Spring Harbor

Bishop M (ed) (1999) Genetic databases. Academic Press, San Diego

Blanchard AP, Hood L (1996) Sequence to array: probing the genome's secrets. Nature Biotech 14:1649

Blattner FR, Plunkett G, Block CA, Perna NT, Burland V et al. (1997) The complete genome sequence of *Escherichia coli* K-12. Science 277:1453–1464

Boguski MS (1995) The turning point in genomic research trends. Biochem Sci 20:295–296

Botstein D, Cherry JM (1997) Molecular linguistics: extracting information from gene and protein sequences. Proc Natl Acad Sci USA 94:5506–5507

Botstein D, White RL, Skolnick M, Davis RW (1980) Construction of a genetic linkage map in man using restriction fragment length polymorphisms. Am J Hum Genet 32:314–331

Botstein D, Chervitz SA, Cherry JM (1997) Yeast as a model organism. Science 277:1259–1260

Braun EL, Halpern AL, Nelson MA, Natvig DO (2000) Large-scale comparison of fungal sequence information: mechanisms of innovation in *Neurospora crassa* and gene loss in *Saccharomyces cerevisiae*. Genome Res 10:416–430

Brenner SE (1999) Errors in genome annotation. Trends Genet 15:132–133

Brocklehurst SM, Hardman CH, Johnston SJT (1999) Creating integrated computer systems for target discovery and drug discovery. PharmaInformatics. Trends Suppl 1999:12–15

Broda PP, Birch RJ, Brooke PR, Sims PRFG (1996) Lignocellulose degradation by *Phanerochaete chrysosporium*: gene families and gene expression for a complex process. Mol Microbiol 19:923–932

Brown APJ, Gow NAR (1999) Regulatory networks controlling *Candida albicans* morphogenesis. Trends Microbiol 7:333–338

Brown DW, Yu J-H, Kelkar HS, Fernandes M, Nesbitt TC et al. (1996) Twenty five coregulated transcripts define a sterigmatocystin gene cluster in *Aspergillus nidulans*. Proc Natl Acad Sci USA 93:1418–1422

Burke DT, Carle GF, Olson MV (1987) Cloning of large segments of exogenous DNA into yeast by means of artificial chromosome vectors. Science 236:806–812

Burset M, Guigo R (1996) Evaluation of gene structure prediction programs. Genomics 34:353–367

C. elegans Sequencing Consortium (1998) Genome sequence of the nematode *C. elegans*: a platform for investigating biology. Science 282:2012–2018

Carlson M (2000) The awesome power of yeast biochemical genomics. Trends Genet 16:49–51

Casadevall A, Perfect JR (1998) *Cryptococcus neoformans*. ASM Press, Washington, DC

Clutterbuck AJ (1973) Gene symbols in *Aspergillus nidulans*. Genet Res 21:291–296

Clutterbuck AJ (1994) Linkage map and locus list. In: Martinelli SD, Kinghorm JR (eds) *Aspergillus* fifty years on. Elsevier, Amsterdam, pp 791–805

Clutterbuck (1997) The validity of the *Aspergillus nidulans* linkage map. Fungal Genet Biol 21:267–277

Cohen J (1997) The genomics gamble. Science 275:767–775

Collins FS, Gallas D (1993) A new five-year plan for the US Human Genome Project. Science 262:43–44

Collins FS, Patrinos A, Jordan E, Chakrevartie A, Gesteland R, Walters L and the members of the DOE and NIH planning groups (1998) New goals for the US Human Genome Project: 1998–2003. Science 282:682–689

Cushion MT, Arnold JA (1998) Proposal for a *Pneumocystis* genome project. J Eukaryot Microbiol 44:7s

Cushion MT, Kaselis M, Stringer SL, Stringer JR (1993) Genetic stability and diversity of *Pneumocystis carinii* infecting rat colonies. Infect Immun 61:4801–4813

Cuticchia AJ, Arnold J, Timberlake WE (1992) The use of simulated annealing in chromosome reconstruction experiments based on binary scoring. Genetics 132:591–601

Davis R (2000) *Neurospora*: contributions of a model organism. Oxford Univ. Press, Oxford

DeBacker MD, Magee PH, Pla J (2000) Recent developments in molecular genetics of *Candida albicans*. Annu Rev Microbiol 54:463–498

Debets F, Swart K, Hockstra RF, Bos CJ (1993) Genetic maps of eight linkage groups of *Aspergillus niger* based on mitotic mapping. Curr Genet 23:47–53

Demain A (1972) Riboflavin over synthesis. Annu Rev Microbiol 26:369–388

Demerec M, Adelberg AE, Clark AJ, Hartman PE (1966) A proposal for a uniform nomenclature in bacterial genetics. Genetics 54:61–76

DeRisi JL, Iyer VR, Brown PO (1997) Exploring the metabolic and genetic control of gene expression on a genomic scale. Science 278:680–686

Diaz-Perez SV, Crouch VW, Orbach MC (1996) Construction and characterization of a *Magnaporthe grisea* bacterial artificial chromosome library. Fungal Genet Biol 20:280–288

Dujon B (1996) The yeast genome project: what did we learn? Trends Genet 12:263–270

Dunn-Coleman N, Prade R (1998) Toward a global filamentous fungus genome sequencing effort. Nature Biotech 16:5

Edman JC, Kovacs JA, Masur H, Sanit DV, Elwood HJ, Sogin ML (1988) Ribosomal RNA sequence shows *Pneumocystis carinii* to be a member of the fungi. Nature 334:519–522

Enright AJ, Illopoulos I, Kyrpides NC, Ouzounis CA (2000) Protein interaction maps for comparing genomes based on gene fusion events. Nature 402:86–90

Ewing B, Green P (1998) Base-calling of automated sequencer traces using *phred*. II. Error probabilities. Genome Res 8:186–194

Ewing B, Hillier L, Wendl MC, Green P (1998) Base-calling of automated sequencer traces using *phred*. I. Accuracy assessment. Genome Res 8:175–185

Farman ML, Leong SA (1998) Chromosome walking to the AVR1-CO39 avirulence gene of *Magnaporthe grisea*: discrepancy between the physical and genetic maps. Genetics 150:1049–1058

Fickett JW (1996) Finding genes by computer. The state of the art. Trends Genet 12:316–320

Fitch WM (1970) Distinguishing homologous from analogous proteins. Syst Zool 19:99–113

Fleischmann RD, Adams MD, White O et al. (1995) Whole-genome random sequencing and assembly of *Haemophilus influenzae* RD. Science 269:496–512

Forsburg SL, Nurse P (1991) Cell cycle regulation in the yeasts *Saccharomyces cerevisiae* and *Schizosaccharomyces pombe*. Annu Rev Cell Biol 7:227–256

Forster H, Coffey MD, Elwood H, Sogin ML (1990) Sequence analysis of the small subunit ribosomal RNAs of three zoosporic fungi and implications for fungal evolution. Mycologia 82:306–312

Fraser CM, Gacayne JD, White O et al. (1995) The minimal gene complement of *Mycoplasma genitalium.* Science 270:397–403

Garber ED (1972) Cytogenetics, an introduction. McGraw Hill Book, New York

Gaskell G, Bauer MW, Durant J, Allum NC (1999) Worlds apart? The reception of genetically modified foods in Europe and the US. Science 285:384–387

Geiser DM, Pitt JI, Taylor JW (1998) Cryptic speciation and recombination in the aflatoxin-producing fungus *Aspergillus flavus.* Proc Natl Acad Sci USA 95:388–393

Gene Ontology Consortium (2000) Gene ontology: tool for the unification of biology. Nature Genet 25:25–29

Gilbert W (1991) Toward a paradigm shift in biology. Nature 349:99

Godfrey T, West S (eds) (1996) Industrial enzymology, 2nd edn. Stockton Press, New York

Goffeau A, Barrell BG, Bussey H, Davis RW et al. (1996) Life with 6000 genes. Science 274:546–567

Goffeau A et al. (1997) The yeast genome directory. Nature 387(Suppl):1–105

Gross C, Kelleher M, Iyer VR, Brown PO, Winge DR (2000) Identification of the copper regulon of *Saccharomyces cerevisiae* by DNA microarrays. J Biol Chem 275:32310–32316

Gygi SP, Corthals GL, Zhang Y, Rochon Y, Aebersold R (2000) Evaluation of two-dimensional gel electrophoresis-based proteome analysis technology. Proc Natl Acad Sci USA 97:9390–9395

Hall D, Bhandarkar SM, Wang J (2001) ODS2: a multiplatform software application for creating integrated physical and genetic maps. Genetics 157:1045–1056

Hamer JE, Farrall L, Orbach MJ, Valent B, Chumley FG (1989) Host species-specific conservation of a family of repeated DNA sequences in the genome of a fungal plant pathogen. Proc Natl Acad Sci USA 86:9981–9985

Hamer L (1997) Meeting review. From genes to genomes: sequencing of filamentous fungal genomes. Fungal Genet Biol 21:8–10

Hawksworth DL (1991) The fungal dimension of biodiversity: magnitude, significance and conservation. Mycol Res 95:641–655

Hayles J, Nurse P (1992) Genetics of the fission yeast *Schizosaccharomyces pombe.* Annu Rev Genet 26:373–402

Hegyi H, Gerstein M (1999) The relationship between protein structure and function: a comprehensive survey with application to the yeast genome. J Mol Biol 288:147–164

Heitman J, Casadevall A, Leodge JK, Perfect JR (2000) The *Cryptococcus neoformans* genome sequencing project. Mycopath 148:1–7

Hieter P, Boguski M (1997) Functional genomics: it's all how you read it. Science 278:601–602

Hodgson J (1994) Genome mapping the "easy" way. Bio/Technology 12: 581–584

Hoheisel JD, Maier C, Mott R, McCarthy L, Grigoriev AV (1993) High resolution cosmid and P1 maps spanning the 14 Mb genome of the fission yeast *S. pombe.* Cell 73:109–120

Hughes TR, Marton MJ, Jones AR, Roberts CJ, Stoughton R et al. (2000) Functional discovery via a compendium of expression profiles. Cell 102:109–126

Hull CM, Raisner RM, Johnson D (2000) Evidence for mating of the "asexual" yeast *Candida albicans* in a mammalian host. Science 289:307–310

Hyman RW (2001) Sequence data: posted vs. published. Science 291:827

International Human Genome Sequencing Consortium (2001) Initial sequencing and analysis of the human genome. Nature 409:860–918

Judelson HS (1996) Recent advances in the genetics of oomycete plant pathogens. Mol Plant-Microbe Interact 9:443–449

Kafer E (1977) Meiotic and mitotic recombination in *Aspergillus* and its chromosomal aberrations. Adv Genet 19:33–131

Kamoun S, Hraber P, Sobral B, Nuss D, Govers F (1999) Initial assessment of gene diversity for the oomycete plant pathogen *Phytophthora infestans* based on expressed sequences. Fungal Genet Biol 28: 94–106

Kelkar HS, Griffith J, Case ME, Covert SF, Hall RD, Arnold J (2001) The *Neurospora crassa* genome: cosmid libraries sorted by chromosomes. Genetics 157:979–990

Kennedy D (2001) Accepted community standards. Science 291:789

Kirk TK, Farrell RL (1987) Enzymatic "combustion," the microbial degradation of lignin. Annu Rev Microbiol 41:465–505

Klich MA, Yu J, Mullaney EJ (1987) DNA restriction enzyme fragment polymorphism as a tool for rapid differentiation of *Aspergillus flavus* from *Aspergillus oryzae.* Exp Mycol 11:170–175

Klich MA, Montalbano B, Ehrlich K (1997) Northern analysis of aflatoxin biosynthesis genes in *Aspergillus parasiticus* and *Aspergillus sojae.* Appl Microbiol Biotech 47:246–249

Koonin EV, Mushegian AR, Rudd KE (1996) Sequencing and analysis of bacterial genomes. Curr Biol 6:404–416

Kraemer E, Wang J, Guo J, Arnold J (2001) An analysis of gene-finding approaches for *neurospora crassa.* Bio information (in press)

Kron J, Gow NAR (1995) Budding yeast morphogenesis. Signalling, cytoskeleton and cell cycle. Curr Opin Cell Biol 7:845–855

Kwon-Chung KJ, Bennett JE (1992) Medical mycology. Lea & Febiger, Philadelphia

Kupfer DM, Reece CA, Clifton SW, Roe BA, Prade RA (1997) Multicellular ascomycetous fungal genomes contain more than 8000 genes. Fungal Genet Biol 21:364–372

Kusumoto K-I, Nogata Y, Ohta H (2000) Directed deletions in the aflatoxin biosynthesis gene homolog cluster of *Aspergillus oryzae.* Curr Genet 37:104–111

Latge JP (1999) *Aspergillus fumigatus* and aspergillosis. Clin Microbiol Rev 12:310–350

Lennon GG, Lehrach H (1992) Gene database for the fission yeast *Schizosaccharomyces pombe.* Curr Genet 21:1–11

Leung H, Borromeo ES, Bernardo MA, Notteghem JL (1998) Genetic analysis of virulence in the rice blast fungus *Magnaporthe grisea.* Phytopathology 78:1227–1233

Levitz SM (1991) The ecology of *Cryptococcus neoformans* and the epidemiology of cryptococcosis. Rev Infect Dis 13:1163–1169

Lipman DJ, Pearson WR (1985) Rapid and sensitive protein similarity searches. Science 227:1435–1441

Lucas M, Gwillam R, Lepingle A, Rajndream MA et al. (2000) Sequence analysis of two cosmids from

Schizosaccharomyces pombe chromosome III. Yeast 16:1519–1526

Machida M, Akita O, Kashiwagi Y, Koyama Y, Yamaguchi S et al. (2000) Analyses of ESTs and the promoters of useful expression patterns from *Aspergillus oryzae*. Int Symp Mol Biol Filamentous Fungi Aspergilli, p 3 (Abstr)

Magee BB, Magee PT (2000) Induction of mating in *Candida albicans* by construction of MTLa and MTLα strains. Science 289:310–313

Marcotte EM, Pellegrini M, Thompson MJ, Yeates TO, Eisenberg D (1999) A combined algorithm for genome-wide prediction of protein function. Nature 402:83–86

Maier E, Hoheisel JD, McCarthy LM, Mott R, Grigoriev AV et al. (1992) Complete coverage of the *Schizosaccharomyces pombe* genome in yeast artificial chromosomes. Nature Genet 1:273–277

Martzen MR et al. (1999) A biochemical genomics approach to identify genes by the activity of their products. Science 286:1153–1155

May GS, Adams TH (1997) The importance of fungi to man. Genome Res 7:1041–1044

McClintock B (1945) Neurospora. 1. Preliminary observations of the chromosomes of *Neurospora crassa*. Am J Bot 32:671–678

Mewes HW, Albermann K, Bahr M, Frishman D, Gleissner A et al. (1997) Overview of the yeast genome. Nature (Suppl) 387:7 8

Mewes HW, Heumann K, Kaps A, Mayer K, Pfeiffer F et al. (1999) MIPS: a database for genomes and protein sequences. Nucleic Acids Res 27:44–48

Minto RE, Townsend CA (1997) Enzymology and molecular biology of aflatoxin biosynthesis. Chem Rev 97:2537–2555

Nelson MA, Kang S, Braun EL, Crawford ME, Dolan PL et al. (1997) Expressed sequences from conidial, mycelial and sexual stages of *Neurospora crassa*. Fungal Genet Biol 21:348–363

Nelson RT, Hua J, Pryor B, Lodge JK (2001) Identification of virulence mutants of the fungal pathogen, *Cryptococcus neoformans*, using signature tagged mutagenesis. Genetics 157:935–947

Nurse P (1994) Ordering S phase and M phase in the cell cycle. Cell 79:547–550

O'Brien SJ (ed) (1993) Genetic maps. Locus maps of complex genomes, vol 3. Lower eukaryotes. Cold Spring Harbor Press, Cold Spring Harbor

Oliver S (2000) Guilt-by-association goes global. Nature 403:601–603

Orbach MJ, Vollrath D, Davis RW, Yanofsky C (1988) An electrophoretic karyotype of *Neurospora crassa*. Mol Cell Biol 8:1469–1473

Pandey A, Mann M (2000) Proteomics to study genes and genomes. Nature 405:837–846

Pearson WR (1990) Rapid and sensitive sequence comparison with FASTP and FASTA. Methods Enzymol 83:63–98

Pearson WR (1991) Searching protein sequence libraries: comparison of the sensitivity and selectivity of the Smith-Waterman and FASTA algorithms. Genomics 11:635–650

Pearson WR, Miller W (1992) Dynamic programming algorithms for biological sequence comparison. Methods Enzymol 210:575–601

Pearson WR, Lipman DJ (1988) Improved tools for biological sequence comparison. Proc Natl Acad Sci USA 85:2444–2448

Perkins DD, Radford A, Newmeyer D, Bjorkman M (1982) Chromosomal loci of *Neurospora crassa*. Microbiol Rev 46:426–570

Perkins D, Radford A, Sachs M (2000) Compendium of chromosomal mutations of *Neurospora*. Academic Press, San Diego

Peruski LF, Peruski AH (1997) The Internet and the new biology. tools for genomic and molecular research. American Society of Microbiology Press, Washington, DC

Pla J, Gil C, Monteoliva L, Navarro-Garcia F, Sanchez M, Nombela C (1996) Understanding *Candida albicans* at the molecular level. Yeast 12:1677–1702

Pontecorvo G (1956) The parasexual cycle in fungi. Annu Rev Microbiol 10:393–400

Pontecorvo G, Roper JA, Hemmons LM, Macdonald KD, Bufton AWJ (1952) The genetics of *Aspergillus nidulans*. Adv Genet 5:141–238

Prade RA (1998) Meeting review. Fungal genomics – one per week. Fungal Genet Biol 25:76–78

Prade RA (2000) The reliability of the *Aspergillus nidulans* physical map. Fungal Genet Biol 29:175–185

Prade RA, Griffith J, Kochut K, Arnold J, Timberlake WE (1997) *In vitro* reconstruction of the *Aspergillus* (= *Emericella*) *nidulans* genome. Proc Natl Acad Sci USA 94:14564–14569

Prade RA, Ayoubi P, Misawa E, Garcia F, Ray T, Samad R et al. (2001a) Large-scale genome DNA sequence survey of the *Aspergillus nidulans* chromosome IV (in preparation)

Prade RA, Ayoubi P, Krishnan S, Macwana S, Russell H (2001b) Accumulation of stress and cell wall degrading enzyme associated transcripts during asexual development in *Aspergillus nidulans*. Genetics 157:957–967

Primrose SB (1995) Principles of genome analysis. Blackwell Science, Oxford

Qutob D, Graber P, Sobral B, Gijzen M (2000) Comparative analysis and expressed sequences in *Phytophthora sojae*. Plant Physiol 123:243–253

Raeder U, Thompson W, Broda P (1989) RLFP-based genetic maps of *Phanerochaete chrysosporium* ME44: lignin peroxidase genes occur in clusters. Mol Microbiol 3:911–918

Radford A, Parish JH (1997) The genome and genes of *Neurospora crassa*. Fungal Genet Biol 21:258–266

Romao J, Hamer JE (1992) Genetic organization of a repeated DNA sequence family in the rice blast fungus. Proc Natl Acad Sci USA 89:5316–5320

Randall TA, Judelson HS (1999) Construction of a bacterial artificial chromosome library of *Phytophthora infestans* and transformation of clones into *P. infestans*. Fungal Genet Biol 28:160–170

Rieder MJ, Taylor SL, Tobe VO, Nickerson DA (1998) Automating the identification of DNA variations using quality-based fluorescence re-sequencing: analysis of the human mitochondrial genome. Nucleic Acids Res 26:967–973

Riley M (1993) Functions of the gene products of *Escherichia coli*. Microbiol Rev 57:862–952

Riley M, Serres MH (2000) Interim report on genomics of *Escherichia coli*. Annu Rev Microbiol 54:341–411

Rippon JW (1988) Medical mycology, the pathogenic fungi and the pathogenic actinomycetes, 3rd edn. WB Saunders, Philadelphia

Rubin GM, Yandell MD, Wortman JR, Miklos GLG, Nelson CR et al. (2000) Comparative genomics of the eukaryotes. Science 287:2204–2220

Schatz BR (1997) Information retrieval in digital libraries: bringing search to the net. Science 275:327–334

Schena M, Shalon D, Davis RW, Brown PO (1995) Quantitative monitoring of gene expression patterns with a complementary DNA microarray. Science 270:467–470

Scherer S, Magee PH (1990) Genetics of *Candida albicans*. Microbiol Rev 54:226–241

Schulte U, Becker I, Mewes HW, Mannhaupt G (2001) Large scale analysis of sequences from *Neurospora crassa*. J Biotech (in press)

Sebghati TS, Engle JT, Goldman WE (2000) Intracellular parasitism by *Histoplasma capsulatum*: ungal virulence and calcium dependence. Science 290:1968–1971

Service RF (2000) Can Celera do it again? Science 287:2136–2138

Sherman F (1981) Genetic nomenclature. In: Strathern JN, Jones EW, Broach JR (eds) Molecular Biology of the yeast *Saccharomyces cerevisiae*. Cold Spring Harbor Lab, Cold Spring Harbor, pp 639–640

Shoemaker DD, Lashkari DA, Morris D, Mittmann M, Davis RW (1996) Quantitative phenotypic analysis of yeast deletion mutants using a highly parallel molecular bar-coding strategy. Nature Genet 14:450–456

Skinner DZ, Budde AD, Leong SA (1991) Molecular karyotype analysis of fungi. In: Bennett JW, LasureLL (eds) More Gene Manipulations in Fungi. Academic Press, New York, pp 86–102

Skinner DZ, Budde AD, Farman ML, Leung H et al. (1993) Genome organization of the rice blast fungus, *Magnaporthe grisea*: genetic map, electrophoretic karyotype, and occurrence of repeated DNAs. Theor Appl Genet 87:545–557

Skolnick J, Fetrow JS, Kolinski A (2000) Structural genomics and its importance for gene function analysis. Nature Biotech 18:283–287

Smith JMB (1989) Opportunistic mycoses of man and other animals. CAB International, Wallingford

Smulian AG, Sesterhenn T, Tanaka R, Cushion MT (2001) The *ste3* pheromone receptor gene of the *Pneumocystis carnii* is surrounded by a cluster of signal transduction genes. Genetics 157:991–1002

Stebbins GL (1966) Chromosome variation and evolution. Science 152:1463–1469

Sunkin SM, Stringer JR (1996) Translocation of surface antigen genes to a unique telomeric site in *Pneumocystis carinii*. Mol Microbiol 19:283–295

Sweigard JA, Valent B, Orbach MJ, Walter AM, Rafalski A et al. (1993) Genetic map of the rice blast fungus *Magnaporthe grisea* (n = 7). In: O'Brien SJ (ed) Genetics Maps, 6th edn. Cold Spring Harbor Lab Press, Cold Spring Harbor, pp 3.112–3.114

Sweigard JA, Carrol A, Kang S, Farrall L, Chumley FC et al. (1995) Identification, cloning and characterization of PWL2, a gene for host species specificity in the rice blast fungus. Plant Cell 7:1221–1233

Swindell SR, Miller RR, Myers GAS (eds) (1995) Internet for the molecular biologist. Horizon Sci Press, Norfolk

Sybenga J (1972) General cytogenetics. Am Elsevier Publ, New York

Tait E, Simon MC, King S, Brown AJ, Gow NAR et al. (1997) A *Candida albicans* genome project: cosmid contigs, physical mapping, and gene isolation. Fungal Genet Biol 21:308–314

Tatusov RL, Koonin EV, Lipman DJ (1997) A genomic perspective on protein families. Science 278:631–637

Taylor JW, Hass T, Kerp H (1999) The oldest fossil ascomycetes. Nature 399:648

Thompson KA (1999) Useful informatics for industrial microbiologists. Soc Ind Microbiol News 49:5–10

Timberlake WE (1990) Molecular genetics of *Aspergillus* development. Annu Rev Genet 24:5–36

Timberlake WE (1991) Cloning and analysis of fungal genes. In: Bennett JW, Lasure LL (eds) More gene manipulations in fungi. Academic Press, San Diego, pp 51–85

Tobin MB, Peery RB, Skatrud PL (1997) An electrophoretic molecular karyotype of a clinical isolate of *Aspergillus fumigatus* and localization of the MDR-like genes AfuMDR1 and AfuMDR2. Diagn Microbiol Infect Dis 29:67–71

Tooley PW, Carras MM (1992) Separation of chromosomes of *Phytophthora* species using CHEF gel electrophoresis. Exp Mycol 16:188–196

Tooley PW, Therrein CD (1987) Cytophotometric determination of the nuclear DNA content of 23 Mexican and 18 non-Mexican isolates of *Phytophthora infestans*. Exp Mycol 11:19–26

Uetz P, Giot L, Cagney G, Mansfield TA, Judson RS, Knight JR et al. (2000) A comprehensive analysis of protein-protein interactions in *Saccharomyces cerevisiae*. Nature 403:623–627

Van der Lee T, De Witte I, Drenth A, Alfonso C, Govers F (1997) AFLP linkage map of the oomycete *Phytophthora infestans*. Fungal Genet Biol 21:278–291

Van der Lee T, Robold A, Testa A, Van't Klooster JW, Govers F (2001) Mapping of avirulance genes in *Phytophthora infestans* with AFLP markers selected by bulked segregant analysis. Genetics 157:949–956

Venter JC, Smith HO, Hood L (1996) A new strategy for genome sequencing. Nature 381:364–366

Venter JC, Adams MD, Sutton GG, Kerlavage AR, Smith HO, Hunkapiller M (1998) Shotgun sequencing of the human genome. Science 280:1540–1542

Venter JC, Adams MD, Myers EW et al. (2001) The sequence of the human genome. Science 291:1304–1351

Verdoes JC, Calil MR, Punt PJ, Debets F, Swart K et al. (1994) The complete karyotype of *Aspergillus niger*: the use of introduced electrophoretic mobility variation of chromosomes for gene assignment studies. Mol Gen Genet 244:75–80

Vos P, Hogers R, Bleeker M, Reijans M, van de Lee T et al. (1995) AFLP: a new technique for DNA fingerprinting. Nucleic Acids Res 23:4407–4414

Walz M (1995) Electrophoretic karyotyping. In: Kück U (ed) Genetics and biotechnology, vol 2. The Mycota. Springer, Berlin Heidelberg New York, pp 63–73

Watson AJ, Fuller LJ, Jeenes DJ, Archer DB (1999) Homologs of aflatoxin biosynthesis genes and sequence of aflR in *Aspergillus oryzae* and *Aspergillus sojae*. Appl Environ Microbiol 65:307–310

Waugh M, Hraber P, Weller J, Wu Y, Chen G et al. (2000) The *Phytophthora* genome initiative database: informatics and analysis for distributed pathogenic research. Nucleic Acids Res 28:87–94

Weidner G, d'Enfert C, Koch A, Mop PC, Brakhage AA (1998) Development of a homologous transformation system for the human pathogenic fungus *Aspergillus fumigatus* based on the PyrG gene encoding orotidine 5'-monophosphate decarboxylase. Curr Genet 33: 378–385

Wendland J, Ayad-Durieux Y, Knechtle P, Rebischung R, Philippsen P (2000) PCR-based gene targeting in the filamentous fungus *Ashbya gossypii*. Gene 242:381–392

Whelan WL (1987) The genetics of medically important fungi. Crit Rev Microbiol 14:99–170

Whelan WL, Kwon-Chung KJ (1986) Genetic complementation in *Cryptococcus neoformans*. J Bacteriol 16:924–928

Woloshuk CP, Prieto R (1998) Genetic organization and function of the aflatoxin B$_1$ biosynthetic genes. FEMS Microbiol Lett 160:169–176

Xiang Z, Moore K, Wood V, Rajandream MA, Barrell BG et al. (2000) Analysis of 114 kb of DNA sequence from fission yeast chromosome 2 immediately centromere-distal to his5. Yeast 16:1405–1411

Xiong M, Chen HJ, Prade RA, Wang Y, Griffith J, Timberlake WE, Arnold J (1996) On the consistency of physical mapping method to reconstruct a chromosome in vitro. Genetics 142:267–284

Young RA (2000) Biomedical discovery with DNA arrays. Cell 102:9–15

Yuan G-F, Liu C-S, Chen C-C (1995) Differentiation of *Aspergillus parasiticus* from *Aspergillus sojae* by random amplification of polymorphic DNA. Appl Environ Microbiol 61:2384–2387

Zhu H, Choi S, Toleston AT, Wing RA, Dean RA (1997) A large-insert (130 Kbp) bacterial artificial chromosome library of the virublast fungus *Magnaporthe grisea*: genome analysis, configuration, assembly, and gene cloning. Fungal Genet Biol 21:337–347

Zhu H, Nowrousian M, Kupfer D, Colot HV, Berrocaltito G et al. (2001) Analysis of ESTs from two starvation, time of day-specific libraries of *Neurospora crassa* reveals novel clock-controlled genes. Genetics 157:1057–1065

Zolan ME (1995) Chromosome-length polymorphisms in fungi. Microbiol Rev 59:686–698

Biosystematic Index

Subject Index